AD1995: NW Europe's Hydrocarbon Industry

AD1995: NW Europe's Hydrocarbon Industry

EDITED BY

K. Glennie & A. Hurst

Department of Geology & Petroleum Geology,
University of Aberdeen, UK

1996
Published by
The Geological Society
London

THE GEOLOGICAL SOCIETY

The Society was founded in 1807 as The Geological Society of London and is the oldest geological society in the world. It received its Royal Charter in 1825 for the purpose of 'investigating the mineral structure of the Earth'. The Society is Britain's national society for geology with a membership of around 8000. It has countrywide coverage and approximately 1000 members reside overseas. The Society is responsible for all aspects of the geological sciences including professional matters. The Society has its own publishing house, which produces the Society's international journals, books and maps, and which acts as the European distributor for publications of the American Association of Petroleum Geologists, SEPM and the Geological Society of America.

Fellowship is open to those holding a recognized honours degree in geology or cognate subject and who have at least two years' relevant postgraduate experience, or who have not less than six years' relevant experience in geology or a cognate subject. A Fellow who has not less than five years' relevant postgraduate experience in the practice of geology may apply for validation and, subject to approval, may be able to use the designatory letters C Geol (Chartered Geologist).

Further information about the Society is available from the Membership Manager, The Geological Society, Burlington House, Piccadilly, London W1V 0JU, UK. The Society is a Registered Charity, No. 210161.

Published by The Geological Society from:
The Geological Society Publishing House
Unit 7, Brassmill Enterprise Centre
Brassmill Lane
Bath BA1 3JN
UK
(*Orders:* Tel. 01225 445046
Fax 01225 442836)

First published 1996

The publishers make no representation, express or implied, with regard to the accuracy of the information contained in this book and cannot accept any legal responsibility for any errors or omissions that may be made.

British Library Cataloguing in Publication Data
A catalogue record for this book is available from the British Library.

ISBN 1-897799-79-9

Typeset by Bath Typesetting, Bath, UK

Printed by The Alden Press, Osney Mead, Oxford, UK

Distributors

USA
AAPG Bookstore
PO Box 979
Tulsa
OK 74101-0979
USA
(*Orders:* Tel. (918) 584-2555
Fax (918) 584-2652)

Australia
Australian Mineral Foundation
63 Conyngham Street
Glenside
South Australia 5065
Australia
(*Orders:* Tel. (08) 379-0444
Fax (08) 379-4634)

India
Affiliated East-West Press PVT Ltd
G-1/16 Ansari Road
New Delhi 110 002
India
(*Orders:* Tel. (11) 327-9113
Fax (11) 326-0538)

Japan
Kanda Book Trading Co.
Tanikawa Building
3-2 Kanda Surugadai
Chiyoda-Ku
Tokyo 101
Japan
(*Orders:* Tel. (03) 3255-3497
Fax (03) 3255-3495)

Contents

Preface

This book is the outcome of a symposium held at the University of Aberdeen in April 1995 to celebrate the Quincentenary of its foundation. Aberdeen is a port that during its long history has been involved with trade throughout NW Europe. And because of Aberdeen's status as the North Sea's operational 'oil capital', it was felt that the topic of the symposium should combine these two features and concern itself with wide ranging aspects of the upstream hydrocarbon industry over the whole of NW Europe. As such, the book contrasts with many thematic volumes by presenting a broad range of topics side-by-side. The book is divided into four sections. As is only proper for a book celebrating five hundred years of history, the first section by **Trewin** is devoted entirely to the history of Aberdeen University and its tradition within the geological sciences.

Part Two looks back at the history of geological exploration and production; **Glennie & Hurst** give an overview of hydrocarbon exploration across NW Europe, although they fail to reach back 500 years! **Archer** provides a personal view of developments in petroleum technology, and takes us back to Georgius Agricola, who was drilling into the sub-surface at the time of Aberdeen University's foundation. **Spencer *et al.*** review North Sea hydrocarbon plays: where the hydrocarbons were found and how the plays developed. The next two papers concern Ireland's role in the oil industry, first **Naylor** focuses on the history of exploration, largely onshore, followed by **Shannon** who describes the country's greater, but still limited success offshore. Crossing the North Sea, **Epting** discusses the Permian Rotliegend gas play in The Netherlands, and emphasizes how the use of 3D seismic in exploration has improved the rate of discovery since the late 1980s. The section ends with **Dean's** paper on the 'Undiscovery' wells of the UK continental shelf: how and why earlier drilling failed to find the Nelson, Scott and Morecombe fields, and how good geological practice led to the discovery of the Wytch Farm Field.

Part Three is much more concerned with the state of the art in hydrocarbon exploration and production. **Iliffe & Dawson** give an update on computer-based basin modelling whereas **Holm** considers overpressure systems in the Central Graben. Still with the Central Graben, **Eggink *et al.*** show how 3D seismic can enhance our understanding of the structural evolution of the Central Graben and relate that to the occurrence of major oil fields. **Larter *et al.*** describe features of the most critical, but perhaps least easily constrained, event in oil-field prospectivity – the geochemical aspects of secondary migration. **Davies & Niko** describe many of the innovations in reservoir engineering that are helping lead reservoir technology and improve recovery toward the twentieth century. **Henriquez & Jourdan** summarize their experience and problems with reservoir modelling in the North Sea and Norwegian Shelf. **Corbett *et al.*** demonstrate the importance of the successful integration of geology and engineering, and **Skopec *et al.*** discuss new, fundamental innovations in petrophysical technology and its application to reserve estimation.

Part Four of the book takes a look into the future. **Daly *et al.*** take an up-beat look at the remaining hydrocarbon resource of the North Sea by introducing comparisons with other long-lived oil provinces, while **Kassler** considers both pessimistic and optimistic scenarios for global energy over the next 50 years. **Fassom** presents how a moderate-sized company must trade its assets in order to keep it viable in terms of cash flow at all stages of the business, while **Bruce *et al.*** consider how the risk involved in oil field development should be managed. **Kemp & Stephen** summarize the economic framework that is critical to controlling the rate of exploration and production in the hydrocarbon industry and present some predictions of future activity. Finally, pollution and the environment are covered by **Tromp**, who describes the history of the international agreements that led to the control of pollution in the North Sea – land-based industry as well as the oil industry – how pollution is being reduced within the different national sectors of the North Sea, and some aspects that still need to be improved.

Although this collection of papers cannot claim to be a comprehensive documentation of all aspects of NW Europe's upstream oil industry, we hope that windows into several key areas of interest are summarized. For the University and City of Aberdeen, where prosperity is closely related to the activity and success of the oil industry, the volume represents a small contribution from an environment within which there is substantial professional, research and educational expertise. Perhaps the collection of papers should encourage readers to ask themselves, 'are we prepared for the challenges that oil exploration and production offer in NW Europe in the new millennium and have we learned from past experience?' Only the future can answer this question.

Andrew Hurst and Ken Glennie

Part One: Geology at Aberdeen University

A brief history of geology in Aberdeen University

Nigel H. Trewin

Department of Geology & Petroleum Geology, King's College, University of Aberdeen, Aberdeen AB24 3UE, UK

ABSTRACT: This brief contribution traces the history of geological teaching and research in Aberdeen University. Activities in the nineteenth century were dominated by the scholarship of Professor James Nicol and Professor H. A. Nicholson. The establishment of a separate Geology Department in 1922 was followed by a period of teaching excellence and the production of a steady supply of graduates to overseas geological survey and mining industries. The last 30 years have seen increasingly successful collaboration with the hydrocarbon industry in both teaching and research, whilst maintaining broad geological interests essential for the continued well-being of the department.

KEYWORDS: *Aberdeen, geology, history, Nicol, Nicholson*

On the Quincentenary of Aberdeen University it is of interest to review the history of geological teaching and research within our institution. A starting date for geology at Aberdeen University can be defined according to one's preference. This could be the establishment of a department in 1922 with the first appointment of the Kilgour Professor of Geology, or maybe the institution of the first 'graduation course' of Geology under the 'New Ordinances' in 1892 and taught by Professor H. A. Nicholson – then Professor of Natural History.

Neither date is satisfactory since prior to Nicholson's period, James Nicol (1810–79) was Professor of Natural History from 1860 to 1878. Nicol was involved in the great Highland Geology controversy with Murchison on the structure and succession of the rocks of the NW Highlands. Murchison, with his superior political clout and influence, appeared at the time to have won the day, but Peach and Horne, working for the Geological Survey, later confirmed many of Nicol's views. The full story is related by D. R. Oldroyd (1990) in his excellent book *The Highlands Controversy*. Geological scholarship was clearly an important part of the Natural History Department. Some specimens in the department were collected by Nicol and we have a superb copy of Griffith's Geological Map of Ireland (1855) which was presented to Nicol by the author. Even prior to Nicol's time some geology was taught as a part of Natural History; a student's notebook of 1820 contains crystal drawings, analyses of minerals and notes on the nature of fossils.

Henry Alleyne Nicholson was Professor of Natural History from 1882 to 1899. A palaeontologist and zoologist of high renown, he published textbooks on zoology and palaeontology for schools and for students. He also wrote monographs on graptolites, and stromatoporoids, and numerous papers on corals and bryozoa. Part of his fossil collections from Britain, Canada and Europe formed the basis of our extensive palaeontological collection, which contains many of Nicholson's type and figured specimens.

Documented history held by the department starts with the Class Record, listing students, their attendance record and achievements from the first graduation class of 1892–3. Twenty students are listed in that year of whom only five were in the Science Faculty. Alfred W. Gibb MA is listed in that class and in 1895 he was appointed as an assistant to teach geology, a task he continued as a lecturer from 1898 until he was appointed Kilgour Professor of Geology in 1922. Gibb was an excellent and well respected teacher, and built up the highly varied departmental collections of rocks, minerals and fossils. The most important acquisition, in 1922, was the Gordon Bequest of the Rev. J. M. Gordon in memory of his father, Harry Gordon, who had been a student at Aberdeen. Apart from scientific instruments and books, the bequest included a magnificent mineral collection, which forms the basis of our departmental mineral displays. Following Gibb's death in 1937, his former students subscribed to a memorial plaque which is now mounted on the wall of the map room, together with the earlier plaques commemorating Nicol and Nicholson.

In 1937 the Kilgour Chair of Geology passed to Professor T. C. Phemister (younger brother of James Phemister of the Geological Survey). Phemister's mineralogical expertise and crystallographic interests influenced all students, and many remember, with mixed emotions, the hours spent at the optical goniometer! The class list for 1936–7 contains the names of F. H. Stewart (now Sir F. H. Stewart FRS) who became head of the Edinburgh Department, and also W. E. Fraser (Bill) who later joined the staff and became personal advisor and friend to generations of students.

Major technological advances took place whilst Phemister occupied the Chair, and he keenly acquired, and frequently had the technicians build, the latest items in our workshops. Teaching was still the paramount activity in the Department but in the late 1960s and early 1970s an increase in academic teaching staff, as part of the general expansion of the universities, resulted in a rapid increase in research output. Following the retirement of Professor Phemister in 1971, Professor E. A. Tait held the Kilgour Chair until 1981.

The department, with its traditional academic outlook, was slow to respond to the arrival of the oil industry in the early 1970s. Most graduates had previously found jobs in mining and overseas geological surveys, but these opportunities were being rapidly replaced with oil industry jobs and their different requirements. Furthermore, finance for research

From K. Glennie & A. Hurst (eds), 1996, *AD1995: NW Europe's Hydrocarbon Industry*, Geological Society, London, pp.3–4

and students was available, and the flavour of the department began to change. James T. C. Hay was instrumental in starting a postgraduate taught course in Petroleum Geology in 1972, and this was continued with the appointment of Arthur J. Whiteman as Professor of Petroleum Geology in 1974. In the early days the concentration was naturally on exploration, but with the maturing of the North Sea oil province, the emphasis on production geology aspects increased. Graduates from this MSc course, together with BSc and PhD graduates, form a world-wide Aberdeen 'Mafia' within the oil industry.

The Government-instigated Earth Sciences Review (1988) resulted in major changes for the department. Departments were largely judged on academic research output and research council grant income. Thus our profile of high industry-related research involvement and teaching counted for little and, at one stage, it was recommended that the department be reduced to two staff for support teaching only! After a vigorous campaign from the department and our friends in industry and politics, it was eventually conceded that a Geology Department with industrial relevance was useful in Aberdeen. We survived, but with funded staff numbers reduced from 18 to 9 and, with our research remit more closely focused on petroleum geology, the department name changed to Geology and Petroleum Geology. The appointment of Professor Brian P.J. Williams to the Chair of Petroleum Geology in 1988 following the retirement of Professor Whiteman, resulted in increased industrial sponsorship including lectureships from Mobil and Conoco, and a Chair of Production Geoscience from Shell in 1992 to which Professor Andrew Hurst was appointed.

The major event which occupied the summer of 1990 was the move of the department from Marischal College to occupy the present premises within the Meston Building at King's College. Our accommodation is smaller, but of higher quality and it is a distinct advantage to be amongst the other science departments on the main University campus.

The Kilgour Chair was held by Professor P.E. Brown from 1981–89, but since that time the chair has been vacant since the original endowment only provides a small fraction of the money required to support the post. The Headship of the Department has been held recently by Professor Ian Parsons (1987–8), now Head of the Geology Department at Edinburgh, Dr Clive M. Rice (1990–3), and is currently held by Dr Gordon M. Walkden.

The department maintains a balance between academic and industry-related research whilst providing a lively teaching and learning environment for undergraduates. Postgraduate student numbers have increased rapidly in recent years, and we now have 37 research students (MSc & PhD) and another 20 take the Petroleum Geology MSc Course. Research output by staff is spread across the geoscience spectrum from hydrocarbon reservoir studies to pure palaeontology and igneous petrology. At present we are increasing the range of our undergraduate courses to provide both vocational and non-vocational strands to the degree. A greater emphasis on Environmental Geology will enable us to produce graduates who understand the scientific basis of both the environmental and industrial arguments surrounding such emotive issues as rig abandonment and radioactive waste disposal.

Our major research income and output is connected with the petroleum industry on a world-wide basis. Important overseas links are in place involving teaching and research personnel in places as diverse as Brunei, Qatar, Canada, Bulgaria, South America, Australia and Nigeria. The current department lecturing staff complement is 15 with several other research and honorary posts; all ably supported by 10 technical and 4 secretarial staff. The range and quality of our technical equipment, and strong links with the Scottish Universities Research and Reactor Centre in East Kilbride, greatly enhance our research output.

The future of Geology and Petroleum Geology is well assured, and we present a surprisingly broad portfolio in relation to the size of the department. Diversification into the international petroleum industry is increasing rapidly, and we now have specialists in petrophysics, risk-analysis, computer-based applications and chemostratigraphy, as well as traditional petroleum geology skills. Metalliferous mining research initiatives in South America and Bulgaria are supported by Research Council and Royal Society finance, and mapping projects for BGS have recently been completed. Productive academic areas of research include early terrestrial environments, with studies in Scotland and Australia; climatic change evidenced in Devonian, Carboniferous Jurassic and Quaternary rocks; generation of magmas at plate margins, and evolution of the Andean Forearc in Chile. A future challenge is to maintain the essential academic geological base required for teaching and research.

The department is committed to high quality research of general applicability to the oil industry. The present high level of collaboration between industry and the Department of Geology and Petroleum Geology must be maintained for the mutual benefit of both parties.

REFERENCE

Oldroyd, D. R. 1990. *The Highlands Controversy.* University of Chicago Press.

Part Two: Historical Development of Exploration and Production

Hydrocarbon exploration and production in NW Europe: an overview of some key factors

K. W. Glennie and A. Hurst

Department of Geology & Petroleum Geology, King's College, University of Aberdeen, Aberdeen AB24 3UE, UK

ABSTRACT: The hydrocarbon industry in NW Europe can be considered to have begun in 1851 when 'Paraffin' Young began to retort Carboniferous oil shales in Scotland, a small industry that lasted just over a century until it was killed off by the taxman. Colonel Drake did not discover oil at Titusville, Pennsylvania, until 1859. Germany was at the centre of Europe's oil industry at the start of the twentieth century. Nationalization of mineral rights in Germany in 1934, put the country back at the forefront of hydrocarbon exploration in NW Europe until the discovery of the Groningen gas field in 1959 and, more importantly, realization in 1963 of its great size. The sale of Groningen gas has been a major factor in the economics of The Netherlands ever since it came on stream.

Following the discovery of Groningen, a second phase of exploration and exploitation began. The first success in the discovery of Permian Rotliegend gas in the UK offshore almost ground to a halt within about four years because the monopoly market became saturated. The resulting availability of cheap seismic encouraged the industry to extend their surveys further north. In 1969, the first offshore oil field to produce 10^9 barrels was found in a Chalk reservoir, the Norwegian Ekofisk Field. Two more discoveries, the giant Forties Field in 1970 and the small but prolific Auk Field in 1971 extended the span of the reservoirs to the Palaeocene and Permian, respectively. The discovery of Brent with a mid-Jurassic reservoir, also in 1971, led to several hectic years of exploration seismic and drilling in UK waters both to find oil and to prove-up oil-free acreage that could be returned to the state under the rules of acreage relinquishment.

From about 1980, the oil industry has consolidated its early successes to the extent that hydrocarbons are now discovered by about 50% of exploration wells in the North Sea area, although not all are commercial at today's prices. Beyond the North Sea, however, success is not so obvious. France has gas and oil in Acquitaine but little oil in the Paris Basin. To date, there is limited oil and gas on the Irish continental shelf. The Norwegian Barents Sea had no successes to match the gas fields of the Russian Pechora Basin. In the Baltic Sea area there are only small oil fields with Palaeozoic reservoirs. The failure to find commercial abiogenic hydrocarbons in the Precambrian rocks of Sweden surprised nobody within the oil industry.

KEYWORDS: *hydrocarbons, exploration, production, history, NW Europe*

INTRODUCTION

The oil industry in NW Europe can be considered as a case of 'before and after'. The early and rather small-scale activity before the discovery of the giant Groningen gas field in 1959 or more significantly, the realization in 1963 of its immense size, and the rapid build-up of a major thriving hydrocarbon industry following the first discovery of offshore gas in the North Sea in 1965. Although the North Sea has been at the heart of this post-Groningen activity, and contains by far the largest reserves of oil within NW Europe, there has been a steady expansion of exploration outwards into the English Channel and around the shores of Ireland, north along the Norwegian margin of the Atlantic Ocean and into the Barents Sea, and eastward to the Baltic Sea (Figs 1 and 2).

These forays have had very variable success in terms of hydrocarbon discoveries, but commercial success or not, the enormous addition to our geological knowledge from seismic and well logs is easily matched by advances in offshore and down-hole engineering, which seem constantly to push development to the limit of human ingenuity. At every turn, engineers have evolved their practices to provide innovative solutions for allowing the economic recovery of hydrocarbons. The background for engineering innovation is the generation of new geological play concepts that have led to the discovery of hydrocarbons in new areas and in new reservoirs. Similarly, as the quality of geological and geophysical data have improved, the accuracy with which exploration targets are defined has increased, thus keeping the

From K. Glennie & A. Hurst (eds), 1996, *AD1995: NW Europe's Hydrocarbon Industry*, Geological Society, London, pp. 7–16

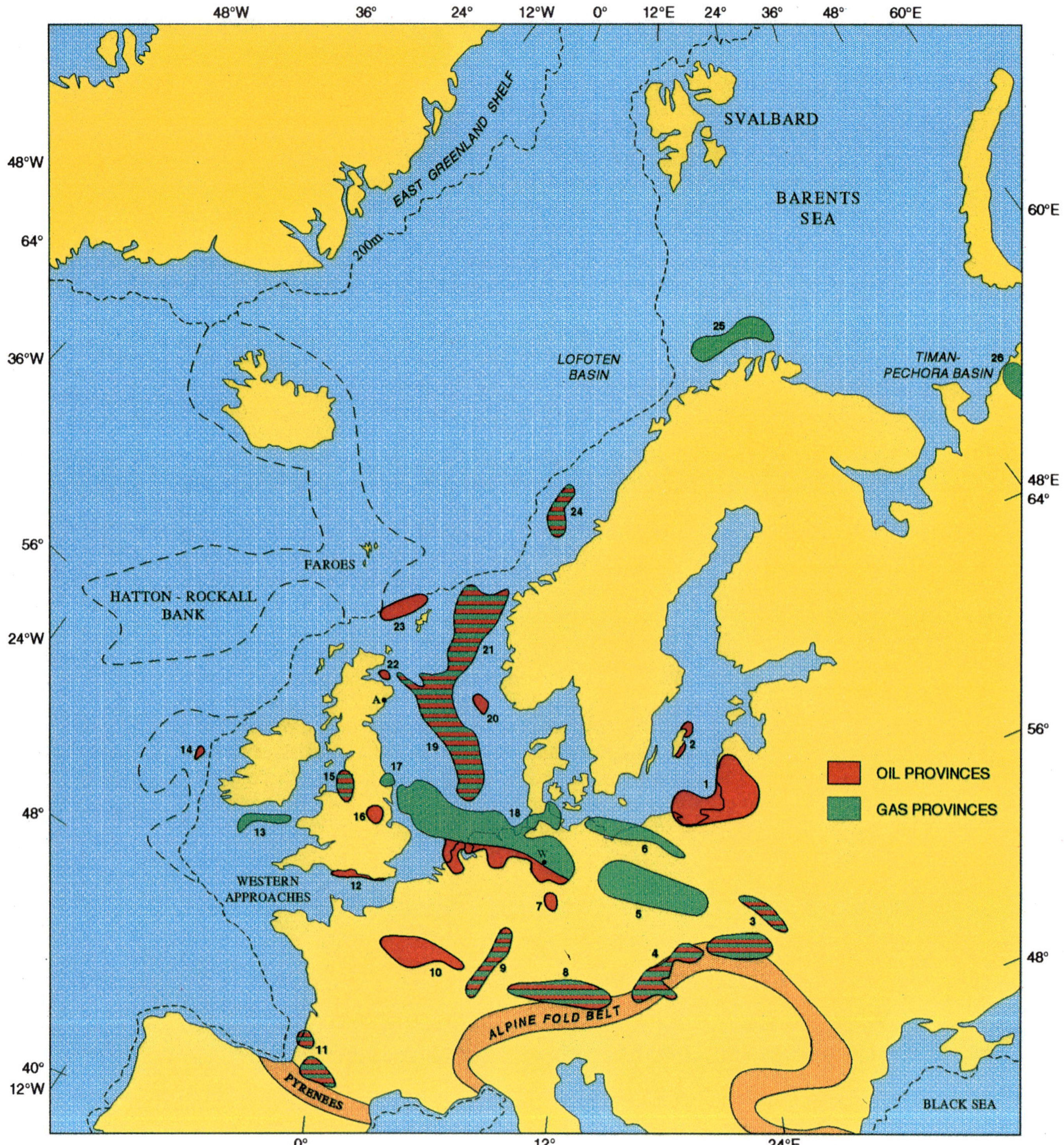

Fig. 1. Major hydrocarbon-producing areas of NW Europe: 1. Baltic States, 2. Gotland, 3. North Carpathian Basin, 4. Vienna Basin, 5. East German-Polish Basin, 6. Mecklenburg-Pommeranian Platform, 7. Thuringian Basin, 8. Molasse Basin, 9. Rhine Graben, 10. Paris Basin, 11. Aquitaine Basin, 12. Isle of Wight Basin, 13. Celtic Sea Basin, 14. Porcupine Basin, 15. East Irish Sea, 16. East Midlands Basin, 17. Eskdale-Lockton area, 18. Anglo-Dutch/Lower Saxony Basin, 19. Central & Wytch Ground Grabens, 20. Brisling-Bream area, 21. Viking Graben, 22. Inner Moray Firth, 23. West Shetland Basin, 24. Mid-Norway Basin, 25. SW Barents Sea, 26. Timan-Pechora Basin, A. Aberdeen, W. Wietze.

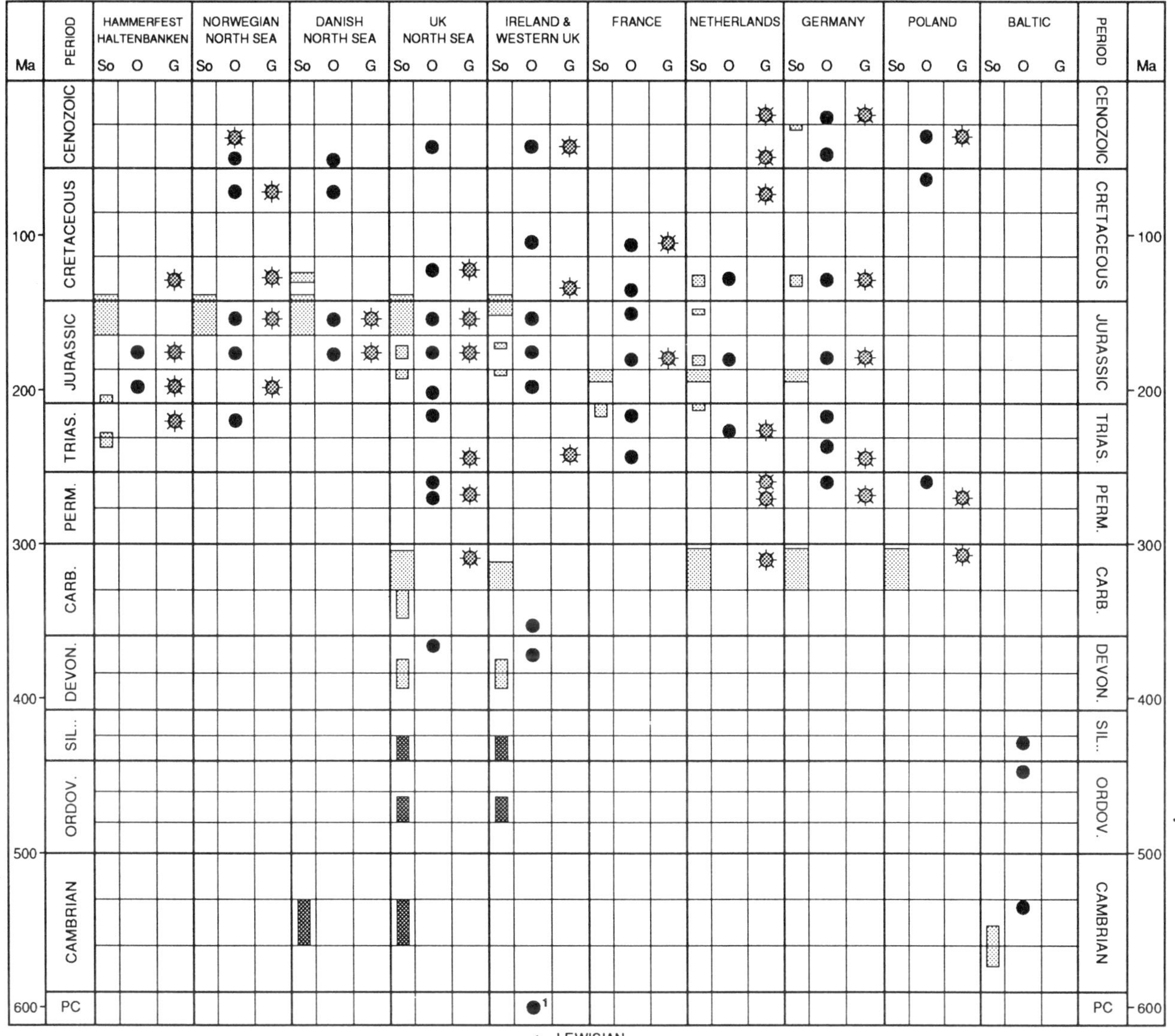

Fig. 2. Ages of main source rocks (So) and oil- and gas-bearing reservoirs (O and G) in NW Europe.

discovery rate of commercial reserves acceptably high.

This paper is an outline of some of the key features in this two-phase development of the up-stream hydrocarbon industry in NW Europe. A third phase, focusing on the deep waters of the Atlantic seaboard may have just begun, with exploration and the discovery of large oil reserves west of Shetland.

HYDROCARBONS IN NW EUROPE PRIOR TO GRONINGEN

Because of the occurrence of natural seeps, oil, gas and asphalt have been known and used in different parts of Europe for several millennia. For instance, according to Carozzi (1992), Neolithic lake dwellers along the shores of Lake Neuchâtel, Switzerland, used asphalt as an adhesive to fasten flint implements to handles of horn or wood. Places where oil seeps provided the basis for small industries in the 18th and 19th centuries included Pechelbronn, Lampersloch and Lobsan in Alsace and at Wietze, in Germany, shafts to recover tar had been sunk before 1800. Asphalt has also been used in recent times to caulk the seams of wooden boats, a practice that is rapidly dying out because of the modern development of tough, lightweight plastic (a product of the modern hydrocarbon industry) for moulding their seamless hulls. Liquid oil from seeps found a use as an axle lubricant and as a medicine or a salve, but was used as lamp oil in Europe for only about a century before being replaced by coal gas and then electricity. None of these early uses, however, can be considered as the basis for a modern economic enterprise; both production and demand were far too small, and the potential market was insufficiently educated in the possible uses of hydrocarbons to drive production.

The establishment of the first economically viable oil industry in Europe can be considered to date from 1851 (Table 1), when the Scot, James 'Paraffin' Young, set up a plant for distilling oil from Lower Carboniferous (Visean) oil shale at Broxburn, in central Scotland (Butt 1983). This event preceded by eight years what most people consider as the founding of the modern oil industry: 'Colonel' Drake's discovery of oil at Titusville, Pennsylvania. A trained chemist,

'Paraffin' Young established the ground rules, which are still in use within the industry today, for refining oils by fractional distillation. His production of paraffin wax on a commercial scale, for instance, led to the marketing by a London firm in 1857 of wax candles which became very popular (Forbes & O'Beirne 1957). The oil-shale industry in Scotland was to operate continuously for just over 100 years, a byproduct, nitrate fertilisers, just enabling it to compete economically with the liquid mineral-oil industry, an estimated total of 10×10^6 tonnes of oil being produced (Cameron & McAdam 1978). It was finally put out of business by the taxman, thereby sorely testing Britain's balance-of-payments through the need to replace oil production by imports and to import agricultural fertilisers from Chile. In the late nineteenth and early twentieth centuries, other distillation plants were established to retort oil shales of differing ages in southern Sweden, Estonia and France. In Estonia, oil shale is still being mined and retorted each year (Gérard *et al.* 1993).

Table 1. *Significant events in the evolution of NW Europe's hydrocarbon industry*

1851	James 'Paraffin' Young started oil-shale industry in Scotland
1859	First geologically selected well at Wietze, Germany
1934	Nationalization of Mineral Rights in Germany
1954	Discovery of oil at Parentis, France
1959	Slochteren-1: discovery of Groningen gas field, The Netherlands
1963	Ten Boer-1 – Confirmation of the great size of the Groningen gas field
1965	First offshore gas find – West Sole
1969	Ekofisk (Norway) – $\times 10^9$ barrel field (Chalk)
1970	Forties (UK) – another $\times 10^9$ barrel oil field (Palaeocene)
1971	Brent (UK) – $\times 10^9$ barrel oil field (Mid Jurassic)
1971	Kinsale Head (Ireland) $\times 10^{12}$ scf gas field (Lower Cretaceous)
1972	Piper (UK) – giant oil field (Upper Jurassic)
1973	Frigg (Norway) – multi $\times 10^{12}$ scf gas field
1974	Discovery of oil West of Shetland
1975	Devonian oil in small Buchan Field (UK)
1975	Troll (Norway) – biggest offshore gas field (Upper Jurassic) long migration path
1981	Second phase of UK Southern North sea gas field development began – price rise
1984	Snorre (Norway) – first giant Triassic oil field
1986	Oil glut and world-wide drop in oil price
1987	Global stock-market collapse – many take-overs
1990	Saddam Hussein invades Kuwait – oil prices strengthened
1992	Foinaven (UK) – first commercial oil West of Shetland

The first deliberately drilled well[1] which discovered oil in Europe was in 1859 at Wietze, in northern Germany (Fig. 1, Table 1), reaching a total depth at 35.5 m. It was designed to test for the presence of coal, but found oil that flowed at a rate of one and a half buckets per day for the next twelve months (Plein 1979). A follow-up well, the first specifically drilled to find oil, produced tar and was drilled in 1860. Deeper wells were drilled at Wietze; by about 1900, a further 100 wells had been drilled, 60 of which were producers giving an annual production between 1908 and 1910 of 11 000 tonnes. This field eventually produced over 18×10^6 bbl of oil. With four other fields elsewhere in Germany (Heide 1856 with 18×10^6 bbl; Eddesse-Nord 1876 with 8.4×10^6 bbl; the small Tegemsee, 1883 and Nienhagen, 1899 with 39×10^6 bbl of recoverable oil), the total annual production was around 50 000 tonnes (Schröder *et al.* 1991).

[1] Pits and simple drilling are reported from the Bóbrka-Rogi oil field in Poland in 1854

Germany had been the centre of Europe's oil industry for some years, with a production in 1882 at Ölheim, for instance, of around 6000 tonnes. Few fields were discovered between 1910 and 1934. In that latter year, all mineral rights in Germany were nationalized to bring some order to an otherwise very complex concession situation. Nine hundred wildcats were drilled between 1935 and 1945, which resulted in the discovery of 28 new fields, mostly with Jurassic and Cretaceous reservoirs. By the end of World War II, 26×10^6 bbl of oil had been produced from these new fields. Today, over 90% of production is from the Northwest German Basin (Schröder *et al.* 1991), the remainder being divided between the Molasse Basin flanking the Alps (Bachmann & Müller 1991) and the Rhine Graben (Durst 1991). It is worth mentioning here that in Austria the giant Matzen Field (discovered 1949; recoverable reserves of over 500×10^6 bbl oil) was the largest onshore hydrocarbon discovery in Europe prior to the discovery in 1959 of the Groningen gas field (K. Chew, pers. comm. 1995).

The bulk of France's oil production had long been obtained from Eocene and Oligocene strata in the Pechelbronn area of the Rhine Graben (Fig. 1, area 9), some of it by underground mining rather than drilling. This area has produced an estimated 23×10^6 bbl over the years. With later discoveries, rates of production eventually reached 500 000 bbl a^{-1} in the immediate past-World War II years. Oil was discovered in Jurassic–Cretaceous carbonates at Lacq in the Acquitaine Basin (Fig. 1, area 11) in 1949 (Bonnard *et al.* 1958), and some 7×10^{12} scf of H_2S-rich gas somewhat later from a deeper horizon. Oil was also found at several other localities during the 1950s, notably at Parentis (Table 1), with reserves of 210×10^6 bbl, and smaller fields in the Paris Basin (Pages 1987).

The first oil discovery in Britain, was at Hardstoft in the West Midlands, which produced from 1919 to 1945. Gas was discovered at Eskdale, in Scotland, in 1938 and oil at Eakring in 1939. Between 1937 and 1954, over 500 wells were drilled in the East Midlands of England (337 producers, 101 dry holes and 69 stratigraphic tests). Over the next 18 years, 150 more exploration wells were drilled, but production still did not amount to more than 1×10^6 bbl of oil per year. Eakring, with multiple Carboniferous reservoirs, had an estimated 25.6×10^6 stb of oil in place, but cumulative recovery amounted to only about 6.5×10^6 stb when production finally ceased in 1971 (Storey & Nash 1993).

In Denmark, 35 deep wells drilled between 1935 and 1959 resulted only in non-commercial shows, two of gas and three of oil (Sorgenfrei & Buch 1964; Nielsen & Japsen 1991). This drilling, however, also resulted in the definition of the E–W trending Ringkøbing-Fyn High, which was later recognized as part of a major structural feature of the North Sea.

On the south side of the Baltic Sea, the nationalized oil industry in Poland comprised 38 small oil fields and 5 small gas fields at the end of World War II. Small quantities of new oil were to be found in Zechstein carbonate reservoirs. The Zechstein was one of the main post-war targets throughout the greater North German Plain (Holland, Germany, Poland) and was the main target in The Netherlands when the Groningen discovery well, Slochteren-1, was drilled in 1959.

Prior to the German invasion of The Netherlands in 1940, only 12 dry holes had been drilled, although, significantly, an oil show was noted in basal Tertiary sands when a demonstration well was drilled at Mient, near The Hague, during the 1938 World Petroleum Congress. Under German pressure, exploration activity was resumed in 1941, and the Schoonebeek oil field was discovered in 1943 close to the

German border although it did not go into production until the war was over. By 1954, there were 6 gas and oil fields in the east and 3 oil fields in the west, including Ijsselmonde-Ridderkerk (75×10^6 bbl), beneath the centre of Rotterdam.

In 1952, a wildcat in the Groningen area found 180 m of the Lower Permian Rotliegend Sandstone to be water bearing. Four years later, a little gas was discovered in Zechstein dolomites in the Ten Boer well. In 1959, however, the Slochteren-1 well found the Rotliegend fully gas bearing (Table 1). Further seismic interpretation and the drilling of four appraisal wells indicated that the gas had a common water table belonging to one major structure (Stäuble & Milius 1970; Van Kesteren 1973). This was confirmed when the Ten Boer well was deepened in 1963 to find the gas–water contact at a depth of about 2900 m. With recoverable reserves then estimated at 58×10^{12} scf (now known to be nearer 97×10^{12} scf; Petroleum Geological Circle 1993), it was realized that there was enough gas in Groningen to alter the fuel economy of much of the adjacent parts of NW Europe. The source of all this gas was considered to be the underlying Carboniferous Coal Measures.

Realization of the size of the Groningen gas field immediately led to the question: are there more 'Groningens' to be found and, if so, where? An unpublished sedimentological study of cores from Groningen by E. Oomkens (acknowledged in Glennie, 1972) indicated that the gas-bearing Rotliegend sandstones were probably desert dune sands. Uniform dune-like foresetting of the gas-bearing Rotliegend sandstones indicated that the reservoir probably extended down-wind from Groningen toward the west. Salt diapirs within the Zechstein cap-rock sequence had been recognized on early offshore seismic lines crossing the Southern North Sea; and Carboniferous Coal Measures existed across the North Sea in central and northern England and even southern Scotland. There seemed to be a good chance that the three requirements of source, reservoir and seal occurred not only at Groningen but probably also within that part of NW Europe's continental shelf that lay beneath the North Sea.

POST-GRONINGEN EXPLORATION FOR NORTH SEA ROTLIEGEND GAS

Following rules established by the Continental Shelf Convention of 1958, agreement was reached by 1965–6 between the UK, Norway, Denmark and The Netherlands for national sovereignty in the North Sea area to extend to the median line half way between them; a boundary dispute between Denmark, Germany and The Netherlands was settled in 1970. Offshore exploration could begin as soon as each country settled its own rules governing the award of concession areas, and regulations to control seismic surveys, drilling and, eventually, it was hoped, production of oil and gas. For further details see Glennie *et al.* (1987), and especially UK regulations, Brennand *et al.* (1990) and The Petroleum Geological Circle (1993) for The Netherlands. Slow enactment of appropriate legislation, prevented early offshore exploration in Dutch waters, but with a free market, once started it has had steady success ever since.

West Germany was not so fortunate in their offshore venture, most gas finds being rich in N_2 and CO_2 rather than CH_4. These gasses were also widespread in onshore Rotliegend reservoirs of the former East Germany, whereas most of West Germany's onshore gas was eminently marketable. There was much more success in land locations, however, the Rotliegend having a total production of natural gas up to 1987 amounting to 330×10^6 m^3. Associated exploration for oil in other reservoirs (Upper Permian Zechstein to Tertiary; Fig. 2) resulted in discovering some 57×10^6 tonnes of recoverable oil by the end of 1987 (Schröder *et al.* 1991).

Offshore drilling in the shallow waters of the UK Southern North Sea began in December 1964 using jack-up rigs. After initial disappointment with wells on the largest structure in the North Sea, the Mid North Sea High (no Rotliegend reservoir), there was a string of successes during the next two years at more southerly locations: West Sole (Table 1), Viking, the small Ann Field, Leman, Indefatigable and Hewett (which, unusually at that stage of exploration, had a Lower Triassic reservoir of Bunter Sandstone). The distribution of fields showed that potential gas-bearing Rotliegend reservoirs were confined to an E–W belt between the London–Brabant Platform to the south and a broad area of non-reservoir evaporitic silty shales (later called the Silverpit Formation of the Rotliegend) south of the Mid North Sea High.

The Netherlands, meanwhile, with its enormous reserves of Groningen gas, had no urgency to develop additional higher-cost offshore fields. Served by a nationwide grid of pipelines distributing Groningen gas, every coal-gas plant in the country could be bought out and the domestic stoves of every household could be replaced with little effect on the profit margins. After a slower enactment of offshore regulations by the Dutch Government, offshore drilling began only in 1968, with the first discovery of gas in the same year; a string of successes in the K and L quadrants was to follow (Oele *et al.* 1981), but Groningen still accounts for more than 50% of all Rotliegend gas reserves in NW Europe. The first offshore oil discovery (F18) was made in 1970 (Petroleum Geological Circle 1993) and was followed from 1978 by a string of small oil discoveries in Lower Cretaceous continental sands in quadrant Q1 (Roelofsen & de Boer 1991).

For the sale of the new-found UK gas, which was brought ashore by submarine pipeline to terminals at Easington in Yorkshire and Bacton in Norfolk, British industry had a monopoly market (British Gas), whose needs for the foreseeable future were satisfied by the discoveries made up to the end of 1967. When realized, this effectively stopped exploration in the UK 'gas province', although with 'obligation wells' to drill, the peak of this activity was not reached until 1969. Seismic ships now became available at reduced rates, which encouraged operators to extend their surveys further to the north.

POST-GRONINGEN DISCOVERY OF OIL AND GAS IN THE CENTRAL AND NORTHERN NORTH SEA

Although seismic ships began their surveys in northern British waters following the collapse of the Rotliegend gas market, the Norwegians and Danes had no such problems. Their waters seemed to contain a different geological world with other targets to explore. The first signs of this different geology came in 1966 with the discovery of oil and gas in a Chalk reservoir on Denmark's Anne Field (later Kraka; Damtoft *et al.* 1987) with apparently uneconomic reserves.

The first discovery of oil in offshore Norway was made in Palaeogene deep-water sandstones of the Balder Field, Well N25/11-1 (Esso) in 1967 (Hanslien 1987). Because of the complexity of its reservoir, Balder has remained undeveloped but, more than a quarter of a century later the Palaeogene sandstone trend constitutes an important exploration play in both the Norwegian and UK North Sea as evidenced by the recent discovery of the Hermod Field (well N25/11-15, Norsk

Hydro) and well N25/8-5 (Esso/Enterprise). In 1968, gas was discovered in deep-water Palaeocene sandstones in the Cod Field, while across the median line in UK waters, oil shows were reported in 1969 from Lower Tertiary sands in what, much later, was to be developed as the Montrose Field. In December of the same year, however, oil and gas were found in a Chalk reservoir, although a Palaeogene target, at Ekofisk (Table 1), in Norwegian waters, with what later turned out to be recoverable reserves of nearly 1500×10^6 bbl oil and 4.4×10^{12} scf gas (Pekot and Gersib 1987).

From now on, the search for oil was to dominate exploration in the central and northern waters of the North Sea. Success in UK waters came with two important finds: the Forties Field late in 1970 with 2400×10^6 bbl reservoired in Palaeocene sandstones (Wills 1991) and, early in 1971, the small but highly productive Auk Field with a Zechstein reservoir (the field reserves were later almost doubled to nearly 90×10^6 bbl when oil was found to extend laterally into the underlying Rotliegend reservoir) (Trewin & Bramwell 1991). The search for oil did not prevent the discovery of some major gas fields. In mid-1971, the giant Frigg gas field was discovered in Lower Eocene turbidites on the Norwegian side of the median line with recoverable reserves, including the later proven extension into British waters, of almost 7×10^{12} scf (Mure 1987; Brewster 1991). Late in 1974, more gas was discovered at Sleipner Vest (Ranaweera 1987). During the early stages of North Sea exploration, prospectivity was, in the minds of some explorationists, restricted to the Palaeogene and several wells were terminated above potential deeper targets. Perhaps most notably well N7/12-1 (Gulf 1969) was plugged and abandoned in the Lower Cretaceous only about 100 m above the Upper Jurassic reservoir of the Ula field that was subsequently 'discovered' in 1976 by well N7/12-2 (Conoco) (Home 1987) with 157×10^6 bbl (25×10^6 m^3).

Arguably the most important North Sea discovery came in mid-1971 (Table 1) with Brent Field (3rd Round licence block; not announced until August 1972 after the award of 4th Round licences). Brent had reserves of 1800×10^6 bbl of oil and 4×10^{12} scf of associated gas trapped in Middle and Lower Jurassic sandstones in a tilted fault block (Bowen 1975). Other similar structures were distributed over a wide area of the Northern North Sea, and formed the targets for an intensive programme of drilling from the award of their 4th Round licences in 1974 until the end of the decade. Almost every structure proved to be oil-bearing, characterized by well performances that were exceeded only by the most prolific Middle Eastern reservoirs. During this period, many giant fields of Brent type were discovered in the UK East Shetland Platform marginal to the Viking Graben, including Brent, Cormorant and Ninian, together with several smaller finds (e.g. Dunlin, Eider, NW Hutton, Murchison, Heather; see creaming curve fig. 8 in Bowen 1991). The largest field of them all, Statfjord which straddles the Norway–UK median line, was discovered in 1972 by well N33/12-1 (Roberts *et al.* 1987). To the south and southwest of the Brent area, oil and gas discoveries were made concurrently in Middle and Upper Jurassic reservoirs in the Outer Moray Firth (Piper Claymore, Tartan). Towards the end of that decade, some companies almost prayed for the occasional dry hole so that they would not have to hand proven oil back to the Government under the rules whereby 50% of each concession had to be relinquished within six years. As a consequence of finite investment resources, the Central North Sea was 'put onto the back burner' until the Brent Province had been effectively delineated and drilled up.

West of Shetland (Fig. 1, area 23), similar Brent Group reservoirs in tilted fault blocks were prognosed and exploration drilling began in 1974 (well 205/21-1A, Shell). Although the first well discovered oil (Table 1), and many of the subsequent wells in the initial exploration period up to 1977 proved hydrocarbons (Verstralen & Hurst 1994), commercial test rates were elusive. Thus, the West of Shetland area gained a bad name with many explorationists who had become accustomed to the phenomenal successes experienced in the North Sea. Discovery of the 4 to 6×10^9 bbl (oil in place) Clair Field (BP) in 1975 is, to date, the largest field on the UKCS (Corey *et al.* 1993). Poor reservoir quality and the existence of less difficult targets for development elsewhere, have meant that Clair remains undeveloped over twenty years after its discovery.

By the late 1970s, oil and gas was known to be trapped in reservoirs that ranged in age from Devonian (Old Red Sandstone) to Lower Tertiary (Eocene turbidites). Out of that stratigraphic spread, one rock unit, the Kimmeridge Clay (now more precisely defined in the Norwegian and Danish waters as the latest Jurassic–earliest Cretaceous Draupne or Borglum formations respectively) became recognized in the later 1970s as the dominant source rock for all the oil and associated gas in the Central and Northern North Sea (minor additional source rocks were recognized in the Devonian and Liassic shales in the Moray Firth, and Liassic shales for oil in the Southern North Sea, and onshore Holland and Germany; Bodenhausen & Ott 1981; Fleet *et al.* 1987). More recently, Patience *et al.* (1995) have demonstrated the local source significance of oil- and gas-prone shales in the Heather Formation and gas- and light oil-prone shales and coals in the mid Jurassic Sleipner and Hugin formations in the South Viking Graben. But how could the one source, the Kimmeridge Clay, charge such a stratigraphic spread of reservoirs? Seismically derived 3D geometry of the North Sea graben system provided much of the answer.

The quality of the available seismic that led to the discovery of Brent had begun to improve dramatically from the end of the 1960s when analogue recording was replaced by digital systems that enabled the data to be processed by computers. Enhanced seismic quality, calibrated by the wells that were already drilled in the central and northern North Sea, soon led to the definition of a major, roughly north–south trending graben system, the Viking and Central grabens, which died out southwards in The Netherlands. Other smaller grabens were discovered further to the east in Denmark and Germany while another in western Britain had long been known. The Central and Viking grabens began to develop during the Permian and early Triassic and ceased to be active early in the Cretaceous. Regional subsidence resulted in deposition of a Cretaceous and Tertiary cover that reached 3 km over the axis of the North Sea graben system.

The level of detail that could be interpreted from improved 3D seismic data not only enhanced understanding of the relationships between source rocks and reservoirs, but also enabled greater accuracy of sub-surface prediction prior to drilling. In 1983 a 3D survey was acquired prior to the award of a Norwegian exploration (called production) licence that later discovered the southern extension of the Snorre Field. Although the scale of data acquisition prior to the award of a licence has hardly set a precedent in exploration practice, 3D seismic surveys are now increasingly common investment features of exploration programmes. With reference to the discovery of the Snorre Field, it is significant that, despite the probable enhancement of the definition of the structure allowed by the 3D survey, the stratigraphy of the reservoir was a surprise; rather than discovering a prognosed Middle Jurassic Brent Group reservoir the Snorre Field proved to be of Lower Jurassic and Triassic age (Hollander 1987).

Burial of the Kimmeridge Clay Formation to depths of the order of 3 km within the North Sea grabens was sufficient to bring about generation of its oil. The oil either migrated vertically into younger reservoir/seal couplets, or laterally into a variety of other reservoirs whose age depended on the local geometric relationship between the graben and its flanks. The higher temperatures associated with deeper burial of the source rock led to the generation first of associated gas and, below about 4 km, of gas only. Discovery of the Troll Field (well N31/2-1, Shell 1979, Table 1) gave proof of the migration of hydrocarbons over substantial distances (> 15 km), from mature source rocks within the graben area into Upper Jurassic reservoirs on an adjacent platform. In the UK sector of the North Sea, recognition that hydrocarbons could migrate in commercial volumes beyond the lateral confines of the graben, and that deep-water Eocene sandstones were potential reservoirs (Alba, Gannet), stimulated exploration along the edges of previously unprospective acreage. West of the Gannet complex, beyond the confines of the Central Graben, wells 21/27-2 and 21/27-4 (Fina) discovered oil in the Eocene approximately 25 km from the nearest probable mature source rocks (Staffurth *et al.* 1995). More recently, and only a few kilometres to the north, on acreage originally awarded in the 4th Round in 1972, relinquished in 1978 without drilling, re-awarded in the 12th Round in 1991 to the same operator, and subsequently farmed out without drilling, well 21/16-1 (Amerada Hess 1993) discovered oil in the Upper Jurassic at least 30 km from the nearest mature source rocks. This trend has been documented by Farquharson (1995) and Nicholson *et al.* (1995) who demonstrate oil migration into a series of salt-supported Upper Jurassic sandstone reservoirs west of the Kittewake Field proving migration distances of approximately 40 km. At the southern tip of the Central Graben, a similar trend is identified where discovery of the Fife Field (well 31/26a-9A, Amerada Hess), again situated on a platform area beyond the graben, has proved lateral oil migration of some 20 km (Mackertich 1996).

By the late 1970s hydrocarbons had been discovered in the North Sea in rocks from the Devonian to the Eocene. West of Shetland, additional reserves were discovered in the Carboniferous and Pre-Cambrian (Lewisian). The major oil discoveries were along the western flank of the North Viking Graben. Exploration along the eastern flank of the North Viking Graben provided discovery of several large Brent Group fields (e.g. Oseberg 1276×10^6 bbl, 1979; Brage 300×10^6 bbl, 1980; Veslefrikk, 310×10^6 bbl, 1980). To date, the eastern flank of the graben seems not to reservoir hydrocarbons in such a broad stratigraphic range as the western flank.

Using the experience gained in the North Sea, offshore exploration was extended to other areas of NW Europe's continental shelf. The English Channel and Western Approaches turned out to be very disappointing, although the onshore Wytch Farm Field (Colter & Howard, 1981) has recoverable reserves of 240×10^6 bbl (Bowman *et al.* 1993). The Morecambe gas field in the UK sector of the Irish Sea has ultimate reserves of 4.5×10^{12} scf in the Triassic Sherwood Sandstone (Stuart & Cowan 1991). In Irish waters, small oil and gas discoveries give hope of bigger finds, but so far only the Kinsale Head Ballycotton gas fields, with ultimate recoverable reserves of 1.5×10^{12} scf and 85.9×10^9 scf respectively are in production (Naylor, this volume).

It was natural for exploration in Norway to extend from the North Sea northwards along their partially collapsed Atlantic margin. This met with moderate success, Jurassic reservoirs with clear commercial potential containing oil, condensate and gas discovered within the Haltenbanken area (Fig. 1, area 24), including the Draugen (Ellenor & Mozetic 1986), Heidrun (Koenig 1986), Smørbukk (Aasheim *et al.* 1986), Tyrihans (Aasheim & Larsen 1984; Larsen *et al.* 1987) and Midgard (Ekern 1987) fields. In contrast, the Arctic Shelf of the Hammerfest Basin proved to be gas-prone (at least $206 \times 10^6\,m^3$, Campbell & Ormaasen 1987) where early to middle Jurassic sandstones appear to have undergone substantial inversion after initial hydrocarbon emplacement (Olaussen *et al.* 1983). Unlike the huge Russian gas fields in the Pechora Basin sector of the Barents Sea, the few wells drilled within its Norwegian waters have so far met with no success.

The failure to find commercial hydrocarbons in the Norwegian–Danish Basin has been attributed to the Kimmeridge Clay not being in a source-rock facies. In the Baltic Sea area, the limited production is mostly on the surrounding land areas, with Lower Palaeozoic reservoirs in Latvia, Lithuania, the Kaliningrad area of Russia (Brangulis *et al.* 1992), three small gas condensate pools offshore Poland (Brangulis *et al.* 1993) and on the island of Gotland, while, as we have already seen, production in Poland and Germany consists of gas of Carboniferous origin but now in Permian Rotliegend reservoirs (Fig. 2). In Estonia two-thirds of the annual hydrocarbon production comes not from conventional reservoirs, but from the distillation of the mid Ordovician algal-rich Kukersite oil shale. About 23×10^6 tons are being mined currently and remaining reserves are estimated to amount to 6.6×10^9 tons of oil shale (Gérard *et al.* 1993). The small quantities of Zechstein oil were probably derived from adjacent intra-Zechstein source rocks.

Apart from Gotland, Sweden and Finland, with their preponderance of Lower Palaeozoic and Precambrian crystalline rocks, have no production of liquid oil or gas. Presumably following a philosophy that there was nothing to lose but money, a deep well was drilled in Sweden during the 1980s to test a hypothesis that all hydrocarbons were of an abiogenic nature and emanated from the mantle; it had been claimed that the oil industry exploited only shallow structures where such hydrocarbons had been trapped. A potential sealing horizon had been identified on seismic deep within the Precambrian. No one in the oil industry was surprised at the complete lack of success of this enterprise, but the exercise did provide the Swedish government with a public airing of its conscience with respect to the need to invest further in nuclear energy in the absence of any natural combustible fuels.

The new frontier for exploration and production in NW Europe is undoubtedly the Atlantic margin of the British Isles and the Faeroe Islands. The first hint of commercial success came in 1990 when, exploring in the West Shetland Basin, Amerada Hess discovered the Strathtay Field by drilling well 205/26a-3, and the Solan Field in the following year (well 205/26a-4). More recently, the much publicized discovery of the Foinavon and Schiehallion fields within the deep waters of the Faeroes Basin by the BP/Shell group has accelerated exploration activity along this area of the Atlantic margin (Table 1). With a projected start of production from the Foinavon Field in 1996, a rapid development of production technology has been necessary in this climatically harsh, environmentally sensitive area where no production infrastructure exists. With the commercial successes West of Shetland, recent geophysical work on the sedimentary fill of the Porcupine Basin (Fig. 1, area 14), west of Ireland (Shannon *et al.* 1993, 1994) becomes all the more interesting as a new exploration frontier is defined.

Since drilling the Slochteren-1 well in Groningen in 1959, some 150×10^{12} scf (14.25×10^{12} m^3) of recoverable Permian gas and an additional 123×10^{12} scf (3.5×10^{12} m^3) of Jurassic gas has been discovered in the North Sea graben system. Jurassic reservoirs also contain 50×10^9 bbl (8×10^9 m^3) of oil, and another 14×10^9 bbl (2.2×10^9 m^3) have been found in Palaeogene reservoirs (Leckie & Chew 1991). More has yet to be discovered, but the chances of emulating the bigger finds of the North Sea are slim, although not impossible, especially in some of the frontier areas.

TECHNICAL ACHIEVEMENTS OF THE OFFSHORE OIL INDUSTRY

It is far beyond the scope of this paper to do justice to all the innovations that have stimulated the present and future status of NW Europe's hydrocarbon exploration and production. It is pertinent, however, to mention a few that, through the eyes of a petroleum geoscientist, seem to have been important. A short summary is given in Table 2.

1. Drilling: The continental shelf of NW Europe can experience some of the worst weather in the world. The first wells to be drilled in the Southern North Sea were in shallow water (*c.* 30 m) where jack-up rigs were ideal for the job. As exploration extended northward (and to the far west), drilling had to be carried out from relatively unstable drill ships or the more stable early semi-submersible mobile rigs. Initially, drilling in northern waters had to be undertaken during the summer 'fair-weather window'. By the mid-1970s, however, technical advances and the use of larger semi-submersibles permitted drilling to take place throughout the year, except during the most violent storms.

   The ability to complete wells by remote control, or to re-enter them through sub-sea manifolds at depths that are too great for divers to be active, smacks of space-age technology. Equally, the ability to control massive overpressure when drilling in some parts of the Central North Sea, and to continue to drill through the cuplrit horizon to deeper targets has enabled several otherwise inaccessible hydrocarbon pools to be explored. Production from these HP/HT reservoirs will represent a further major advance in technology.

   Application of horizontal and highly deviated wells to hydrocarbon recovery has allowed production from thin oil legs, economic production from satellite fields too small to warrant production from a dedicated sub-sea template, development of fields with too low well performance from conventional, near-vertical wells, and revitalization of mature fields by accessing by-passed oil. Planning and controlling of horizontal wells has only been possible by use of logging and measurement while drilling technology (LWD and MWD) and careful interpretation of geological data (e.g. Dobb *et al.* 1995; McClure *et al.* 1995).
2. Navigation: Apart from the technical advances in computer science that have made 3D seismic such an important exploration and development tool, it needed the precision of satellite navigation to make it viable. In the early days of seismic acquisition, navigation depended on 'line of sight from scattered shore stations'. The precise location of some lines were questioned, and more than one well 'located on the line' failed to match the associated seismic with anything like the expected accuracy. Together with accurate surface navigation, sea-bottom transponders now enable production platforms weighing tens of thousands of tons, to be placed to within a metre of their preferred location. With such precise location, 3D seismic and improved definition of sub-surface features and targets, greater recovery efficiency and recovery of by-passed hydrocarbons are achievable.

These technical advances have been a great boon to the modern oil industry. They result from the ingenuity of humanity, and require further ingenuity to apply them to their maximum advantage. Such developments usually resulted from a technical need, and as further needs become recognized, we will no doubt rise to the challenge and produce something that meets them. And here, it is perhaps worthwhile stating that today, many advances within the oil industry result from a very healthy interaction between oil companies, consultants, contractors and academia.

Table 2. *Some technical advances that helped offshore exploration and production*

Late 1960s	Change from analog to digital recording of seismic data
1960s to 1990s	Continued rapid improvement in computer technology
Early 1960s	Air gun energy source for seismic
Early 1970s	Satellite navigation offshore
Early 1980s	Water gun energy source for seismic
Early 1980s	3D seismic
Early 1960s	Development of marine jack-up rigs
Mid 1960s?	Early semi-submersibles in North Sea
Mid 1970s	Semi-submersibles capable of year-round drilling in North Sea
Mid 1970s	Exposed Location Single-Buoy Mooring (ELSBM) (oil storage) for tankers
Mid 1970s?	Modified tankers for offshore oil storage
Early 1980s	Sea-bottom completion manifolds
Late 1980s	Horizontal drilling revolutionized development of complex reservoirs and thin oil legs and satellite developments.

CONCLUSIONS

The history of hydrocarbon exploration in NW Europe dates back over a century to the mining and retorting of oil shale, beginning in Scotland in 1851, and the development of small oil fields in continental areas, especially Germany. Although oil was found on land in limited quantities up until about 1960, hydrocarbon production was never sufficient to meet national demand (except for gas retorted from Carboniferous coal) until offshore exploration was spurred by a realization of the huge size of the Groningen gas field: Permian (Rotliegend-reservoired) gas of Carboniferous coal-measure origin. Since then, hydrocarbon exploration in NW Europe has been a story of phenomenal success for those nations bordering the North Sea graben system. Different countries developed their resources at different rates, depending in part on their national economic needs, but mostly on the geological framework of their own offshore sectors. There was a natural progression from exploration for gas in the Southern North Sea in the wake of the Groningen discovery, to more northern waters where the geological sequence and trapping configuration differed from that in the south; oil, derived mostly from an Upper Jurassic (Kimmeridge Clay) source rock, became the dominant target, and was found to be reservoired in rocks ranging in age from Devonian (Old Red Sandstone) to Eocene. The progression in exploration ventures extended beyond the North Sea to areas south and west of the British Isles, the Norwegian continental shelf and the Baltic Sea, but with mixed success.

The rate of discovery was helped by the relative freedom to shoot seismic almost anywhere in these marine areas, a freedom not experienced in land concessions. The rate was also helped by an increasingly open approach to the exchange of geological information, first by the exchange of well information between interested companies, and later by the publication of geological data at numerous conferences. And this information also resulted in renewed exploration in some onshore areas, especially in Germany, Austria and Poland, with varying success.

Successful exploration for, and exploitation of, offshore oil and gas fields was also aided by enormous technical advances, whether in seismic acquisition and processing, year-round offshore drilling, the building of huge platforms or the pipelines and offshore loading facilities for bringing the hydrocarbons ashore. Sometimes working almost at the limit of human ingenuity, engineers continued to extend the frontiers of their capabilities provided there was a sufficient economic incentive.

Ken Chew is thanked for his critical review of the paper that included much valuable additional information.

REFERENCES

AASHEIM, S. M. & LARSEN, V. 1984. The Tyrihans discovery – preliminary results from well 6407/1-2. *In:* Spencer, A. M. *et al.* (eds) *Petroleum Geology of the North Europen Margin*, Graham & Trotman, London, pp. 285–292.

———, DALLAND, A., NETLAND, A. & THON, A. 1986. The Smørbukk gas/ condensate discovery, Haltenbanken, *In:* Spencer, A. M. *et al.* (eds), *Habitat of hydrocarbons on the Norwegian Continental Shelf*, Graham & Trotman, London, pp. 299–305.

BACHMANN, G.H. & MÜLLER, M. 1991. The Molasse Basin, Germany: evaluation of a classic petroliferous foreland basin. *In:* Spencer, A. M. (ed.) *Generation, Accumulations and Production of Europe's Hydrocarbons.* European Association of Petroleum Geoscientists Special Publication No. ?. Oxford University Press, pp. 263–276.

BODENHAUSEN, J. W. R. & OTT, W. F. 1981. Habitat of the Rijswijk Oil Province, Onshore The Netherlands. *In:* Illing, L. V. & Hobson, G. D. (eds) *Petroleum Geology of the Continental Shelf of North-West Europe.* Institute of Petroleum, London, pp. 301–309.

BONNARD, E., DEBOURLE, A, HLAUSCHEK, H., MICHEL, P., PEREBASKINE, V. *et al.* 1958. The Acquitaine Basin, south-west France. *In:* Weeks, L. G. (ed). *Habitat of Oil.* American Association of Petroleum Geologists Memoir, 4, 1091–1122.

BOWEN, J. M. 1975. The Brent Oil Field. *In:* Woodland, A. W. (ed.). *Petroleum and the Continental Shelf of N.W. Europe. Vol. 1: Geology.* Elsevier Applied Science Publishers, Barking, 353–363.

——— 1991. 25 years of North Sea Exploration. *In:* Abbots, I. L. (ed.) *United Kingdom Oil and Gas Fields 25 Years Commemorative Volume.* Geological Society, London, Memoir, **14**, 1–7.

BOWMAN, M. B. J., McCLURE, N. M., & WILKINSON, D. W. 1993. Wytch Farm oilfield: deterministic reservoir description of the Triassic Sherwood Sandstone. *In:* Parker, J. R. (ed.) *Petroleum Geology of Northwest Europe: Proceedings of the 4th Conference.* Geological Society, London, pp. 1513–1517.

BRANGULIS, A. P., KANEV, S. V., MARGULIS, A. S. & HASELTON, T. M. 1992. *Hydrocarbon geology of the Baltic Republics and the adjacent Baltic Sea.*

———, ———, MARGULIS, L. S. & PEMERANTSEVA, R. A. 1993. Geology and hydrocarbon prospects of the Paleozoic in the Baltic Region. *In:* Parker, J. R. (ed.) *Petroleum Geology of Northwest Europe: Proceedings of the 4th Conference.* Geological Society, London, pp. 651–656.

BRENNAND, T. P., VAN HOORN, B. & JAMES, K. H. 1990. Historical review of North Sea Exploration. *In:* Glennie, K. W. (ed.) *Introduction to the Petroleum Geology of the North Sea.* Blackwell Scientific Publishers, 1–33.

BREWSTER, J. 1991. The Frigg Field, Block 10/1 UK North Sea and 25/1 Norwegian North Sea. *In:* Abbots, I. L. (ed.). *United Kingdom Oil and Gas Fields, 25 Years Commemorative Volume.* Geological Society, London, Memoir, **14**, 117–126.

BUTT, J. 1983. *James 'Paraffin' Young: founder of the Mineral Oil Industry.* Scotland's Cultural Heritage, Edinburgh.

CAMERON, I. B. & McADAM, A. D. 1978. *The oil-shales of the Lothians, Scotland: present resources and former workings.* Institute of Geological Sciences Report **78/28**.

CAMPBELL, C. J. & ORMAASEN, E. 1987. The discovery of oil and gas in Norway: an historical synopsis. *In:* Spencer, A. M. *et al.* (eds). *Geology of the Norwegian Oil and Gas Fields.* Norwegian Petroleum Society, Graham & Trotman, London, 1–37.

CAROZZI, A. V. 1992. The history of Petroleum Geology and the Geneva Naturalists (1790–1815). *Journal of Petroleum Geology*, **15**, v-viii.

COLTER, V. S. & HOWARD, D. J. 1981. The Wytch Farm Oil Field, Dorset. *In:* Illing, L. V. & Hobson, G. D. (eds) *Petroleum Geology of North-West Europe.* Institute of Petroleum, London, 494–503.

COREY, D., FYFE, T. B., RETAIL, P. & SMITH, P. J. 1993. Chair appraisal – the benefits of a co-operative approach. *In:* Parker, J. R. (ed.) *Petroleum Geology of Northwest Europe: Proceedings of the 4th Conference.* Geological Society, London, 1409–1432.

DAMTOFT, K., ANDERSEB, C. & THOMSEN, E. 1987. Prospectivity and hydrocarbon plays of the Danish Central Trough. *In:* Brooks, J. & Glennie, K. (eds) *Petroleum Geology of North West Europe.* Graham & Trotman, London, 403–417.

DOBB, A., ROBERTSON, G., REILLY, G. M. & PURPICH, A. J. 1995. The use of well seismic profiles and MWD to steer a horizontal well. *Petroleum Geoscience*, **1**, 3–11.

DURST, H., 1991. Aspects of exploration history and structural style in the Rhine Graben area. *In:* Spencer, A. M. (ed.) *Generation, Accumulations and Production of Europe's Hydrocarbons.* European Association of Petroleum Geoscientists Special Publication No. 1. Oxford Science Publications, pp. 247-261.

EKERN, O. F. 1987. Midgard. *In:* Spencer, A. M. (ed.), *Geology of the Norwegian Oil and Gas Fields.* Graham & Trotman, London, pp. 403–410.

ELLENOR, D. W. & MOZETIC, A. 1986. Draugen oil discovery. *In:* Spencer, A. M. *et al.* (eds), *Habitat of Hydrocarbons on the Norwegian Continental Shelf*, Graham & Trotman, London, pp. 313–316.

FARQUHARSON, G. E. 1995. From source to seal, hydrocarbon tracking in the western Central North Sea. *In:* Cayley, G. T., Fraser, A. J. & Erratt, D. (eds), In pursuit of subtle traps. Geological Society (abstract).

FLEET, A. J., CLAYTON, C. J., JENKYNS, H. C. & PARKINSON, D. N., 1987. Liassic source-rock deposition in Western Europe. *In:* Brooks, J. & Glennie K. (eds), *Petroleum Geology of North West Europe.* Graham & Trotman, London, pp. 57–70.

FJAERAN, T. & SPENCER, A. M. 1991. Proven hydrocarbon plays, offshore Norway. *In:* Spencer, A. M. (ed.), *Generation, Accumulations and Production of Europe's Hydrocarbons.* European Association of Petroleum Geoscientists Special Publication No. 1. Oxford Science Publications, pp. 25–48.

FORBES, R. J. & O'BEIRNE, D. R. 1957. *The Technical Development of the Royal Dutch/Shell 1890–1940.* Brill, Leiden.

GÉRARD, J., WHEATLEY, T. J., RITCHIE, J. S., SULLIVAN, M. & BASSETT, M. G. 1993. Permo-Carboniferous and older plays, their historical development and future potential. *In:* Parker, J. R. (ed.) *Petroleum Geology of Northwest Europe: Proceedings of the 4th Conference.* Geological Society, London, pp. 641–650.

GLENNIE, K. W., 1972. Permian Rotliegendes of Northwest Europe interpreted in light of Modern desert Sedimentation Studies. *AAPG Bulletin*, **56**, 1048–1071.

———, BROOKS, J. & BROOKS J. R. V. 1987. Hydrocarbon exploration and geological history of north west Europe. *In:* Brooks, J. & Glennie, K. (eds), *Petroleum Geology of Northwest Europe*, Graham & Trotman, London, pp. 1–10.

HANSLIEN, S. 1987. Balder. *In:* Spencer, A. M. *et al.* (eds). *Geology of the Norwegian Oil and Gas Fields.* Graham & Trotman, London, pp. 193–201.

HOLLANDER, N. B. 1987. Snorre. *In:* Spencer, A. M. *et al.* (eds) *Geology of the Norwegian Oil and Gas Fields.* Norwegian Petroleum Society and Graham & Trotman, London, pp. 307–318.

HOME, P. C. 1987. Vla. *In:* Spencer, A. M. *et al.* (eds) *Geology of the Norwegian Oil and Gas Fields.* Graham & Trotman, London, pp. 143–151.

KOENIG, R. H. 1986. Oil discovery in 6507/7, an initial lok at the Heidrun Field. *In:* Spencer, A. M. *et al.* (eds) *Habitat of Hydrocarbons on the Norwegian Continental Shelf*, Graham & Trotman, London, pp. 307–311.

LARSEN, V., MØRKESETH, P. O. & RASHEIM, S. M. 1987. Tyrihans. *In:* Spencer, A. M. *et al.* (eds) *Geology of the Norwegian Oil and Gas Fields.* Norwegian Petroleum Society and Graham & Trotman, London, pp. 411–418.

LECKIE, G. G. & CHEW, K. J. 1991. The discovered hydrocarbon reserves in western Europe. *In:* Spencer, A. M. (ed.) *Generation, Accumulations and Production of Europe's Hydrocarbons.* European Association of Petroleum

Geoscientists Special Publication No. ?. Oxford University Press, pp. 1–23.

MACKERTICH, D. 1996. The Fife Field, Blocks 31/26a, 31/27a, 39/1, 39/2. UK North Sea. *Petroleum Geoscience*, in press.

McCLURE, N. M., WILKINSON, D. W., FROST, D. P. & GEEHAN, G. W. 1995. Planning extended reach wells in Wytch Farm Field, UK. *Petroleum Geoscience*, **1**, 115–127.

MURE, E. 1987. Frigg. *In:* Spencer, A. M. *et al.* (eds) *Geology of the Norwegian Oil and Gas Fields.* Norwegian Petroleum Society and Graham & Trotman, London, 203–213.

NAYLOR, D. 1996. History of oil and gas exploration in Ireland. *This volume.*

NICHOLSON, P. H., CAYLEY, G. T., CONVERSE, D. R. & GRAY, G. G. 1995. Geologic controls on subsurface pore pressure distribution: North Sea. *In:* Cayley, G. T., Fraser, A. J. & Erratt, D. (eds) *In pursuit of subtle traps.* Geological Society, London (abstract).

NIELSEN, L. H. & JAPSEN, P. 1991. *Deep wells in Denmark 1935–1990.* Danmarks Geologiske Undersøgelse. III Raekke 36.

OELE, J. A., HOL, A. C. P. J. & TIEMENS, J. 1981. Some Rotliegend Gas Fields of the K and L Blocks, Netherlands Offshore (1968–1978) – A Case History. *In:* Illing, L. W. & Hobson, G. D. (eds) *Petroleum Geology of the Continental Shelf of North-West Europe.* Institute of Petroleum, London, pp. 289–300.

OLAUSSEN, S., DALLAND, A., GLOPPEN, T. G. & JOHANNESSEN, E. 1983. Depositional environment and diagenesis of Jurassic reservoir sandstones in the eastern part of Troms I area. *In:* Spencer, A. M. *et al.* (eds) *Petroleum Geology of the North European Margin.* Graham & Trotman, London, pp. 61–79.

PAGES, L., 1987. Exploration in the Paris Basin. *In:* Brooks & Glennie, *Petroleum Geology of Northwest Europe.* Graham & Trotman, London, pp. 87–93.

PATIENCE, R. L., VAN GRAAS, G., ISAKSEN, G. H. & JENSEN, A. I. 1995. High-resolution reservoir geochemistry; Greater Sleipner area, South Viking Graben. *In:* Cayley, G. T., Fraser, A. J. & Erratt. D. (eds) In pursuit of subtle traps. Geological Society, London (abstract).

PEKOT, L. J. & GERSIB, G. A. 1987. Ekofisk. *In:* Spencer, A. M. *et al.* (Eds) *Geology of the Norwegian Oil and Gas Fields.* Norwegian Petroleum Society and Graham & Trotman, London pp. 73–83.

Petroleum Geological Circle 1993. Synopsis: Petroleum Geology of The Netherlands – 1993. *Geologie en Mijnbouw*, pre-print AAPG Conference, The Hague, October 1993.

PLEIN, E. 1979. Das deutsche Erdöl und Erdgas. *Jahrbuch Ges. Naturkde. Wurtemburg*, **134**, 5–33.

RANAWEERA, H. K. A., 1987. Sleipner Vest. *In:* Spencer, A. M. (ed.), *Geology of the Norwegian Oil and Gas fields.* Norwegian Petroleum Society and Graham & Trotman, London pp. 253–263.

ROBERTS, J. D., MATHIESON, A. S. & HAMPSON, J. M. STATFJORD. 1987. *In:* Spencer, A. M. *et al.* (eds) *Geology of the Norwegian Oil and Gas Fields.* Norwegian Petroleum Society and Graham & Trotman, London, pp. 319–340.

ROELOFSEN, J. W. & DE BOER, W. D. 1991. Geology of the Lower Cretaceous Q/1 oil fields, Broad Fourteens Basin, The Netherlands. *In:* Spencer, A. M. (ed.) *Generation, Accumulation and Production of Europe's Hydrocarbons.* European Association of Petroleum Geoscientists., Special publication. Oxford University Press: 203–216.

SCHRÖDER, L., LÖSCH, J., SCHÖNEICH, H., STANCU-KRISTOFF, G. & TAFEL, W.-D. 1991. Oil and gas in the north-west German Basin. *In:* Spencer, A. M. (ed.) *Generation, Accumulation and Production of Europe's Hydrocarbons.* European Assocation of Petroleum Geoscientists Special Publication No. ?. Oxford University Press, pp. 139–148.

SHANNON, P. M., JACOB, A. W. B., MAKRIS, J., O'REILLY, B., HAUSER, F. & VOGT, U. 1994. Basin evolution of the Rockall region, North Atlantic. *First Break*, **12**, 515–522.

———, MOORE, J. G., JACOB, A. W. B. & MAKRIS, J., 1993. Cretaceous and Tertiary basin development west of Ireland. *In:* Parker, J. R. (ed.) *Petroleum Geology of Northwest Europe: Proceedings of the 4th Conference.* Geological Society, London, pp. 1057–1066.

SPENCER, A. M. *et al.* (eds) 1987. *Geology of the Norwegian Oil and Gas Fields.* Norwegian Petroleum Society and Graham & Trotman, London.

SPENCER, A. M. (ed.), 1991. *Generation, Accumulation and Production of Europe's Hydrocarbons.* European Assocation of Petroleum Geoscientists, Special publication No ?. Oxford University Press.

SORGENFREI, T. & BUCH, A. 1964. *Deep tests in Denmark, 1935–1959.* Geological Survey Denmark 111(36), pp. 12–146.

STAFFURTH, J. H., LAVER, R. M. & EVANS, D. G. 1995. Eocene stratigraphic trapping in UKCS Block 21/27. In: Cayley, G. T., Fraser, A. J. & Erratt, D. (eds) In pursuit of subtle traps. Geological Society, London (abstract).

STÄUBLE, A. J. & MILIUS, G. 1970. Geology of the Groningen gas field. *In:* Weeks, L. G. (ed.) *Habitat of Oil.* American Association of Petroleum Geologists, Memoir **14**, 359–369.

STOREY, M. W. & NASH, D. F. 1993. The Eakring Dukeswood Field: an unconventional technique to describe a field's geology. *In:* Parker, J. R. (ed.) *Petroleum Geology of Northwest Europe: Proceedings of the 4th Conference.* Geological Society, London, pp. 1513–1517.

STUART, I. A. & COWAN, G. 1991. The South Morecambe Field, Blocks 110/2a, 110/3a, 110/8a, UK Irish Sea. *In:* Abbots, I. L. (ed.) *United Kingdom Oil and Gas Fields, 25 Years Commemorative Volume.* Geological Society, London, memoir, **14**, 527–541.

TREWIN, N. H. & BRAMWELL, M. G. 1991. The Auk Field Block 30/16, UK North Sea. In: Abbots, I.L. (ed.) *United Kingdom Oil and Gas Fields, 25 Years Commemorative Volume.* Geological Society, London, Memoir, **14**, 227–236.

VAN KESTEREN, J. 1973. The analysis of future subsurface subsidence resulting from gas production in the Groningen Field. *Verhandelingen van het Koninklijk Nederlands Geologisch Mijnbouwkundig Genootschap*, **28**, 11–18.

VERSTRALEN, I. & HURST, A. 1994. Sedimentology and reservoir characteristics of the Rona Sandstone, West of Shetland (UKCS). *First Break*, **12**, 11–20.

WILLS, J. M. 1991. The Forties Field, block 21/10, 22/6a, UK North Sea. *In:* Abbots, I. L. (ed.) *United Kingdom Oil and Gas Fields, 25 Years Commemorative Volume.* Geological Society, London, Memoir, **14**, 301–308.

Petroleum reservoir engineering – a personal perspective

John S. Archer

Imperial College of Science, Technology & Medicine, London SW7 2AZ, UK

ABSTRACT: This paper was invited as part of the AD 1995 – NW Europe's Hydrocarbon Industry Symposium to mark Aberdeen University's 500th Anniversary. The author has taken the opportunity to recall, from a highly personal and selective perspective, some of the events of the last 25 years in reservoir engineering, through his own experiences of North Sea fields. The first part of the paper sets the background to reservoir engineering through some of the key contributions to the literature. The second part recalls the early North Sea reservoir planning and the emergence of multidisciplinary approaches to reservoir characterization and asset management. The third part of the paper focuses on reservoir engineering research at Imperial College, and the final part poses a number of questions on the future of reservoir engineering in a UK North Sea context. The author does not intend to provide a critical appraisal of the reservoir engineering literature that has emerged in the last 25 years, therefore, the reservoir engineering with which the author has been most closely associated naturally receives the greatest prominence in a paper of this type. In no way should this be construed as a claim to invention.

KEYWORDS: *reservoir engineering, North Sea, multidisciplinary*

INTRODUCTION

This paper has been prepared specially for the Symposium to mark the 500th Anniversary of King's College, the forerunner of the University of Aberdeen. I was invited by the organizers to reflect over the time that I have been associated with the petroleum industry and, particularly, the last 25 years of North Sea activity. The paper does not pretend to be a balanced review, but rather represents the things that I have found interesting from a personal perspective and some areas in which I and my colleagues have been able to make a modest contribution.

On occasions such as this it is tempting to seek a connection in the subject matter over the period of the anniversary. It will be virtually impossible to make the connection in petroleum engineering, but the flow of fluids in underground rock systems was of practical interest to the early minerals engineers. Almost exactly at the time of the founding of this University in Aberdeen, another significant event took place. On the 24 March 1494, in Glauchau in Saxony, Georgius Agricola was born. He subsequently joined the students from all over Europe who flocked to Italian universities to be at the centre of the newly awakened learning and, like his contemporaries, he returned to his native country to spread ideas and thinking. In about 1525 Agricola visited Basel and met Erasmus who became his great friend and patron. Around this time, Agricola visited the Universities of Bologna, Venice and probably Padua and there became interested in working in the sciences. In 1527 he was appointed physician in a little Bohemian town in the middle of a prolific mining region. There he spent much of his time visiting the mines and smelters and in researching all he could find in history about the art and science of mineral exploitation. Until his death in 1555 Agricola published profusely on the subjects of minerals, mining and physical geology and these works, in particular *De Re Metallica* (Agricola 1556), were the basis of knowledge for the next two centuries. His book on subterranean waters and gases was part of a series of five published under the title *De Ortu et Causis Subterraneorum*. It is tempting to think that his works were studied in Aberdeen by those early students of the geosciences.

Neither petroleum engineering nor petroleum reservoir engineering are, in my view, strict disciplines. Both are applications of the physical and natural sciences to the extraction of petroleum fluids from the pore spaces of permeable rocks. In order for such processes to be of benefit to humankind and to be of commercial interest they need to be safe, economic and environmentally sound. Petroleum reservoir engineering is, for me, essentially a multidisciplinary field, encompassing the engineering application of physics, mathematics, the earth sciences and economics. It is natural philosophy in the broad – a way of harnessing aspects of the natural world for the improvement of human life. Our forefathers of the Renaissance would have recognized the breadth of this description.

It may be surprising to know that the UK has always been at the leading edge of education in petroleum engineering, and that its strategic importance has long been recognized by politicians. In my own university the first courses in petroleum engineering were started in 1913 as a four-year programme entitled 'Oil Technology', with the aim of providing 'an urgent need for a course of training suitable for those who intend subsequently to follow occupations connected with the oil industry'. The first graduates were amongst the earliest in the world to receive degrees in this field. By the 1920s the course's reputation had been firmly established and the Royal School of Mines (RSM) at Imperial College was recognized as one of the leading centres of training for petroleum technologists in the world. In 1933 the RSM was selected as the venue for the First World Petroleum Congress. The political setting and history at the RSM is described in the excellent commentary by Douglas Hobson

From K. Glennie & A. Hurst (eds), 1996, *AD1995: NW Europe's Hydrocarbon Industry*, Geological Society, London, pp. 17–23

(Hobson 1990), written for the 75th Anniversary Symposium held at Imperial College in July 1988.

THE EARLY DEVELOPMENT OF RESERVOIR ENGINEERING

The science of reservoir engineering was established during the 1930s and 1940s and it built upon an understanding of water flow in porous soils and rocks. The concept of Darcy flow in porous media had been developed as a civil engineering principle by Henri Darcy in the middle of the nineteenth century and was described in his classic work *Les Fontaines Publiques de la Ville de Dijon* (Darcy 1856). From the end of the 1800s until the 1930s, the production of oil from shallow underground porous rocks utilized the natural compressive energy of petroleum fluids and exploited the potential difference between the underground reservoir and the surface. The idea of using a displacing fluid such as water in the recovery of petroleum was both accidental and revolutionary.

The early use of water as a displacement fluid is described by Smith in his classic text on the mechanics of secondary recovery (Smith 1966). He reported that in the Second Geological Survey of Pennsylvania, 1875–9, J. F. Caril noted that water seeped into a wellbore from a level above the oil zone and that, although it reduced productivity in the well, the water found its way to nearby wells and increased their production of oil. In 1888, J. D. Dinsmoor noted the beneficial effect on oil production of using a higher pressure gas sand to communicate with a lower pressure oil sand. He made use of this observation to design a gas injection project in another well in about 1890. A patent by William Richards in 1884 described how maintenance of pressure on an injection well by use of compressed air, gas or fluids could force oil towards adjoining wells, where it could be recovered. In 1917, James Lewis wrote in the *US Bureau of Mines Bulletin* that oil production was being improved by the injection of air and gas into shallow wells and that water injection was proving beneficial in the Bradford oil field. Smith has commented that Lewis appeared to appreciate at this time the heterogeneous character of the oil sands and something of the role of capillary pressure and the viscosity of reservoir fluids on the oil recovery processes. The practice of water injection provided an opportunity to increase the fraction of oil recovered from the underground accumulation from less than 5% to more than 40%. Optimal recovery of petroleum fluids thus became a matter of design.

The scientific framework for modern petroleum reservoir engineering was laid down by giants such as Morris Muskat in his books *Flow of Homogeneous Fluids through Porous Media* (Muskat 1937) and in *Physical Principles of Oil Production* (Muskat 1949). Schilthuis described the physical basis for the distribution of connate water in oil and gas reservoirs (Schilthuis 1938) and thus allowed insight into the estimation of fluids initially in place in reservoirs. Capillary behaviour in porous rocks (Leverett 1941) and the classic contribution to the understanding of oil displacement by water in sandbodies (Buckley & Leverett 1942) were landmarks of knowledge in the field. The recognition of permeability heterogeneity and its role in influencing oil recovery was the subject of another key publication (Stiles 1949) entitled the 'Use of permeability distribution in waterflood calculations'. The post-war era saw consolidation and improvement to the physical and analytical understanding of the flow of petroleum fluids in idealized reservoir rocks. The foundations of well test interpretation were laid with the classic paper (Horner 1951) to the third World Petroleum Congress on pressure build-up in wells, and with Miller, Dyes and Hutchinson's paper (Miller *et al.* 1950) on the estimation of permeability and reservoir pressure from bottom hole pressure build-up characteristics. The books *The Physics of Flow through Porous Media* (Scheideggar 1950) and *Handbook of Natural Gas Engineering* (Katz *et al.* 1959) made significant contributions to knowledge. The classic analysis of Darcy's Law and the field equations of the flow of underground fluids was published in 1956 (Hubbert 1956) in which it was shown how the Darcy equations were related to the Navier–Stokes equations.

In the early 1950s, reservoir engineering interest was largely concerned with the possibility of using laboratory sandpack models and somehow scaling the results obtained from them to field usefulness. The development of the general equations for the displacement of oil by water or gas for each fluid flowing in the direction of the three principal axes thus made a valuable contribution (Rapoport 1955). He used the Darcy equations for each fluid phase, together with the continuity equation, and assumed that the fluids were incompressible and that the process was isothermal. Richardson described how Rapoport's equations could be used with scaling criteria to design sandpack experiments which might help understand field recovery processes (Richardson 1960). Laboratory experimentation provided much of the physical insight into fluid displacement processes, and understanding about the competing displacement forces of capillarity, gravity and viscous flow began to be established. The recognition of the role of displacement stabilization by gravity forces was first described by Hill (Hill 1952) in work on critical rates for channelling in sugar refining, and a year later by Dietz (Dietz 1953). By 1959 the ability to use numerical methods to solve the differential equations of multiphase multidimensional immiscible flow in porous media began to make an impact, and the paper by Douglas, Peaceman and Rachford (Douglas *et al.* 1959) was significant. Advances in high speed computing over the following 35 years have seen the increasing domination of numerical methods over analytical approaches in reservoir engineering. The year 1959 also saw the publication of the classic paper by Johnson, Bossler and Naumann (Johnson *et al.* 1959) on the calculation of relative permeability from unsteady state immiscible displacement experiments. It finally provided individual phase relative permeabilities instead of ratios, and allowed the work of Welge (Welge 1952) and of Buckley and Leverett (Buckley & Leverett 1942) to be used more effectively.

In reservoir engineering, the 1960s were the years in which laboratory experiments were refined, in which reservoir heterogeneity became more important, in which data acquisition in the field became more widespread, and in which computing and numerical methods started to be used in the area of reservoir simulation. Warren & Price (Warren & Price 1961) wrote their paper entitled 'Flow in heterogeneous porous media'. The first monograph commissioned by the Society of Petroleum Engineers in 1967 was the landmark volume entitled *Pressure Build-up and Flow Tests in Wells* (Matthews & Russell 1967). The same year saw a useful contribution (Havlena 1967) entitled 'Interpretation, averaging and use of the basic geological engineering data'.

NORTH SEA DISCOVERIES

The late 1960s and the early 1970s were particularly important years for the UK as the significance of oil and gas discoveries and developments in the North Sea became apparent. Glennie & Hurst (Glennie & Hurst 1996) have overviewed this period as a contribution to the symposium. The First UK Licence Round was in 1964 and 51 companies were awarded 53 licences in 348 blocks. In the Second Round a year later, 44 companies were awarded 37 Licences in 127 blocks. The largest round in those years was the Fourth Round in 1971/1972 in which 213 companies were awarded 118 licences in 282 blocks.

The first gas discoveries in the southern part of the North Sea were made in West Sole in September 1965 and in Viking in December the same year. Production started from West Sole in March 1967. The year 1966 saw discoveries of gas in April at Leman, in June at Indefatigable, and in October at Hewett. A North Sea map from 1970 would show a North Sea petroleum industry centred on the East Anglian coast with no hint of the more northerly activity to come. The first reported oil discovery in UK Continental Shelf (UKCS) waters was the Montrose Field in December 1969 in the central area of the North Sea. In the same area it was closely followed by the discovery of Forties in November 1970, by Auk in February 1971 and Argyll in August 1971. In the Viking Graben area of the Northern North Sea, the first major discovery was the Brent Field in July 1971. There followed Frigg (May, 1972), Beryl, Deveron and Cormorant (September, 1972), Piper (January, 1973), Maureen (February, 1973), Thistle (July, 1973), Hutton and Heather (December, 1973), Statfjord and Ninian (April, 1974), Claymore (May, 1974), Magnus (July, 1974) and Buchan (August, 1974).

RESERVOIR ENGINEERING EXPERIENCES 1970–1990

Reservoir engineering during the last 25 years has been very much influenced by the practices developed in the North Sea arena and by the recognition of 'asset management'. In the following paragraphs I will trace some of these influences and will relate them to events in my own career. This is not to claim invention, but merely to put into a time sequence some of the changes which I regard as memorable. The topics are not presented as critical reviews and no attempt has been made to reference the many valuable papers in the fields.

In 1972 and 1973 the SPE ran two European Spring Meetings. Some of the 'hot issues' of those days are still current. In Amsterdam in 1972 the programme included 'two-phase flow through well head chokes', 'numerical coning applications', 'analysis of statistical methods', 'critical water-free production rates', and 'economics of underground storage of oil and gas'. In London in 1973 the programme included 'flow in fissured reservoirs', 'environmental instrumentation for offshore installations', 'new developments in platform production safety systems', 'top reservoir mapping problems', 'computerized storage and retrieval of well data' and 'economics of oil exploration and production'.

In 1973, the *Journal of Petroleum Technology* published a review entitled 'A quarter of a century of progress in the application of reservoir engineering' (Richardson & Stone 1973). Until 1973, I was with Imperial Oil, part of the Exxon Group, in Calgary, Canada, as a research reservoir engineer. Much of my own work was related to understanding the multiphase flow behaviour of reservoir rocks, and in designing experimental and interpretative procedures for reservoir characterization (Archer & Wong 1973). In those days, our research laboratory was of modest size and it was very easy to have informal contact between engineers and geoscientists concerned with field development and operations. I found it perfectly natural to work on a diversity of projects in loose, project-orientated, multidisciplinary teams. The projects included building reservoir models of fields like Leduc and Golden Spike; understanding the reservoir character of wildcat wells in the Arctic; finding solutions for casing collapse in permafrost sections; and arguing for reserves, and the pipeline capacity share that went with them, at the Oil and Gas Conservation Board hearings in the Calgary Courthouse. It was only much later that I discovered that this was called 'synergy', and that it would be a buzzword for multidisciplinary collaboration in the industry in the late 1970s and early 1980s.

When I returned to the UK in May 1973, I had the good fortune to work with Gas Council (Exploration), the E&P part of British Gas. Multidisciplinary approaches to problem solving were encouraged and I worked as an engineer alongside geoscientists. It was at this time that the mysteries of the Wytch Farm Field were being unravelled (Colter & Havard 1980). There was intense debate in 1973 and 1974 between engineers and geoscientists concerning the significance of the exploration wells drilled on the Isle of Wight at Arreton, which gave a clue to the oil migration route, and hinted at reservoir potential in the Sherwood and Bridport sands. I was the petroleum engineer on the Wytch Farm discovery well in December 1973. I remember being surprised that we shut down over Christmas, and that we had to observe restrictions on well testing so that the house guests of the local landowner would not have their view over Poole Harbour obstructed by smoke. Wytch Farm turned out to be one of the largest onshore oil reservoirs in Europe and the reservoir engineering design was a challenge in understanding heterogeneity.

One of my most interesting tasks in 1973/4 was the analytical design for a gas storage scheme. It was based on gas injection into the reservoir from nearby fields during low-demand months, followed by high deliverability production for peak load shaving during the winter. There was considerable uncertainty regarding the reservoir volume and its variable distribution, and much of the design was based on deliverability predictions based on production well tests. This project eventually became the Rough Field storage scheme (Hollis 1982).

During 1974 I joined the embryonic European office of a Canadian consulting company, where my job was to develop and conduct reservoir engineering and simulation modelling. Reservoir simulation was conducted using decks of punched cards and mainframe computers. One job was for the, then, Burmah Oil Company, to evaluate their interests in the recently discovered Ninian oil field and to conduct a reserves estimation and a development plan. Forecasting of long-term production using reservoir simulation was a time-consuming activity, and it involved a lot of customizing of the 'rates' routine and the output files. The end result may not have been much better than one calculated from fractional flow and material balance methods (which of course I did as a routine check in any case!). The power of reservoir simulation was not really harnessed until sufficient well data became available to understand the flow paths between proposed wells. It was clear to a substantial part of the community that the real challenge lay in making the most of all the available data – and that meant a more structured and multidisciplinary approach to evaluation of the reservoir database. Reservoir engineers set out to form liaisons with geoscientists and sought to use geological models in a more quantitative manner than had been commonplace. Reserves estimates for secondary recovery processes in North Sea oil fields began to be represented in a probabilistic rather than a deterministic manner (Archer 1985), and this was extended to the economic appraisals. The approach initially shocked several of the petroleum engineers employed by US banks, who were engaged in putting together the financing of North Sea field development. The other thing that took some of the analysts a while to accept was that the oil fields in the North Sea would not, except while waiting for water-injection wells to be drilled, have a primary depletion stage. The design of a full-pressure maintenance scheme by water or gas injection from the beginning, in fields like Brent, Dunlin, Forties and Statfjord, was something fairly new.

During that period I also had an interesting contract with the Abu Dhabi Petroleum Company for the design of a pressure maintenance scheme for a reservoir in the Bu Hasa Field. It was in Abu Dhabi that I first came across the use of the Repeat Formation Tester Tool, which was later to make a huge impact on reservoir interpretation in the North Sea. That tool, in conjunction with good geological hypotheses in a dynamic reservoir environment, was capable of allowing great insight into reservoir heterogeneity and compartmentalization (Van Rijswiik *et al.* 1980).

A landmark in 1978 was the publication of the now classic textbook in reservoir engineering by Laurie Dake (Dake 1978), which provided an excellent base for a whole generation of UK-trained engineers. The last 15 years or so have seen remarkable developments in computing and these have influenced the way in which design studies for reservoir development and management are conducted. The old mainframe and the 'overnight run' of a massive deck of punched cards have long gone, and there are now distributed systems of personal workstations. It is currently possible to represent the geometry and heterogeneity of geological facies within reservoirs much more faithfully – at least we have more data and possibly more imagination. We have learnt that nature is more complex than we would have wished to believe. Now many engineers go out and look at reservoir analogues in outcrops, and some are more keenly aware of the limitations of their representations. Twenty-five years ago it was thought that newly discovered North Sea oil fields were very simple and their reservoir engineering could be modelled in simulators with coarse-grid cell dimensions and little heterogeneity between cells. Those early representations gave rise to performance predictions which indicated late-time water-cut development and high recovery factors. The development strategy of the time was based on the earliest possible return on investment and the highest daily rate in the early years. Reservoir management appeared to be something that came along later – and it focused the mind when oil prices plummeted after 1986. What saved the day for many investment decisions based on those simple models was the increasing value of oil. It will be a long time before we see $US 40 per barrel of oil again (Tuft 1994).

In 1977, with ERC-Robertson Research, I helped lead a multiclient study of all the wells drilled in the Viking Graben area of the North Sea (RRI-ERC 1980). This database of reservoir information was interpreted by an interdisciplinary team to show the added value that might be obtained from these data if a closely integrated geoscience/engineering study was made. To that end, over the subsequent three years, structural geology, geochemistry, sedimentology, petrophysics and reservoir engineering was integrated. I do not think the sales of the report resulted in a huge net profit, but it was an enormously exciting project to do and on a technical level we learnt a great deal. One spin-off was the initiation, in 1979, of one of the first integrated field courses in the UK for geoscientists and engineers entitled 'North Sea Jurassic Reservoirs'. It was a combination of classroom sessions and discussions at selected outcrop sites along the Yorkshire coast between Whitby and Scarborough. The course was put together and led by Nigel Hancock and me, and later Nigel's role was taken on by Jim Harris. It was a great success in promoting multidisciplinary contact and ran for many years, long after I left ERC. The outcrops have become extremely well-known analogues for North Sea reservoirs (Archer & Hancock 1980; Rudkiewicz *et al.* 1989) and there is now a body of knowledge about them which has helped to guide many people in the industry. We also developed a similar course for the Moray Firth area which helped the understanding of the Beatrice Field. At a later date, I transported the concept to the Bridport Sands at West Bay, to assist students at Imperial College in visualizing the upper reservoir at Wytch Farm.

The late 1970s and early 1980s saw a number of significant privatizations and sales of oil and gas interests. The break-up of the British National Oil Corporation and the sale of BGC's oil interests to form Enterprise Oil were major exercises which required extensive multidisciplinary co-ordination, and ERC performed the technical audit for all the reservoirs involved.

One of the outcomes of demonstrating the 'added value' of a multidisciplinary approach to reservoir management was that in 1979, Kyrre Nese of Statoil asked me to help him develop the approach with his young Statfjord team. My job was to catalyse group working between geoscientists, petrophysicists and reservoir engineers through periodic visits, teaching and setting project related work assignments. The work was goal-orientated, and targets were defined and progress monitored. Initially, we met for two days every two or three weeks and, in between times, the members of the group worked on their assignments. Later on, as the group gelled, the period between my visits lengthened to five or six weeks. The group became used to working closely together, and my role became much more one of helping technology transfer, aiding communication, bringing in relevant expertise and being an 'explainer'. Over the years some people moved out of the group, new ones moved in and the cycles repeated themselves. The culture spread and, after some 14 years and the development of several fields, it has been considered reasonably successful.

RESERVOIR ENGINEERING AT IMPERIAL COLLEGE

At Imperial College I have had an opportunity to look at the research needs of the industry and to help shape some of the graduates who have assisted in the development of North Sea resources.

The industry now concentrates its attention on obtaining the best possible value for money when selecting new well locations in reservoirs already in production. This involves a greater interest in the prediction of fluid displacement flow paths and in understanding the heterogeneity of reservoirs. I have always found it valuable to make use of the full-fractional-flow equation in which the balance between viscous, capillary and gravity forces can be appreciated, and which remains a powerful analytical tool (Archer & Wall 1986). Once again, the ability to represent physical processes through the mathematics of reservoir simulation has advanced more quickly than our ability to describe the reservoir and the complexity of its rock–fluid interactions. At Imperial College, my research team devoted its energy to progress on both fronts. We developed one of the first really portable field minipermeameters (Hamp 1985–6; Daltaban *et al.* 1989) and we used it to map permeability variations in analogue reservoir rocks. Our approach to tool design went hand-in-hand with numerical modelling of the flow behaviour, so that we were able to understand not only the calibration process, but also what the 'signal' really meant in terms of a resultant permeability (Wang 1992; Wang *et al.* 1993). Some of the early field testing of our prototype devices were on the Brora sandstones and involved Andy Hurst, then with Statoil, and Jon Lewis, then at Imperial College. It became very obvious to many in the field that the revealed permeability distributions could help condition geostatistical models of reservoir heterogeneity, at least so long as the underlying 'rock-type' differences were respected (Al-Rumhy *et al.* 1991). We interpreted rock-type in the petrophysical and fluid-flow behaviour sense, rather than in the sense of the sedimentary unit. As a result of field portable minipermeameters becoming widely used in the UK and the USA, there was a rapid availability of information describing permeability variation in analogue reservoirs, which helped in establishing geostatistical mapping as a useful precursor to reservoir simulation modelling. Geostatistical realizations have resulted in a growth in stochastic modelling of reservoirs, although deterministic methods reflecting a base case, together with sensitivity analyses, still represent common usage (Archer 1990).

Permeability distributions obtained from analogue formations provide only part of the information required for reservoir characterization, and even that can be difficult to relate to real reservoir data. A significant limitation in the use of minipermeameter measurement appears to be that it is best as hemispherical single phase 'absolute permeability' in dry rock. When a second phase is present the interpretation of 'effective phase permeability' as a function of different but poorly defined saturation conditions, remains difficult. The average saturation in the 'tested' region may only be one of the parameters which may influence 'relative permeability'. The influence of rock type and pore geometry on residual saturations and on relative permeability has been indicated from work on several Jurassic reservoirs (Archer 1987), but the ability to use the relationships in reservoir simulation models was, and probably is still, not widely utilized. Other influences on saturation distributions at reservoir conditions which control effective phase permeability is emerging from work on electrical characteristics of a variety of rock types (Jing *et al.* 1992; Elashahab *et al.* 1995).

Interest in reservoir heterogeneity and the modelling of fluid displacement pathways in producing reservoirs (Archer & Wall 1986) has been fuelled by the need to locate wells effectively. It has also revived research interests in the community in reservoir simulation concerning grid orientation and grid-block aspect ratios, as well as on modelling the connection between cells containing well-bore perforations and adjacent grid blocks. In order to utilize at the scale of grid cells reservoir data which come from measurements at a variety of scales there has been great interest, by various groups around the world including my own, in the subject of upscaling and downscaling to achieve effective grid-block values. The ability to understand effective properties in simulation has also benefited in the last ten years from wholesale changes in the way in which computer simulations are performed. The large mainframe devices have been largely superseded by powerful workstations. Pre-processing, post-processing, and graphic visualizations of fluid displacement simulations are now highly sophisticated and have replaced the reams of numerical output that reservoir engineers used to deal with (see for example Pink 1992; Caamano *et al.* 1994; Balough *et al.* 1994). There is also an increasing awareness amongst reservoir engineers that parallel and vector parallel devices can offer advantages in modelling particular processes. At Imperial College, in the Centre for Advanced Parallel Computing, we are exploring this potential with the Fujitsu AP1000 system.

The ability to model the behaviour of highly deviated and horizontal wells has been recognized around the world as of particular importance in reservoir engineering. My first introduction to the use of horizontal wells was in 1970, when I explored the potential for a horizontal well to reduce coning restrictions on production rates in the Leduc Field in Canada. In the mid-1980s, at Imperial College, we embarked on a programme to model inflow behaviour of horizontal wells and to represent it effectively in conventional reservoir simulation well blocks. This work has been regarded as largely successful (Folefac & Archer 1990; Padua *et al.* 1991) and has provided an ability to model horizontal well performance in quite complex heterogeneous environments in order to help select preferred well locations. The use of horizontal wells in a North Sea field like Gullfaks has been an example of effective reservoir management (Tollefsen *et al.* 1992). A variation on the theme of horizontal wells, with which we have been associated since 1986, is the multilateral trunk and branch concept in which multiple horizontal or high-angle wells emanate from a single wellbore. It takes a long time for an idea to catch on, but several operators in the North Sea now seem interested in exploiting the concept.

Tensorial representation of permeability and effective-phase permeability has received considerable attention in recent years, and effort has been directed towards modelling flow in directions which are not orthogonal to the orientation of grid cells. This is quite a powerful technique, and some of our own research has focused on modelling fracture systems within a permeable matrix, and for highly contrasting permeable systems. Fluid flow paths are, of course, influenced by structural effects as well as by sedimentary character. Under the auspices of NERC and with the co-operation of BP, we have utilized tensorial permeability fracture model concepts (Folefac *et al.* 1991, 1992; Wang *et al.* 1992*b*) to model flow paths where fault dislocations are anticipated at a scale below that which can be interpreted from conventional seismic methods. An example of the description of fault patterns and characteristics is the work emerging from the Fault Modelling Group at Liverpool University (Yielding *et al.* 1992; Gillespie *et al.* 1993). At present, the mathematical ability to compute the flow behaviour appears to be in advance of the ability to characterize real fault systems. The resolution of the problem is unlikely be absolute, but where the characterization may be important, there are opportunities to add value through multidisciplinary interpretation. For example, multi-phase tracer tests are becoming more widely utilized, particularly in the Norwegian sector of the North Sea (Rogde & Ljosland 1992), there are significant advances in measuring downhole pressure using fibre optics (Geisler 1993), and my group has been active in the subsequent interpretation of well tests in complex reservoir systems (Kumoluyi 1994).

There is great interest in exploring the potential for modelling flow through permeable reservoir regions with arbitrary geometry and with or without internal flow barriers. In collaboration with colleagues in the Aeronautics Department at Imperial College, we have developed a technique involving mathematics derived from 'panel methods' which was originally used to compute flow around surfaces with complex geometry, such as the space shuttle. The method has now been extended for application to non-zero permeability surfaces (Wong *et al.* 1990). We have also included particle tracking methods in the vicinity of permeability variation and it appears that the method could, in some instances, have significant computational advantages over fine-grid finite-difference methods in tracking fluid flow paths.

FUTURE DIRECTIONS

The oil production in the UKCS area is now higher than it has been for several years and the DTI is optimistic (Coleman 1994) about improved recovery techniques.

So far as present challenges in North Sea reservoir engineering are concerned, my own team has two particular areas of interest, namely improved gas reservoir engineering and the modelling of dynamic permeability change. The first area includes the performance of gas condensate reservoirs, as well as recognizing the enormous gas potential still remaining in the North Sea. It is particularly appropriate to focus on issues of gas reservoir management in a deregulated market and in the context of CRINE (Tuft 1994). Our interest in dynamic permeability change has been long lasting, but is of current significance in the context of blowdown of major North Sea oil reservoirs and in the development of new high-pressure reservoirs. For many years (Marsden *et al.* 1989) we have studied the geomechanical behaviour of a wide range of reservoir rocks at elevated temperatures and pressures, and have developed appropriate experimental and interpretative techniques. More recently, these interests have extended to the understanding of petrophysical and rock–fluid interaction in these systems at high pressures and high temperatures in order to devise physical models of the observed behaviour. We are developing more rigorous mathematical models which go beyond empirical correlations, and we find that tensorial effective-phase permeability appears to provide a necessary bridge. The models allow for permeability variation during the processes of oil and gas production according to defined physical criteria. These pore-space modification processes have not yet all been described, but are likely to include the consequences of pore-pressure decline, fluid interactions during production, and the result of stimulation treatment.

What of the more distant future? As ever, that will be influenced by demand for petroleum products, by environmental considerations and by oil and gas prices. We might imagine that the trends set by CRINE will continue, and I hope that the role of reservoir engineers will continue to be pivotal. We still recover only some 50% of the oil in place in UKCS oil reservoirs. The current knowledge base will inevitably become more accessible through the use of 'expert systems' and 'artificial intelligence'. Will that knowledge base be renewed and extended? Scientific expertise must be nurtured in industrial laboratories as well as in the universities. Although such a situation might just be tolerated for a while by the industry, the need for an environment in which innovation and continuing accumulation of knowledge can thrive seems to me to be essential if new challenges in oil and gas recovery are to be met. The major oil companies now appear to be less research intensive than major international companies in other fields (Pink 1992). New partnerships will need to be forged between the operating companies, the service and contract companies and the universities if there is to be a new knowledge base with long-term value. There is already a shift towards contractor companies commissioning research from the universities and they are hiring many of the trained petroleum engineers coming from the universities. In 25 years time will we find asset owners managing natural resources, or will they have abrogated those responsibilities to contractors? Will industry remain competent to commission reservoir engineering research from the universities, and will the universities still have the resources to maintain their reservoir engineering expertise and training facilities? Probably not, at least without a considerable investment from industry and government. The Rothschild Principle for an 'effective market forces' economy will break down if there is not an informed purchaser to commission research from an informed supplier. The future for reservoir engineering in the next 25 years hangs very much in the balance.

REFERENCES

AGRICOLA, G. 1556. *De Re Metallica*, translated 1950 (Hoover, H. C. & Hoover, L. H.), Dover, NY.

AL-RUMHY, M., ARCHER, J. S. & DALTABAN, T. S. 1991. A Synergistic approach to characterisation of reservoir permeability: a conditional kriging method. Paper SPE 21446, Proceedings of MEOGC, Abu Dhabi.

ARCHER, J. 1985. Reservoir volumetrics and recovery factors. *In:* Dawe, R. A. & Wilson, D. C. (eds) *Developments in Petroleum Engineering*. Academic Publisher, London, 11.

——— 1987. Some factors influencing fluid flow in petroleum reservoir models. *In: North Sea Oil and Gas Reservoirs*. Graham & Trotman, London ,327.

——— 1988. Heterogeneity of permeability in oil reservoirs. *In:* Ala, M. (ed.) *75 Years of Progress in Oilfield Science and Technology*. Balkema, Rotterdam.

ARCHER, J. S. 1990. Reservoir uncertainties. In *Proceedings of the Third Conference on Reservoir Management in Field Development and Production, Kristiansand, Norway.*

——— & HANCOCK, N. 1980. An appreciation of middle Brent sand reservoir features by analogy with Yorkshire coast outcrops. *EUR Paper 197, Proceedings of Europec*, 501.

——— & WALL, C. 1986. *Petroleum Engineering – Principles and Practice*. Graham & Trotman, London.

——— & WONG, S. 1973. Use of a simulator to interpret laboratory waterflood data. *Society of Petroleum Engineering Journal*, 343.

BALOUGH, S. *et al.* 1994. Managing oilfield data management. *Oilfield Review* (July), Schlumberger, 32.

BUCKLEY, S. & LEVERETT, M. 1942. Mechanism of fluid displacement in sand. *Trans AIME*, **146**, 107.

CAAMANO, E. *et al.* 1994. Integrated Reservoir Interpretation. *Oilfield Review* (July), Schlumberger, 50.

COLEMAN, B. 1994. IOR: the new challenges on the UKCS. *SPE Rev* (Oct), 11.

COLTER, V. S. & HAVARD, D. J. 1980. The Wytch Farm oilfield, Dorset. *In:* Illing and Hobson (eds) *Pet Geol Cont Shelf NW Europe*. (Publ) Heydon and Son, London, 494.

DAKE, L. 1978. *Fundamentals of Reservoir Engineering*. Elsevier, Amsterdam.

DALTABAN, T. *et al.* 1989. Field Minipermeameter Measurements – their collection and interpretation. *Proc Eur Symp IOR, Budapest*, 671–682.

DARCY, H. 1856. *Les fontaines publiques de la ville de Dijon*. Dalmint, Paris.

DIETZ, D. 1953. A theoretical approach to the problem of encroaching and by-passing edge water. *Proc Acad van Wetenschappen, Amsterdam*, **56B**, 83.

DOUGLAS, J., PEACEMAN & RACHFORD 1959. A method for calculating multidimensional immiscible displacement. *Trans AIME*, **216**, 297.

ELASHAHAB, B. M. *et al.* 1995. Resistivity index and capillary pressure hysteresis for rock samples of different wetability characteristics. *SPE 29888 Proc 9th MEOS, Bahrain*.

FOLEFAC, A. & ARCHER, J. 1990. Modelling of horizontal well performance to provide insight in coning control. *Rev IFP*, **45**, 51.

——— *et al.* 1991. Effect of pressure drop along horizontal wellbores on well performance. *Proc Conf Offshore Europe, Aberdeen*.

——— *et al.* 1992. Explicit representation of flow in a fractured porous medium with boundary fitted coordinates. *Eur Conf Math Oil Prod, Delft*.

GEISLER, W. 1993. Fiber optics improves BHP monitoring reliability. *World Oil* (Oct), 89.

GILLESPIE, P. A. *et al.* 1993. Measurement and characterisation of spatial distributions of fractures. *Technophysics* (**226**), 113.

GLENNIE, K. W & HURST, A. 1995. Hydrocarbon exploration and production in NW Europe: an overview of some key factors. *This volume*.

HAMP, R. 1985/6. The design and calibration of a field portable minipermeameter. *MEng Dissertation, Imperial College, London*.

HAVLENA, D. 1967. Interpretation, averaging and use of the basic geological-engineering data. *J Can Pet Tech (6)*, 236.

HENRIQUES, A. *et al.* 1990. Novel simulation techniques used in a gas reservoir with a thin oil zone. *SPE 21181, LAPEC*, Oct.

HILL, S. 1952. Channelling in packed columns. *Chem Eng Sci*, **I** (6), 246.

HOBSON, G. D. 1990. The history of the oil technology course and its offshoots. *In:* Ala, M. (ed.) *75 Years of Progress in Oilfield Science and Technology*, Balkema, Rotterdam.

HOLLIS, A. 1982. Some petroleum engineering considerations in the change over of the Rough gas field to the storage mode. *EUR Paper 295, Proc Europec*, 175.

HORNER, D. 1951. Pressure build up in wells. *In:* Brill (ed) *Proc 3rd World Petroleum Congress, Leiden*, **11**, 503.

HOWARTH, G. 1994. Decision making and mature fields. *Pet Rev* (Sept), 401.

HUBBERT, M. K. 1956. Darcy's Law and the field equations of the flow of underground fluids. *Trans AIME*, **207**, 222.

JING, X. *et al.* 1992. Laboratory study of the electrical and hydraulic properties of rocks under simulated reservoir conditions. *Marine and Pet Geol*, (9), 115.

JOHNSON, E., BOSSLER & NAUMANN 1959. Calculation of relative permeability from displacement experiments. *Trans AIME*, **216**, 370.

KATZ, D. *et al.* 1959. *Handbook of Gas Reservoir Engineering*. McGraw Hill, NY.

KUMOLUYI, A. 1994. Well test modelling identification using higher order neural networks. *SPE 27558, Proc Eur Pet Comp Conf, Aberdeen*.

LEVERETT, M. 1941. Capillary behaviour in porous solids. *Trans AIME*, **142**, 152.

MARDSEN, J. *et al.* 1989. An investigation of peak rock strength behaviour for wellbore stability application. *Proc Ann Conf Int Soc Rock Mech mtg: Rock at Great Depth, Pau*.

MATTAX, C. & DALTON, R. 1990. *Reservoir Simulation*. SPE Monograph, SPE, Richardson, Tx.

MATTHEWS, C. & RUSSELL, D. 1967. *Pressure build up and flow tests in wells*. SPE Monograph No 1, SPE, Dallas, Tx.

MILLER, C., DYER & HUTCHINSON 1950. The estimation of permeability and reservoir pressure from bottom hole pressure build up characterisation. *Trans AIME*, **189**, 91.

MUSKAT, M. 1937. *Flow of homogeneous fluids through porous media*. McGraw Hill, NY.

——— 1949. *Physical principles of oil production*. McGraw Hill, NY.

PADUA, K. *et al.* 1991. Horizontal well modelling in numerical simulation. *Proc Int Symp Horiz Well Tech, Indonesia*, 73.

PINK, M. 1992. Exploration and Appraisal Technology. *Selected Paper, Shell Int Pet Co, London*.

RAPOPORT, L. 1955. Scaling laws for use in design and operation of water–oil flow models. *Trans AIME*, **204**, 143.

RICHARDSON, J. 1960. Flow through porous media. *In:* Streeter (ed.) *Handbook of Fluid Dynamics*, Sect 16.

RICHARDSON, J. & STONE, H. J. 1973. A quarter century of progress in the application of reservoir engineering. *J. Pet Tech*, **25**, 1371.

ROGDE, S. & LJOSLAND, E. 1992. Tracer testing: – field examples – radioactive tracers. *In:* Skjaeveland, S. and Kleppe, J. (eds) *Recent advances in improved oil recovery methods for N Sea sandstone reservoirs*. NPD, Stavanger as SPOR Monograph.

RRI-ERC. 1980. *The Brent Sand in the North Viking Graben, N. Sea – A Sedimentological and Reservoir Engineering Study*. RRI, Llandudno.

RUDKIEWICZ, J. L. *et al.* 1989. An integrated software for stochastic modelling of reservoir lithology and property with an example from the Yorkshire Middle Jurassic. *In: North Sea Oil and Gas Reservoirs II*. Graham & Trotman, London.

SCHEIDEGGAR, A. 1957. *The physics of flow through porous media*. (Publ) Macmillan, NY.

SCHILTHUIS, R. 1938. Connate water in oil and gas sands. *Trans AIME*, **127**, 199.

SEYMOUR, R. & ARCHER, J. 1990. Some requirements from seismic methods for use in reservoir simulation models. *In: North Sea Oil and Gas Reservoirs 11*. (Publ) Graham and Trotman, London.

SHERWOOD, J. 1993. 3D Seismic and geostatistical methods provide more accurate reservoir models. *Pet Eng Intl* (Dec), 15.

SKJAEVELAND, S. & KLEPPE, J. (eds). 1992. *Recent advances in improved oil recovery methods for N Sea sandstone reservoirs*. NPD, Stavanger as SPOR Monograph.

SMITH, C. R. 1966. *Mechanics of Secondary Oil Recovery*. Litton Educ Pub, NY.

STILES, W. 1949. Use of permeability distribution in waterflood calculation. *Trans AIME*, **186**, 9.

TOLLEFSEN, S. *et al.* 1992. The Gullfaks field development challenges and perspectives. *SPE 25054, Europec, Cannes*.

TUFT, V. 1994. CRINE – a drive for cost reduction offshore. *Offshore Technology*, (Sept), 15.

VAN RIJJSWIIK, J. *et al.* 1980. The Dunlin field – review of field development and performance to date. *EURPaper 168, Proc Europec*, 217.

WANG, J. 1993. Permeability description in the presence of small scale heterogeneity. *PhD Thesis, University of London*.

——— *et al.* 1992. The role of permeability tensors in modelling fluid flow in fractured reservoirs. *OSEA 92214, Proc OSEA, Singapore*.

——— *et al.* 1992. The use of permeability tensor in modelling heterogeneity and fractured flow media. *SPE 24503, Proc ADNOC/SPE, Abu Dhabi*.

WARREN, J. & PRICE, A. 1961 Flow in heterogeneous porous media. *Trans AIME*, **222**, 153.

WELGE, H. 1952. A simplified method for computing oil recovery by gas or water drive. *Trans AIME*, **195**, 91.

WONG, D. *et al.* 1990. Modelling flow through heterogeneous porous media with boundary integrals using higher-order surface singularities. *Proc 2nd Eur Conf Math Oil Prod, Technip (ed), Paris*.

YIELDING, G. *et al.* 1992. The prediction of small scale faulting in reservoirs. *First Break*, **10**, 449.

North Sea hydrocarbon plays and their resources

A. M. Spencer[1], G. G. Leckie[2] and K. J. Chew[3]

[1] *Statoil, 4035 Stavanger, Norway*

[2] *Petroconsultants, 266 Upper Richmond Road, London SW15 6TQ, UK*

[3] *Petroconsultants, 24 Chemin de la Mairie, 1258 Perly, Geneva, Switzerland.*

ABSTRACT: Two main systems of hydrocarbon plays occur in the North Sea basin when grouped according to their source rock. The most important and varied system comprises the 403 oil and gas finds in six plays related to Upper Jurassic marine shale source rocks of the Jurassic rift province of the central and northern North Sea. These are the plays reservoired in Triassic to Palaeozoic, Lower–Middle Jurassic, Upper Jurassic, Lower Cretaceous, Upper Cretaceous and Palaeogene strata in the UK, Norwegian and Danish sectors. The second main system comprises 299 gas finds offshore and 67 onshore in three plays related to Upper Carboniferous coal source rocks in the southern North Sea region. Finds occur in Upper Carboniferous, Permian and Triassic plays in the UK and Dutch sectors and onshore in The Netherlands. A third, minor, system comprises 30 oil finds onshore and offshore The Netherlands sourced from Lower Jurassic marine mudstones. In total 100×10^9 barrels of oil equivalent resources have been discovered in the North Sea basin, 71% in the Upper Jurassic sourced system, 29% in the Carboniferous gas system and less than 1% in the third minor play.

KEYWORDS: *North Sea, hydrocarbon plays, resources, finding rates*

INTRODUCTION

Exploration in the North Sea since 1964 has involved the drilling of 2617 new field wildcat wells and resulted in 722 discoveries. These discoveries, plus the 77 accompanying onshore finds, have originally recoverable hydrocarbon resources totalling 100×10^9 barrels oil equivalent (BOE). This article reviews the petroleum geology and finding patterns of these offshore discoveries, including also the related finds if the plays continue onshore. There are two main systems of hydrocarbon plays when grouped according to their source rocks. In the centre and north of the North Sea, oil and gas finds are intimately associated with the presence of a late Jurassic to early Cretaceous rift system buried beneath a Cretaceous and Tertiary cover. Thick, organic-rich, syn-rift marine mudstones were laid down throughout most of the rift system and provide the main source rocks. Reservoir rocks, principally sandstones, are present in every system from Devonian to Oligocene. In the south of the North Sea, gas finds occur in a broad E–W belt related to the presence of a thick Upper Carboniferous coal sequence. The gas finds occur principally in sandstone reservoirs in Carboniferous, Permian and Triassic strata. In the centre and north of the North Sea, except for a tiny faulted strip in northeast Scotland, the hydrocarbon plays do not extend onshore. In the south, however, the hydrocarbon plays do extend onshore. Nine minor gas finds occur onshore in Yorkshire. The gas finds onshore Netherlands include the largest find in the whole North Sea region – the Groningen gas field – the discovery of which in 1959 was the impetus which provoked the offshore exploration. A third, minor, hydrocarbon play occurs in The Netherlands, where oil finds in Jurassic and Cretaceous sandstones have been sourced from Lower Jurassic marine mudstones.

This article first outlines the main features of the geology which have been important in creating the favourable conditions for the hydrocarbon plays. Following this the plays are then reviewed, particularly their controlling factors and the discovery pattern of their resources.

GEOLOGICAL HISTORY

The geological history of the North Sea region is well known (Andrews *et al.* 1990; Ziegler 1990; Cameron *et al.* 1992; Cope *et al.* 1992; Johnson *et al.* 1993; Gatliff *et al.* 1994). A few particular episodes have had most effect on the development of the petroleum finds (Figs 1, 2).

Devonian continental deposits, which accumulated in intramontane grabens and basins at the end of the Caledonian Orogeny, are known from Scotland, Norway and the central and northern North Sea. Marine conditions were probably widespread in the southern North Sea in early Carboniferous times gradually changing over by Westphalian times to widespread coal swamps in a major sedimentary trough north of the developing Hercynian mountains.The pattern in Scotland was different, with Carboniferous sedimentation in separate basins of small extent, and this is likely also to have been the case beneath the central and northern North Sea. In the south the Carboniferous strata were gently arched and faulted and a major unconformity was developed, above which come the desert deposits of the Rotliegend. These are widespread in both the southern and central North Sea. This continental basin was flooded, from the Arctic Ocean far to the north, by the Zechstein Sea, the repeated evaporation of which gave rise to a bulls-eye facies pattern with peripheral carbonates and thick evaporite sequences in the centre. Two main depositional basins formed, in the south and centre, separated by the Mid North Sea High. In Triassic times,

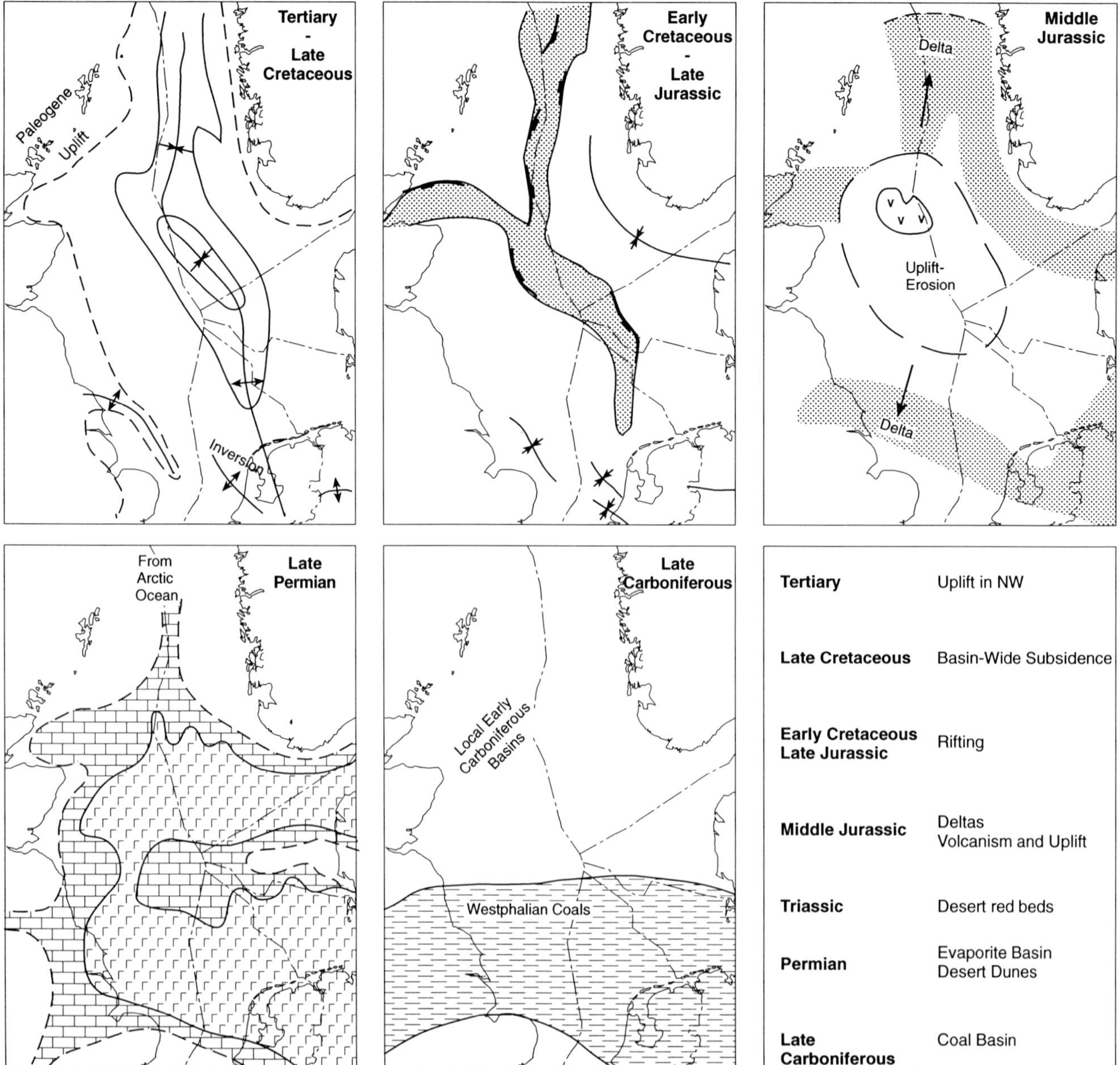

Tertiary	Uplift in NW
Late Cretaceous	Basin-Wide Subsidence
Early Cretaceous Late Jurassic	Rifting
Middle Jurassic	Deltas Volcanism and Uplift
Triassic	Desert red beds
Permian	Evaporite Basin Desert Dunes
Late Carboniferous	Coal Basin

Fig. 1. Geological history of the North Sea: the events most relevant to the petroleum prospectivity.

continental red beds accumulated throughout the basin and, especially in the centre and north, N-trending rifting occurred.

Shallow marine strata accumulated throughout the region in the early Jurassic. An important volcanic episode took place in the central North Sea in middle Jurassic times and was accompanied by major regional doming and the accumulation of deltaic sequences on the periphery of the dome (Underhill & Partington 1993). The most important event with respect to the hydrocarbon finds in the centre and north was the ensuing rifting episode there in late Jurassic to early Cretaceous times (Fig. 3). These movements created the tilted, eroded, fault blocks – the most important traps – and the grabens, in which restricted marine circulation caused anoxic conditions allowing the deposition of source rocks. In Cretaceous times, active rifting ceased but basinal subsidence and marine sedimentation occurred throughout the region with lacustrine and shallow marine conditions in the south giving way to deeper, turbiditic basins in the north; late Cretaceous times saw chalk sequences everywhere, except in the far north where thick mudstones accumulated.

In the southern North Sea and The Netherlands important inversion movements affected the Mesozoic troughs in latest Cretaceous to Tertiary times. Except for these movements, broad subsidence and marine conditions affected the whole of the North Sea basin thoughout the Tertiary. Atlantic opening in early Eocene times was preceded by major volcanic activity from Scotland to East Greenland. The accompanying uplift of northern Scotland shed a flood of clastic detritus into the north and centre of the subsiding North Sea basin in the Palaeocene and Eocene times, providing the youngest major reservoir sequence there.

Neogene uplifts of the Norwegian and Scottish landmasses have affected some hydrocarbon finds and the prospectivity of the adjacent edges of the North Sea basin.

SOURCE ROCKS AND MATURITY

In the north and centre, late Jurassic to early Cretaceous rifting resulted in marine strata being deposited throughout

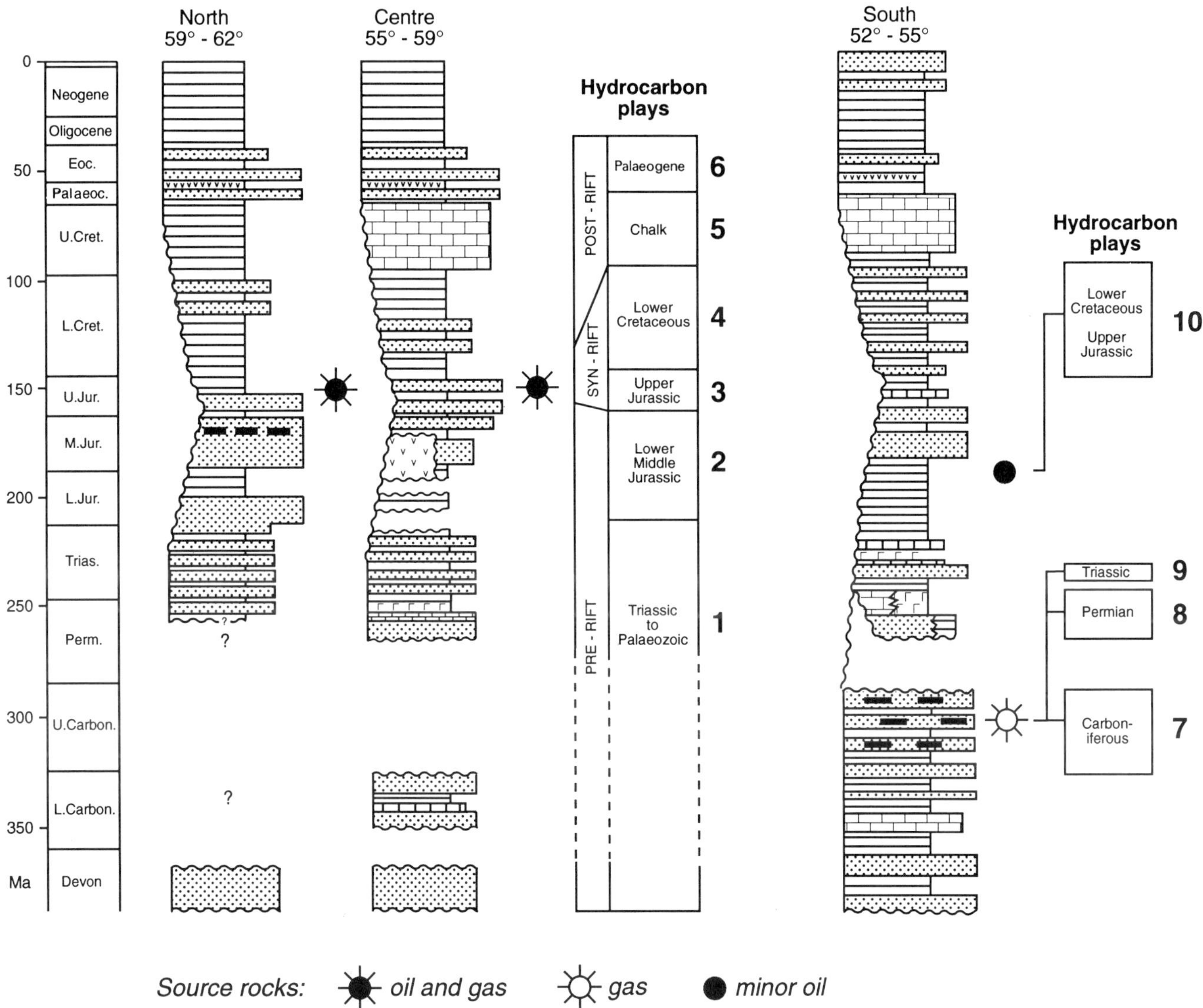

Fig. 2. Generalized stratigraphy of the North Sea and classification of the hydrocarbon plays by source rock and reservoir age.

the basin. The thickest sequences exceed 3 km, whilst incomplete sequences, sometimes only a few metres thick, cover highs and fault blocks. These strata are predominantly mudstones and range from Callovian to Ryazanian in age (Doré *et al.* 1985). At many levels there are black shales with high radioactivity ('hot shales') and total organic carbon contents ranging up to 15%. Many studies have indicated that these 'hot' shales are the principal source rocks of the northern North Sea (e.g. Schou *et al.* 1985; Mackenzie *et al.* 1987). The 'hot' shales achieved maturity during their continuous burial from Cretaceous to recent times. Oil generation began in late Cretaceous times and gas generation was achieved in Neogene times. Due to the continuous subsidence, the widest area of generation and the maximum rank of generation occur at the present day. The regions over which these source rocks have achieved maturity for oil generation are outlined on Figs 4 to 8.

In the southern North Sea region, from eastern England to Germany, a Westphalian A–B coal measure sequence is present everywhere. It reaches 1000 m or more in thickness and contains many coal seams and carbonaceous shales. Throughout the region, burial to present day depths of 1 to 6 km, although in several areas interrupted by inversion, has allowed gas to be generated over wide areas beginning in Jurassic times (Cope 1986; Cornford 1990).

In parts of the Netherlands sector of the North Sea and onshore in the West Netherlands Basin Toarcian marine oil shales a few tens of metres thick have generated minor quantities of oil.

HYDROCARBON PLAYS

A hydrocarbon play is an association of finds, prospects and leads which share the same reservoir, hydrocarbon source system and regional top seal. Application of this concept is subjective. We believe the petroleum geology of the North Sea can be analysed by identifying a series of ten hydrocarbon plays (Fig. 2). Many of the ten plays may be groups of plays but in this review we prefer not to subdivide too finely so as to be able to give a concise overview.

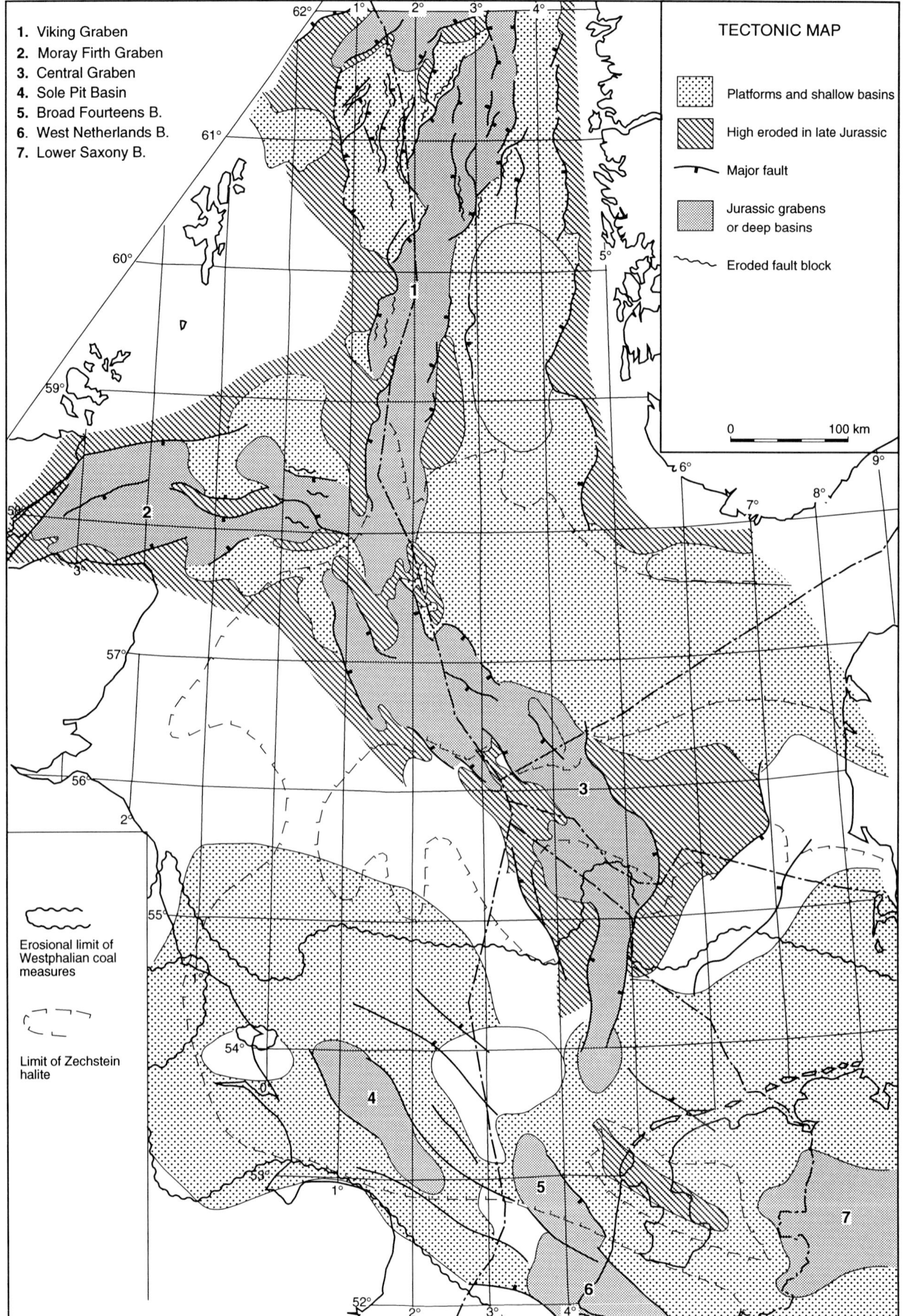

Fig. 3. Tectonic elements of the North Sea of most relevance to the petroleum prospectivity: the three-arm late Jurassic rift system in the centre and north and the Carboniferous - Permian basin in the south.

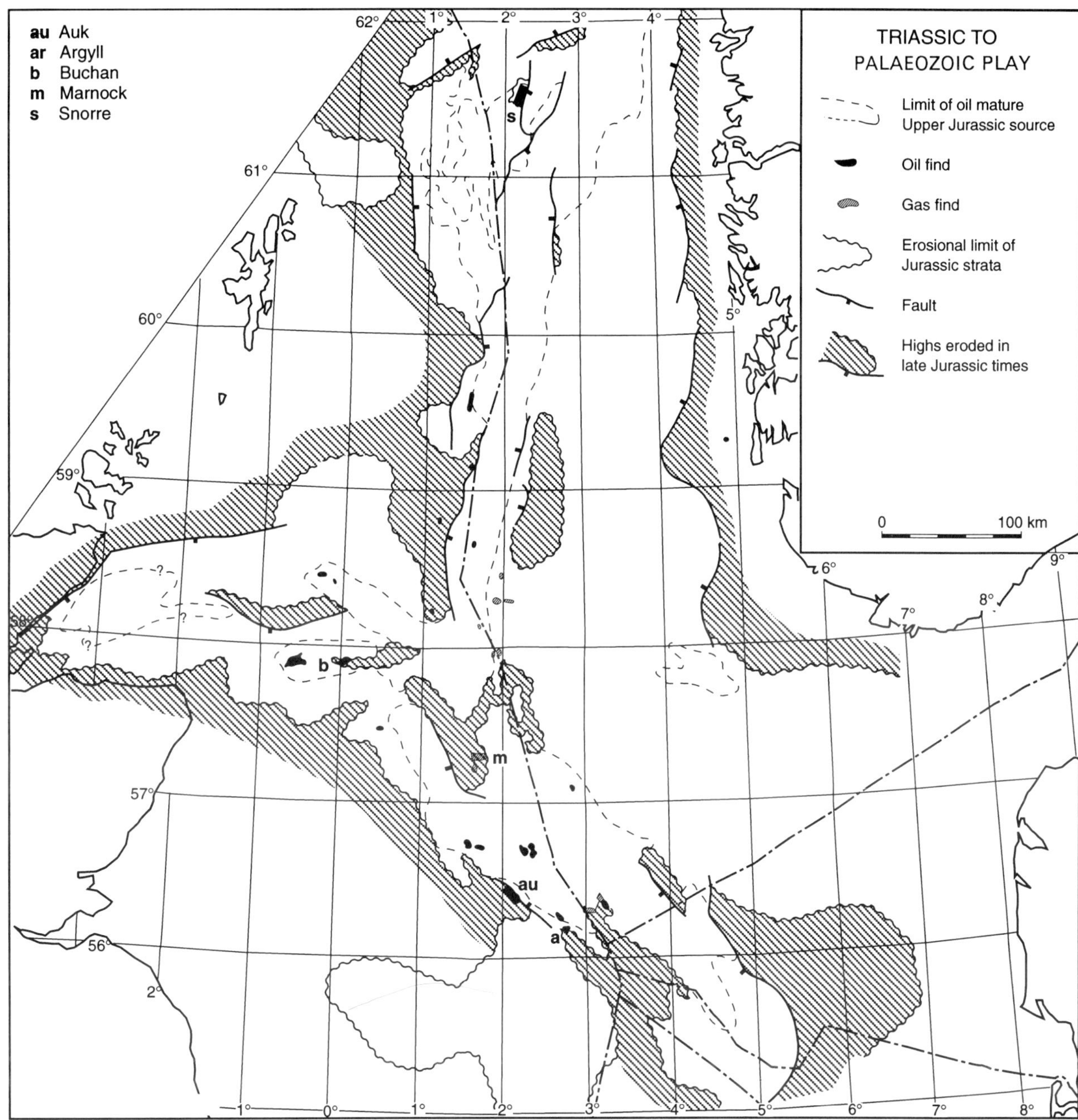

Fig. 4. Triassic to Palaeozoic play map. All the finds lie in areas where Jurassic rocks are absent or thin.

The hydrocarbon geology of the north and centre of the North Sea can be analysed by identifying a series of six groups of hydrocarbon plays, with reservoir ages: Triassic to Palaeozoic, Lower–Middle Jurassic, Upper Jurassic, Lower Cretaceous, Chalk and Palaeogene (Pegrum & Spencer 1990). The tectonic history of the area allows these plays to be grouped into 'Pre-rift', 'Syn-rift' and 'Post-rift'. This also emphasizes that there are close links between the hydrocarbon geology and the late Jurassic rifting. The regions where the source rocks have achieved maturity are approximately coincident with the main rift grabens. These grabens are also the regions where fault-block structures and thick reservoir sections are best developed. This fortunate coincidence of mature source rocks, traps and reservoirs explains the prolific character of these hydrocarbon plays.

In the southern North Sea region, three main groups of hydrocarbon plays can be recognized, grouped by reservoir age: Carboniferous, Permian and Triassic (Figs 9–11). The plays are related by the presence or absence of regional seals: Permian Rotliegend shales trap the gas within the Carboniferous sequence; Zechstein halite caps the Rotliegend reservoirs. These two sealing horizons, particularly the Zechstein halite, are so effective that only minor gas quantities have migrated up to the Triassic reservoir.

The tenth hydrocarbon play occurs in the Netherlands sector of the North Sea and onshore in the west Netherlands Basin where minor quantities of oil are reservoired in Upper Jurassic to Lower Cretaceous sandstones. The play is restricted to three separate basins (Figs 6, 7) and is made complicated by the inversion movements which affected the areas in late Cretaceous to Tertiary times.

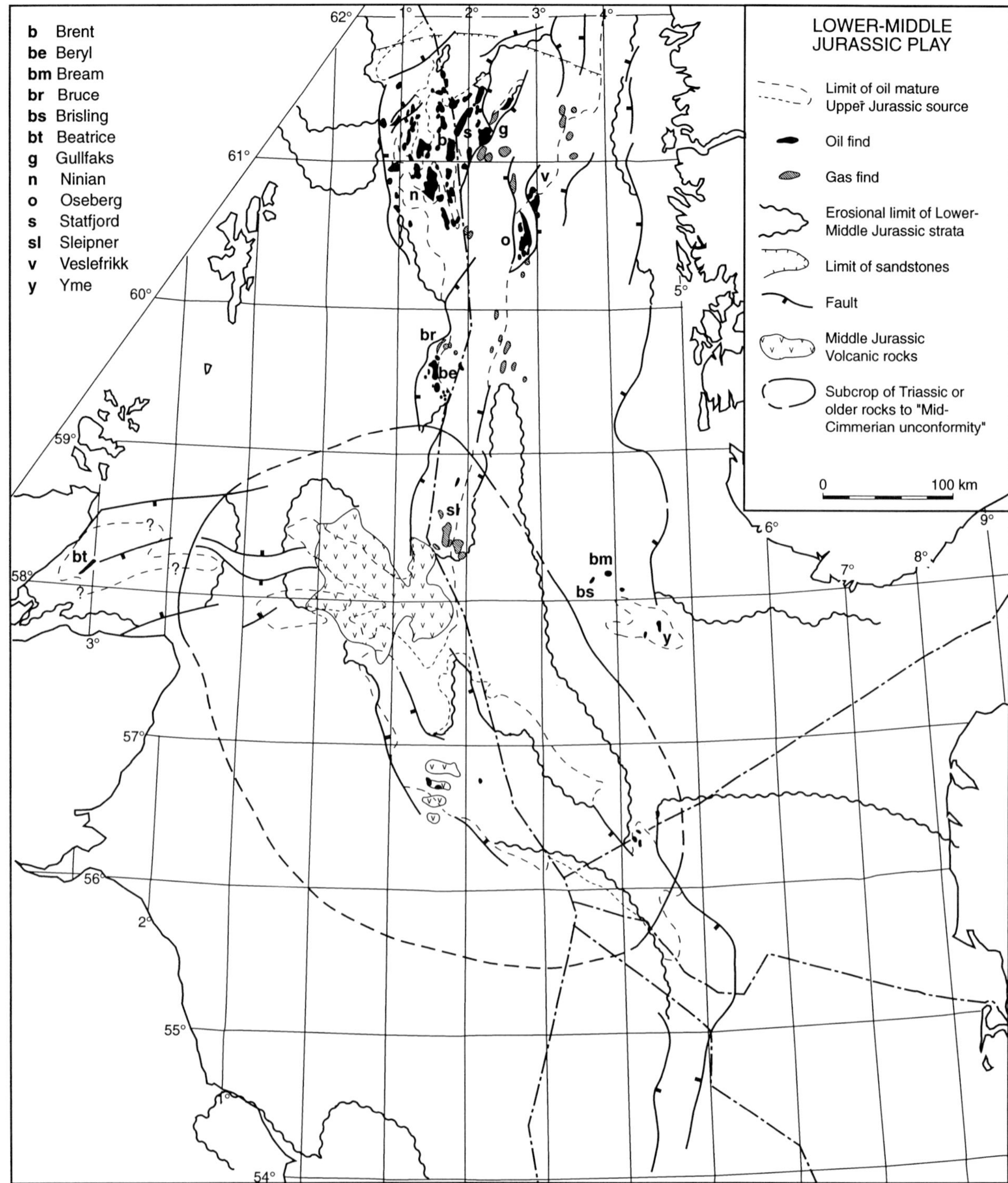

Fig. 5. Lower–Middle Jurassic play map. Middle Jurassic volcanism and uplift in the centre were linked with peripheral deposition of deltaic reservoirs. Prolific finds occur in the Viking Graben and minor finds in the east (Bream, Brisling, Yme) and west (Beatrice).

JURASSIC RIFT PROVINCE

Pre-rift plays

Play 1 includes all finds with pre-Jurassic reservoirs, including Devonian (Buchan), Carboniferous (Claymore), Rotliegend (Auk, Argyll), Zechstein (Auk) and Triassic (Marnock, Snorre) reservoirs (Fig. 4). In all the finds the reservoir is overlain unconformably by Upper Jurassic or Cretaceous cap rocks, for they are located in the faulted, deeply eroded highs formed during the late Jurassic rifting. They are close to areas with mature source rocks and have short migration routes.

Play 2 includes finds with Lower–Middle Jurassic reservoirs (Fig. 5). In the north Viking Graben these contain many of the largest fields (Spencer *et al.* 1987; Abbotts 1991) and constitute the most important hydrocarbon play in the northern North Sea. The oldest reservoirs are non-marine to

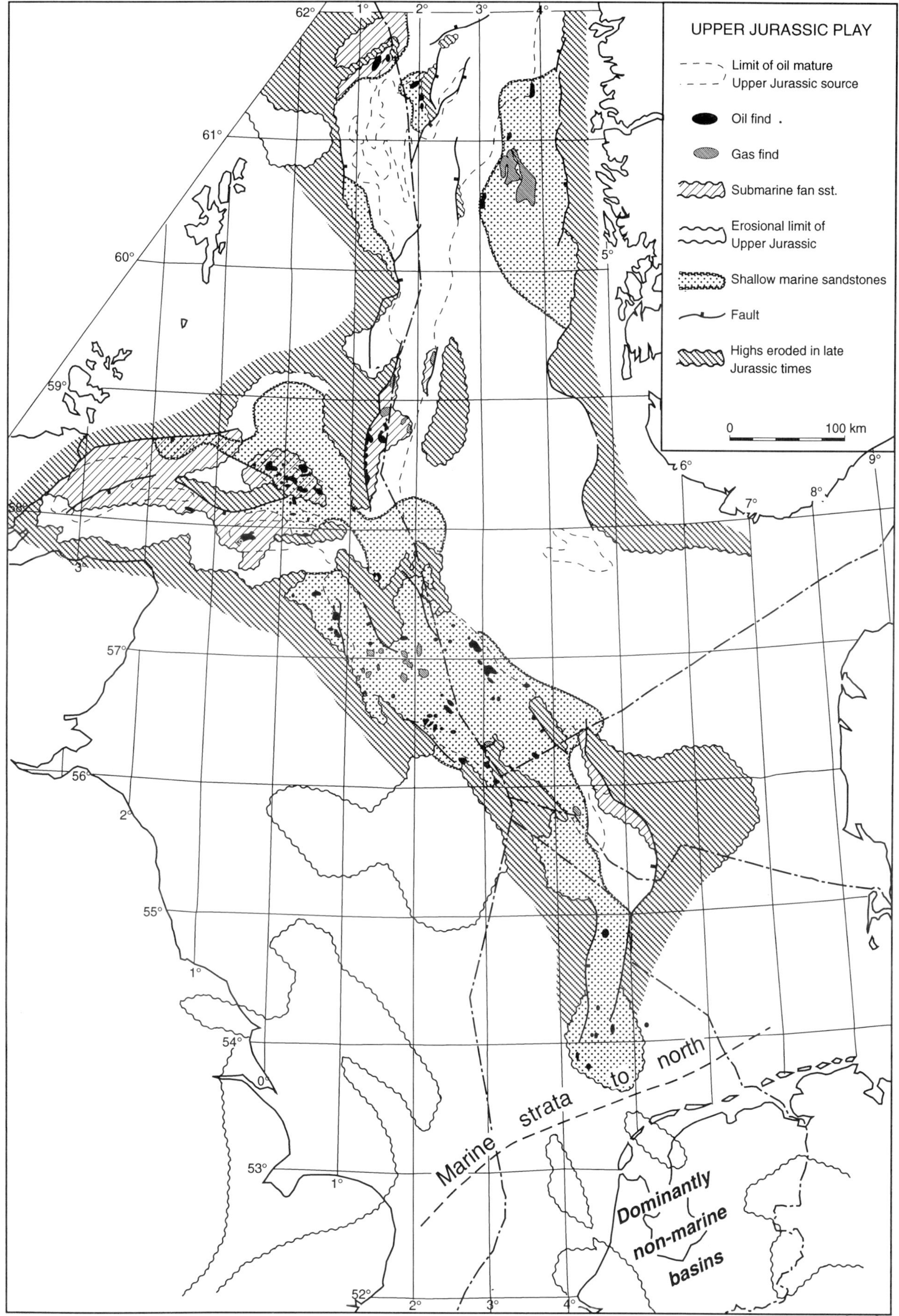

Fig. 6. Upper Jurassic play map. These rocks are present widely but vary much in thickness and lithology due to the rift movements occuring during deposition.

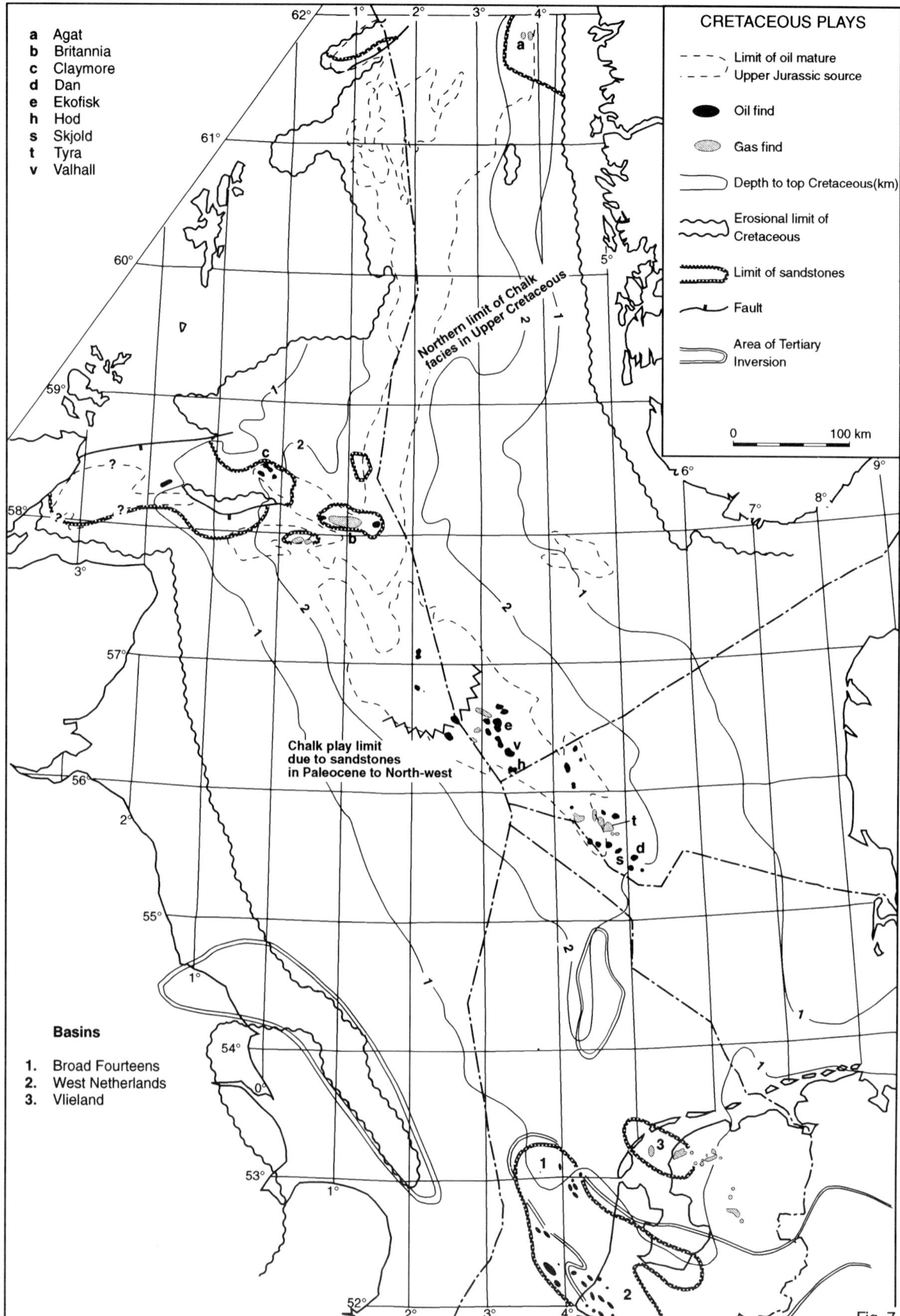

Fig. 7. Cretaceous play map. Cretaceous rocks are present throughout the basin but finds occur only in restricted areas where reservoirs are developed and are in communication with mature source rocks: Upper Cretaceous chalks in the Central Graben and Lower Cretaceous sandstones elsewhere.

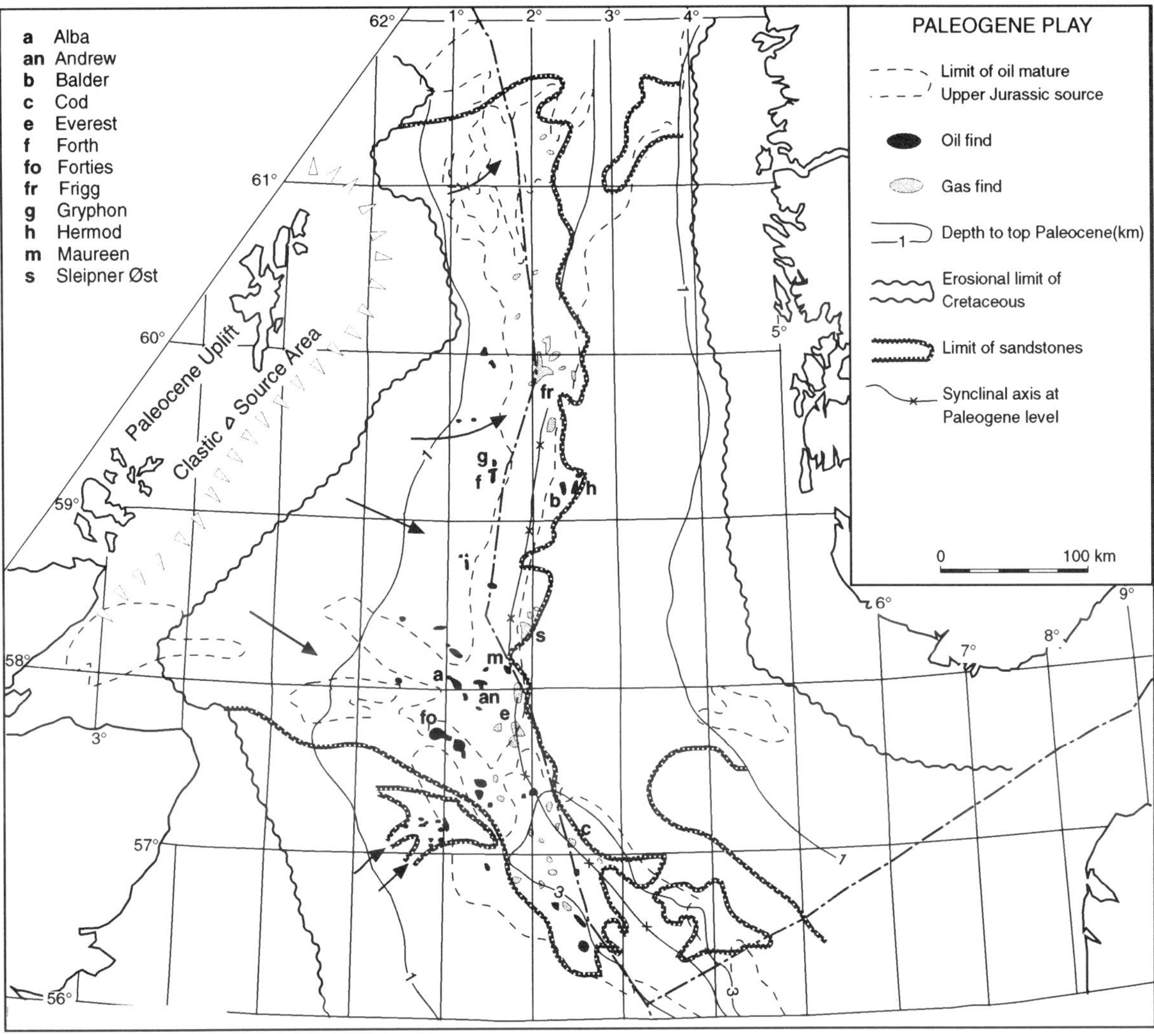

Fig. 8. Palaeogene play map. Finds are located in the centre and north of the North Sea in Palaeocene and Eocene sandstones derived from erosion of the north Scottish landmass. N.B. Forth Field is now re-named Harding.

marginal marine sandstones of the Statfjord Formation. The most important are the sandstones of the Brent Group, deposits of a delta system which reached its maximum northward extent in late Bajocian times and retreated southwards because of marine transgression in Bathonian and Callovian times. Most traps are tilted fault blocks with truncated reservoirs due to footwall uplift and erosion. The traps are sealed by unconformable Upper Jurassic or Cretaceous shales. Subsidence in Cretaceous and Tertiary times buried the fault blocks without disrupting them. Migration routes from the adjacent mature source rocks are often short.

In the south of the Viking Graben, Callovian marine sandstones of the Hugin Formation are the reservoir in the Sleipner Vest area. In the east, Middle Jurassic sandstones provide the reservoirs in the small Bream, Brisling and Yme finds: the play here is restricted by the low maturity of the source rocks. In the west, in the Moray Firth Graben close to the Scottish mainland, Lower–Middle Jurassic shallow marine and coastal sandstones form the reservoir in the Beatrice Field.

Syn-rift plays

Play 3 comprises the Upper Jurassic sandstone reservoirs that accumulated just before and during the main rifting movements (Fig. 6). These clastic marine strata show great variations in thickness and lithology, containing both source rocks and sandstone reservoirs. Two groups of sandstones may be recognized: older units (Magnus, Troll, Piper), which were afterwards affected by trap-forming faulting; younger units (Claymore, Brae, Ula) deposited during the main fault movements.

For simplicity, we group all these into our Upper Jurassic play. In the Central Graben, shallow-marine sandstone reservoirs of Oxfordian to Kimmeridgian ages accumulated inside the active graben margins (e.g. Ula and Fulmar fields). In the southern Viking Graben, Volgian, syn-rift, submarine fan conglomerates and sandstones, 1–3 km thick, accumulated along the giant Brae-trend fault system. In the Outer Moray Firth there are pre-rift, deltaic to shallow-marine sandstones of Oxfordian to Kimmeridgian age (Piper Field) and syn-rift turbiditic sandstones of Kimmeridgian to Volgian

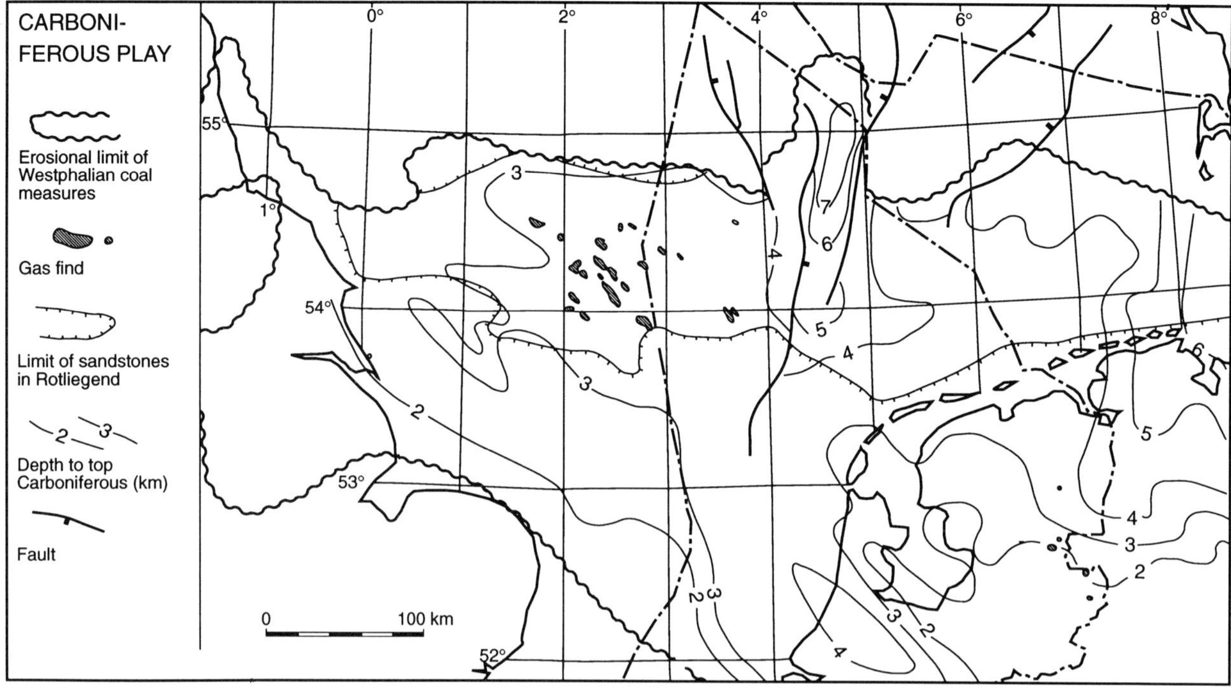

Fig. 9. Carboniferous gas play map. Finds are limited to the area where the unconformably overlying Rotliegend strata are in a shale facies and act as a regional seal.

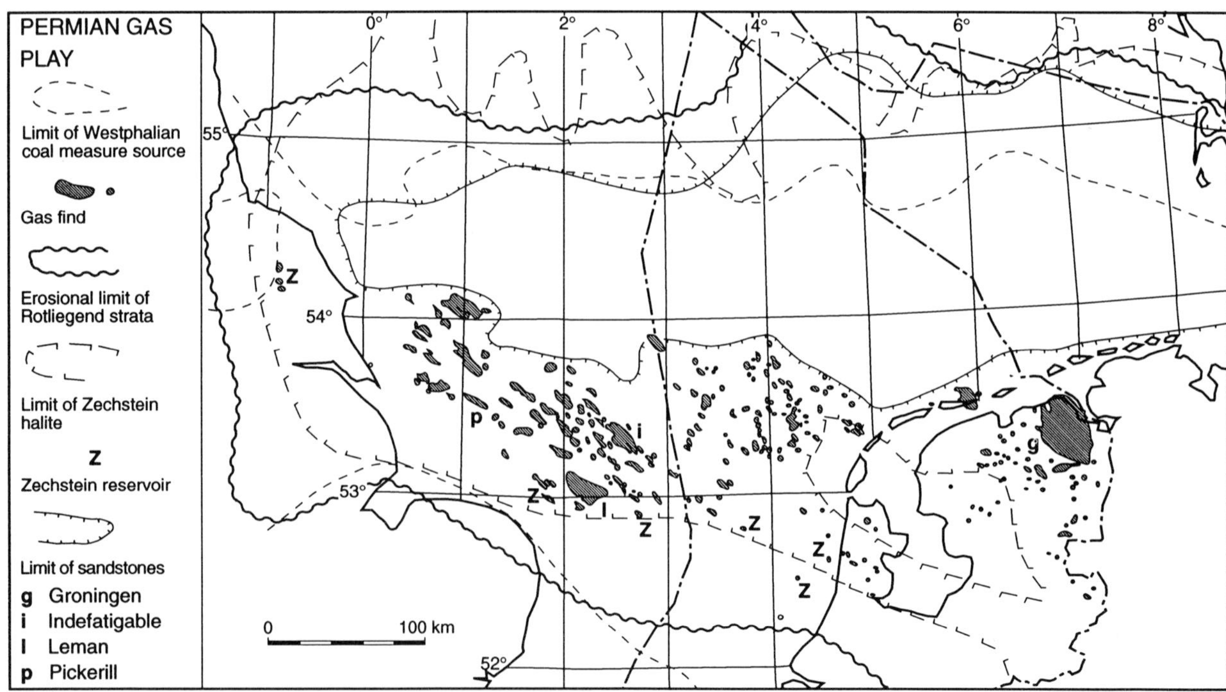

Fig. 10. Permian gas play map. Most finds occur in Rotliegend sandstones. In the south and west the Zechstein halites are replaced laterally by carbonates and the gas has migrated up into those reservoirs.

age (Claymore). In the northeast, shallow-marine sheet sandstones of Bathonian to Kimmeridgian age occur in a sedimentary wedge produced by the uplift of the Norwegian mainland (Troll). In the northwest, Kimmeridgian basin-floor fan sandstones are the reservoir in the Magnus Field.

The traps of these Upper Jurassic fields are also varied. The Troll Field is shallow (1000 m beneath the sea bed) but enormous (770 km^2) and comprises gently east-dipping fault blocks. The Brae 'trend' continues for 100 km inside the down-faulted margin of the South Viking Graben: the fields are classic syn-rift hydrocarbon accumulations, trapped in the hanging wall of a major fault, on which footwall uplift was responsible for the supply of clastic material. The main fields in the Outer Moray Firth are tilted, eroded fault-block traps, but with a more complex faulting history than is the case in the Middle Jurassic play in the north. The Ula Field lies in a trend containing over ten finds with traps that vary from domes above salt structures to closures involving stratigraphic truncations.

Play 4 comprises oil and gas finds in Lower Cretaceous

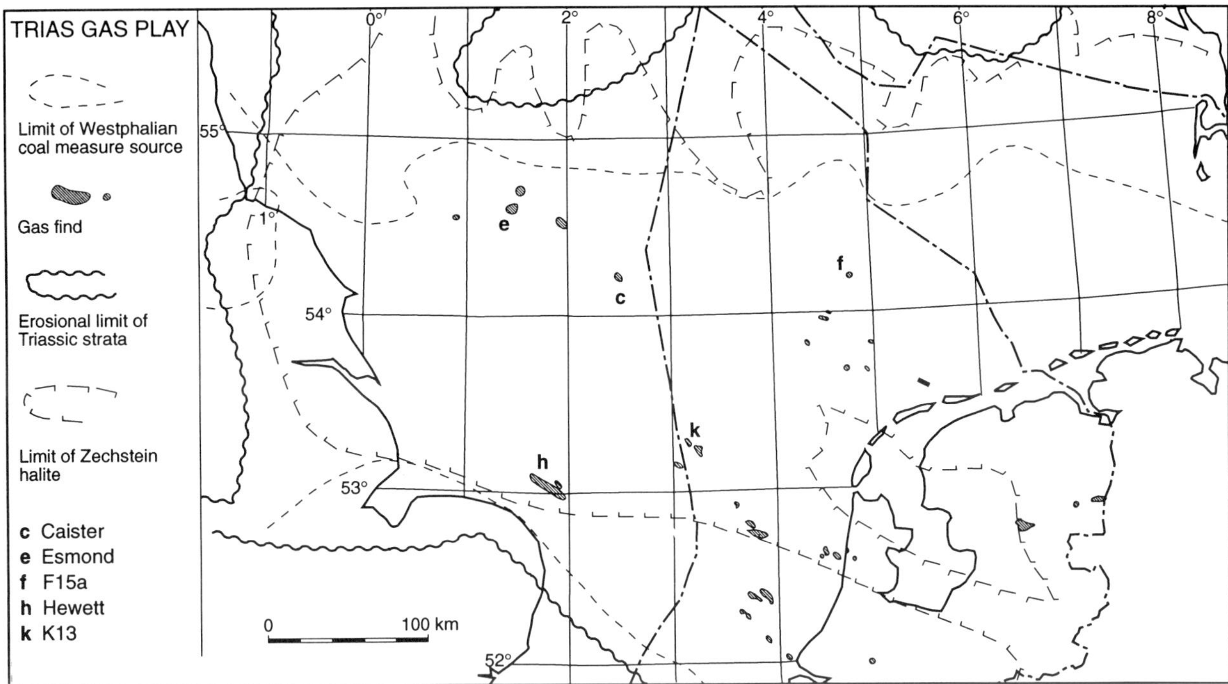

Fig. 11. Triassic gas play map. Most finds occur in areas where migration upwards from the Westphalian source is possible because the Zechstein halite is thin or absent.

submarine-fan sandstones (Fig. 7). Rifting continued into early Cretaceous times but was less extensive and movements on many of the major faults had ceased, so Lower Cretaceous marine strata – dominantly mudstones – usually drape and infill the post-rift relief. In the Moray Firth rifting movements continued so that Valanginian to Hauterivian submarine-fan sandstone reservoirs formed as aprons flanking the eroding fault blocks (Claymore). Submarine gravity-flow sandstones of Aptian–Albian age are an important reservoir further east (Britannia). In the Norwegian sector such reservoirs are developed only in the northeast where minor gas finds occur in stratigraphically-trapped, westerly-thickening, submarine-fan sandstones (Agat).

Post-rift plays

Play 5 comprises profilic finds reservoired in Upper Cretaceous to Danian chalks. Rifting had now ceased but subsidence was greatest along the axes of the earlier rift grabens, and there Upper Cretaceous strata locally exceed 1000 m. Upper Cretaceous and Danian strata are in a chalk facies south of about 59°N but only in a small area in the Central Graben does the chalk contain economically producible hydrocarbons (Fig. 7). There it is an unexpectedly good reservoir due to early migration of oil and the development of overpressure. Also, the most porous intervals are debris-flow chalks formed as a result of re-sedimentation near rising salt and inversion structures and along the oversteepened flanks of the trough. All the chalk fields have four-way dip closure but stratigraphic variations control the distribution of the productive reservoir zones. Most fields overlie salt structures but Valhall and Hod occupy local culminations on an inversion ridge. The Chalk play is bounded by the pinch-out line of the Palaeocene sandstones, which provide an escape route for hydrocarbons.

Play 6 is provided by Palaeogene reservoirs, principally Palaeocene and Eocene submarine-fan, mass-flow and turbidite sandstones (Fig. 8). These were derived from the uplift and erosion of the north of Scotland area. The basin axis exerted an important control on the submarine fan sands, which pass east and south into marine mudstones. Despite a lack of subsequent tectonic movements many fields are structural four-way dip closures, caused mainly by drape over older buried highs (Forties). Other fields lie near the pinch-out line of the sandstones (Balder, Sleipner Øst, Everest). Domal traps occur above Zechstein salt diapirs (Cod, Andrew, Maureen). Other fields have strong stratigraphic elements to their trap geometrys: submarine-fan relief (Frigg); submarine channel/canyon (Alba); depositional mounds (Gryphon, Harding). All of these Palaeogene fields have been sourced from Jurassic source rocks. Fields located near the graben axes have been supplied by vertical migration paths. Others, such as Balder, Sleipner Øst and Forties have no mature Jurassic source rocks vertically below and have been fed laterally by secondary migration of hydrocarbons within the Palaeogene sandstones.

CARBONIFEROUS GAS PROVINCE

Play 7 comprises gas finds in Westphalian coal-measure sandstones in the UK and Dutch sectors of the southern North Sea and minor gas finds onshore Netherlands (Fig. 9). The offshore finds are restricted to the area where the overlying Rotliegend is in a shale facies and acts as a regional seal. Reservoir beds include 20–50 m thick, stacked, braided floodplain sandstones in the Westphalian B in Caister Field (Ritchie & Pratsides 1993). Although sandstones are abundant in the Westphalian A–B (Collinson *et al.* 1993), they are often thin and of limited lateral extent (Besley 1990) and have undergone considerable diagenetic alteration (Leeder & Hardman 1990). The traps are NW-trending Variscan fold and fault-block structures sealed by mudstones within the Carboniferous sequence. It is likely that these structures have been modified by post-Variscan movements: during Permian

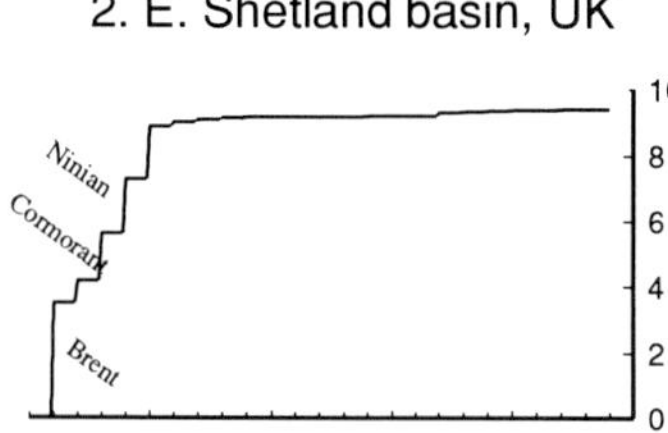

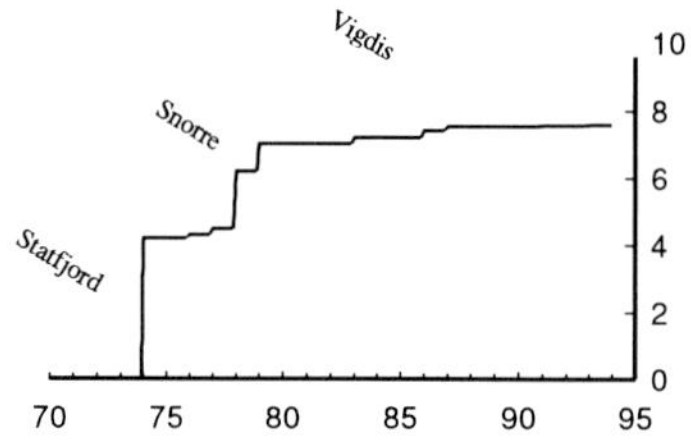

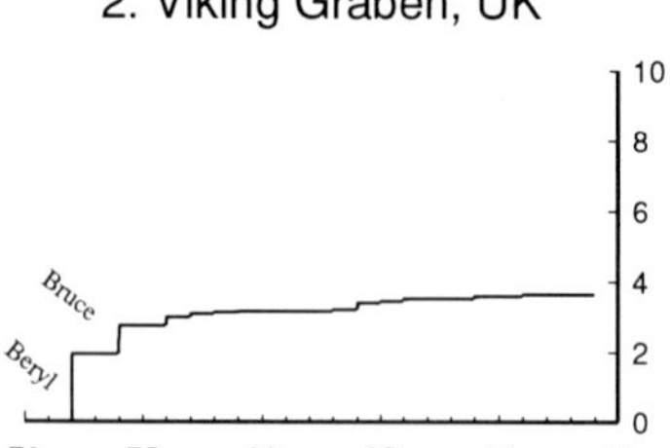

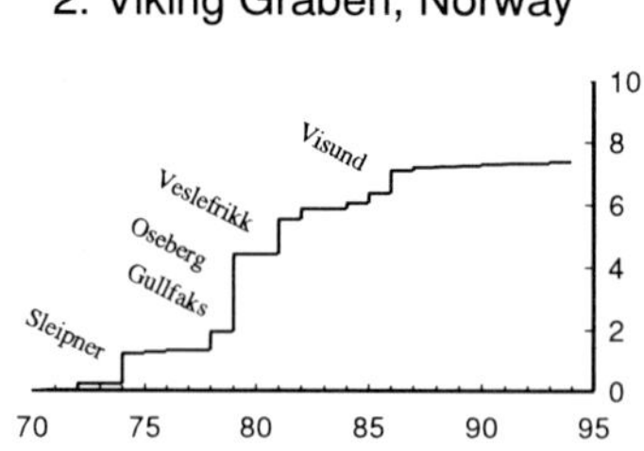

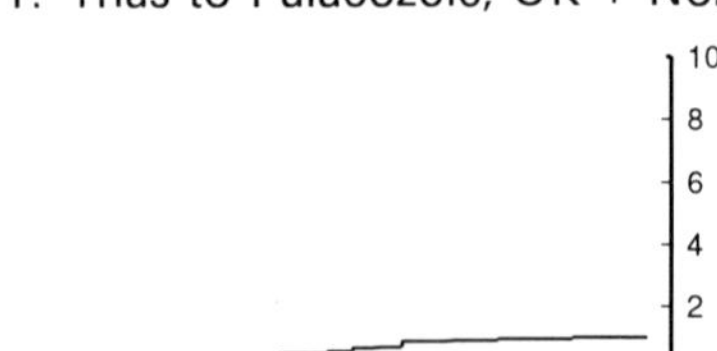

Fig. 12. Cumulative resources discovered in the Jurassic rift province in the pre-rift plays (a 'finding curve'). Horizontal axis: 1970 to 1994. Vertical axis: recoverable resources of oil and gas expressed as $\times 10^9$ BOE.

to early Jurassic burial, during Jurassic extension and uplift and during late Cretaceous to Tertiary inversions (Hollywood & Whorlow 1993). This has not affected the entrapment of the gas, however, for in the area of the offshore finds it is likely that the Westphalian coal-measure source became mature to generate its gas only during Tertiary times (Leeder & Hardman 1990).

Play 8 includes the prolific gas finds in the Lower Permian Rotliegend sandstones (Fig. 10). The Rotliegend reservoirs formed in a desert as wadi deposits and dune sands around a lake which had sabkha margins and in which shales and evaporites accumulated. The reservoirs total as much as 300 m in gross thickness and often have excellent properties (porosity up to 25%; permeability up to 3 darcies) (Glennie & Provan 1990). The traps are fault blocks and faulted highs sealed by Zechstein halites. There is structural detachment of the faults below and above the Zechstein halite and so the age of individual traps is often uncertain. The faults trend dominantly NW, like the Variscan structures in the underlying Westphalian rocks. They formed mostly during broad Mesozoic extension and uplift: movements which affected the region in Jurassic and early Cretaceous times (Arthur 1993). Further structuration at Rotliegend level resulted from later inversion of the Mesozoic basins. These inversion movements, during the late Cretaceous, Palaeocene and mid-Tertiary times, are well known (Glennie & Boegner 1981; Walker & Cooper 1987; van Hoorn 1987; Hillier & Williams 1991; Oudmayer & de Jager 1993).

The Westphalian coal source rocks had started to generate gas by Jurassic to Cretaceous times as a result of burial to depths of some 4 km beneath the regions of the deep Mesozoic basins (Fig. 3). This 'early' phase of gas generation has been suggested in several, separate studies (van Wijhe *et al.* 1980; Oele *et al.* 1981; Kirby *et al.* 1987; van Wijhe 1987; Leeder & Hardman 1990; Hillier & Williams 1991). The inversions of these basins in late Cretaceous times, involving uplifts of some 2 km, 'switched off' gas generation. Outside the inverted zones, however, continued Tertiary subsidence (up to 2 km) allowed further gas generation. Although this history of trap formation and gas generation is complex, there are few areas with unfilled Rotliegend traps (Alberts & Underhill 1991). This fortunate circumstance is probably due to the excellence of the regional Zechstein seal, which has not allowed gas to escape but has caused the gas to remigrate as structures changed geometry and position (van Wijhe 1987; Hillier & Williams 1991).

Play 8 also includes minor finds in Upper Permian carbonates (Fig. 10). These are poor to medium quality reservoirs in Zechstein carbonates which fringe the Zechstein halite basin (Fig. 1). As the Zechstein halites form the seal to the Rotliegend finds, these Zechstein finds occur close to or beyond the 'outer' margin of the Zechstein halite (Yorkshire: Scott & Colter 1987; UK offshore: Clark 1986; Taylor 1990; Netherlands: Roos & Smits 1983).

Play 9 includes the gas finds in the Lower Triassic Bunter Sandstone (Fig. 11). These sandstones formed as braided stream and sheet-flood and aeolian dune sand deposits in a desert; they are present thoughout the basin and have gross

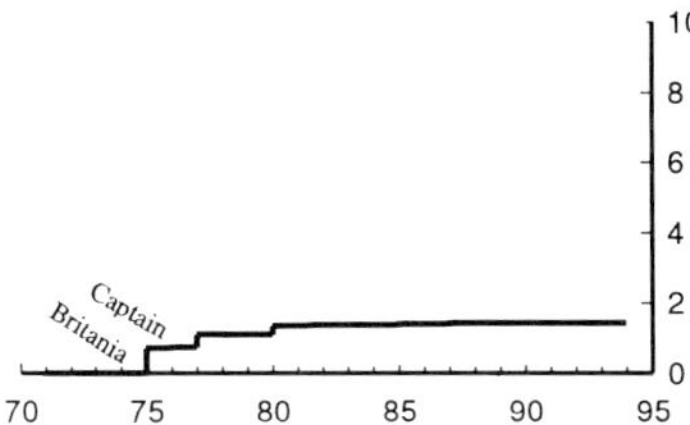

3. E. Shetland Basin, UK + Nor.

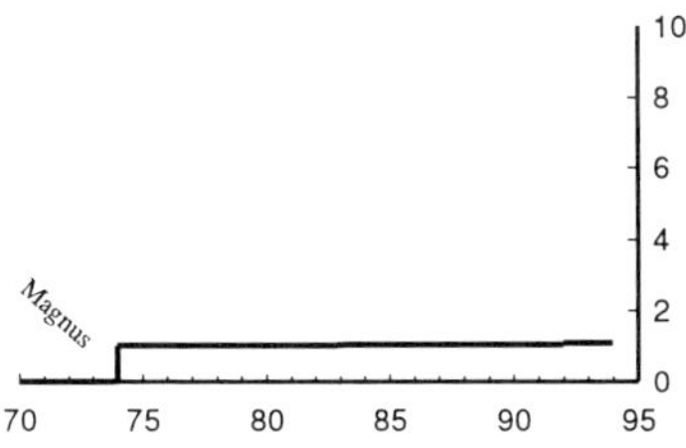

3. Horda Platform, Norway

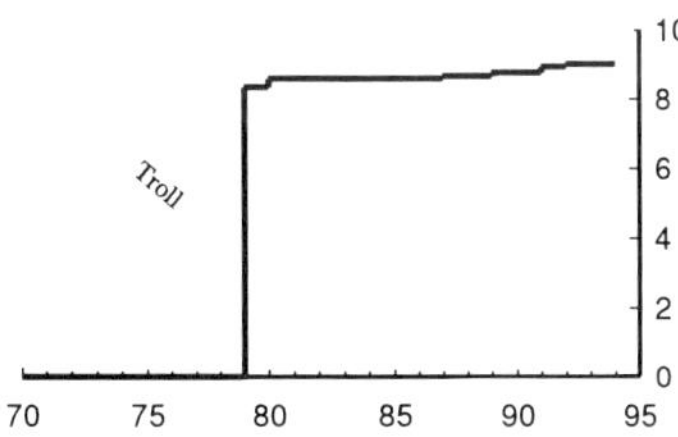

3. S. Viking Graben, UK + Nor.

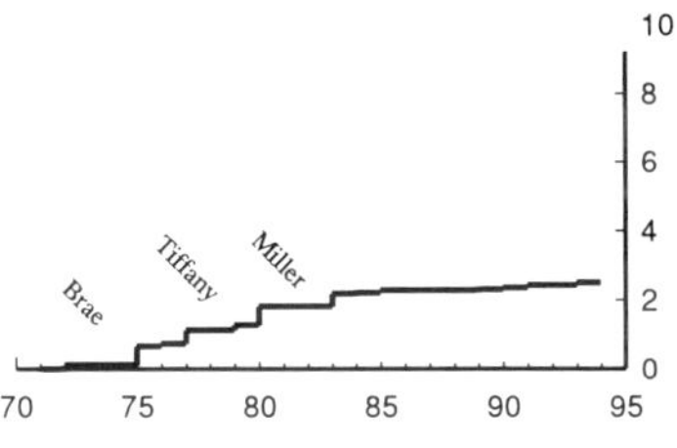

3. Outer Moray Firth, UK

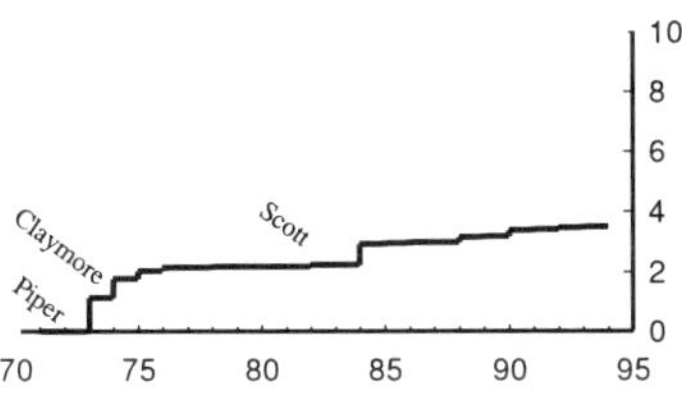

3. Central Graben, UK

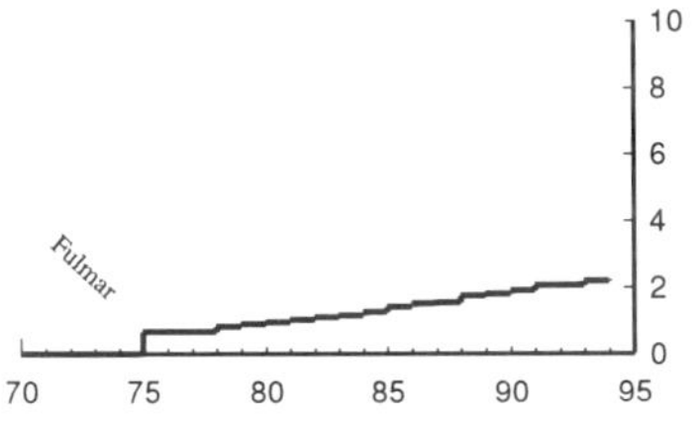

3. Central Graben, Norway

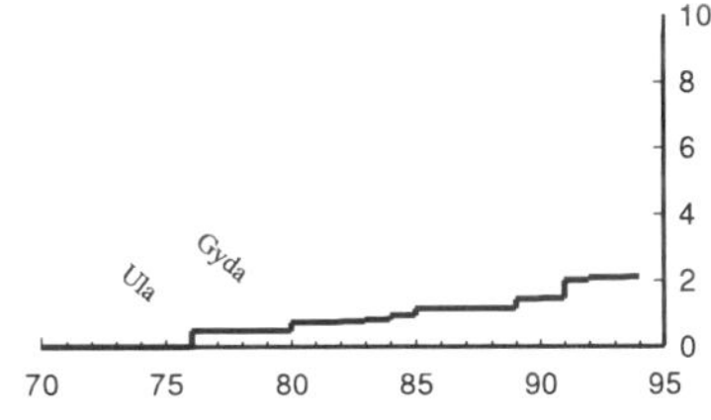

Fig. 13. Cumulative resources discovered in the Jurassic rift province in the syn-rift plays. Axes as Fig. 12.

thicknesses of 50–300 m. They often have excellent reservoir qualities and are capped by a regional seal, the Röt Halite. In addition, Bunter strata occur in large domal closures above salt swells and in inversion anticlines. Despite these attractive features, this play is of minor significance because of the difficulty of migration. In UK Quadrants 43–44 gas has had to traverse two regional seals – Rotliegend shale and Zechstein halite – and finds are few (Esmond fields: Ketter 1991; Caister: Ritchie & Pratsides 1993). The Zechstein seal has been traversed at the faulted margins of the Dutch Central Graben (F15a finds: Fontaine *et al.* 1993) and as a result of remigration due to inversion movements (K13 finds: Roos & Smits 1983). Most other Triassic finds lie close to the pinch-out line of the Zechstein halites (Hewett Field: Cooke-Yarborough 1991) or beyond it (southern Netherlands sector).

MESOZOIC OIL IN THE NETHERLANDS

Play 10 includes finds with Upper Jurassic or Lower Cretaceous reservoirs in The Netherlands. The play is present in four separate basins: the Dutch Central Graben (Fig. 6) and the Broad Fourteens, West Netherlands and Vlieland basins (Fig 7). In these basins, non-marine and marine clastic sequences, with thickness of up to 2 km, accumulated in late Jurassic to early Cretaceous times. This subsidence was

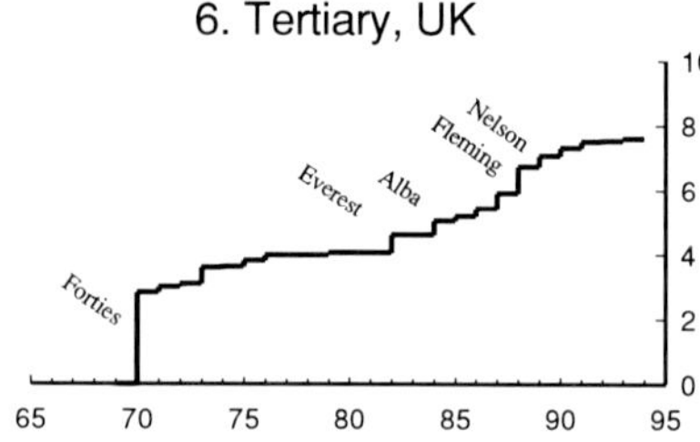

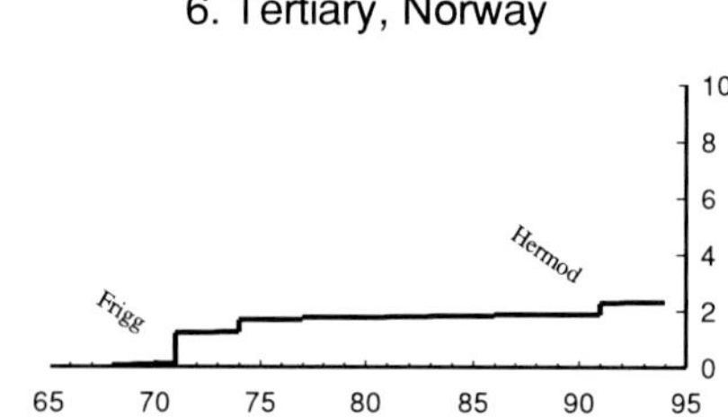

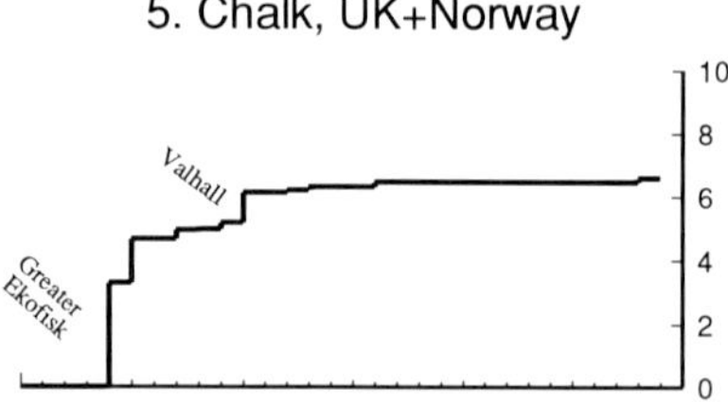

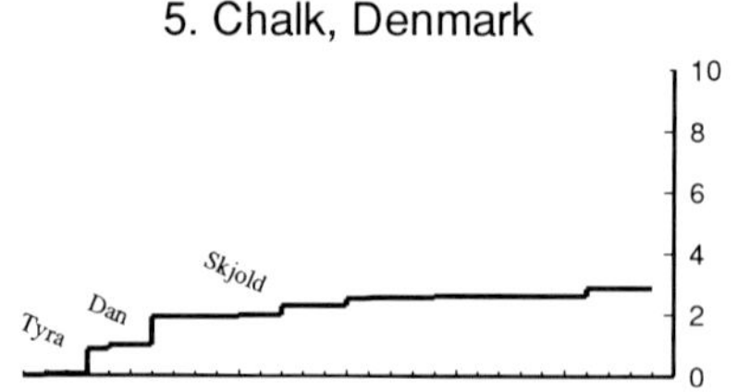

Fig. 14. Cumulative resources discovered in the Jurassic rift province in the post-rift plays. Axes as Fig. 12.

sufficient to allow the start of oil generation from the underlying Toarcian Posidonia Shale (Bodenhausen & Ott 1981; Roelofsen & de Boer 1991; Binot *et al.* 1993). The late Cretaceous and Tertiary inversion movements, which were so intense as to produce overthrusts, disrupted early-formed traps and caused hydrocarbons to escape or to re-migrate (Dronkers & Mrozek 1991; Goh 1994). Subsidence during Tertiary times in The Netherlands has allowed the Posidonia Shale to again source oil. Oil is now trapped in anticlines and fault blocks of late Cretaceous (inversion) age. One structure is so complicated that the reservoir is repeated four times by thrusting (Hoorn Field: Roelofsen & de Boer 1991).

FINDING RATES

The rate at which the resources of each play were found has been analysed (Figs 12–15). These data have been compiled on a find-by-find basis as follows. The original recoverable resources of oil, condensate and gas contained in each find were converted to BOE by summing the oil and condensate figures and adding the gas, converted at a rate of 6000 SCF per barrel. The volumes were taken from the Petroconsultants database as at July 1994, which in turn was derived from a variety of sources. Official figures are published annually for fields in production in the UK (these often give estimates for only one type of hydrocarbon, i.e. oil fields may not neccessarily include an associated gas figure) and for fields, both developed and undeveloped in Denmark and Norway. Individual field figures are not available for producing fields in The Netherlands. For the fields, the finds and the hydrocarbon phases not covered by official figures, estimates have been derived from academic and trade literature and from press releases by operating and partner companies. Otherwise, estimates have been made by Petroconsultants' staff.

The original recoverable resources for a field have been recalculated by Petroconsultants in the cases where production data indicate that the published estimate for a field is out of date. Such revisions have been backdated to the time of the discovery. Figures 12 to 15 thus show the volumes added by New Field Wildcat wells each year, although the full volumes of hydrocarbons being added may not have been understood at the time of discovery. Similarly, the age of the reservoir in each field has been derived from official sources: where no official statement of age has been released, an estimate has been made by Petroconsultants.

Five plays contain only small resources: Triassic to Palaeozoic (Play 1), Lower Cretaceous (4), Carboniferous (7), Triassic (9) and Jurassic–Cretaceous in The Netherlands (10). The other five plays contain major resources and distinctive patterns to their finding rates (Figs 12–15).

The Lower-Middle Jurassic play(2), with its simple fault-block traps and widespread reservoirs, was revealed rapidly. In the UK, the first exploration well drilled in the East Shetland Basin found the largest field there (1971, Brent) and over 90% of the resources were discovered in the following 5 years, giving a steep finding curve. In Norway the first valid test in the north found the largest field in the whole play (1974, Statfjord). Subsequent finds there were spread out over the next 20 years (Fig. 12), because of the different licensing policy in Norway compared with the UK.

The Upper Jurassic play (3) varies in geological make up from area to area (Fig. 13). The Magnus and Troll fields are isolated giants and there has been little growth of resources in the play around them. In the South Viking Graben and Outer Moray Firth the finding curves were steep for the first 10 years. In the Central Graben, the finding curves have an almost constant upward trend over 20 years, reflecting the difficulty of the exploration: varying stratigraphy, complex structures and deep targets with high overpressures.

The Chalk play (5) in Norway has a steep finding curve (Fig. 14). Once the existence of the play was known, exploring the domal closures was straightforward, apart from problems with gas clouds obscuring the seismic image. Again, the largest find came first (1969, Ekofisk) and over 95% of the resources were found in 5 years. The pattern in Denmark is similar.

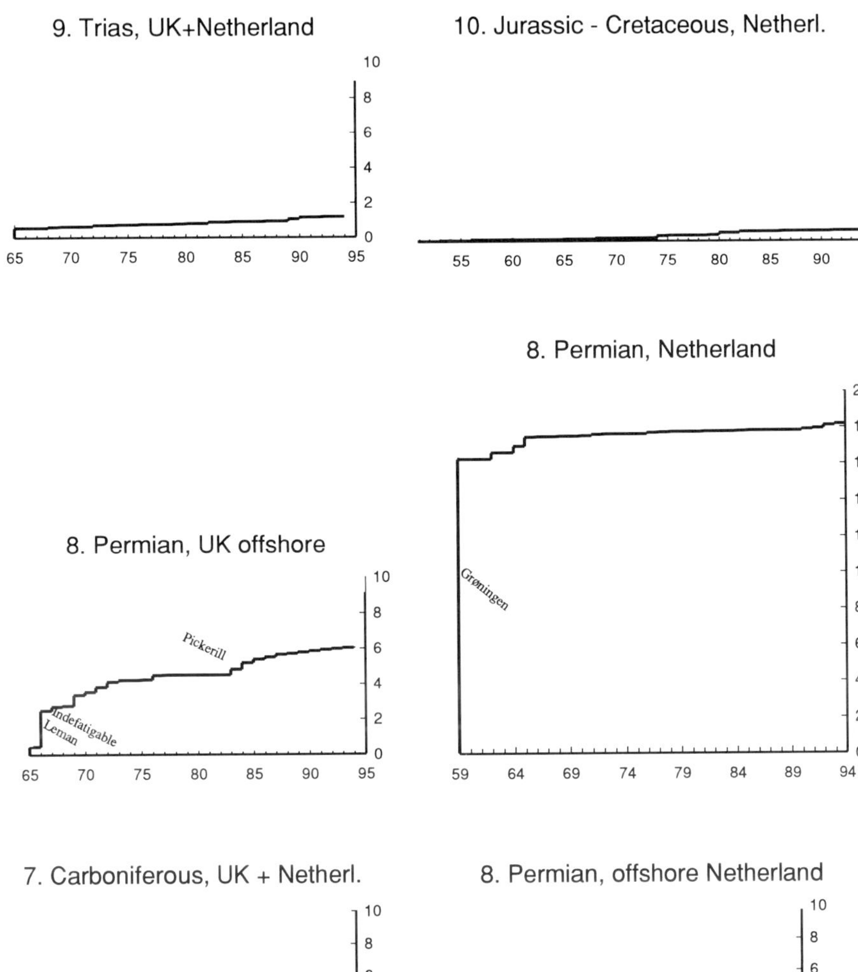

Fig. 15. Cumulative gas resources discovered in the plays in the southern North Sea. Note that the vertical axis is again $\times 10^9$ BOE, to allow direct comparison with Figs 12–14.

Table 1. *Petroleum resources and finds in the hydrocarbon plays of the North Sea region*

	UK offshore	Norway offshore	Denmark offshore	Netherlands Onshore	Netherlands Onshore	Total
Jurassic Rift Province						
6. Palaeogene	7.60 (72)	2.30 (20)		0.05 (5)		9.95 (97)
5. Chalk	0.45 (5)	6.15 (15)	2.85 (18)			9.45 (38)
4. L. Cretaceous	1.15 (11)	0.25 (1)				1.40 (12)
3. U. Jurassic	9.05 (98)	10.80 (25)	n (1)			19.85 (124)
2. L–M. Jurassic	13.30 (61)	16.45 (55)	0.10 (3)			29.85 (119)
1. Trias-Palaeozoic	0.60 (8)	0.25 (5)		0.85 (11)		
Subtotal	32.15 (255)	36.20 (121)	2.95 (22)	0.05 (5)		71.35 (403)
10. L. Cretaceous				0.35 (15)	0.05 (10)	0.40 (25)
U. Jurassic				0.15 (5)		0.15 (5)
Subtotal				0.50 (20)	0.05 (10)	0.55 (30)
Carboniferous Gas Province						
9. Trias	0.15 (7)			0.80 (29)	0.05 (4)	1.00 (40)
8. Permian	6.05 (112)			2.9 (120)	18.20 (62)	27.15 (294)
7. Carboniferous	0.70 (25)			0.10 (6)	n (1)	0.80 (32)
Subtotal	6.90 (144)			3.80 (155)	18.25 (67)	28.95 (366)
Total	39.05 (399)	36.20 (121)	2.95 (22)	4.35 (180)	18.30 (77)	100.85 (799)

Recoverable resources expressed as $\times 10^9$ BOE. Number of finds shown thus: (15). n = negligible

The Palaeogene play (6) in the UK has a finding curve with two steps. The largest field was again found very early (1970, Forties) and by 1975 to 1980 the curve was flat; the finds in this early phase were structural traps with dip closure. Afterwards, from 1981 to 1990, a new period with a steep finding rate is related to successes in stratigraphic traps.

The Permian gas play (8) in The Netherlands is totally dominated by the discovery in 1959 of Groningen – the find that started off the exploration of this play and of the whole basin. In the UK offshore the largest finds were made early (1966, Leman, Indefatigable) but new finds have continued over a 25-year period. Steady resource growth is clearer still in The Netherlands (Fig. 15). The capping Zechstein salt has provided an almost perfect seal to the gas: few traps are dry. However, this same irregular salt layer and the often complicated structure of the overlying strata mean that seismic imaging and depth conversion of the Rotliegend traps are often difficult.

PETROLEUM RESOURCES

The North Sea hydrocarbon plays have petroleum resources totalling 100×10^9 BOE (Table 1). The six hydrocarbon plays of the Jurassic rift province are most important, with 403 finds containing a total of 71×10^9 BOE, all offshore. The three plays of the Carboniferous gas province contain 299 finds offshore and a further 67 finds onshore The Netherlands, giving total resources, respectively, of 11 and 18×10^9 BOE. The third minor system comprises 30 oil finds onshore and offshore The Netherlands sourced from Lower Jurassic rocks and with total resources of 0.5×10^9 BOE.

We wish to thank Statoil and Petroconsultants for allowing us the facilities to prepare this article. Kent Høgseth helped compile the maps and finding curves. Wenche Frisvold and Inger Oppedal are thanked for the fine drawings.

REFERENCES

ABBOTTS, I. L. 1991. *United Kingdom Oil and Gas Fields, 25 Years Commemorative Volume.* Geological Society, London, Memoir **14**.

ALBERTS, M. A. & UNDERHILL, J. 1991. The effect of Tertiary structuration on Permian gas prospectivity, Cleaver Bank area, southern North Sea, UK. *In:* Spencer, A. M. (ed.). *Generation, Accumulation and Production of Europe's Hydrocarbons.* European Association of Petroleum Geoscientists. Special Publication 1, Oxford University Press, 161–173.

ANDREWS I. J. and six others. 1990. *United Kingdom offshore regional report: the geology of the Moray Firth.* HMSO–British Geological Survey, London.

ARTHUR, T. J. 1993. Mesozoic structural evolution of the Southern North Sea: insights from analyses of fault systems. *In:* Parker, J. R. (ed.) *Petroleum Geology of Northwest Europe, Proceedings of the 4th Conference.* Geological Society, London, 1269–1279.

BESLEY, B. M. 1990. Carboniferous. *In:* Glennie, K. W. (ed.) *Introduction to the Petroleum Geology of the North Sea.* Blackwell, Oxford, 90–119.

BINOT, F., GERKING, P., HILTMAN,W., KOCKEL, F. & WEHNER, H. 1993. The petroleum system in the Lower Saxony basin. *In:* Spencer, A. M. (ed.) *Generation, Accumulation and Production of Europe's Hydrocarbons III.* European Association of Petroleum Geoscientists, Special Publication, **3**, Springer, Heidelberg, 121–139.

BODENHAUSEN, J. W. A. & OTT, W. F. 1981. Habitat of the Rijswijk oil province, onshore, the Netherlands. *In:* Illing, L. V. & Hobson, G. D. (eds) *Petroleum Geology of the Continental Shelf of North-West Europe.* Heyden, London, 301–309.

CAMERON, T. D. J., CROSBY, A., BALSON, P. S., JEFFERY, D. H., LOTT, G. K., BULAT, J. & HARRISON, D. J. 1992. *United Kingdom offshore regional report: the geology of the southern North Sea.* HMSO–British Geological Survey, London.

CLARK, D. N. 1986. The distribution of porosity in Zechstein carbonates. *In:* Brooks, J., Goff, J. C. & van Hoorn, B. (eds) *Habitat of Palaeozoic Gas in N. W. Europe.* Geological Society, London, Special Publication, **23**, 121–149.

COLLINSON, J. D., JONES, C. M., BLACKBOURN, G. A., BESLEY, B. M., ARCHARD, G. M. & McMAHON, A. H. 1993. Carboniferous depositional systems of the Southern North Sea. In: Parker, J. R. (ed.) *Petroleum Geology of Northwest Europe: Proceedings of the 4th Conference*, Geological Society, London, 677–687.

COOKE-YARBOROUGH, P. 1991. The Hewett Field, Blocks 48/28-29-30, 52/4a-5a, UK North Sea. In: Abbotts, I. L. (ed.) *United Kingdom Oil and Gas Fields, 25 Years Commemorative Volume.* Geological Society, London, Memoir, **14**, 433–442.

COPE, J. W. C., INGHAM, J. K. & RAWSON, P. F. 1992. *Atlas of Paleogeography and Lithofacies.* Geological Society, London, Memoir, **13**.

COPE, M. J. 1986. An interpretation of vitrinite reflectance data from the Southern North Sea basin. *In:* Brooks, J., Goff, J. & van Hoorn, B. (eds) *Habitat of Paleozoic Gas in N. W. Europe.* Geological Society, London, Special Publication, **23**, 85–98.

CORNFORD, C. 1990. Source rocks and hydrocarbons of the North Sea. *In:* Glennie, K. W. (ed.) *Introduction to the Petroleum Geology of the North Sea.* Blackwell, Oxford, 294–361.

DORÉ, A. G., VOLLSET, J. & HAMAR, G. P. 1985. Correlation of the offshore sequences referred to the Kimmeridge Clay Formation – relevance to the Norwegian sector. *In:* Thomas, B. M. *et al.* (eds.). *Petroleum Geochemistry in Exploration of the Norwegian Shelf.* Graham & Trotman, London, 27–37.

DRONKERS, A. J. & MROZEK, F. J. 1991. Inverted basins of the Netherlands. *First Break*, **9**, 409–425.

FONTAINE, J. M., GUASTELLA, G., JOUAULT, P. & DE LA VEGA, P. 1993. F15-A: a Triassic gas field on the eastern limit of the Dutch Central Graben. *In:* Parker, J. R. (ed.) *Petroleum Geology of Northwest Europe: Proceedings of the 4th Conference.* Geological Society, London. 583–593.

GATLIFF, R. W. and 14 others. 1994. *The geology of the central North Sea.* HMSO: British Geological Survey, London.

GLENNIE, K. W. & BOEGNER, P. L. E. 1981. Sole Pit inversion tectonics. *In:* Illing, L. V. & Hobson, G. D. (eds) *Petroleum Geology of the Continental Shelf of North-West Europe.* Heyden, London, 110-120.

——— & PROVAN, D. M. J. 1990. Lower Permian Rotliegend reservoirs of the Southern North Sea gas province. *In:* Brooks, J. (ed.) *Classic Petroleum Provinces.* Geological Society, London, Special Publication, **50**, 399–416.

GOH, L. S. 1994. The Logger Field: geology and reservoir characterization. *In:* Aasen, J. O., Berg, E., Buller, A. T., Hjelmeland, O., Holt, R. M., Kleppe, J. and Torsæther, O. (eds). *North Sea Oil and Gas Reservoirs III.* Kluwer, Dordrecht, 75–93.

HILLIER, A. P. & WILLIAMS, B. P. J. 1991. The Leman Field, Blocks 49/26, 49/27, 53/1, 53/2, UK North Sea. *In:* Abbotts, I. L. (ed.) *United Kingdom Oil and gas fields, 25 Years Commemorative Volume*, Geological Society, London, Memoir, **14**, 451–458.

HOLLYWOOD, J. M. & WHORLOW, C. V. 1993. Structural development and hydrocarbon occurrence of the Carboniferous in the U. K. Southern North Sea. *In:* Parker, J. R. (ed.) *Petroleum Geology of Northwest Europe: Proceedings of the 4th Conference.* Geological Society, London, 689–696.

JOHNSON, H., RICHARDS, P. C., LONG, D. & GRAHAM, C. C. 1993. *United Kingdom offshore regional report: the geology of the northern North Sea.* HMSO–British Geological Survey, London.

KETTER, F. J. 1991. The Esmond, Forbes and Gordon fields, Blocks 43/8a, 43/13a, 43/15a, 43/20a, UK North Sea. *In:* Abbots, I. L. (ed.) *United Kingdon Oil and Gas fields, 25 Years Commemorative Volume.* Geological Society, London, Memoir, **14**, 425–432.

KIRBY, G. A., SMITH, K., SMITH, N. J. P. & SWALLOW, P. W. 1987. Oil and gas generation in eastern England. *In:* Brooks, J. & Glennie, K. W. (eds). *Petroleum Geology of Northwest Europe.* Graham & Trotman, London, 171–180.

LEEDER, M. R. & HARDMAN, M. 1990. Carboniferous geology of the Southern North Sea basin and controls on hydrocarbon prospectivity. *In:* Hardman, R. F. P. & Brooks, J. (eds) *Tectonic Events Responsible for Britain's Oil and Gas Reserves.* Geological Society, London, Special Publication, **55**, 87–105.

MACKENZIE, A. S., LEYTHAUSER, D., MULLER, P., RADKE, M. & SHAEFER, R. G. 1987. The expulsion of petroleum from Kimmeridge clay source rocks in the area of the Brae oilfield, U.K. continental shelf. *In:* Brooks, J. & Glennie, K. W. (eds). *Petroleum Geology of Northwest Europe.* Graham & Trotman, London, 865–878.

OELE, J. A., HOL, A. C. P. J. & TIEMENS, J. 1981. Some Rotliegend gas fields of the K and L Blocks, Netherlands offshore (1968–1978) a case history. *In:* Illing, L. V. & Hobson G. D. (eds) *Petroleum Geology of the Continental Shelf of North-west Europe.* Heyden, London, 289–300.

OUDMAYER, B. C. & DE JAGER, J. 1993. Fault reactivation and olique slip in the Southern North Sea. *In:* Parker, J. R. (ed.) *Petroleum Geology of Northwest Europe: Proceedings of the 4th Conference.* Geological Society, London, 1281–1290.

PEGRUM, R. M. & SPENCER, A. M. 1990. Hydrocarbon plays in the northern North Sea. *In:* Brooks, J. (ed). *Classic Petroleum Provinces.* Geological Society, London, Special Publication, **50**, 441–470.

RITCHIE, J. S. & PRATSIDES, P. 1993. The Caister Fields, Block 44/23a, UK North Sea. *In:* Parker, J. R. (ed.) *Petroleum Geology of Northwest Europe: Proceedings of the 4th Conference.* Geological Society, London, 759–769.

ROELOFSEN, J. W. & DE BOER, W. D. 1991. Geology of the Lower Cretaceous Q/1 oil-fields, Broad Fourteens basin, The Netherlands. *In:* Spencer, A. M. (ed.) *Generation, Accumulation and Production of Europe's Hydrocarbons.* European Association of Petroleum Geoscientists, Special Publication, **1**, Oxford University Press, 203–216.

ROOS, B. M. & SMITS, B. J. 1983. Rotliegend and Main Buntsandstein gas fields in Block K/13–a case history. *Geologie en Mijnbouw*, **62**, 74–82.

SCHOU, L., EGGEN, S. S. & SCHOELL, M. 1985. Oil-oil and oil-source rock correlation, northern North Sea. *In:* Thomas, B. M. *et al.* (eds). *Petroleum Geochemistry in Exploration of the Norwegian Shelf.* Graham & Trotman, London, 101–117.

SCOTT, J. & COLTER, V. S. 1987. Geological aspects of current onshore Great Britain exploration plays. *In:* Brooks, J. & Glennie, K. W. (eds) *Petroleum Geology of Northwest Europe*, Graham & Trotman, London, 95–107

SPENCER, A. M. *et al.* (eds) 1987. *Geology of the Norwegian Oil and Gas Fields.* Graham & Trotman, London.

TAYLOR, J. C. M. 1990 Upper Permian–Zechstein. In: Glennie, K. W. (ed.) *Introduction to the Petroleum Geology of the North Sea*, Blackwell, Oxford, 153–190

UNDERHILL, J. R. & PARTINGTON, M. A. 1993. Jurassic thermal doming and deflation in the North Sea: implications of the sequence stratigraphic evidence. *In:* Parker, J. R. (ed.). *Petroleum Geology of Northwest Europe: Proceedings of the 4th Conference.* Geological Society, London, 337–345.

VAN HOORN, B. 1987. Structural evolution, timing and tectonic style of the Sole Pit inversion. *Tectonophysics*, **137**, 239–284.

VAN WIJHE, D. H. 1987. The structural evolution of the Broad Fourteens Basin. *In:* Brooks, J. & Glennie, K. W. (eds). *Petroleum Geology of Northwest Europe.* Graham & Trotman, London. 315–323.

———, LUTZ, M. & KAASSCHIETER, J. P. H. 1980. The Rotliegend in the Netherlands and its gas accumulations. *Geologie en Mijnbouw*, **59**, 3–24.

WALKER, I. M. & COOPER, W. G. 1987. The structural and stratigraphic evolution of the northeast margin of the Sole Pit Basin. *In:* Brooks, J. & Glennie, K. W. (eds) *Petroleum Geology of Northwest Europe.* Graham & Trotman, London, 263–275.

ZIEGLER, P. A. 1990. *Geological Atlas of Western and Central Europe.* Shell International Petroleum Maatschappij B. V.

History of oil and gas exploration in Ireland

D. Naylor

ERA-Maptec Ltd., 5 South Leinster Street, Dublin 2, Ireland

ABSTRACT: Onshore exploration in the Republic of Ireland began in 1959 and, during 1962–63, a consortium of Ambassador, Conoco and Marathon drilled six exploration wells in the main Carboniferous basins. The first designation of the Irish offshore took place in 1969, when a large concession was granted to Marathon. Further offshore designations were made in 1970, 1974, 1976 and 1977, extending the Irish designated area westwards over the Porcupine and south Rockall regions.

Marathon commenced offshore drilling in 1970 and, with the third well (48/25-2) in 1971, discovered the Kinsale Head gas field. The First Round of offshore Licensing in 1976 was followed by an extension of drilling to the Fastnet Basin (1976), the Kish Bank Basin (1977) and the deep waters of the Porcupine Basin (1977). The Aran/BP 26/28-1well in 366 m of water in the northern part of the Porcupine Basin discovered the Connemara Field (undeveloped), with proven reserves in place of 120×10^6 bbl of 32–38° API oil.

Further Licensing Rounds were held – a Second in 1982 and a Third in 1985. Drilling continued at reduced levels, but without notable success except for a small undeveloped oil discovery (Helvick) off the southeast coast in Block 49/9 (1983).

Low oil prices, lack of success, and deep water in the promising western regions, were all contributing factors to the low levels of exploration in the last decade. The Kinsale Head gas field (1.5×10^{12}scf) was brought on stream in 1978. A small gas accumulation immediately to the northwest – the Ballycotton gas field – has been linked to the Kinsale facility and brought onto production in 1991. These are the only commercial developed discoveries in the Irish offshore. Recent attempts to stimulate exploration have been made through the holding of Frontier Licensing Rounds in Slyne-Erris (1993) and Porcupine (1994), and through improvement of the financial terms (1992).

KEYWORDS: *Ireland, exploration history, onshore/offshore*

Petroleum exploration in the Republic of Ireland has followed a quite different path to that in the United Kingdom. There was no equivalent in Ireland to the exploration effort onshore UK between the First and Second World Wars, and in the immediate post-War period (Lees & Taitt 1946; Falcon & Kent 1960). Geological factors, such as an almost total lack of Mesozoic rocks onshore and the presence of anthracitic rank coals in the Leinster coalfields, acted as deterrents. The areas of Coal Measure rocks in Ireland are relatively small, and the workings did not have the frequent showings of oil which were a feature of some of the northern England coalfields. For these and other reasons, Ireland remained completely unexplored for hydrocarbons during the first half of this century.

There is now an extensive literature covering the geological results of oil and gas exploration in Ireland. Reference is made here to only a handful of regional geological papers or recent reviews, to published accounts covering the historical aspects of exploration, and to recent pronouncements regarding possible field developments.

ONSHORE EXPLORATION

Onshore exploration finally began through an Irish American initiative in 1959. The impetus behind the American oilmen was as much to do with doing something for the 'old country' as with the perception of large areas of unexplored and prospective Carboniferous rocks. An entertaining account of this entry into the realm of oil exploration is found in Collins (1976). An agreement in 1959 between the Minister for Industry and Commerce and Ambassador Irish Oil Ltd gave the latter an Exploration Licence for 20 years from 29 March, 1960 applying to the whole onshore area of the Republic and any seas under Irish jurisdiction, subject to the surrender of 25% of the original area every five years. The apparent generosity of these terms has to be judged against the fact that there had been no previous exploration, and against a general climate of low expectation in the period before any North Sea development had taken place.

Conoco and Marathon Petroleum Ireland Ltd farmed into the licence in 1961. This consortium drilled 6 exploration wells in the main onshore Carboniferous basins during 1962–63. The first reflection seismic data within the Republic had also been acquired across the prospective areas, with S.S.L. as contractor using vibroseis for the first time outside the United States. Two wells were drilled in 1962 at Trim, Co. Meath and Doonbeg, Co. Clare (Fig. 1). A third well, drilled at Dowra, Co. Cavan in the Northwest Carboniferous Basin

From K. Glennie & A. Hurst (eds), 1996, *AD1995: NW Europe's Hydrocarbon Industry*, Geological Society, London, pp. 43–52

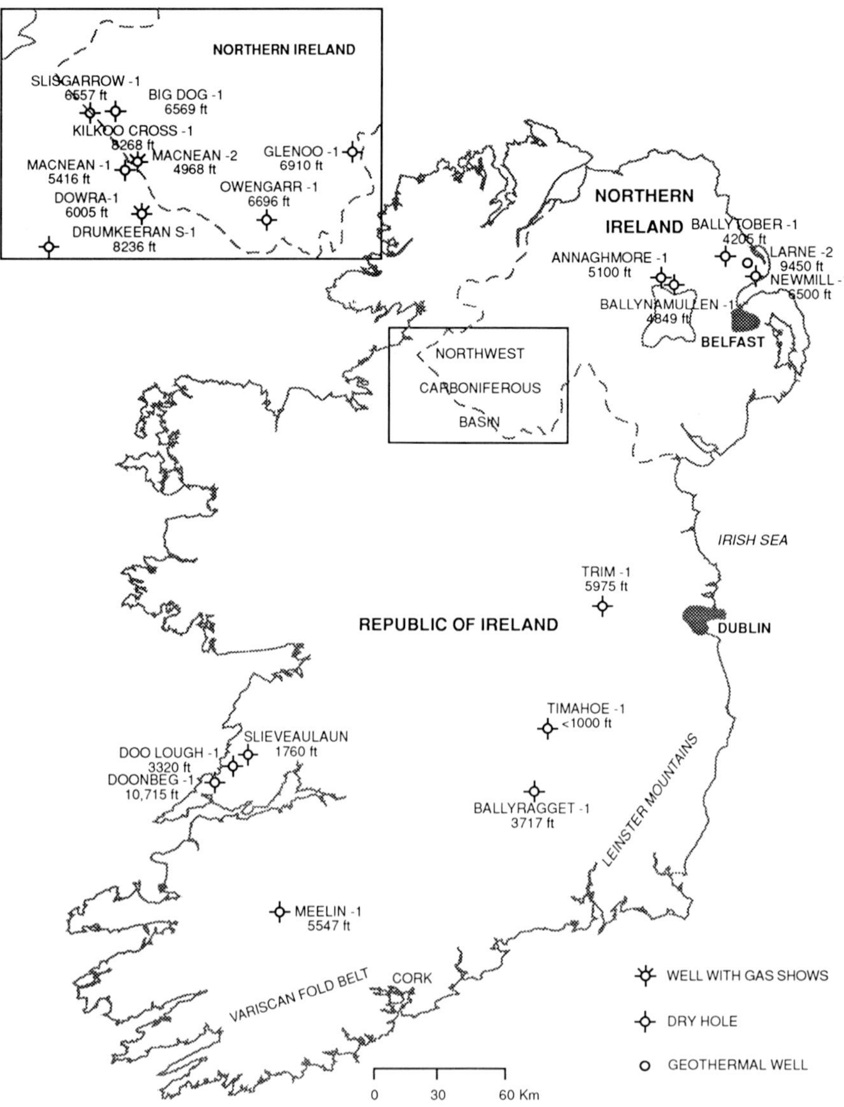

Fig. 1. Onshore wells in Ireland. The inset shows wells drilled in the Northwest Carboniferous Basin.

during 1963, flowed gas at a rate of about 31 000 scf per day from Lower Carboniferous sandstones (Sheridan 1977). Two further wells drilled at Meelin, Co. Kerry and Ballyragget, Co. Kilkenny were dry. However, a sixth, Macnean-1, about 12 km from Dowra and close to the border with Northern Ireland, also tested gas at 15 000 scf per day (Fig. 1 inset). The outline results of the exploration programme were given by Sheridan (1972, 1977), and details of individual wells were released in other publications. This exploration programme produced the first regional overview of the Irish Carboniferous in terms of basin development, regional isopachytes and reservoir potential.

The shows of gas in the northwest encouraged the consortium to apply for an Exploration Licence over a contiguous area in Northern Ireland. A licence covering 3500 km^2 was issued to Marathon Petroleum (UK) Ltd in May 1965 under the Petroleum (Production) Act (Northern Ireland), 1964 (Griffith 1983). Three wells were drilled for Carboniferous targets in the border area of Northern Ireland–Big Dog-1, Owengarr-1 and Glenoo-1–but only traces of gas were encountered (Fig. 1).

In the Republic, after withdrawal of two partners (Conoco in 1964 and Ambassador in 1966), and a first relinquishment of area (1965), Marathon was left as the single licensee holding the whole of the remaining concession. However, by this time the company had become interested in the offshore areas, about which it was negotiating with the government. Onshore exploration became severely curtailed during the 1970s, and drilling was restricted to two exploration coreholes by Irish Petroleum Prospecting Ltd. in 1980 to test the Carboniferous in Co. Clare–Slieveaulaun to 1760 ft, and Doo Lough to 3320 ft–both without recorded shows (Fig. 1).

The first onshore oil exploration programme in Ireland therefore met with little success. With the benefit of hindsight, and particularly with improved understanding of maturation and source rocks, the disappointing results in the onshore can be placed in perspective. The region was affected by the Variscan orogeny, and the putative trace of the Variscan 'front' is normally depicted as crossing the southern part of the country (Fig. 2). Recent maturation studies (Clayton *et al.* 1989) and fission track data (Keeley *et al.* 1993) demonstrate that the Palaeozoic rocks of the southern part of onshore Ireland are overmature for hydrocarbon sourcing and preservation. The Carboniferous rocks generally have higher maturation levels than those encountered beneath the offshore basins, probably due to

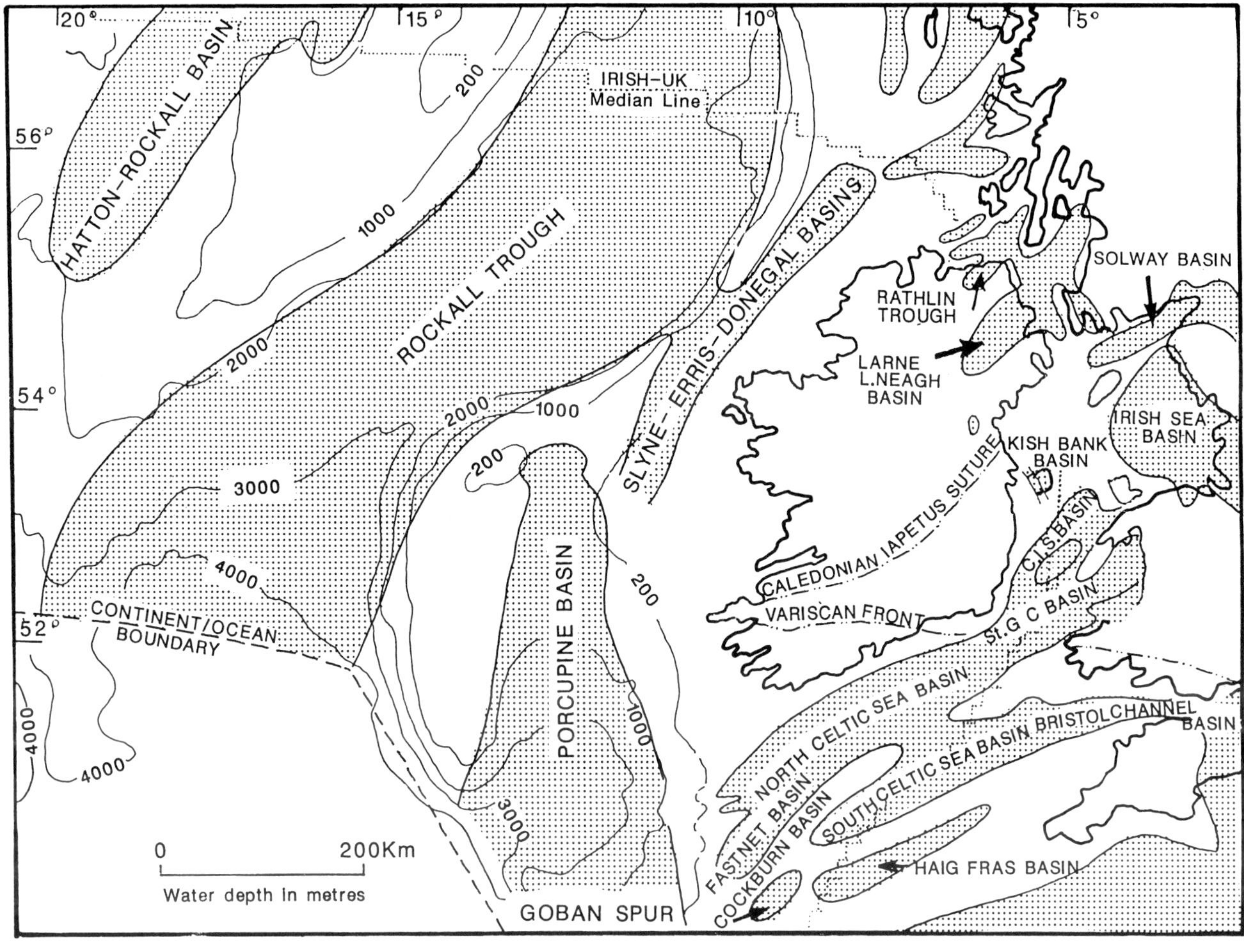

Fig. 2. Mesozoic–Tertiary sedimentary basins around Ireland (after Naylor & Shannon 1982).

higher heat flows. Onshore maturation values record decreasing maturation from the greenschist meta-argillite regime of the Munster Basin in the south of the country to the more promising oil-window maturation values of the northeast.

Naylor (1992) emphasized the strong contrast between the development of the onshore and offshore regions. Onshore Ireland has been a positive and largely emergent feature for much of post-Variscan time. The Early Jurassic and Late Cretaceous periods were probably the main periods of post-Variscan marine submergence onshore, but evidence of these incursions has been removed by later erosion.

After a period during which the onshore area was largely ignored, licences were taken up in 1980 over the Northwest Carboniferous Basin, and adjoining areas by a consortium consisting of Aran Energy, Marinex Petroleum and North West Oil & Gas. This group, operated by Marinex, re-entered and attempted to stimulate the old Dowra-1 well (with a resultant flow rate of 289 000 scf per day), and carried out a regional seismic programme. Santa Fe joined the group and, with Aran Energy now as operator, four wells were drilled in the period 1984–85. All were plugged and abandoned without encountering significant shows. One well was drilled in Co. Leitrim (Drumkeeran South-1), one in Co. Cavan (Macnean-2), and two in the immediate border area of Co. Fermanagh, Northern Ireland (Kilkoo Cross-1 and Slisgarrow-1: Fig. 1 inset).

Philcox *et al.* (1992) have outlined the development of the Northwest Carboniferous Basin, based in part on the results of the Aran Energy consortium drilling. Vitrinite reflectance studies indicate that the Carboniferous rocks are generally supra-mature for oil generation, but mature for gas generation.

A licence was also held by Tullow Oil in the 1980s in east central Ireland over the crop of the Leinster Coalfield. A slimhole exploration test was drilled in December 1987 at the northern end of the coalfield (Timahoe-1, Fig. 1) but was terminated at less than 1000 ft due to mechanical difficulties.

In consequence of the perceived low potential for oil and gas exploration, there has been no significant exploration activity in the Republic in recent years. However, maturation levels and reservoir quality improve northwards, and Upper Palaeozoic rocks in the northernmost areas of the Republic and in Northern Ireland have some exploration potential.

The thick Permo-Triassic reservoir sequences of the Ulster Basin (comprising the Larne-Lough Neagh Basin and the Rathlin Trough : Fig. 2) retain interest for hydrocarbon traps sourced from underlying Carboniferous strata. The Permo-Triassic to Palaeogene subsurface geology of the Ulster Basin has been outlined by Thompson (1979), McCann (1988, 1990) and McCaffrey & McCann (1992). Parnell (1991a,b, 1992) reported on reservoir potential in the Carboniferous and Permo-Triassic sandstones of the Ulster Basin, and discussed hydrocarbon source rocks in northern Ireland. Parnell (1991a,b) concluded that Carboniferous source rocks are probably the most promising and widely distributed.

Recent exploration in Northern Ireland has centred around

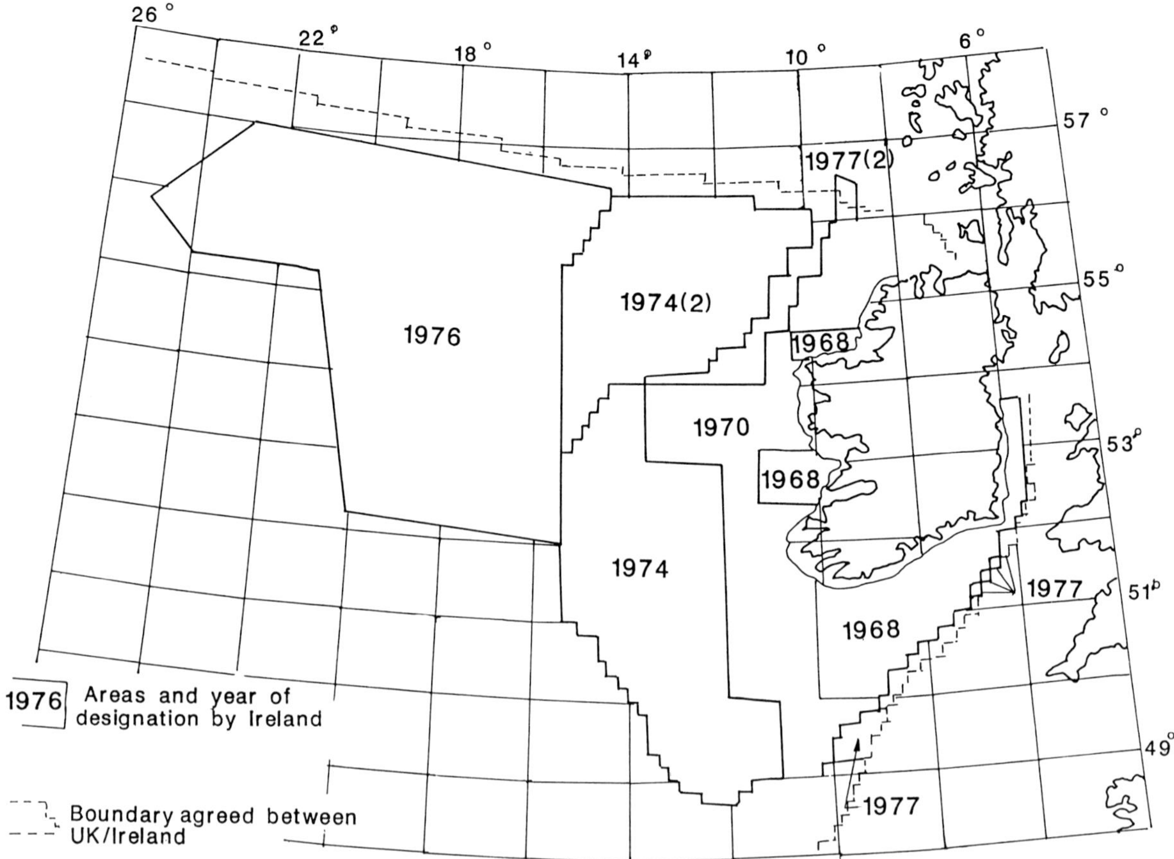

Fig. 3. Successive stages in the designation of the Irish Continental Shelf (after Naylor & Shannon 1982).

the potential of the Permo-Triassic basin although, as we have seen, several earlier wells were drilled in pursuit of Carboniferous targets as an extension of the Ambassador and Aran exploration programmes in the Republic. In fact, the earliest exploration of the Permo-Triassic basin was driven by the search for concealed coal fields beneath the Mesozoic cover (Griffith 1983). Coal was formerly mined in the Tyrone coal field, small outliers of Coal Measures west of Lough Neagh, and early tests had been made east of the Lough in an unsuccessful attempt to locate more reserves. Government-sponsored boreholes were drilled in the sub-basins of the Ulster Basin between 1956 and 1980 in search of Carboniferous coal and Permo-Triassic halite and evaporites. A deep geothermal test – Larne No. 2 (Fig. 1) – was also drilled at the coast (Penn 1982). These deep tests, although not drilled in search of hydrocarbons, did confirm the nature of the thick Permo-Triassic succession.

In 1968, Marathon obtained an exploration licence for an area northeast of Belfast, and covering the adjacent offshore. Some offshore seismic lines were shot. An exploration well (Newmill-1, at the head of Larne Lough) was drilled in the Permo-Triassic basin by Shell under a farm-in arrangement with Marathon, but was dry and abandoned. This was the first oil exploration well to be drilled in Northern Ireland.

None of the coreholes or wells drilled in the Ulster Basin prior to 1980 had been located using seismic surveys. Under an initiative from the Department in 1981, CGG shot two vibroseis lines across the Larne–Lough Neagh portion of the basin, and a further line extending from Tertiary lavas onto Lower Carboniferous rocks to the west of Lough Neagh. These lines were made available to the industry (Illing 1982). They showed in particular the presence of promising structures in the Permo-Triassic, and also that acquisition techniques had advanced sufficiently for reasonable seismic results to be obtained through the basalt which covers much of the basin. This latter point had been a source of technical concern. Interestingly, the first acquisition of reflection seismic in Ireland had been by the Geological Survey of Northern Ireland in 1959, using an SSL dynamite crew to attempt, unsuccessfully, to image sub-basalt structure beneath part of the Rathlin Trough.

Three petroleum exploration wells have been drilled during the last four years to test the potential of the Permo-Triassic sequence in the Ulster Basin, each located using vibroseis surveys. The first, Ballytober-1, was drilled northwest of Larne by a consortium with Kirkland as operator, and the other two (Annaghmore-1 and Ballynamullan-1) by Nuevo Energy (as operator for a group including Lough Neagh Exploration and Wiser) on the north shore of Lough Neagh (Fig. 1). The wells were drilled on seismically-defined features and were plugged and abandoned, but no stratigraphic details are available. Press statements from the Nuevo group indicate that shows of oil were encountered in the Annaghmore well. The basin remains under-explored, but the existence of structure, seal and reservoir potential has been demonstrated. However, the distribution of pre-Permian source rocks is unknown (only one of the published deep tests has penetrated beneath the Permian).

OFFSHORE EXPLORATION

General summaries of the geology and exploration history of the Irish offshore basins can be found in Naylor & Shannon (1982), Naylor (1984), and Shannon (1991a, 1993).

Table 1. *Petroleum authorizations*

Petroleum prospecting licence(non-exclusive)
Confers the right to search for petroleum in any part of the Irish offshore which is not subject to an existing exploration licence, reserved area or petroleum lease. Annual rental.

Licensing option
Confers the first right, exercisable at any period of the option, to an exploration licence over all or part of the area covered by the option.

Exploration licence
Confers the exclusive right of searching for petroleum in the area to which the licence applies. There are 3 categories of exploration licence:
Standard: valid for 6 years in areas up to 200 m water depth;
Deepwater: valid for 12 years in water depths exceeding 200 m
Frontier: valid for 15 years minimum in areas of special difficulty related to physical environment, geology or technology.

Lease undertaking
Covers a situation in which a discovery has been made but there is not enough time within the licence period for the licensee to determine whether the discovery is commercial. Involves an undertaking by the Minister to grant a petroleum lease, subject to certain conditions, in relation to that part of the licensed area which contains the discovery.

Petroleum lease
Issued in respect of the development of a commercial discovery. The rental fee is revised annually in line with the Wholesale Price Index.

Reserved area licence
Issued with respect to a specific area adjacent to or surrounding the leased a leased area. Confers the same rights as an exploration licence.

Source: Petroleum Affairs Division (1994)

After negotiations with the Irish Government, Marathon Oil Company was granted three offshore tracts (Robinson & Riddihough 1975; Naylor & Mounteney 1975). In 1968, the first Irish offshore designation claimed areas slightly larger than those granted to Marathon (Fig.3). The following year a revised agreement was negotiated between the Government and Marathon under which the company relinquished any exclusive claim to the Continental Shelf, except for the areas which it already held. Marathon retained claims in these three areas for the second term of the original licence (1965–70). These were further reduced for the third term (1970–75) and again in March 1975, for the final term (1975–80).

The first seismic surveys were shot in the Celtic Sea in 1969. The discoveries in the southern North Sea had by this time imparted a considerable impetus to exploration, together with the growing realization of the extension of the Mesozoic trough system around Britain and Ireland (Fig.2).

A series of further offshore designations took place in 1970, 1974, 1976 and 1977, extending the area westwards over the Porcupine and south Rockall region (Fig.3). There was no agreed boundary between Ireland and the United Kingdom at this time and, in fact, the second Irish designation of 1977 counterclaimed a small area also designated by the United Kingdom. It was not until 1990 that agreement on a boundary was reached by negotiation between the two countries. As in United Kingdom waters, each 1° Latitude × 1° Longitude sector of the Irish offshore was numbered and divided into thirty licence blocks.

In May 1970 the 48/25-1 well (P&A) became the first well to be drilled in the Irish offshore. The well was located in the North Celtic Sea Graben, with Marathon as operator. The company then drilled another unsuccessful well, before returning to drill a second well in Block 48/25. This well, completed in late 1971, was the discovery well of the Kinsale Head gas field. The field, which produces from Lower Cretaceous sandstones, contained recoverable reserves of 1.5×10^{12} scf of gas (Winn 1994).

In 1971, Esso and Marathon agreed a farm-out deal, whereby Esso would earn a 50% interest in just over half of Marathon's Celtic Sea concession, in return for a work obligation involving the drilling of a number of exploration wells and further seismic surveys. The farm-out area covered the western portion of the Celtic Sea concession, but did not include the 48/25 Kinsale Head gas field. At this time, no other concessions had been granted outside the Marathon areas, despite growing interest in the Irish offshore. By 1974, sixty-two companies and groups had been granted non-exclusive exploration licenses (Table 1), which allowed the companies to carry out exploration surveys on open acreage on the Irish Continental Shelf. A list of the companies is given in Robinson & Riddihough (1975) and reveals a substantial oil industry interest in the Irish offshore sector at that time.

Exploratory surveys and drilling were carried out by Marathon and Esso–Marathon on the Celtic Sea acreage thoughout the 1970s. Although no further commercial discoveries were made, the companies were granted a total of 20 petroleum leases (17 to Marathon, 2 to Marathon–Esso, and 1 to Marathon–Esso–Elf) covering 37 blocks, which continue in force for a period of twenty-one years. Production leases awarded on blocks licensed under the 1975 licensing terms and conditions will be for twenty-eight years (Table 1), and beyond this in the case of production blocks, until production ceases.

Legislation governing exploration for hydrocarbons was brought forward in the Petroleum and other Minerals Development Act, 1968. In 1975, the Department of Industry and Commerce published a set of terms and conditions under which exclusive exploration licences would be awarded. Applications were then invited for a First Round of Licensing. The Round covered all blocks within Irish-designated territory, with the exception of those held by Marathon and Esso/Marathon, and 24 other blocks, which at the time were being negotiated for by a number of joint-venture groups. The first exploration licences were granted in 1976 to 11 different groups, for a total of 43 blocks. These gave the holders exclusive rights to carry out exploration, including drilling, on the specified blocks. Licences were issued for an initial six-year period where water depths were less than 600 ft, or nine years in deeper water. A 50% relinquishment was required after four or six years, respectively. Current authorizations are listed in Table 1. The awards in the First Round included blocks in several of the offshore basins. Significantly, the new terms made a provision for State

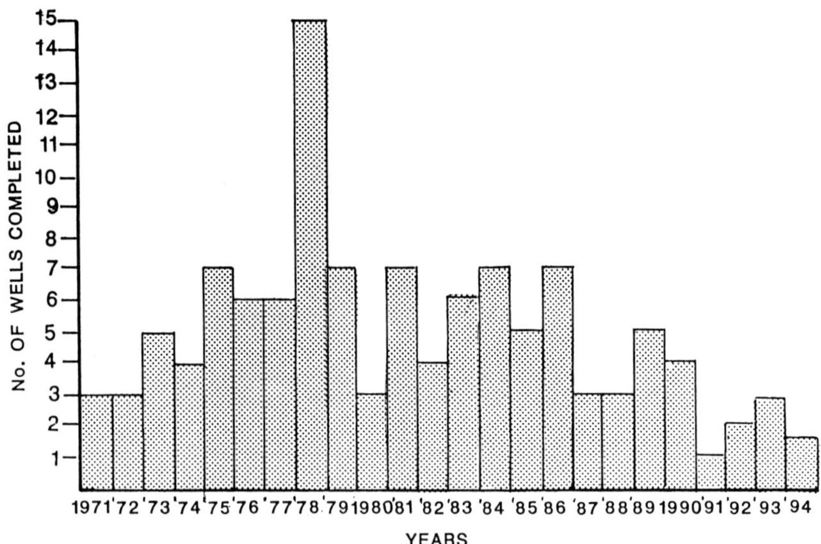

Fig. 4. Histogram of the offshore exploration wells drilled each year since 1971 (source: Department of Transport, Energy and Communications, Dublin. Petroleum Affairs Division 1994).

participation in the event of a commercial discovery.

In 1976, the first drilling took place in the Fastnet Basin, a western extension of the Celtic Sea basins (Fig. 2), in about 130 m of water.

The following year saw further drilling in the Fastnet Basin, and the first well in the Kish Bank Basin, east of Dublin. Drilling also commenced in 1977 in the deeper waters of the Porcupine Basin, off the west coast, when Shell drilled 35/13-1 in 482 m of water. Interest in the western basins was being fuelled by the major discoveries being made at that time in the UK northern North Sea.

Since the First Licensing Round the Irish Department of Energy has, in contrast to the UK, operated an 'open door' policy with respect to licensing. Companies are free to enter into discussions with the Department regarding the work programme which would be required to obtain a licence over a block or blocks in open acreage. This policy resulted in the granting of a small flow of licences over the succeeding years. 'Option' agreements were first signed by exploration groups in 1978. Under the terms of an option (Table 1), the company is required to carry out a work programme, normally a seismic survey, within a specified period. The company then has to decide whether any of the blocks held under the option are to be retained as an exploration license with an agreed work programme.

In November 1980, the Minister announced a Second Licensing Round, with the closing date on 29 January 1982. In all, 108 blocks were on offer (45 in the North Celtic Basin, 46 in the Porcupine Basin, 7 in the Slyne Trough, 6 in the Donegal Basin and 4 in the Kish Bank Basin). Thirty-seven companies (twenty-one new to offshore Ireland) were involved in applications. In June 1982 the Minister announced the award of 24 blocks to 10 consortia (18 in the Celtic Sea basins, 4 in the Porcupine Basin, and 2 in the Kish Bank Basin).

The Second Licensing Round was followed by an intensive phase of seismic acquisition, followed in 1983 by the drilling of further wells in the North Celtic Basin. Due to the outcrop of Chalk over much of the basin, the quality and penetration of the early seismic surveys in the region had been relatively poor. Newer seismic techniques now allowed better imaging of the deeper plays, particularly in the Jurassic, which was judged to have considerable potential. The discovery of hydrocarbons in 1983 by a Gulf–Union–Atlantic Resources consortium within the Middle–Upper Jurassic section in Block 49/9 at the eastern end of the North Celtic Graben, gave rise to considerable industry interest. The well tested at rates up to 6467 bopd of 44°API oil (aggregating 9911 bopd oil and 2.1×10^6 scfd gas), and demonstrated the potential for light oil accumulations in the Celtic Sea area. The Department of Energy suspended the 'open door' policy for licence acquisition in the discovery area, and announced (29 February 1984) a Third Round of Licensing, encompassing most of the open acreage (76 blocks) in the Irish sector of the Celtic and Irish seas, with a closing date of applications of 15 February 1985. Twenty-two companies were involved in the Round and, in October 1985, the Minister announced the award of fifteen blocks to nine consortia. Thirteen of the blocks were in the Celtic Sea and two in the Irish Sea. Unfortunately, follow-up drilling of the 49/9 discovery showed that it was of small extent. This discovery – the Helvick Field – now operated by Arcon Resources plc. remains undeveloped (Caston 1994). In 1985, Gulf drilled a further success when well 50/6-1 flowed 2074 bopd and 1.3×10^6 scfd gas.

The Kinsale Head gas field, about 30 miles offshore, went on stream in 1978 and the Ballycotton satellite structure (89.5×10^9 scf gas in place: Winn 1994) 9 miles to the northwest in Block 48/18, was linked into the Kinsale Head facility in 1991 (Murray 1994). Further east, gas flow rates of 119×10^6 scfd were recorded in the Marathon well 49/13-1, on the Ardmore prospect. An offset well, 49/14-1, flowed gas on test at a rate of 8.81×10^6 scfd, but the accumulation was not regarded as commercial.

A number of non-commercial oil accumulations have been encountered in drilling within the Celtic Sea area. One is the Esso–Marathon Seven Heads prospect, on which two wells flowed hydrocarbons on test. Esso–Marathon 48/28-1 produced at a rate of 1550 bopd from a thin sandstone. A further well, 48/23-1 had oil and gas shows, but failed to flow on test. The final well on the prospect, 48/24-2, was drilled in 1978 and encountered both gas and oil zones. Reserve estimates by the operator are about 100×10^9 scf gas, and $1–2\times10^6$ bbl of oil. The well 48/10-2, drilled in 1992, west of the Ballycotton

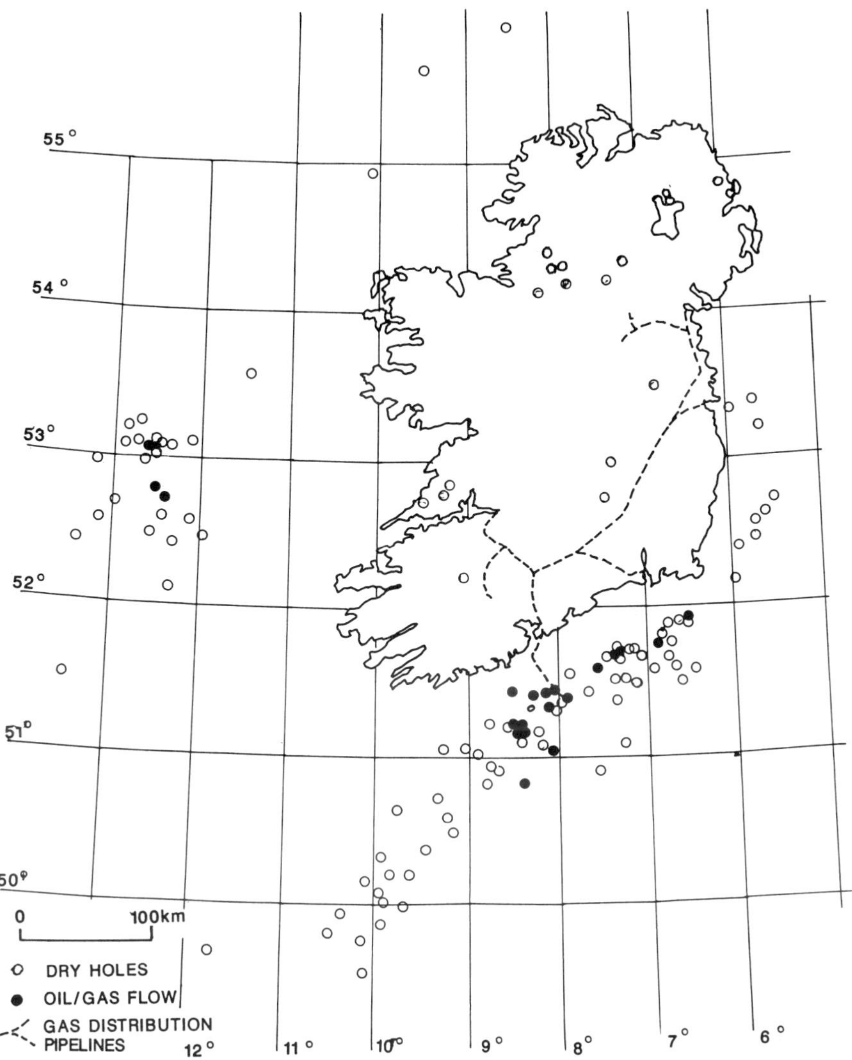

Fig. 5. Distribution of exploration wells, offshore Ireland (source: Department of Transport, Energy and Communications, Dublin. Petroleum Affairs Division 1994).

Field, encountered a large accumulation of heavy biodegraded oil (Howell & Griffiths 1994), similar to the Seven Heads structure, in Greensand and Wealden sandstones, probably sourced from the Liassic. In 1985, a Gulf-operated group tested 2074 bopd and 1.3×10^6 scfd in well 50/6-1, probably from the pre-Cretaceous section.

The general tectonic setting and prospectivity of the Celtic Sea basins has been reviewed by Shannon (1991b), who emphasized the important control exercised on the younger sequences by Caledonian and Variscan structures, and also pointed to the need to explore a wider range of play types. Eighty wells have been drilled in the offshore Celtic Sea region over the past 20 years (Table 2 & Fig. 5), but the Kinsale and Ballycotton gas fields represent the only commercial successes. Contributing reasons for this lack of success have been the relatively poor seismic data quality, and the tectonic complexity of the area. A range of untried plays remain at pre-Cretaceous levels in the Celtic Sea basins. However, due to the structural complexity, individual prospects are likely to be modest in size.

The Porcupine Basin remains an area of exploration promise, with up to 9 km of Cretaceous and Tertiary strata preserved (Naylor & Anstey 1987; Shannon *et al.* 1993). Because of the deep water and low oil prices, there has been little exploration effort there in recent years.

Table 2. *Location of offshore wells (to end 1994)*

Celtic Sea	68
Fastnet/Mizen	12
Porcupine	25
Goban Spur	1
Slyne	1
Erris Trough	2
Donegal Basin	1
Kish Basin	3
Irish Sea	5
Total	118

Source: Petroleum Affairs Division (1994).

Between 1977 and 1982 twenty-one wells were drilled in the Porcupine Basin. Much of this early drilling in the basin was focused on the type of tilted Jurassic horst blocks and

Tertiary targets which had been productive in the North Sea. Although many of the wells encountered hydrocarbons, the lack of thick Jurassic reservoirs, in particular, gave rise to disappointing results. More than 80 000 km of seismic data have been acquired in the basin, and a total of twenty-five wells drilled. Four of the wells flowed significant quantities of petroleum. Phillips 35/8-1(1978) flowed good quality light oil from Lower Cretaceous marine sandstones. A later well on the same block flowed condensate from Upper Jurassic sandstones. No further appraisal wells were drilled and the company subsequently relinquished the licence. Also in 1979, BP 26/28-1 flowed 5589 bopd from Middle and Upper Jurassic fluvial and shallow-marine sandstones. A follow-up well flowed 1500 bopd from similar rocks. However, three further appraisal wells were dry, indicating a relatively small and complex structure. MacDonald *et al.* (1987) report that the field (the Connemara Field) contains approximately 120×10^6 bbl of proven oil (32–38° API) in place, and a further 75×10^6 bbl of possible reserves. The undeveloped Connemara Field (Earls 1994) remains the only likely commercial discovery made to date in the Porcupine Basin. However, the basin is large and underexplored. There is a variety of untried plays, including thick sand-prone units, which have not been drilled under closure. The Department is attempting to encourage exploration in the sector with a current Frontier Licensing Round. This Round involves 172 blocks and one part-block, and applications closed on 15 December, 1994. The recent focus of the industry on the West Shetland area (Knott 1994), and an increased perception of the prospectivity of the Tertiary sections in the western basins, added interest to the Round (Shannon 1996).

A stratigraphic framework for the Goban Spur has been suggested by Colin *et al.* (1992). Basin development was initiated in the Triassic. Only one well has been drilled in the sector to date.

The Northwest Offshore basins were the subject of a Frontier Licensing Round in 1993. Only three wells have been drilled in the area, and it is evident from the published data that the area retains exploration potential. Trueblood (1992) provided a comprehensive review of the petroleum geology of the Slyne Trough based on oil company exploration data. Adequate source, reservoir and seal sequences were demonstrated in the Lower and Middle Jurassic. The Lower Jurassic source rocks were considered to be probably mature in the basin centre.

Murphy & Croker (1992), of the Irish Government's Petroleum Affairs Division, discussed the geological development, exploration history and petroleum potential of the Erris and Slyne troughs (Fig. 2). They released further seismic and geological data for the two basins, revealing fault-controlled structures. A large number of play types were proposed for the area. The Frontier Licensing Round, which offered 128 blocks covering the Slyne and Erris troughs, closed on 15 December,1993. Ten companies, grouped in five consortia, applied for twenty-eight of the blocks. Five licences were awarded to the consortia, led by Enterprise (2 licences), Kerr-McGee Oil, Statoil (UK) Ltd and Texaco (Petroleum Affairs Division 1994).

If the Porcupine Basin requires a willingness on the part of the oil industry to venture back into deeper water, this is doubly true of the Rockall–Hatton region. The promise of this frontier region is becoming increasingly obvious as more data are collected and published. Structure and thick sedimentary sequences are undoubtedly present, together with some of the other ingredients necessary for hydrocarbon generation and entrapment. Makris *et al.* (1992) presented the results of a 700 km reflection/refraction profile across the Erris Trough, Rockall Trough and Rockall Bank, which showed fault-bounded basins with up to 5 km of sediments, probably ranging in age from Late Palaeozoic to Cenozoic, resting on thinned continental crust. To date, no wells have been drilled in the Irish sector of the Rockall–Hatton troughs.

The Irish Sea basins are being re-examined in the light of new developments in the Morecambe Bay area (Trueblood *et al.* 1994). The Permo-Triassic section holds the greatest promise, and has been the target for wells in the central and southern sections of the Irish Sea. However, interest could spread northwards into the area between Northern Ireland and Scotland, where speculative seismic surveys in the past have demonstrated the existence of a basin with faulted structures. The renewed interest in the Irish Sea province following the Hamilton discoveries off North Wales is also likely to extend into the eastern portion of the Celtic Sea. The discovery in 1994 by UK well 103/01-1, of oil in the Jurassic of the St George's Channel Basin (Marathon Press Release) close to the UK/Irish median line, should also spur activity in the region. The Irish Sea prospects are discussed by Shannon elsewhere in this volume.

Three wells have been drilled in the Permian to Liassic Kish Bank Basin (Naylor et al. 1993). Much of the Westphalian section penetrated at the margin of the basin by the first well–Amoco 33/22-1 (1977)–lies within the gas window and has vitrinite values ranging from 0.8% to 1.3%. The two later wells, Shell 33/21-1 (1979) and Fina 33/17-1 (1986), penetrated the prospective Permo-Triassic section but were dry, despite encountering good reservoir sections. The lack of exploration success is possibly due to the absence of Upper Carboniferous source sequences beneath the main part of the basin.

FISCAL TERMS

The main legislation with respect to oil and gas exploration and production in Ireland is the following:

Petroleum and Other Minerals Development Act, 1960
Continental Shelf Act, 1968
Finance Act, 1992.

During the 1980s there were a number of revisions to the 1975 fiscal terms for exploration and development, including the announcement in 1987 of no royalties on production, and no State participation. The continuing low levels of exploration prompted a further radical review, which produced new petroleum taxation measures in the Finance Act, 1992. This provided a special low rate of 25% Corporation Tax – now the only State 'take' from oil and gas production – on production under petroleum leases granted before certain future dates (2003–2013), dependent on the length of petroleum lease tenure. There will be normal 40% Corporation Tax beyond those dates. There will be 100% allowances for exploration, development and operating expenses.

The new fiscal terms were accompanied (November 1992) by revised Licensing Terms from the Minister for Transport, Energy & Communications (Petroleum Affairs Division 1994). These terms apply to all new authorizations and replace the amended 1975 Terms. Existing authorizations (i.e. pre-1992) continue to be subject to the 1975 Terms, but the operating companies may under appropriate circumstances opt for the new Terms. The full range of authorizations (Table 1) were covered under the 1992 Terms and the rental fees were restructured.

CONCLUSION

In general, the level of exploration onshore–offshore Ireland during recent years has been low (Fig. 4), with no major discoveries. Ireland produces an average of 239×10^6 scf (6.77×10^6 m^3) of gas per day from the two offshore gas fields. This represents about 18% of the country's primary energy needs. A pipeline system (Fig. 3) relays gas to most of the major centres of population, and construction of a gas import/export interconnector with the UK has recently been completed. The country has one small refinery, situated on the south coast.

Despite offshore exploration since 1970, most of the prospective offshore sedimentary basins remain underexplored, particularly in the deeper-water areas off the west coast. A total of 118 wells have been drilled to date (Table 2) at a cost probably in the order of £900 million in 1994 values. Seismic acquisition costs are probably approaching £200 million. To date, the only significant discoveries resulting from this expenditure have been the Kinsale–Ballycotton gas fields and the undeveloped Connemara oil accumulation. Recent attempts to stimulate exploration have been made through the holding of Frontier Licensing Rounds in the Northwest basins (1993) and Porcupine (1994), and through improvement of the Financial Terms (1992). Low oil prices, lack of success after early promise, and deep water in the prospective western regions, have all contributed to the low levels of exploration offshore Ireland in the last decade. There is currently renewed interest in Irish waters due to a combination of factors – notably, disenchantment with the financial terms in the North Sea and, conversely, the improved Irish terms, together with the new large discoveries west of Shetland.

I would like to thank Dr K. W. Robinson, Dr P. M. Shannon and Dr M. E. Philcox for providing some of the data contained in this paper.

REFERENCES

CASTON, V. N. D. 1994. The Helvick oil accumulation, Block 49/9, North Celtic Sea Basin. The Petroleum Geology of Ireland's Offshore Basins Conference, Dublin 21–22 April 1994. *Abstracts of Papers*, 20–21.

CLAYTON, G., HAUGHEY, N., SEVASTOPULO, G. D. & BURNETT, R. 1989. *Thermal maturation levels in the Devonian and Carboniferous rocks in Ireland*. Geological Survey of Ireland.

COLIN, J-P., IOANNIDES, N. S. & VINING, B. 1992. Mesozoic stratigraphy of the Goban Spur, offshore south-west Ireland. *Marine and Petroleum Geology*, 9, 527–241.

COLLINS, C. B. 1976. *Wildcats and Shamrocks*. Mennonite Press Inc., Kansas.

EARLS, T. C. 1994. Potential for development of the Connemara Field – Block 26/28. The Petroleum Geology of Ireland's Offshore Basin. Conference, Dublin 21–22 April 1994. *Abstracts of Papers*, 46.

FALCON, N. L. & KENT, P. E. 1960. *Geological results of petroleum exploration in Britain 1945–1957*. Geological Society, London, Memoir, 2

GRIFFITH, A. E. 1983. The search for petroleum in Northern Ireland. *In:* Brooks, J. (ed.) *Petroleum Geochemistry and Exploration of Europe*. Geological Society, London, Special Publication, **12**, 213–222.

HOWELL, T. J. & GRIFFITHS, P. S. 1994. A study of the Blocks 48/18, 48/19 and 48/15 Lower Cretaceous prospectivity, North Celtic Sea Basin. The Petroleum Geology of Ireland's Offshore Basins. Conference, Dublin 21–22 April 1994. *Abstracts of Papers*, 40–41.

ILLING, L. V. 1982. *An assessment of petroleum prospects near Larne, Northern Ireland*. The Department of Economic Development for Northern Ireland.

KEELEY, M. L., LEWIS, S. L. E., SEVASTOPULO, G. D., CLAYTON, G. & BLACKMORE, R. 1993. Apatite fission track data from southeast Ireland: implications for post-Variscan burial history. *Geological Magazine*, **130**, 171–176.

KNOTT, D. J. 1994. West of Shetland emerges as U.K. exploration hotspot. *Oil & Gas Journal*, June 20, 16–20.

LEES, G. M. & TAITT, A. H. 1946. The geological results of the search for oilfields in Great Britain. *Quarterly Journal of the Geological Society of London*, **101**, 255–317.

MacDONALD, H., ALLAN, P. M. & LOVELL, J. P. B. 1987. Geology of an oil accumulation in block 26/28, Porcupine Basin, offshore Ireland. *In:* Brooks, J. & Glennie, K. W. (eds) *Petroleum Geology of Northwest Europe: Proceedings of the 3rd Conference*. Graham & Trotman, London, 643–651.

MAKRIS, J., GINZBURG, A., SHANNON, P. M., JACOB, A. W. B., BEAN, C. J. & VOGT, U. 1992. A new look at the Rockall region, offshore Ireland. *Marine and Petroleum Geology*, **8**, 410–416.

McCAFFREY, R. J. & McCANN, N. 1992. Post-Permian basin history of northeast Ireland. *In:* Parnell, J. (ed.) *Basins on the Atlantic Seaboard*. Geological Society, London, Special Publication, **62**, London, 277–290.

McCANN, N. 1988. An assessment of the subsurface geology between Magilligan Point and Fair Head, Northern Ireland. *Irish Journal of Earth Sciences*, **9**, 71–78.

——— 1990. The subsurface geology between Belfast and Larne, Northern Ireland. *Irish Journal of Earth Sciences*, **10**, 157–174.

MURPHY, N. J. & CROKER, P. E. 1992 Many play concepts seen over wide area in Erris, Slyne troughs off Ireland. *Oil & Gas Journal*, **90**(37) September 14

MURRAY, M. V. 1994. Development of small gas fields in the Kinsale Head area. The Petroleum Geology of Ireland's Offshore Basins. Conference, Dublin 21–22 April 1994. *Abstracts of Papers*, 20–21.

NAYLOR, D. 1984. Petroleum exploration in the Republic of Ireland: a review. *Energy Exploration & Exploitation*, **3**, 5–26.

——— 1992. The post-Variscan history of Ireland. *In:* Parnell, J. (ed.) *Basins on the Atlantic Seaboard*. Geological Society, London, Special Publication, **62**, 255–275.

———, HAUGHEY, N., CLAYTON, G. & GRAHAM. J. R. 1993. The Kish Bank Basin, offshore Ireland. *In:* Parker, J. R. (ed.) *Petroleum Geology of Northwest Europe: Proceedings of the 4th Conference*. Geological Society, London, 845–855.

——— & MOUNTENEY, S. N. 1975. *Geology of the North-West European Continental Shelf*. (Vol 1). Graham & Trotman, London.

——— & SHANNON, P. M. 1982. *The Geology of Offshore Ireland and West Britain*. Graham & Trotman, London.

——— & ANSTEY N. A. 1987. A reflection seismic study of the Porcupine Basin, offshore west Ireland. *Irish Journal of Earth Sciences*, **8**, 187–210

PARNELL, J. 1991*a*. Hydrocarbon potential of Northern Ireland: 1. Burial histories and source rock potential. *Journal of Petroleum Geology*, **14**, 65–78.

——— 1991*b*. Hydrocarbon potential of Northern Ireland: 2. Reservoir potential of the Carboniferous. *Journal of Petroleum Geology*, **14**, 143–160.

——— 1992. Hydrocarbon potential of Northern Ireland: 3. Reservoir potential of the Permo-Triassic. *Journal of Petroleum Geology*, **15**, 51–70.

PENN, I. E. 1982. Larne No.2 *Geological Well Completion Report. Deep Geology Unit*, Institute of Geological Sciences Report, 81/6.

PETROLEUM AFFAIRS DIVISION, DUBLIN.1994. *Summary of hydrocarbon exploration in Ireland* (and accompanying well chart). Department of Transport, Energy and Communications, Dublin.

PHILCOX, M. E., BAILY, H., CLAYTON G. & SEVASTOPULO, G. D. 1992. Evolution of the Carboniferous Lough Allen Basin, Northwest Ireland. *In:* Parnell, J. (ed.) *Basins on the Atlantic Seaboard*. Geological Society, London, Special Publication, **62**, 203–216.

ROBINSON K. W. & RIDDIHOUGH, R. P. 1975. Ireland – Oil and Gas Exploration. *Geological Survey of Ireland Information Circular*, **8**, 11.

SHANNON, P. M. 1991*a*. Irish offshore basins: geological development and petroleum plays. *In:* Spencer, A. M. (ed.) *Generation, Accumulation and Production of Europe's Hydrocarbons*. Special Publication of the European Association of Petroleum Geologists, 1, Oxford University Press, Oxford, 99–109.

——— 1991**b**. The development of Irish offshore sedimentary basins. *Journal of the Geological Society, London*, **148**, 181–190.

——— 1993. Oil and gas in Ireland – exploration, production and research. *First Break*, **11**, 429–433

——— 1996. Current and future potential of oil and gas exploration in Ireland. *This volume*.

———, MOORE, J. G., JACOB, A. W. B., & MAKRIS, J. 1993. Cretaceous and Tertiary basin development west of Ireland. *In:* Parker, J. R. (ed.) *Petroleum Geology of Northwest Europe: Proceedings of the 4th Conference*. Geological Society, London, 1057–1066.

SHERIDAN, D. J. R. 1972. Upper Old Red Sandstone and Lower Carboniferous of the Slieve Beagh Syncline and its setting in the Northwest Carboniferous Basin, Ireland. *Geological Survey of Ireland Special Paper*, **2**, 129.

——— 1977. The hydrocarbons and mineralization proved in the Carboniferous strata of deep boreholes in Ireland. *In:* Garrard, P. (ed.) *Proceedings of Forum on Oil and Ore in Sediments*. Imperial College London, 1975, 113–145.

THOMPSON, S. J. 1979. Preliminary Report on the Ballymacilroy No. 1 Borehole, Aloghill, Co. Antrim. *Geological Survey of Northern Ireland Open File Report*, **63**.

TRUEBLOOD, S. 1992. Petroleum geology of the Slyne Trough and adjacent basins. *In:* Parnell J. (ed.) ***Basins on the Atlantic Seaboard.*** Geological Society, London, Special Publication, **62**, 315–326.

———, BRYAN, C. & PICKERING, S. 1994. The Douglas Oilfield and its implications for exploration on the Irish Continental Shelf. The Petroleum Geology of Ireland's Offshore Basins Conference, Dublin 21–22 April 1994. *Abstracts of Papers*, 4–5.

WINN, R. D. 1994. Shelf sheet-sand reservoir of the Lower Cretaceous Greensand, North Celtic Sea Basin, offshore Ireland. ***American Association of Petroleum Geologists Bulletin***, **78**, 1775–1789.

Current and future potential of oil and gas exploration in Ireland

P. M. Shannon
Department of Geology, University College Dublin, Ireland.

ABSTRACT: The Late Palaeozoic to Cenozoic sedimentary basins in the Irish offshore are among the most lightly explored basins of Northwest Europe. Although the basins in the Irish sector of the Irish Sea have not yet yielded any discoveries, they developed in a similar setting to the oil- and gas-bearing East Irish Sea Basin in UK waters and a range of comparable, albeit smaller, Permo-Triassic plays are predicted. Exploration in the Celtic Sea basins to the south of Ireland has resulted in the discovery of Ireland's two producing fields. A number of unappraised or currently sub-commercial oil and gas fields have also been discovered in the North Celtic Sea Basin. Exploration to date in these basins has concentrated on Early Tertiary inversion structures with Lower Cretaceous reservoirs, and on Jurassic tilted fault blocks. Triassic plays have recently begun to attract attention along the margins of the basins. There are also untested stratigraphic and halokinetic-related plays in the basins. The frontier basins lying to the west of Ireland are typically large but lie in deep waters. Some oil accumulations have been discovered in the Porcupine Basin. Exploration in these basins has generally concentrated on Jurassic tilted fault blocks. There has been a significant renewal of exploration interest in these basins within the last couple of years. A large range of untested plays exist. These include tilted fault blocks in the Rockall Trough and in the Porcupine and NW Offshore Basins, and a large range of stratigraphic plays at Cretaceous and Tertiary levels in the Porcupine and Rockall basins.

KEYWORDS: *Celtic Sea, Irish Sea, Porcupine, Rocall, exploration*

INTRODUCTION

Ireland is surrounded by a series of Late Palaeozoic to Cenozoic sedimentary basins that formed in a general east-west extensional domain during the break-up of the Pangean supercontinent and the subsequent development of oceanic crust in the region to the west of Ireland. These basins have been the subject of periodic phases of exploration during the past 25 years (Croker & Shannon 1995; Naylor 1996). To date approximately 120 exploration and appraisal wells have been drilled in the Irish offshore. Two commercial gas fields (Kinsale Head and Ballycotton) have been discovered in the North Celtic Sea Basin (Fig. 1) and a further gas discovery has recently (1995) been made adjacent to the Kinsale Head Field. A number of unappraised or currently sub-commercial discoveries has been found in the North Celtic Sea and the Porcupine basins. These include the Seven Heads oil and gas accumulation (Naylor & Shannon 1982), the Helvick (Caston 1995) and Connemara accumulations (MacDonald *et al.* 1987; Earls 1995; Fig. 1).

Following a period of relative exploration dormancy during the late 1980s there has been a revival of exploration interest in the Irish offshore since 1994. This is attributed partly to the relatively unexplored nature of most of the basins. They represent some of the largest and most lightly explored offshore basins in NW Europe. The renewal of interest is also due in part to the successes of the exploration industry in UK waters west of the Shetlands and in the Irish Sea. The Irish licensing terms, previously regarded as being harsh by comparative standards, were altered to abolish state participation and royalties. No special petroleum taxes are payable and Corporation Tax at 25% is the only tax likely to be due on most fields. During the past decade the North Sea has developed into a mature province where the likelihood of very large discoveries is rapidly diminishing. There is thus the requirement to find new provinces to replace its declining production. Exploration has begun to move to deeper water and more frontier regions. These include the deep water basins west of the Shetlands where significant discoveries such as the Foinaven, Schiehallion and Stronsay oil fields have provided significant encouragement. These basins lie along strike from a number of the deep water frontier basins of the Irish offshore where comparable stratigraphies and structural development are seen (Trueblood & Morton 1991; Murphy & Croker 1992) and similar plays can be expected. Recent oil discoveries in the UK sector of the Irish Sea have also prompted a renewal of interest in the Irish sector of the region. In addition, the recent, first, discovery in the St George's Channel Basin adjacent to the UK–Ireland median line is likely to further heighten interest in the region. During the past few years a significant number of blocks have been licensed in most of the Irish offshore basins. The First Frontier Round awards of 1994 licensed 28 blocks to 5 consortia in the Slyne, Erris and Rockall basins. The Second Frontier Licensing Round resulted in the award, in 1995, of 32 blocks in the Porcupine Basin.

Ireland produces an average of 255×10^6 scf gas per day, representing almost 20% of the country's primary energy

From K. Glennie & A. Hurst (eds), 1996, *AD1995: NW Europe's Hydrocarbon Industry*, Geological Society, London, pp. 53–64

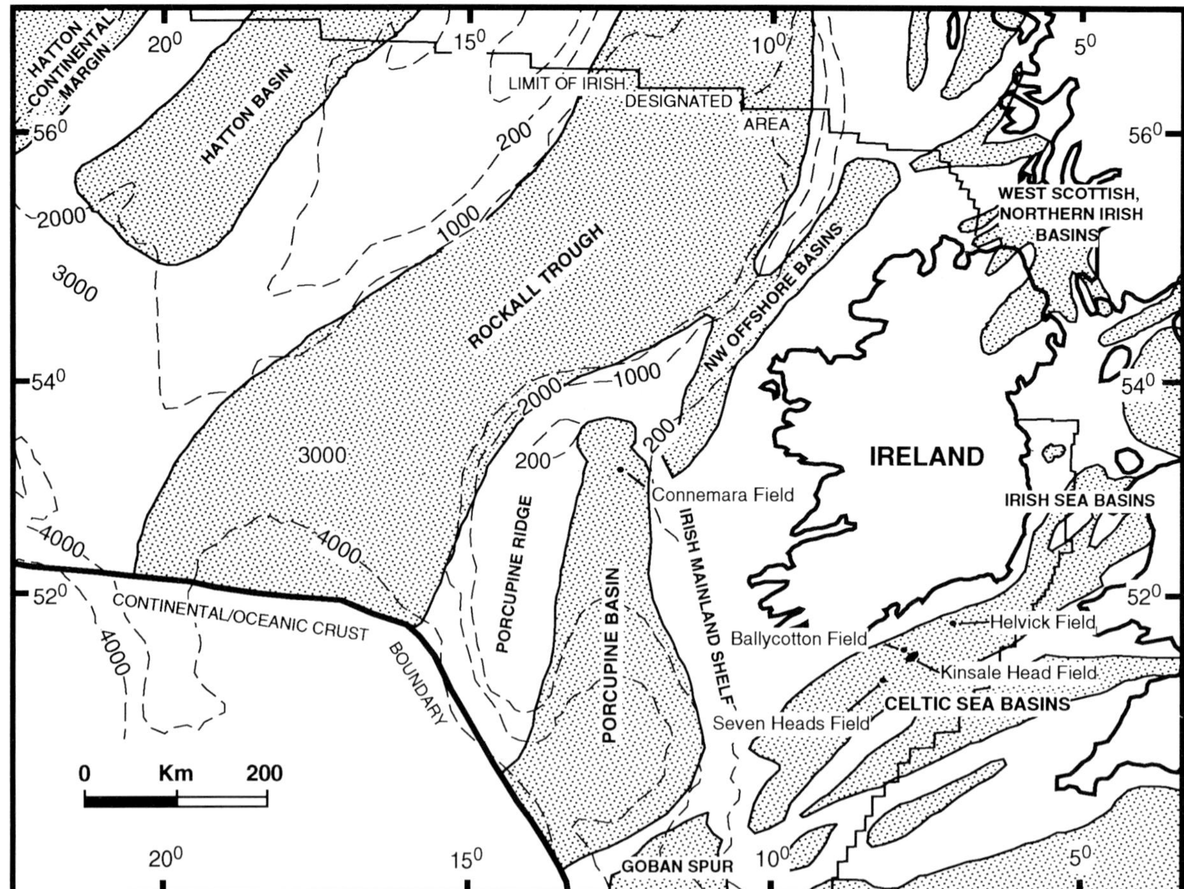

Fig. 1. Distribution of Irish offshore sedimentary basins. The major oil and gas discoveries are shown. Bathymetry is in metres.

requirements, from the two offshore gas fields. A pipeline network provides gas to most of the major centres of population and industry and a 292 km long gas import/export interconnector between Dublin and Scotland was completed in 1994 and the first gas was imported in 1995.

Gas from the Kinsale Head and Ballycotton fields, produced from two fixed platforms and one subsea system respectively, is landed on the south coast. The Whitegate oil refinery, situated on the south coast of Ireland, is operated by the Irish National Petroleum Corporation (INPC). This hydroskimming refinery has a primary distillation capacity of 56 000 bbl per day, naphtha reforming capacity of 14 500 bbl per day and hydrodesulphurization capacity of 6000 bbl per day (Shannon 1993a). It is planned to upgrade the refinery, preferably in conjunction with an investing partner. A large oil terminal at Whiddy Island adjacent to the refinery is also owned by the INPC and has storage capacity of over 7×10^6 bbl.

The objectives of the present paper are to review the geology of the main basins in the Irish offshore and to summarize and speculate upon the major current and future petroleum plays in the basins. The basins are divided for the purposes of the paper into the Irish Sea, Celtic Sea and Frontier basins.

OFFSHORE GEOLOGY

Irish Sea basins

The Irish Sea basins (Fig. 1), lying in shallow (less than 100 m) Irish designated waters between Ireland and the UK, comprise the Kish Bank and the Central Irish Sea basins. Lower Palaeozoic deformed basement, similar to that exposed onshore in eastern Ireland, is overlain locally by Visean limestones. The overlying geology is similar to that of the East Irish Sea Basin which contains significant oil and gas reserves (Ebbern 1981; Knipe *et al.* 1993; Trueblood *et al.* 1995) and comparisons at formation and member level have been made (e.g. Naylor *et al.* 1993). Basin development probably commenced in Late Palaeozoic times along reactivated Caledonian and older lineaments and tectonic fabrics. The Visean is overlain unconformably by Namurian and Westphalian sandstones, shales and coals (Jenner 1981). The Namurian Holywell Shale Formation, while not encountered in drilling to date, may be locally preserved in the Irish basins (Trueblood *et al.* 1995).

The main phase of extensional development took place during the Permian and Triassic and continued locally during the Early Jurassic. The Permo-Triassic strata, up to 3 km thick, are typically developed in red bed facies with continental sandstones and thick evaporites. There is no evidence of Mid-Jurassic to Tertiary strata in the Kish Bank Basin (Naylor *et al.* 1993) with possible preservation of such strata only locally in the Central Irish Sea Basin (Maddox *et al.* 1995). This may be due to non-deposition, to erosion following inversion, or to a combination of effects. The generalized stratigraphy for the Irish Sea basins is shown in Fig. 2.

Celtic Sea basins

The basins lying to the south of Ireland have an ENE–WSW to NE–SW orientation. Two parallel series of generally interlinked basins are bisected by a discontinuous basement

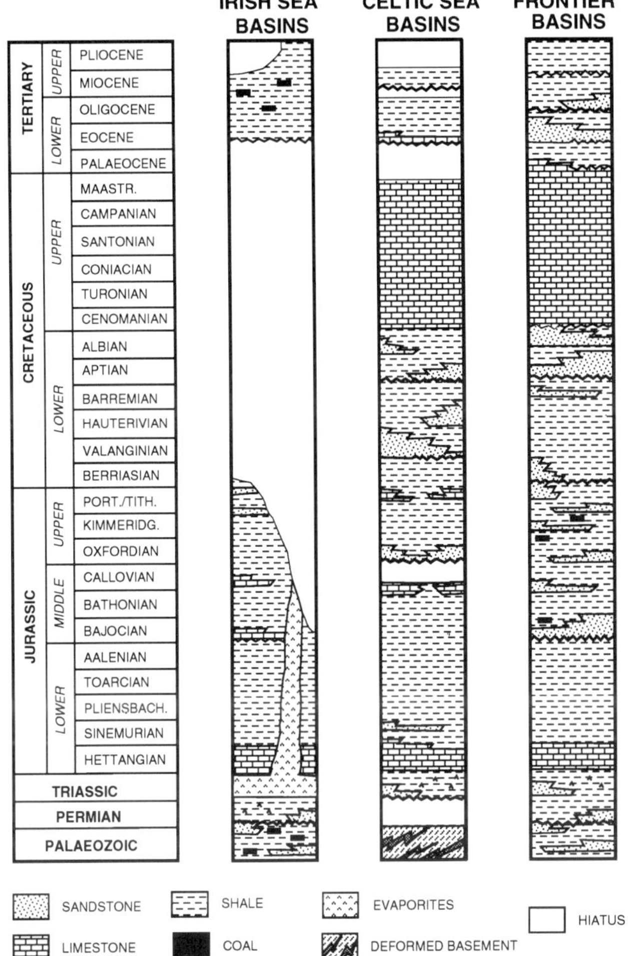

Fig. 2. Generalized stratigraphy of the Irish offshore basins. These are divided into the Irish Sea, Celtic Sea and Frontier basins as discussed in the text.

ridge (Fig. 1). The Fastnet, North Celtic Sea and St George's Channel basins lie to the north of the ridge with the Cockburn, South Celtic Sea and Bristol Channel basins lying to the south. The basin orientation in the region is controlled by Caledonian and Variscan structures (Pinet *et al.* 1987; Petrie *et al.* 1989; Shannon 1991a, b; McCann & Shannon 1994). Water depths in the region increase westwards from 100 m through much of the North and South Celtic Sea basins to 200 m in the Fastnet Basin. The sedimentary succession in the region is up to 9 km thick.

Basement comprises Variscan-deformed Devonian and Carboniferous strata (Robinson *et al.* 1981; Higgs 1983). Permian strata are rarely preserved in the region. Triassic strata typically comprise a lower Sherwood Sandstone Group equivalent and a Mercia Mudstone Group equivalent. The sandstones appear to be only locally developed while the red mudstones, anhydritic in places, are areally extensive. Thick salts are generally absent from the Fastnet and North Celtic Sea basins (except for a narrow band along the southern margin of the North Celtic Sea Basin), but are present in the other basins. The Triassic development was described by Shannon & MacTiernan (1993), Musgrove *et al.* (1995) and Shannon (1995). The Lower and Middle Jurassic marine strata are shale-prone with local sandstone-prone members (Fig. 2) and represent a thermal subsidence succession with little evidence of tectonism. A major unconformity towards the base of the Upper Jurassic marks the onset of Cimmerian tectonism resulting in syn-rift sandstone-prone strata of continental to shallow marine origin. A relative rise in sea-level resulted in the development of more shale- and locally limestone-prone strata during the Late Jurassic throughout the North Celtic Sea Basin. However, there is no evidence of regional Upper Jurassic strata in the Fastnet and Cockburn basins. Late Middle Jurassic basic igneous sills and dykes within the Fastnet Basin (Caston *et al.* 1981) may reflect the presence of a hot spot which caused local doming and prevented the development of a thick cover of Upper Jurassic strata comparable to that in the basins both to the east (North Celtic Sea Basin) and to the west (Porcupine Basin).

Lower Cretaceous strata are typically sandstone-prone (Fig. 2). They represent a succession of fluvial, alluvial, deltaic and lacustrine strata (Robinson *et al.* 1981; Naylor & Shannon 1982; Shannon 1991b; McCann & Shannon 1993) which are comparable to the Wealden of southern England (Allen 1981). The non-marine character of these strata is in marked contrast to coeval deep marine strata in the Porcupine Basin and the North Sea and is thought to represent uplift due to compression in the region associated with the commencement of sea floor spreading in the Bay of Biscay region (Ziegler 1988). Lower Cretaceous strata of pre-Aptian age are generally thin to absent, probably due to non-deposition, in the South Celtic Sea and Cockburn basins.

The Aptian and Albian of the region consist of shallow marine, sandy, Greensand, facies (Colley *et al.* 1981; Robinson *et al.* 1981; Hartley 1995; Taber *et al.* 1995). These are overlain by a progressively deepening succession of shales (Gault Clay) and Upper Cretaceous chalk.

The effects of early Tertiary inversion are pronounced in the North and South Celtic Sea basins, throughout much of the St George's Channel Basin and locally in the Fastnet Basin (Shannon 1991b). Lower Tertiary strata are generally thin and in a non-marine facies in these regions, while in the Fastnet Basin a thicker succession of Tertiary strata contains deep to shallow marine and non-marine lithologies (Robinson *et al.* 1981).

Frontier Basins

A number of large, generally deep-water basins lie to the west, southwest and northwest of Ireland (Fig. 1). These include the Porcupine Basin, the Goban Spur, the NW Offshore basins (Slyne, Erris and Donegal) and the basins of the Rockall Region (Rockall, Hatton and Hatton Continental Margin). These basins have seen little exploration to date and their geology is consequently generally poorly constrained. Only the Porcupine Basin, with 25 wells, is understood to any significant extent. The basins contain a variable thickness of Upper Palaeozoic to Recent strata. The thickest succession is in the Porcupine Basin where in excess of 10 km are preserved. The basins in the Rockall Region contain up to 6 km of strata (Shannon *et al.* 1993, 1994) while the thickness of strata in the NW Offshore and Goban Spur basins is typically of the order of 3 km.

Most of the frontier basins (excepting the Porcupine Basin) have a NE–SW orientation which is thought to reflect the control by reactivated Caledonian structures and fabrics. Basement in the area probably consists of granites and Lower Palaeozoic and older metamorphic rocks. Upper Palaeozoic sediments encountered by drilling in the Porcupine (Croker & Shannon 1987), Erris and Donegal basins (Tate & Dobson 1989) are relatively undeformed. The Upper Carboniferous strata in these basins offer reservoir and source rock potential.

Permo-Triassic to Lower Jurassic basins in the region are thought to have developed along Caledonian structures. They have a relatively widespread development in the NW Offshore basins but are less common in the Porcupine Basin (Croker & Shannon 1987; Shannon 1991b). They are likely to be locally preserved in the Rockall Region (Shannon *et al.* 1995). It is also likely that such strata are present in the Goban Spur Basin (Cook 1987). Permo-Triassic to Lower Jurassic strata are likely to comprise continental to shallow-marine sandstones, evaporites and limestones deposited during the initial rift phase marking the break-up of the Pangean supercontinent.

Middle Jurassic strata have a widespread development throughout the Porcupine Basin, and are also predicted throughout much of the Rockall Region, the Goban Spur and most of the NW Offshore basins (Fig. 2). In the Porcupine Basin the succession comprises a braided fluvial succession which is interpreted (Sinclair *et al.* 1994) as the deposits of an onset warp phase of tectonism prior to the Late Cimmerian rifting. A broadly similar tectono-stratigraphic setting is envisaged for the other frontier basins. Upper Jurassic strata reflect the deposition in a major syn-rift setting, with the development of a range of lithologies and facies. These range from basin-edge alluvial fans and braided to meandering fluvial strata (Croker & Shannon 1987; MacDonald *et al.* 1987) to deep marine submarine fans (Shannon 1992, 1993b).

Rifting waned during the early part of the Cretaceous (Moore 1992) and a major unconformity marks the approximate Jurassic–Cretaceous boundary in most of the basins. The Cretaceous strata throughout most of the frontier basins consist of deepening marine shale-prone strata (Fig. 2). Shale deposition was interrupted, especially in the Porcupine Basin and locally in the NW Offshore basins, by deposition of deltaic sandstones which reflect a minor rift episode of Aptian–Albian age (Shannon *et al.* 1993). Upper Cretaceous chalk was deposited through most of the basins but it is suggested (Shannon *et al.* 1993) that Upper Cretaceous strata in the basins of the Rockall region are marly rather than pure chalk. However, in some of the frontier basins (e.g. Hatton Basin) the Cretaceous succession is generally absent through basin inversion (Shannon *et al.* 1993, 1994).

Tertiary strata in the basins of the Rockall region are thought to be predominantly shale-prone (Shannon *et al.* 1993, 1994, 1995) and relatively thin in comparison with the succession deposited in the Porcupine Basin. Here, a sandy succession in the Eocene is developed in deltaic to submarine fan facies (Croker & Shannon 1987; Moore & Shannon 1992; Shannon 1992) and was interpreted (Shannon *et al.* 1993, 1994) as the result of ridge-push effects due to sea floor spreading and oceanic crustal development in the area to the west of the Rockall region. The post-Eocene succession in the region is generally represented by a deepening marine succession, interrupted locally in early Oligocene times by a number of unconformities (Naylor & Anstey 1987). Fluctuations in relative sea-level resulted in the development of slump structures (Moore & Shannon 1991) and occasional submarine fans (Shannon 1992, 1993b) During the Tertiary, rapid thermal subsidence occurred in the basins west of Ireland. The relative distance of the Rockall Trough from sediment sources led to a comparatively thin sedimentary succession and to the development of a deep bathymetric basin where subsidence outstripped sedimentation. On the basin margins, adjacent to the NW Offshore basins, a series of prograding fan deposits were laid down during the Tertiary. Subsidence in the Porcupine Basin, flanked on three sides by basement ridges which acted as sediment sources, led to the development of a thick sedimentary succession, although sedimentation was outpaced by subsidence in Late Tertiary times.

PETROLEUM PLAYS

A wide range of petroleum plays exists in the Irish offshore basins. A small number of these have been tested, with moderate success, in some of the basins. However, large areas of the offshore basins remain undrilled while many play types have not been tested. The current and future plays of each of the major groups of basins are outlined with a view to illustrating the oil and gas potential of the region.

Irish Sea basins

Only 9 wells have been drilled in the Irish sector of the Irish Sea basins and two of these reported hydrocarbon shows. The others appear to have been dry. Three of the wells were in the Kish Bank Basin, four in the Central Irish Sea and two in the St George's Channel Basin. The region has been traditionally regarded as containing gas plays, although the recent oil discoveries, such as the Douglas Field in the adjacent East Irish Sea Basin (Trueblood *et al.* 1995), have increased the awareness of oil potential in the region.

Source rocks lie in Upper Carboniferous coals and shales and, where encountered in the Kish Bank Basin, are at the peak stage of gas generation (Jenner 1981). The Upper Carboniferous Coal Measures may not be regionally extensive, which puts a significant risk on the source rocks (Naylor *et al.* 1993). There is also a possible Namurian source rock fairway containing the oil-prone Holywell Shale Formation in the Kish Bank and northern part of the Central Irish Sea Basin (Trueblood *et al.* 1995). Recent exploration interest in the region has been stimulated by discoveries associated with this source rock in the East Irish Sea Basin. It is unlikely that this source fairway extends through the southern part of the Central Irish Sea Basin. Reservoir rocks are Permo-Triassic fluvial, alluvial and locally aeolian sandstones which have been drilled in most of the wells in the Kish Bank Basin and in the Central Irish Sea Basin. The Sherwood Sandstone Group in these basins contains up to 1000 m of sandstones, with typical porosity of 14–18%. Thick Permo-Triassic evaporites and marls of the Mercia Mudstone Group cap the sandstones.

The wells drilled to date tested structural traps: either four-way dip-closed structures or tilted fault blocks. With few exceptions the drilling results are encouraging from the viewpoint of reservoir and cap rocks but disappointing with respect to the lack of shows. This probably reflects the absence of regional source rocks although it may also point to late structuration which breached earlier structures or created structures which post-dated petroleum migration. Few wells have been drilled within the deep parts of the basins (most were drilled on basin margins), while a number of these did not penetrate the entire source rock succession. There is, consequently, a poor understanding of the palaeogeographic framework and geometry of pre-Permian strata in the region. It is likely that the Mesozoic basin configuration differs from that of the Late Carboniferous basins. The older basins, containing the source rocks for all the plays, developed in response to a pre-Variscan regional framework while the location and geometry of the Mesozoic basins were influenced by the Variscan topography and

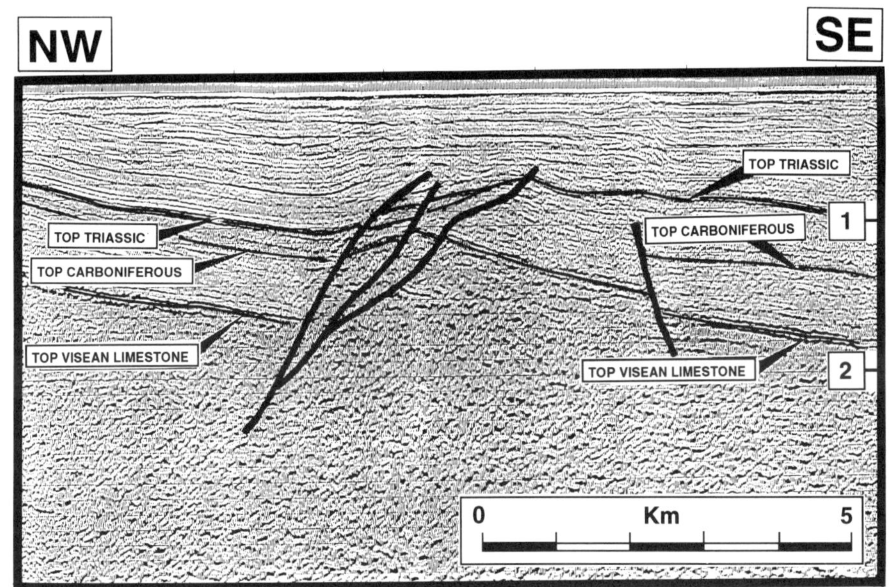

Fig. 3. Seismic profile from the Central Irish Sea Basin. This illustrates a Permo-Triassic stratigraphic play with Lower Triassic clastic reservoirs shed from an intrabasinal Carboniferous high. Cap rocks are prognosed at Upper Triassic level with source rocks in the Upper Carboniferous. Note the thinning of the Triassic section to the NW, away from the high. Vertical scale is in seconds (TWT).

tectonic framework (Musgrove *et al.* 1995).

The most obvious main plays are the undrilled tilted fault block structures, with Permo-Triassic sandstone reservoirs, salt or marl caps, and source rocks in the Upper Carboniferous Coal Measures or the Namurian marine shales. A more adventurous play, as yet undrilled, involves stratigraphic traps (Fig. 3). Permo-Triassic sediments were predominantly sourced from the Irish and Welsh massifs, with local sediment input from the footwall crests of rotated fault blocks. Downlapping of reflectors from the rotated footwall crests into the hanging wall, indicates the presence of local alluvial fans (O' Reilly & Shannon 1994). Cap rocks are Upper Triassic salts or marls with gas-prone source rocks predicted within the Upper Carboniferous. Overall, the Irish sector of the Irish Sea is an underexplored region, with potential gas plays in the Central Irish Sea Basin and in the Kish Bank Basin. The major risks to the region are the possible absence of a regionally extensive source rock and the effects of late structuration.

Celtic Sea basins

Approximately 80 exploration and appraisal wells have been drilled to date in the Irish sector of the Celtic Sea basins (Fastnet, North Celtic Sea, South Celtic Sea, St George's Channel basins), with 15 of these flowing oil or gas on test and many of the others recording shows. Gas production in the North Celtic Sea Basin is from shallow-marine Greensand (Albian) and shoreface to fluvial Wealden (Barremian–Aptian) sandstone reservoirs (Colley *et al.* 1981; Taber *et al.* 1995). Oil flows have also been recorded from these levels and from Lower Wealden (Valanginian–Hauterivian) and Oxfordian fluvial sandstones and Middle Jurassic shelf limestones in the North Celtic Sea Basin. Significant oil shows were recorded from Lower Jurassic deltaic sandstones in a number of wells drilled in the Fastnet Basin. These potential reservoirs are also likely to be locally developed in the North Celtic Sea Basin. The Kinsale Head (*c.* 1.5×10^{12} scf of gas initially in place) and Ballycotton (70×10^{9} scf of gas initially in place) fields are in production, while the Helvick (Caston 1995) and Seven Heads (Naylor & Shannon 1982) accumulations represent the most significant of the currently sub-commercial oil discoveries in the region (Fig. 1). Recently well 103/1-1 (late 1994) in the UK sector of the St George's Channel Basin, adjacent to the Irish/UK boundary, reported the first interesting, and potentially significant, gas discovery in Jurassic sandstones in the basin.

Reservoir quality sandstones have also been encountered in Triassic fluvial–aeolian sandstones in the Fastnet Basin (Robinson *et al.* 1981) and the St George's Channel Basin (Barr *et al.* 1981) and were predicted along the southern margin of the North Celtic Sea Basin (Shannon & MacTiernan 1993). They are also likely in the South Celtic Sea and Cockburn basins. Reservoirs in the latter two basins are likely to be confined to the Permo-Triassic, locally the Lower Jurassic and the Aptian–Albian Greensand. In these basins the Upper Jurassic and Lower Cretaceous successions are generally either absent or thinly developed. In the St George's Channel Basin reservoirs are developed in the Middle and Upper Jurassic and may also be present at Permo-Triassic level, although the latter may be too deep to have preserved reservoir potential. The Cretaceous succession is absent or only thinly developed in this basin.

Source rocks in the Celtic Sea Basins lie in the Lower Jurassic (oil and gas), Middle and Upper Jurassic (oil) and, locally, in the lowermost Lower Cretaceous shales (oil). The Lower Jurassic, with mixed oil and gas potential, is the only source rock in the Fastnet Basin and, probably, also in the South Celtic Sea and Cockburn basins. It is likely to be within the gas window in the central parts of the North Celtic Sea and St George's Channel basins but is likely to be only marginally mature in the Cockburn Basin and much of the Fastnet and South Celtic Sea basins. Burial history analysis suggests that the Lower Jurassic entered the oil window in Late Jurassic times in the central part of the North Celtic Sea Basin and in Early Cretaceous times on the basin shoulder zones of the basin (Shannon & MacTiernan 1993). The Upper Jurassic contains a number of rich oil source zones and entered the oil window in Late Cretaceous times in the basin centre and in Early Tertiary times along the shoulder zone (Shannon & MacTiernan 1993). Cap rocks occur at Upper Triassic, Lower, Middle and Upper Jurassic and Lower Cretaceous levels. The lithologies are evaporites and evaporitic shales in the Upper Triassic and shales and marls at the other levels.

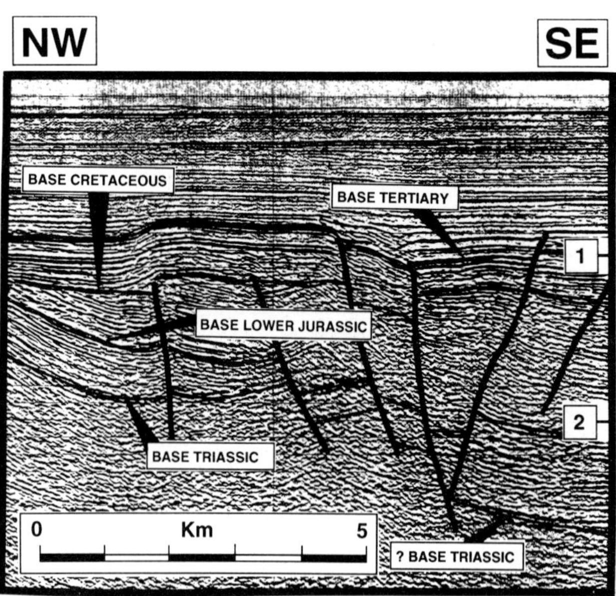

Fig. 4. Early Tertiary inversion structure in the Fastnet Basin. Reservoirs are prognosed at Lower Jurassic and Cretaceous level with source rocks in the Lower Jurassic succession. Vertical scale is in seconds (TWT).

The initial exploration in the Celtic Sea basins, undertaken in the early and middle 1970s, was targetted at four-way dip-closed structures which could be seen at Aptian to Upper Cretaceous levels (Fig. 4). These Early Tertiary inversion structures were found to contain gas in commercial quantities at the Kinsale Head Field (Colley *et al.* 1981; Taber *et al.* 1995) and sub-commercial oil and/or gas accumulations in other structures. Seismic data quality during the initial exploration phase was very poor and deeper structures could not be identified.

Improvements in seismic acquisition and processing techniques in the early 1980s allowed the identification of deeper structures and led to a resurgence of interest in the region, with emphasis on the deeper plays. These were generally tilted fault block structures, similar to those that had been drilled without significant success some years previously in the adjacent Fastnet Basin. Although the play was technically successful in a number of instances no commercial discoveries were made. The main Jurassic reservoir fairway for this play lies along the northern margin of the basin, with the major sediment source area probably lying to the north, onshore in Ireland and in the shelf area south of the Irish coast (Ewins 1992). The Upper Jurassic provides the major source rock for the play, with gas also sourced from the Lower Jurassic depocentre along the basin axis. The syn-rift structures are typically of Jurassic age. All of the accumulations to date appear to be small and this is thought to be due to a combination of the structural complexity of the northern basin margin with the breaching of structures during Early Tertiary basin inversion. The syn-rift Jurassic sandstones appear to decrease in thickness toward the basin centre and this, together with the burial and resulting porosity occlusion, renders the centre of the basin less prospective for this play.

The early 1990s saw a return to interest in remaining undrilled, small inversion structures with Lower Cretaceous reservoirs. The Ballycotton gas field was discovered during this phase. These structures, largely gas-prone, are mainly located towards the centre of the North Celtic Sea Basin. Similar structures which have been drilled were either dry or contained heavy and biodegraded oil. This is probably due to the fact that the timing of structuration and petroleum migration are closely related. The major phase of oil and gas generation and migration took place in Late Cretaceous to Early Tertiary time, while the inversion structuration is of Palaeogene age. The inversion led to a major phase of freshwater flushing which has resulted in the biodegradation of trapped oil (Howell & Griffiths 1995). The petroliferous structures may be the fortuitous consequence of the breaching of older structures and the remigration of hydrocarbons into late inversion structures. Furthermore, many of the structures are shallow and faults run close to the sea floor. To date, little exploration of Triassic potential has taken place in the region. Most of the wells terminated within the Jurassic, and the geometry of Triassic lithofacies is poorly understood. Recent exploration has begun to focus upon Triassic targets in areas away from the inversion axis. The most likely reservoir fairways appear to lie along the southern shoulder zone of the North Celtic Sea Basin (Shannon & MacTiernan 1993), with sediment shed northwards from the Pembrokeshire Ridge, and in the eastern and northeastern parts of the basin, where the basin is relatively narrow and sediment may have been derived from the Lower Palaeozoic Leinster granite-cored massif and from the Pembrokeshire Ridge. Source rocks are likely to be the Upper Jurassic (oil) and Lower Jurassic (gas) shales in the main basin depocentre. The traps within this play are tilted fault blocks (Fig. 5) which formed during the Triassic and Late Jurassic phases of syn-rift development and onlap stratigraphic traps (Figs 5 and 6) at Lower and Upper Jurassic levels.

The Lower Cretaceous succession in the region is sandstone-prone and offers potential stratigraphic traps. A large range of depositional environments from fluvial to shallow marine are indicated (Millson 1987) and these are likely to afford the requisite juxtaposition of Jurassic source with Lower Cretaceous reservoir, seal and stratigraphic trap. In common with all the Irish basins, stratigraphic traps have rarely, if ever, been tested in the Celtic Sea.

The St George's Channel Basin lies along strike from the North Celtic Sea Basin and is largely in UK waters. Until

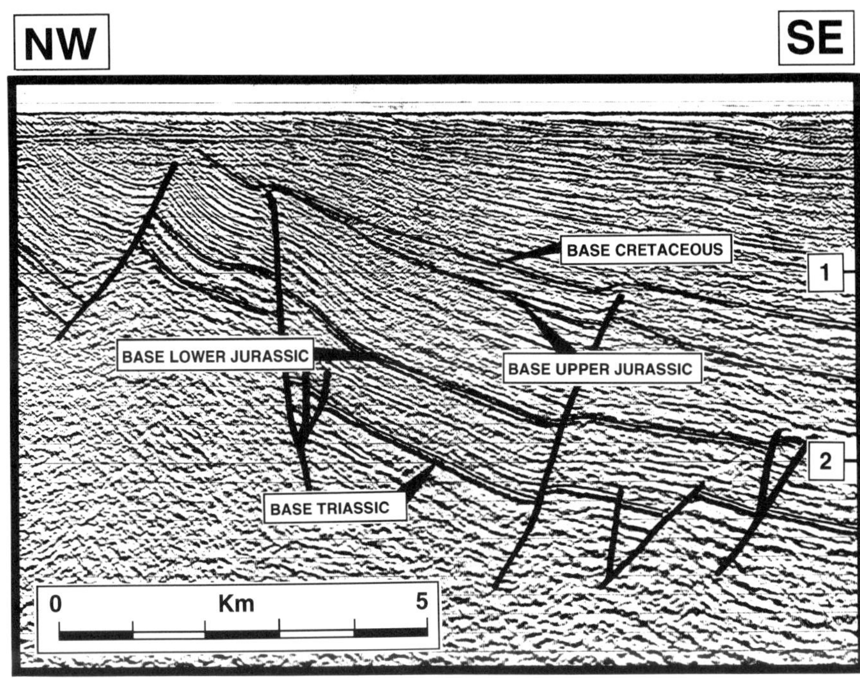

Fig. 5. Jurassic tilted fault block structures on the northern margin of the North Celtic Sea Basin. Upper Jurassic strata onlap the Lower–Middle Jurassic and provide a stratigraphic trap. Reservoirs are predicted at Triassic and Upper Jurassic levels with source rocks in the Lower and Upper Jurassic and cap rocks in the Upper Jurassic. Vertical scale is in seconds (TWT).

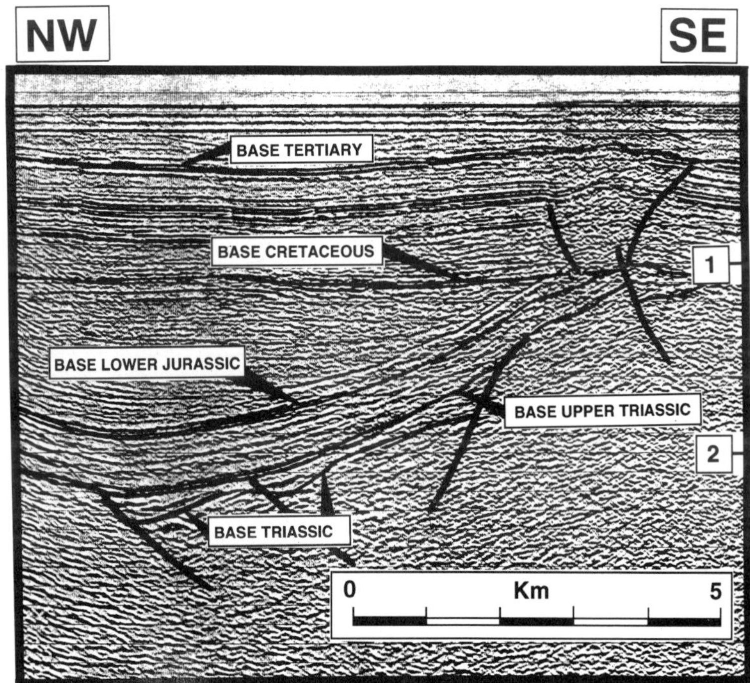

Fig. 6. Small Lower Triassic fault blocks are overlain by Upper Triassic strata that thins southwards onto basement at the southern margin of the North Celtic Sea Basin. Lower Jurassic strata thin and onlap southwards. Reservoirs are prognosed at Lower Triassic and locally at Lower Jurassic levels. The Lower Jurassic shales are gas- and oil-prone. Vertical scale is in seconds (TWT).

recently the lack of success in this basin had led to a downgrading of the perceived prospectivity. Although Triassic, and locally Jurassic, reservoirs were known from the region (Barr *et al.* 1981) relatively few shows were reported and it was considered that source rocks were likely to be generally immature. The recent gas discovery in UK Block 103/1, adjacent to the Irish/UK boundary, with Jurassic sandstone reservoirs, demonstrated the presence of a mature gas-prone Jurassic source rock and has improved the prospectivity of the region in Irish waters. It is likely that drilling of this play, with similar structures, will take place shortly in the Irish sector.

The Fastnet and Cockburn basins are regarded as having less potential than the North Celtic Sea Basin. Although they probably contain reservoirs at Triassic, Lower Jurassic and Lower Cretaceous levels, together with Triassic and Jurassic tilted fault block structures, the major problem appears to be the lack of a regionally mature source rock (Robinson *et al.* 1981). The Upper Jurassic succession is largely absent from the basins and the Lower Jurassic shales are only locally within the oil window. The South Celtic Sea Basin is likewise viewed as being less prospective than the North Celtic Sea and St George's Channel basins. Once again, the Upper Jurassic is not regionally developed while the Lower Cretaceous succession is thinner than in the North Celtic Sea Basin. On the positive side, however, the presence of halokinetic structures, due to the extensive Triassic salt, has resulted in some dramatic structural and stratigraphic structures (Shannon 1991a).

Frontier basins

These basins are very lightly explored. Twenty-five wells have been drilled in the Porcupine Basin. Four of these flowed good quality oil (35–45° API) or gas, with many others recording shows. Four wells have been drilled in the NW Offshore basins (Slyne Trough, Erris Trough and Donegal Basin) with the one in the Slyne Trough being the only one which recorded shows (Trueblood 1992). No wells have been drilled in the Irish sector of the Rockall Trough or in the Hatton Basin. One well has been drilled in the Goban Spur (Cook 1987). Water depths in the Porcupine Basin range from approximately 300 m in the north, to in excess of 2000 m in the south of the basin. Water depths in the other basins further west vary from less than 200 m to more than 2000 m.

Source rocks in the Porcupine Basin lie at Upper Carboniferous (gas), locally Lower Jurassic (oil), Middle and Upper Jurassic (oil) and Lower Cretaceous (oil) levels (Croker & Shannon 1987). In particular, the Upper Jurassic source rocks, which are mature throughout most of the basin, have pyrolysis yields sometimes in excess of 7 kg/tonne (Croker & Shannon 1987). Source rocks in the NW Offshore basins occur in the Upper Carboniferous (gas) and in the Lower Jurassic (oil). The latter, described by Trueblood (1992), are likely to be mature within the central parts of the basins. The geology of the basins in the Rockall region is poorly known but it has been suggested (Shannon *et al.* 1995) that oil-prone source rocks may be present in the Jurassic and in the Cretaceous of the Rockall Trough.

Sandstone reservoirs have been encountered in the Middle and Upper Jurassic (fluvial), and Upper Jurassic and Lower Cretaceous (sediment gravity flows) of the Porcupine Basin (Croker & Shannon 1987; MacDonald *et al.* 1987; Shannon 1992, 1993b). Other reservoir quality sandstones have been drilled in Triassic (non-marine), Lower Cretaceous (deltaic), Lower Tertiary (deltaic) and locally Upper Carboniferous (deltaic) strata (Croker & Shannon 1987; Moore & Shannon 1992). Similar reservoirs, albeit often limited in thickness and areal distribution, are present in the NW Offshore basins (Murphy & Croker 1992). The Mesozoic and Tertiary succession in the basin contains adequate thicknesses of shales to provide efficient cap rocks for the sandstone reservoirs. The distribution of likely reservoir rocks in the Rockall region is speculative. Shannon *et al.* (1995) suggested that localized Permo-Triassic sandstones and extensive Jurassic sandstones may occur in the Rockall Trough and Hatton Basin, while occasional submarine sandstones may occur in the Tertiary succession of the Rockall Trough.

Exploration drilling commenced in the Porcupine Basin in 1977 and most of the wells were drilled in the late 1970s and early 1980s. These were concentrated in the northern, shallower, waters of the basin and most were drilled on tilted fault block structures with prognosed Middle and Upper Jurassic reservoirs. This phase of exploration met with limited success, notably the Connemara oil field (Fig. 1). This was discovered in 1979 and four appraisal wells were drilled on the blocks. The results indicated that the field is complex with fault-compartmentalized segments. It contains close to 200×10^6 bbl oil in-place (MacDonald *et al.* 1987). Apart from a small number of wells in the deeper southern parts of the basin, no other play types appear to have been tested and most companies had withdrawn from the basin by 1990. The general exploration belief at the time appeared to have been that the basin contained thinner reservoirs than were expected while the deep water also inhibited exploration. However, there has been a renewal of interest in the basin within the past couple of years with the licensing of 32 blocks to groups representing 15 companies in 1995.

The Porcupine Basin contains a large number of plays at various levels. These include tilted fault blocks at Middle and Upper Jurassic levels (Fig. 7). Although most of the obvious fault blocks appear to have been tested in the northern part of the basin, others remain undrilled in the deeper, southern region. It is likely that the reservoir sandstones will be best developed close to the basin edges, having been derived from the Palaeozoic basement platforms of the Porcupine Ridge or the Irish Mainland Shelf. The basin is also likely to contain significant potential for stratigraphic plays at Cretaceous (Moore & Shannon 1995) and Tertiary levels. Thick Cretaceous (Aptian–Albian) deltaic structures occur especially in the northern and eastern parts of the basin and could be sourced from Jurassic or Cretaceous shales in the central part of the basin. The major risk to this play is the up-dip seal against the basin edge. Thick deltaic sandstones occur within the Lower Tertiary in the northern part of the basin (Croker & Shannon 1987; Moore & Shannon 1992). They are sealed by overlying shales and could be sourced from the basin depocentre. Again, however, the major risk lies in the updip seal of the deltas.

Submarine fan deposits are interpreted at various levels within the Jurassic, Cretaceous and Tertiary of the Porcupine Basin (Shannon 1992, 1993b) and a number of these are thought to offer significant potential. It is thought likely that some of these will be the centre of exploration attention during the next phase of exploration in the basin. While those at Jurassic and Lower Cretaceous levels are difficult to image clearly on seismic sections, the quality of seismic data allows the mapping of a variety of mounded and probably sand-prone stratigraphic structures within the Tertiary (Shannon 1992, 1993b). These are likely to be analogous in age and nature to recent stratigraphic discoveries in the North Sea (Shannon 1992). Their location is often topographically controlled with sand-prone sediment gravity flows preferentially focused into the topographic residual footwall lows associated with the early Cretaceous thermal subsidence. In Tertiary times, the submarine fans are coeval with the late Palaeocene–Eocene deltaic development in the northern part of the basin and occur as stacked mounds (Fig. 8). Smaller mounded structures, possibly somewhat more shale-prone, occur at Miocene level within the basin. Because many of the mounds are developed a distance from the basin edge, they are likely to be sealed by shales and contain a smaller risk of up-dip leakage than the deltaic prospects. They are likely to be charged from the basin centre by mature Jurassic and Cretaceous shales. Hydrocarbon migration may be facilitated along small faults that occur through the Cretaceous and Tertiary succession

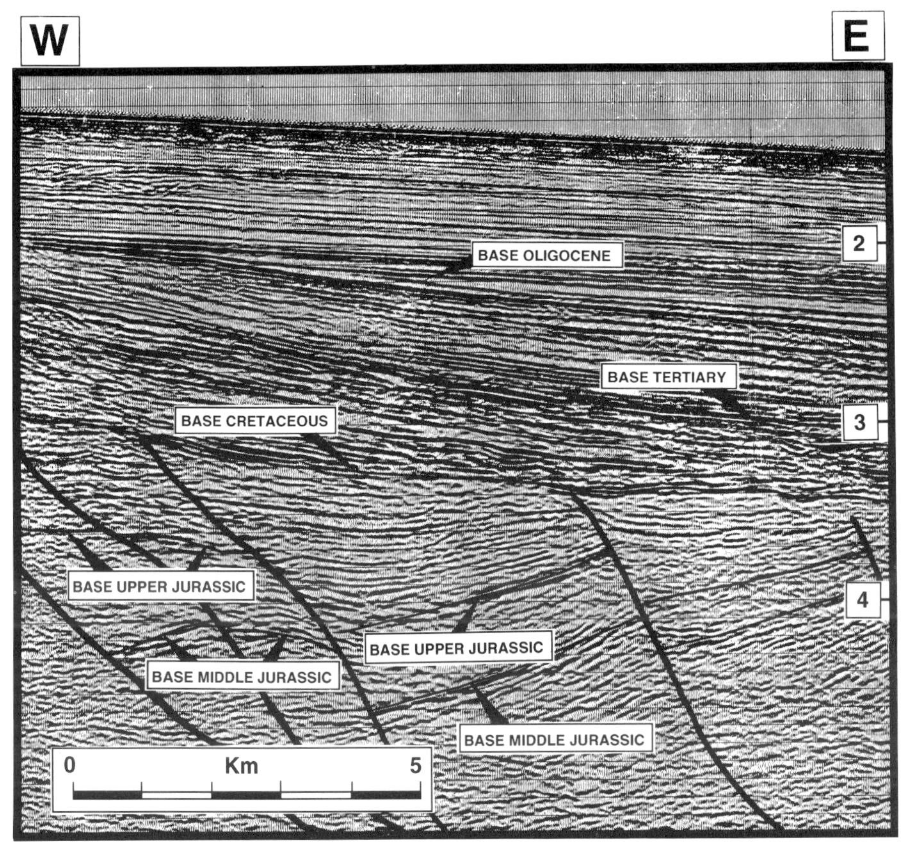

Fig. 7. Tilted fault block structures on the western margin of the Porcupine Basin showing growth faulting at Upper Jurassic level. The Lower Cretaceous downlaps and thins eastwards. Reservoirs are predicted at Jurassic and locally Cretaceous level with mature source rocks occurring in the Upper Jurassic. Vertical scale is in seconds (TWT).

(McCann *et al.* 1995) or by seismic pumping along microfractures.

In a number of areas of the basin, a series of mounded, circular to slightly elongate structures occur at the top of the Cretaceous chalk succession. These structures, which show evidence of bidirectional downlap on seismic sections, are onlapped and draped by uppermost Cretaceous and lowermost Tertiary strata (Fig. 9). They are interpreted as possible biohermal reefal build-ups, and may offer reservoir potential. The Upper Cretaceous chalk succession also contains structures which are interpreted as calciturbidites (Moore & Shannon 1995) and these structures, which are capped by Lower Tertiary shales, also offer some reservoir potential. They are likely to be sourced by migration along small faults or microfractures by seismic pumping from deeper, mature Jurassic or Lower Cretaceous shales.

Exploration in the NW Offshore basins had a brief phase of exploration in the late 1970s and early 1980s. The 4 wells drilled appear to have been targetted on tilted fault block structures. While these were generally encouraging in terms of thickness of reservoirs at various levels, the general lack of shows were interpreted as pointing to either lack of mature source rocks or to reactivated and breached structures. The last well was drilled in 1982 and most companies withdrew from the basins, although some interest lingered in the Slyne Trough where good quality source rocks and some oil shows were encountered. However, a renewal of exploration interest in the basins has taken place in the last few years, in tandem with the exploration interest and successes (e.g. Foinaven, Schiehallion and Stronsay) of the West Shetland region which lies along regional strike. In 1994, a total of 28 blocks in the Slyne, Erris and eastern Rockall troughs were licensed following the First Frontier Round.

It is likely that future exploration in the Slyne and Erris basins will concentrate upon tilted fault blocks with Jurassic reservoirs. There are also likely to be exploration targets in Cretaceous submarine fan sandstones in the northern part of the region. A number of large tilted fault blocks are likely to occur on the eastern margin of the Rockall Trough and these are likely to represent targets for Jurassic reservoirs. A number of stratigraphic traps are also likely in this area with westward-dipping prograding submarine fans of Cretaceous and Tertiary age shed westwards from the Erris Ridge or the Irish mainland during thermal subsidence of the Rockall Trough. Finally, a play somewhat analogous to the Clair Field play could be anticipated in the Erris Ridge region. This intermittent basement ridge separating the Rockall Trough from the Erris Trough is covered by probable Cretaceous and Tertiary strata. Reservoir could comprise either fractured Carboniferous, or older, basement, or sandy drape structures which are sealed by overlying shales. The play could be sourced from either the Erris or Rockall troughs.

CONCLUSIONS

1. The basins offshore Ireland are lightly explored. Petroleum accumulations occur in the North Celtic

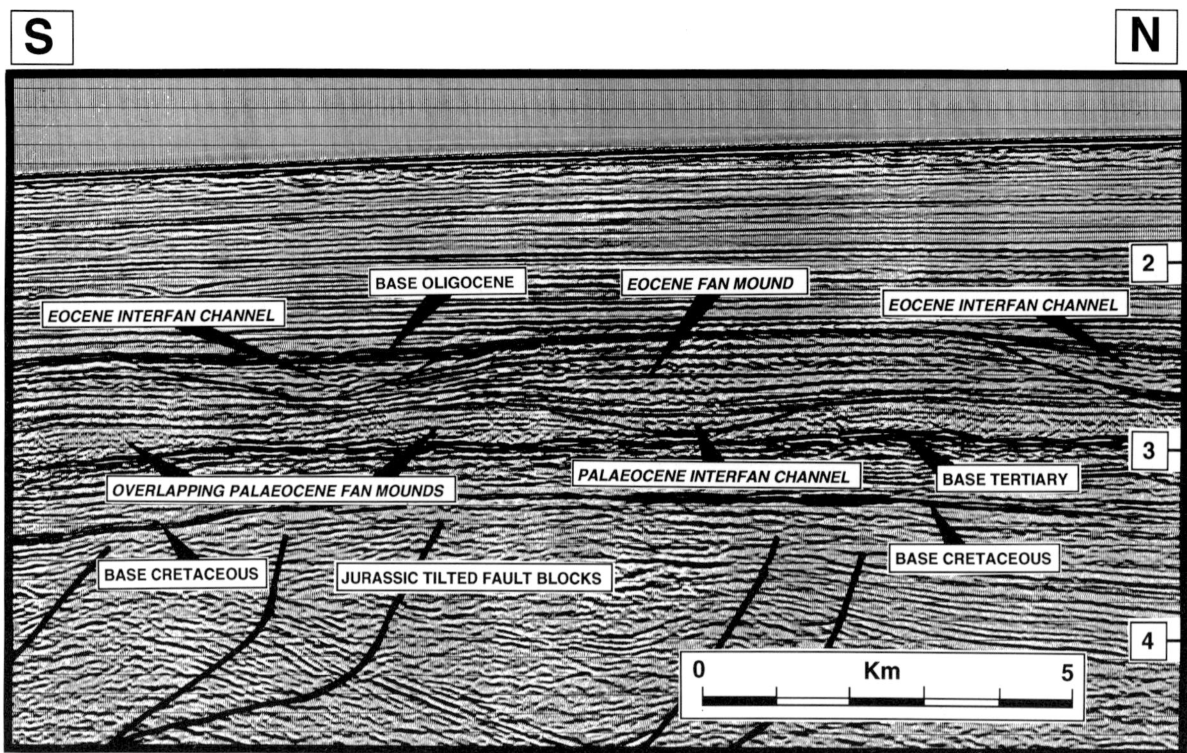

Fig. 8. Lower Tertiary submarine fan structures on the western margins of the Porcupine Basin. Palaeocene fans sometimes overlap and are overlain by larger Eocene mounds. Interfan channel structures are also indicated. Reservoirs are predicted at Palaeocene and Eocene levels with source rocks in the Jurassic and Lower Cretaceous. Vertical scale is in seconds (TWT).

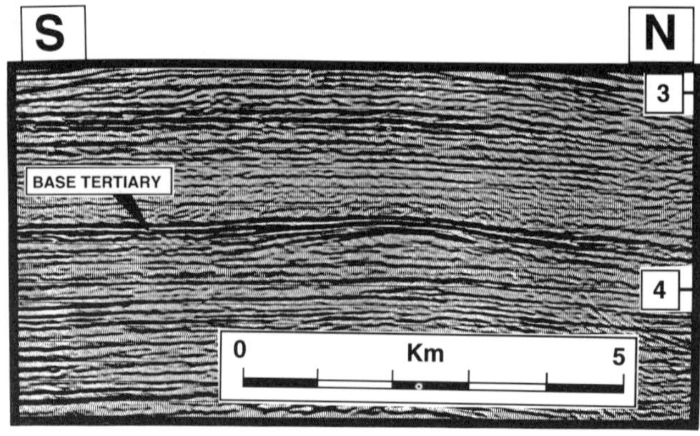

Fig. 9. Upper Cretaceous biohermal build-up structure in the southeast of the Porcupine Basin. Carbonate reservoirs are predicted at this level and are capped by draped Lower Tertiary shales. Source rocks lie in the Upper Jurassic. Vertical scale is in seconds (TWT).

Sea and Porcupine basins. Two gas fields are currently in production in the former basin while oil accumulations in both basins are either unappraised or are currently sub-commercial.

2. The past few years have seen a renewal of exploration interest in the Irish offshore basins, especially the frontier basins lying to the west of Ireland. This is due to a combination of discoveries in adjacent basins in UK waters, to improvements in production technology for marginal and deep water areas, and to attractive licensing terms. Recent licensing rounds have had considerable success in attracting companies to the frontier basins.
3. The basins contain a wide range of plays at different levels. Carboniferous, Permo-Triassic and Jurassic structural plays occur, while the Cretaceous and Tertiary targets offer significant potential for stratigraphic plays.

REFERENCES

ALLEN, P. 1981. Pursuit of Wealden models. *Journal of the Geological Society, London*, **138**, 375–405.

BARR, K. W., COLTER, V. S. & YOUNG, R. 1981. The geology of the Cardigan Bay–St George's Channel Basin. *In:* Illing, L. V. & Hobson, G.D. (eds) *Petroleum Geology of the Continental Shelf of North-West Europe.* Heyden, London, 432–443.

CASTON, V. N. D. 1995. The Helvick oil accumulation, Block 49/9, North Celtic Sea Basin. *In:* Croker, P. F. & Shannon, P. M. (eds) *The Petroleum Geology of Ireland's Offshore Basins.* Geological Society, London, Special Publication, **93**, 209–225.

———, DEARNLEY, R., HARRISON, R. K., RUNDLE, C. C. & STYLES, M. T. 1981. Olivine–dolerite intrusions in the Fastnet Basin. *Journal of the Geological Society, London*, **138**, 31–46.

COLLEY, M. G., McWILLIAMS, A. S. F. & MYERS, R. C. 1981. Geology of the Kinsale Head gas field, Celtic Sea, Ireland. *In:* Illing, L.V. & Hobson, G.D. (eds) *Petroleum Geology of the Continental Shelf of North-West Europe.* Heyden, London, 504–510.

COOK, D. R. 1987. The Goban Spur–exploration in a deep-water frontier basin. *In:* Brooks, J. & Glennie, K. (eds) *Petroleum Geology of North-West Europe.* Graham & Trotman, London, 623–632.

CROKER, P. F. & SHANNON, P. M. 1987. The evolution and hydrocarbon prospectivity of the Porcupine Basin, Offshore Ireland. *In:* Brooks, J. & Glennie, K. W. (eds) *Petroleum Geology of North-West Europe.* Graham & Trotman, London, 633–642.

——— & ——— 1995. *The Petroleum Geology of Irelands' Offshore Basins.* Geological Society, London, Special Publication, **93**.

EARLS, T. C. 1995. Potential for development of the Connemara field - Block 26/28. *In:* Croker, P. F. & Shannon, P. M. (eds) *The Petroleum Geology of Ireland's Offshore Basins.* Geological Society, London, Special Publication, **93**, 343.

EBBERN, J. 1981. The geology of the Morecambe Gas Field. *In:* Illing, L. V. & Hobson, G. D. (eds) *The Geology of the Continental Shelf of North-West Europe.* Heyden, London, 485–493.

EWINS, N. P. 1992. *Stratigraphy, sedimentology and diagenesis of the Jurassic and Cretaceous of the North Celtic Sea and Fastnet Basins.* PhD thesis, National University of Ireland.

HARTLEY, A. 1995. Sedimentology of the Cretaceous Greensand, Quadrants 48 and 49, North Celtic Sea Basin: a progradational shoreface deposit. *In:* Croker, P. F. & Shannon, P. M. (eds) *The Petroleum Geology of Ireland's Offshore Basins.* Geological Society, London, Special Publication, **93**, 245–257.

HIGGS, K. 1983. Palynological evidence for the Carboniferous strata in two wells drilled in the Celtic Sea area. *Bulletin of the Geological Survey of Ireland*, **3**, 107–112.

HOWELL, T. J. & GRIFFITHS, P. 1995. A study of the hydrocarbon distribution and Lower Cretaceous Greensand prospectivity in Blocks 48/15, 48/17, 48/18 and 48/19, North Celtic Sea Basin. *In:* Croker, P. F. & Shannon, P. M. (eds) *The Petroleum Geology of Ireland's Offshore Basins.* Geological Society, London, Special Publication, **93**, 261–275.

JENNER, J. K. 1981. The structure and stratigraphy of the Kish Bank Basin. *In:* Illing, L. V. & Hobson, G. D. (eds) *Petroleum Geology of the Continental Shelf of North-West Europe.* Heyden, London, 426–431.

KNIPE, R., COWAN, G. & BALENDRAN, V. S. 1993. The tectonic history of the East Irish Sea Basin with reference to the Morecambe Fields. *In:* Parker, J. R. (ed.) *Petroleum Geology of Northwest Europe: Proceedings of the 4th Conference.* Geological Society, London, 857–866.

MacDONALD, H., ALLAN, P.M. & LOVELL, J.P.B. 1987. Geology of oil accumulation in Block 26/28, Porcupine Basin, offshore Ireland. *In:* Brooks, J. & Glennie, K. W. (eds) *Petroleum Geology of North-West Europe.* Graham & Trotman, London, 643–651.

MADDOX, S.J., BLOW, R. & HARDMAN, M. 1995. Hydrocarbon prospectivity of the Central Irish Sea Basin with reference to Block 42/12, offshore Ireland. *In:* Croker, P.F. & Shannon, P.M. (eds) *The Petroleum Geology of Ireland's Offshore Basins.* Geological Society, London, Special Publication, **93**, 59–77.

McCANN, T. & SHANNON, P. M. 1993. Lower Cretaceous seismic stratigraphy and fault movement in the Celtic Sea Basin, Ireland. *First Break*, **11**, 335–344.

——— & ——— 1994. Late Mesozoic reactivation of Variscan Faults in the North Celtic Sea Basin, Ireland. *Marine and Petroleum Geology*, **11**, 94–103.

———, ——— & MOORE, J. G. 1995. Fault styles in the Porcupine Basin, offshore Ireland: tectonic and sedimentary controls. *In:* Croker, P. F. & Shannon, P. M. (eds) *The Petroleum Geology of Ireland's Offshore Basins.* Geological Society, London, Special Publication No, **93**, 371–383.

MILLSON, J. A. 1987. The Jurassic evolution of the Celtic Sea basins. *In:* Brooks, J. & Glennie, K. W. (eds) *Petroleum Geology of North West Europe.* Graham & Trotman, London, 599–610.

MOORE, J. G. 1992. A syn-rift to post-rift transition sequence in the Main Porcupine Basin, offshore western Ireland. *In:* Parnell, J. (ed.) *Basins on the Atlantic Seaboard: Petroleum Geology, Sedimentology and Basin Evolution.* Geological Society, London, Special Publication, **62**, 333–349.

——— & SHANNON, P.M. 1991. Slump structures in the Late Tertiary of the Porcupine Basin, offshore Ireland. *Marine and Petroleum Geology*, **8**, 184–197.

——— & ——— 1992. Palaeocene–Eocene deltaic sedimentation, Porcupine Basin, offshore Ireland: a sequence stratigraphic approach. *First Break*, **10**, 461–469.

——— & ——— 1995. The Cretaceous succession in the Porcupine Basin, Offshore Ireland: Facies distribution and hydrocarbon potential. *In:* Croker, P. F. & Shannon, P. M. (eds) *The Petroleum Geology of Ireland's Offshore Basins.* Geological Society, London, Special Publication, **93**, 345–370.

MURPHY, N. J. & CROKER, P. F. 1992. Many play concepts seen over wide area in Erris, Slyne troughs off Ireland. *Oil & Gas Journal*, Sept. 14, 92–97.

MUSGROVE, F. W., MURDOCH, L. M. & LENEHAN, T. 1995. The Variscan fold-thrust belt of southeast Ireland and its control on early Mesozoic extension and deposition: a method to predict the Sherwood Sandstone. *In:* Croker, P. F. & Shannon, P. M. (eds) *The Petroleum Geology of Ireland's Offshore Basins.* Geological Society, London, Special Publication, **93**, 81–100.

NAYLOR, D. 1996. History of oil and gas exploration in Ireland. *This volume.*

——— & ANSTEY, N. A. 1987. A reflection seismic study of the Porcupine Basin, offshore west Ireland. *Irish Journal of Earth Sciences*, **8**, 187–210.

——— & SHANNON P. M. 1982. *The Geology of Offshore Ireland and West Britain.* Graham & Trotman, London.

———, HAUGHEY, N., CLAYTON, G. & GRAHAM, J. R. 1993. The Kish Bank Basin, offshore Ireland. *In:* Parker, J. R. (ed.) *Petroleum Geology of Northwest Europe: Proceedings of the 4th Conference.* Geological Society, London, **845–855**.

O'REILLY & SHANNON 1994. Fault Analysis and Modelling–An Example from the Central Irish Sea–St George's Channel Region. *Extended Abstracts of Papers, European Association of Petroleum Geoscientists & Engineers, Vienna, Austria*, P505.

PETRIE, S. H., BROWN, J. R., GRANGER, P. J. & LOVELL, J. P. B. 1989. Mesozoic history of the Celtic Sea Basins. *In:* Tankard, A. J. & Balkwill, H. R. (eds) *Extensional Tectonics and Stratigraphy of the North Atlantic Margins.* American Association of Petroleum Geologists, Memoir, **46**, 433–444.

PINET, B., MONTADERT, L., MASCLE, A., CAZES, M. & BOIS, C. 1987. New insights on the structure and formation of sedimentary basins from deep seismic profiling in Western Europe. *In:* Brooks, J. & Glennie, K. W. (eds) *Petroleum Geology of North West Europe.* Graham & Trotman, London, 11–31.

ROBINSON, K. W., SHANNON, P. M. & YOUNG, D. G. G. 1981. The Fastnet Basin: An integrated analysis. *In:* Illing, L. V. & Hobson, G. D. (eds) *The Geology of the Continental Shelf of North-West Europe.* Heyden, London, 444–454.

SHANNON, P. M. 1991a. Tectonic framework and petroleum potential of the Celtic Sea, Ireland. *First Break*, **9**, 107–122.

——— 1991b. The development of Irish offshore sedimentary basins. *Journal of the Geological Society London*, **148**, 181–189.

——— 1992. Early Tertiary submarine fan deposits in the Porcupine Basin, offshore Ireland. *In:* Parnell, J. (ed.) *Basins on the Atlantic Seaboard: Petroleum Geology, Sedimentology and Basin Evolution.* Geological Society, London, Special Publication, **62**, 351–373.

——— 1993a. Oil and gas in Ireland – exploration, production and research. *First Break*, **11**, 429–433.

——— 1993b. Submarine Fan Types in the Porcupine Basin, Ireland. *In:* Spencer, A. M. (ed.) *Generation, Accumulation and Production of Europe's hydrocarbons III.* European Association of Petroleum Geoscientists, Special Publication, **3**. Springer-Verlag, Berlin, 111–120.

——— 1995. Permo-Triassic development of the Celtic Sea region, offshore Ireland. In: Boldy, S.A.R. (ed.) *Permian and Triassic Rifting in Northwest Europe.* Geological Society, London, Special Publication, **91**, 215–237.

——— & MacTIERNAN, B. 1993. Triassic prospectivity in the Celtic Sea, Ireland – a case history. *First Break*, **11**, 47–57.

———, MOORE, J. G., JACOB, A. W. B. & MAKRIS, J. 1993. Cretaceous and Tertiary basin development west of Ireland. *In:* Parker, J. R. (ed.) *Petroleum Geology of Northwest Europe: Proceedings of the 4th Conference.* The Geological Society, London, 1057–1066.

———, JACOB, A. W. B., MAKRIS, J., O'REILLY, B., HAUSER, F. & VOGT, U. 1994. Basin evolution in the Rockall region, North Atlantic. *First Break*, **12**, 515–522.

———, ———, ———, ———, ——— & ——— 1995. Basin development and petroleum prospectivity of the Rockall and Hatton region. 1995. *In:* Croker, P. F. & Shannon, P. M. (eds) *The Petroleum Geology of Ireland's Offshore Basins.* Geological Society, London, Special Publication, **93**, 435–457.

SINCLAIR, I. K., SHANNON, P. M., WILLIAMS, B. P. J., HARKER, S. D. & MOORE, J. G. 1994. Tectonic control on sedimentary evolution of three North Atlantic borderland Mesozoic basins. *Basin Research*, **6**, 193–218.

TABER, D. R., VICKERS, M. K. & WINN, R. D. Jr. 1995. The definition of the Albian 'A' Sand reservoir fairway and aspects of associated gas accumulations in the North Celtic Sea Basin. *In:* Croker, P. F. & Shannon, P. M. (eds) *The Petroleum Geology of Ireland's Offshore Basins.* Geological Society,

London, Special Publication, **93**, 227–244.

TATE, M. P. & DOBSON, M. R. 1989. Pre-Mesozoic geology of the western and northwestern Irish continental shelf. *Journal of the Geological Society, London*, **146**, 229–240.

TRUEBLOOD, S. 1992. Petroleum geology of the Slyne Trough and adjacent basins. *In:* Parnell, J. (ed.) *Basins of the Atlantic Seaboard: Petroleum Geology, Sedimentology and Basin Evolution.* Geological Society, London, Special Publication, **62**, 315–326.

——, BRYAN, C. & PICKERING, S. 1995. The Douglas oil field and its implications for exploration on the Irish Continental Shelf. *In:* Croker, P. F. & Shannon, P. M. (eds) *The Petroleum Geology of Ireland's Offshore Basins.* Geological Society, London, Special Publication, **93**, 39–40.

—— & MORTON, N. 1991. Comparative sequence stratigraphy and structural styles of the Slyne Trough and Hebrides Basin. *Journal of the Geological Society, London*, **148**, 197–201.

ZIEGLER P. A 1988. *Evolution of the Arctic–North Atlantic and the Western Tethys.* American Association of Petroleum Geologists, Memoir, **43**.

The Rotliegend play in the northeast Netherlands: an exploration evergreen

Manfred Epting

Nederlandse Aardolie Maatschappij (NAM, Business Unit Exploration), Postbus 28 000, 9400HH Assen, The Netherlands.

ABSTRACT: The vast majority of Dutch gas reserves is contained in Rotliegend sandstone reservoirs deposited as desert sediments in the Southern Permian Basin. Likewise, these reservoirs form the objective of some 60% of NAM's prospect portfolio. Based on 3D seismic, acquired on a regional scale, new structures can be mapped. Seismic anomalies and their modelling in relation to regional lithofacies trends assist in defining drillable locations that resulted in a sustained rate of commercial discoveries in the early nineties.

KEYWORDS: *NE Netherlands, Rotliegend, gas exploration, 3D seismic*

EXPLORATION HISTORY

The northeast Netherlands lies above a belt of Permian desert sediments which stretches from eastern England across the southern North Sea to northern Germany and Poland (Glennie 1972; Van Wijhe *et al.* 1980; Ziegler 1990). The Rotliegend reservoirs, called Slochteren Sandstone, have been charged with gas sourced from the underlying Carboniferous coal measures and are sealed by Late Permian Zechstein evaporites. These reservoirs contain the vast majority of the Dutch gas reserves and form the objective of two thirds of NAM's present prospect portfolio. The first gas was discovered in 1959 with the giant Groningen gas field and was followed in the early 1960s by several other large fields such as Annerveen, Marum and Ameland (Fig. 1). While the main emphasis shifted to offshore Rotliegend exploration in the late 1960s and 1970s, several small discoveries were made onshore with ultimate recoveries ranging from 1 to $5\times10^9\,m^3$ (Knaap & Coenen 1987). Only minor gas volumes were added in the 1980s. With total reserves exceeding $3\times10^{12}\,m^3$ (*c.* 100×10^{12} scf), the northeastern Netherlands is the most prolific onshore gas province in Western Europe.

RESERVOIR CHARACTERISTICS

The width of the east–west trending reservoir belt is restricted to some 100 km (Fig. 2). In the south, alluvial conglomerates prevail while the northern boundary is marked by desert lake claystones. In the intermediate zone, dune and interdune deposits constitute the bulk of the Rotliegend reservoirs. The highest permeability and productivity is encountered in the aeolian dune and dry sandflat deposits where there is a reduced content of clay minerals and early carbonate cements. These mineral phases deterio-

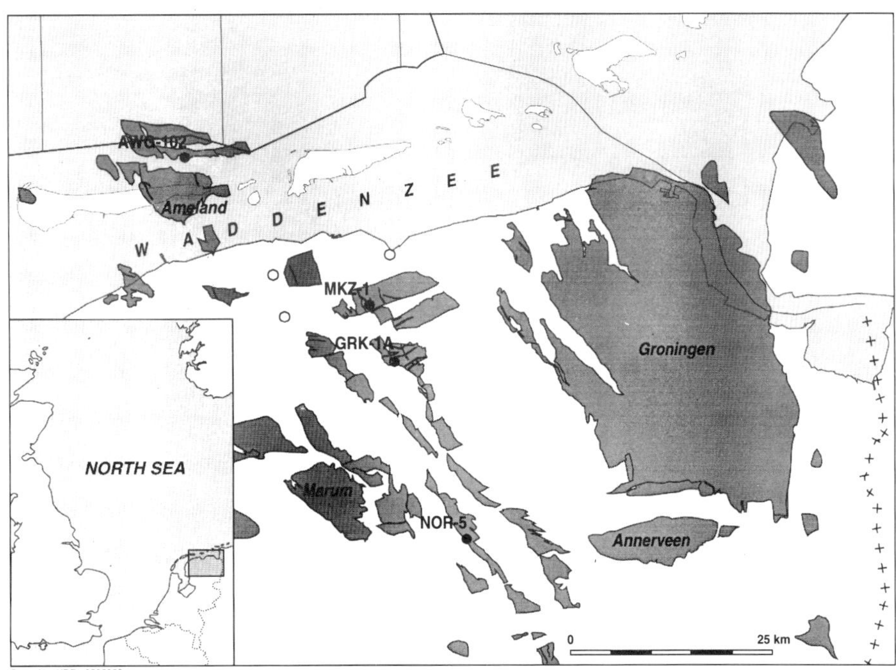

Fig. 1. Map of the northeast Netherlands gas province. Depth to objective Rotliegend level varies between 2500 m (in the south) and more than 4000 m (in the north). Locations are indicated for gas wells NOR-5, GRK-1A, AWG-102 (Figs 2, 3) and for the 1992 Munnekezijl discovery (MKZ-1, Fig. 5). Three recent discoveries near the coastline are shown as open circles.

From K. Glennie & A. Hurst (eds), 1996, *AD1995: NW Europe's Hydrocarbon Industry*, Geological Society, London, pp. 65–67

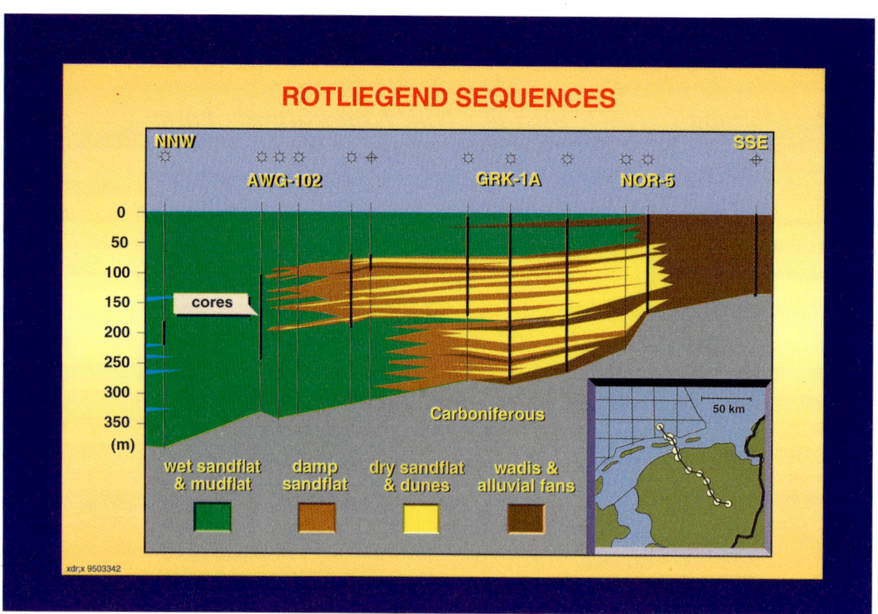

Fig. 2. Cross-section perpendicular to the strike of major facies belts. Well and core control is indicated. Logs of the Rotliegend from wells NOR-5, GRK-1A and AWG-102 and their correlation are displayed on Fig. 3.

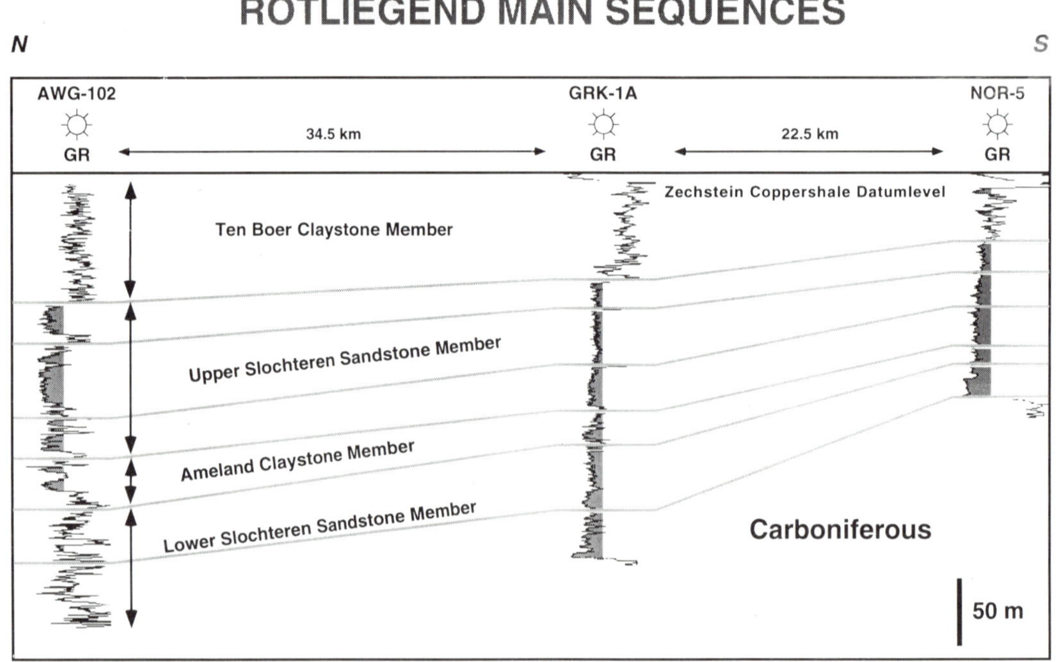

Fig. 3. Gamma ray logs of three gas-bearing wells with major subdivision into depositional sequences showing general thickening towards the north. Well NOR-5 in the south is dominated by dry sandflat and dune deposits. In well AWG-102 the lower sequences are dominated by mudflat facies while the Upper Slochteren interval is mainly developed as cemented wet sandflat sequences. See Fig. 1 for location of wells.

rate the reservoir quality of the adjacent damp and wet sandflat deposits where higher groundwater levels prevailed.

During short, occasional periods of heavy rainstorm, the desert lake transgressed southwards over large parts of the relatively flat desert area. Thin mudflat deposits testify to such lake incursions and subdivide the Slochteren Formation into sequences (Figs 2, 3). The mudflat deposits are underlain by thin dune and dry sandflat layers representing periods of maximum dryness. These high porosity/permeability sand sheets extend far into the basin and are the main producing intervals in wells recently drilled in the north.

REVIVAL BASED ON 3D SEISMIC

Until the late 1980s, the fields discovered had been defined on 2D seismic only. The remaining prospects were perceived to be too small to be drilled, which explains the low volumes found during the 1980s (Fig. 4). Obviously, the first 3D surveys were acquired over producing fields, while exploration acreage was covered only in the late 1980s. Improved seismic definition of some of the prospects revived exploration drilling. The well Grijpskerk-1 (GRK-1A, Figs 1–3), drilled in 1990, was the first success based on 3D interpretation. Some $10 \times 10^9 \, m^3$ of gas were added to the reserves base.

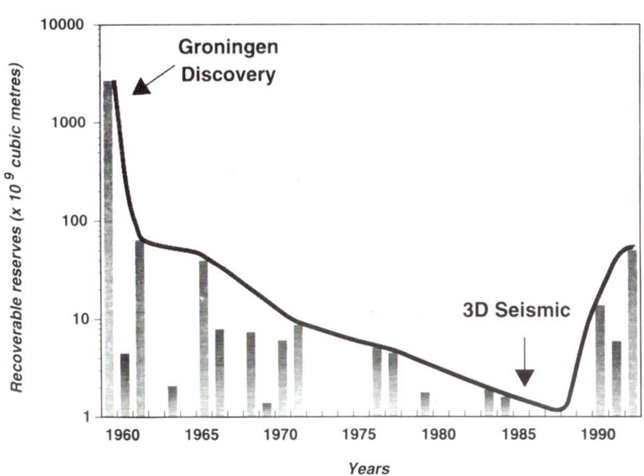

Fig. 4. Rotliegend gas discoveries in the northeast Netherlands 1959-1992. The dramatic decline in the seventies and eighties was reversed by the availability of 3D seismic data in the early nineties.

This discovery spurred exploration activity in adjacent areas.

RECENT RESULTS AND FUTURE ACTIVITY

Essentially, the entire northeastern Netherlands area is now covered by more than $6000\,km^2$ of contiguous 3D seismic, with some areas still to be acquired in the shallow offshore domain of the Waddenzee. In the early 1990s, a number of fault blocks such as Munnekezijl (MKZ-1, Fig. 5) were drilled and resulted in a sustained rate of commercial discoveries amounting to some $100 \times 10^9\,m^3$ of gas (Fig. 4). Further north, more structures await interpretation and drilling. Future wells are expected to eventually delineate the northern extension of this productive Slochteren reservoir trend.

This short paper is published by permission of the Nederlandse Aardolie Maatschappij BV (NAM), Shell Internationale Petroleum Maatschappij BV (SIPM) and Exxon Exploration Company (EEC). Special thanks are extended to the numerous staff members who have built up the knowledge of the Rotliegend Group during the last 25 years, and to all the colleagues who continue to unravel the last secrets of NAM's major gas play.

REFERENCES

GLENNIE, K. W. 1972. Permian Rotliegendes of Northwest Europe interpreted in the light of modern desert sedimentation studies. *American Association of Petroleum Geology Bulletin*, **56**, 1048–1071.

KNAAP, W. A. & COENEN, M. J. 1987. Exploration for oil and natural gas. *In:* Visser, W. A., Zonneveld, J. I. S. & van Loon, A. J. (eds) *Seventy-five years of geology and mining in The Netherlands (1912–1987)*. Royal Geological and Mining Society of the Netherlands (KNGMG), The Hague, 207–242.

VAN WIJHE, D. H., LUTZ, M. & KAASSCHIETER, J. P. H. 1980. The Rotliegend in the Netherlands and its gas accumulations. *Geologie en Mijnbouw*, **59**, 3–24.

ZIEGLER, P. A. 1990. *Geological Atlas of Western and Central Europe (2nd edn)*. Shell International Petroleum Maatschappij, The Hague.

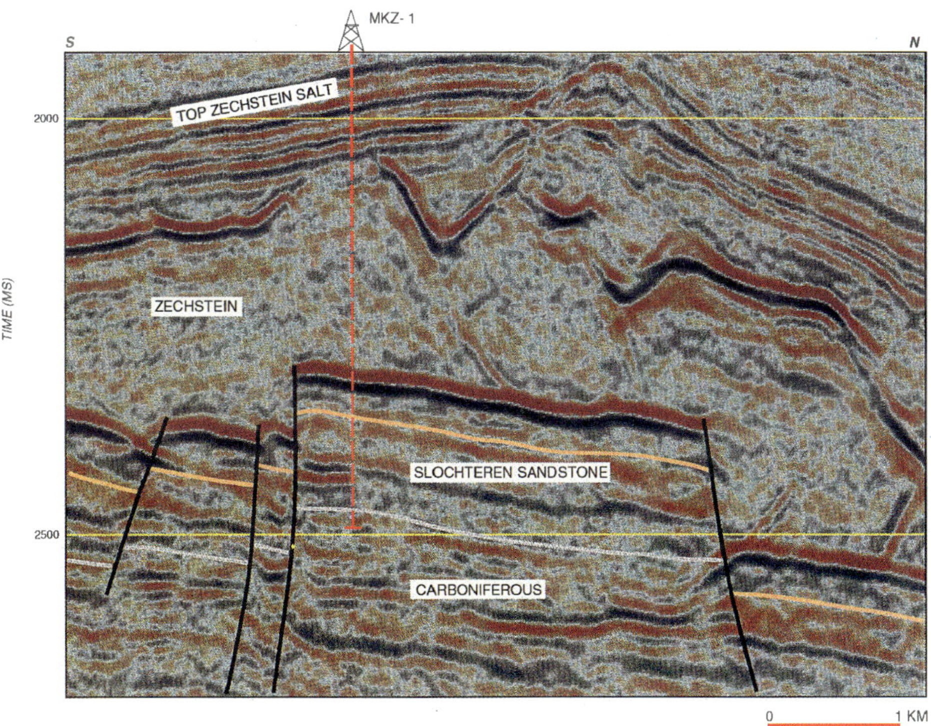

Fig. 5. 3D seismic line across the gas-bearing Munnekezijl horst block. Gas migrated into the Slochteren Sandstone from the underlying Carboniferous coal measures. The Zechstein evaporites provide an efficient top seal. See Fig. 1 for location of well MKZ-1.

Undiscovery wells of the UK Continental Shelf

Graham Dean
Amerada Hess Limited, 33 Grosvenor Place, London SW1X 3PL, UK

ABSTRACT: Several of today's most important fields were missed when they were first drilled. This paper looks at some of these Undiscovery wells and examines how they were mis-interpreted. Some of the examples show how the wells were reinterpreted and how this led to successful field developments. The examples in the paper are all from old UK wells which are in the public domain. However, the paper suggests that Undiscovery wells are still being drilled today.

KEYWORDS: *Undiscovery, Nelson, J Block, Scott, Wytch Farm, South Morecambe*

UNDISCOVERY WELLS IN THE UKCS

The search for oil and gas in the UK Continental Shelf (UKCS) over the last thirty years has been extremely successful and active. However, with such active exploration, there will also be many disappointments. This paper discusses Undiscovery wells. Undiscovery wells are exploration wells which were originally thought to be disappointing but were later found to have drilled into an oil or gas field.

It takes many years and a great deal of work before an oil company can drill a prospect. The company must first decide to participate in a licensing round and then appraise blocks on offer in order to identify prospects. Once the block is offered more detailed seismic mapping and other studies have to be done before the first well can be drilled. It is most unfortunate if the oil company drills the well but does not realize that it has found a significant oil or gas field. The examples in this paper are chosen because they are old wells where the information about the wells is in the public domain. No offence or criticism is intended. Instead it is the intention of this paper to emphasize the work of the individuals who turned some of these wells from Undiscovery into discovery wells.

The first oil field to be discovered in the North Sea was the Kraka Field in Denmark. It was a long time later before oil was finally discovered in the UK section of the North Sea. Over a dozen wells were drilled in the Central North Sea before Amoco announced an oil discovery in December 1969. This well, 22/18-1, discovered the Arbroath/Montrose fields. About the same time Phillips discovered the Ekofisk Field in Norway. These discoveries coincided with the 3rd UKCS licensing round so perhaps it is not surprising that about a third of the 3rd Round blocks were awarded to groups associated with these discoveries. Amoco, Phillips, Amerada Hess, British Gas/Enterprise Oil and Mobil became major players in the North Sea thanks, in part, to these early discoveries (Abbotts 1991). It could all have been very different if there had not been Undiscovery wells.

Nelson

One of the first Undiscovery wells was drilled over a year before the Arbroath well. This was the Nelson Field Undiscovery well, 22/11-1. The Nelson Field lies close to the Arbroath Field, (Fig. 1). Like the adjacent fields, Montrose, Arbroath and Forties, the Nelson Field is situated on the Forties–Montrose High and the reservoir rock is the Forties sandstone which is a turbidite sandstone sequence. 22/11-1 lies in the centre of the field between the two discovery wells 22/11-5 and 22/11-6Z.

The operator of 22/11-1 abandoned the well as a well with oil shows. Enterprise Oil seems to have had a very different view of the well because they drilled their first two wells not as exploration wells but as development wells through a seabed template positioned next to 22/11-1. So why was the potential of 22/11-1 not recognized at the time? Figure 2 compares the wireline logs from 22/11-1 with the eventual Nelson discovery well 22/11-5 (Bilsland *et al.* 1989). The two wells appear to have similar amounts of oil from the logs. Figure 3 is the composite log from 22/11-1. The high gas readings from the cuttings indicated that the well had found hydrocarbons and the operator set casing to just above the reservoir. With casing above the reservoir, the operator was able to run a bare-foot drill stem test. Before a test could be performed the total depth had to be reached with a smaller bit. The reservoir section had been drilled with a 12 1/4' bit, the hole had then been filled with cement when setting the casing and then the section had been redrilled with the smaller bit. The smaller bit leaves a sheath of cement around the walls of the borehole. When the total depth was reached, the well was plugged back with cement and the cement redrilled to just above the oil–water contact. Figure 4 shows the test results taken from the operator's reports (DTI Released Well Records). Despite testing through the cement sheath the well flowed a total of 84 bbl fluid in a total flowing time of only 37 minutes, equivalent to 3270 bopd. This was a higher flowrate than the Arbroath discovery well which only flowed at 2160 bopd despite having a lower bottom-hole flowing pressure of 2198 psig.

Today it is common to spend several days to test a discovery well and it is unusual to flow the well for less than eight hours at a time. So compared to modern practices the testing of 22/11-1 appears inadequate in the extreme.

22/11-1 was abandoned as a dry hole. Enterprise oil had a very different interpretation of the well results. The first two Nelson discovery wells were drilled as possible production wells from a seabed template close to 22/11-1. The first well, 22/11-5, was rather disappointing, finding oil but not the channel sands present in 22/11-1. 22/11-6Z eventually found oil-bearing channel sands and proved Nelson to be a large and highly productive oil field.

From K. Glennie & A. Hurst (eds), 1996, *AD1995: NW Europe's Hydrocarbon Industry*, Geological Society, London, pp. 69–80

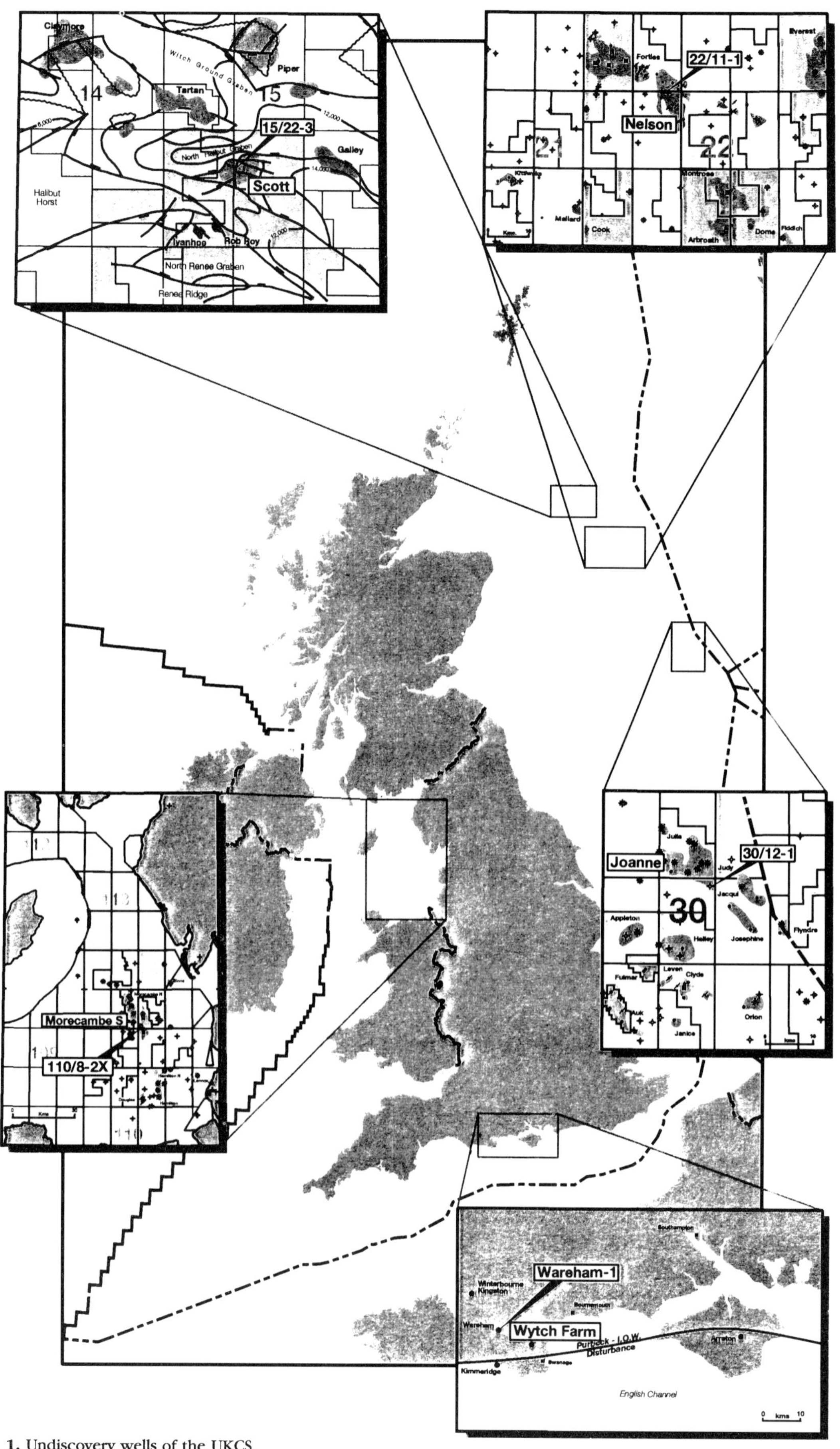

Fig. 1. Undiscovery wells of the UKCS.

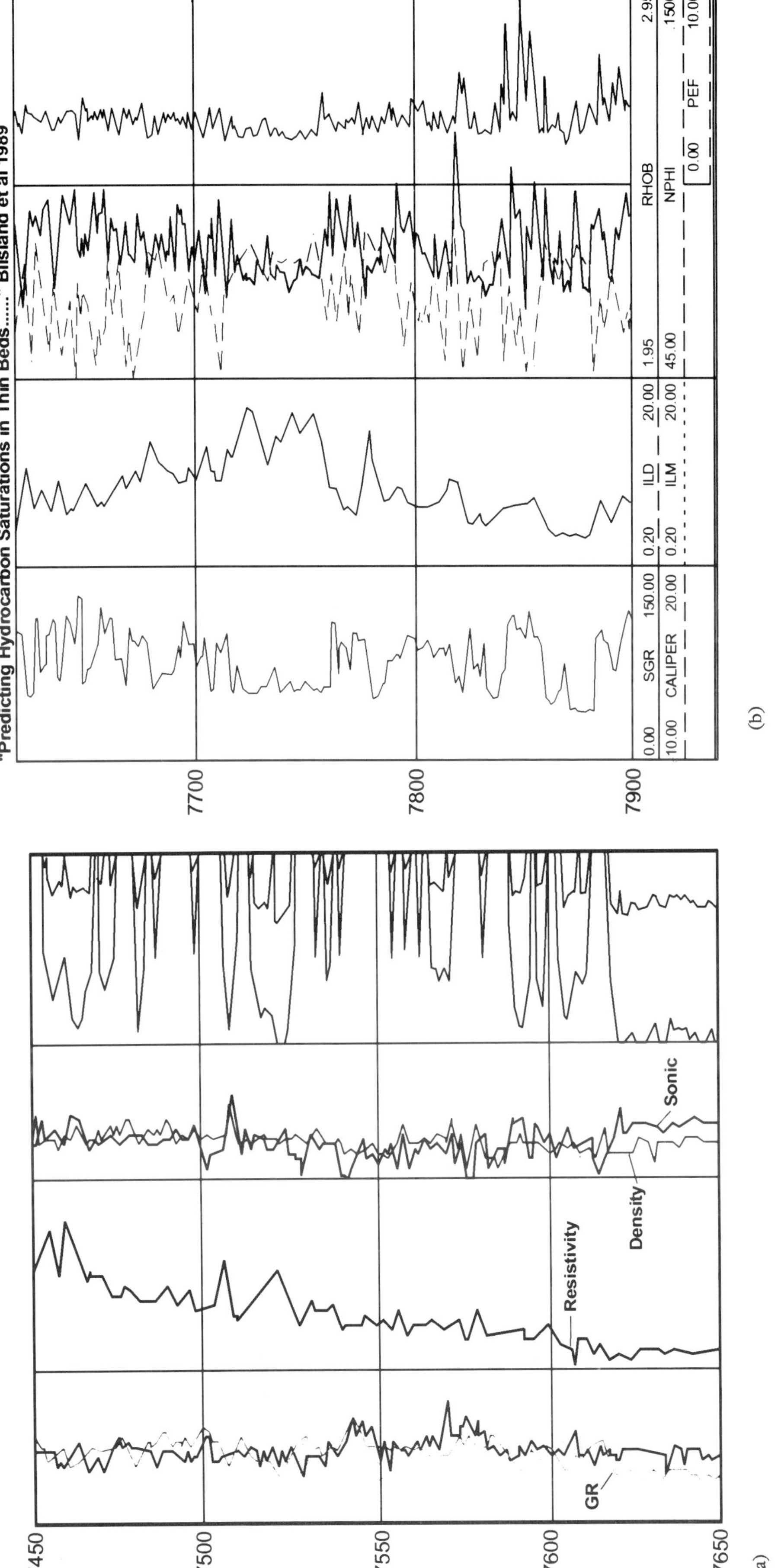

Fig. 2. Wireline logs from (**a**) the Nelson Undiscovery well, 22/11-1; (**b**) the Nelson discovery well, 22/11-5, the first of two wells drilled from a subsea production template positioned adjacent to 22/11-2. The thin sands at the top of (a) are oil-bearing. There is a thick channel sand with excellent reservoir properties below 7620 ft, but in this Undiscovery well it is below the oil–water contact.

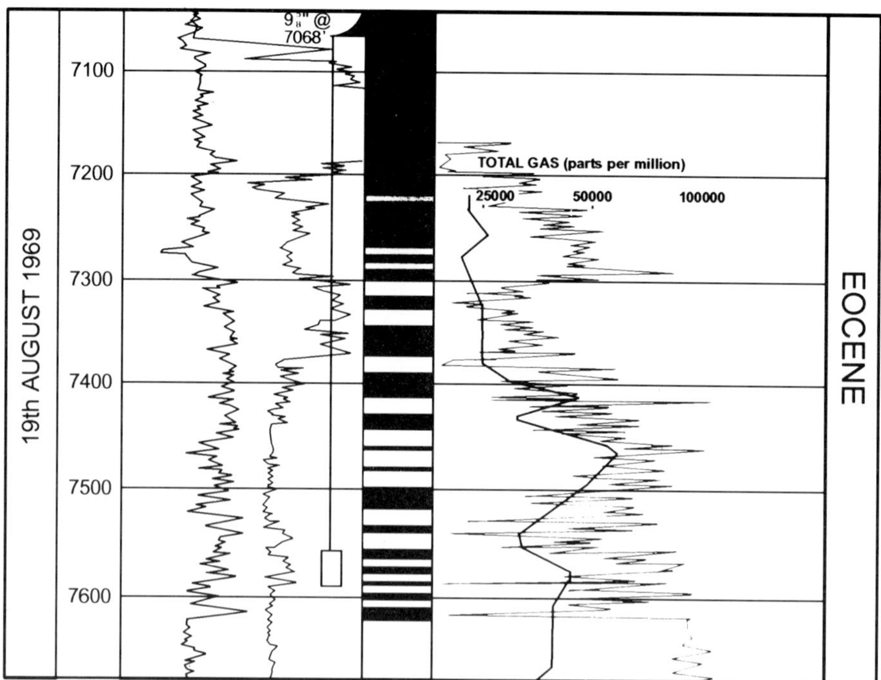

Fig. 3. Section of the operator's composite log over the Nelson reservoir in well 22/11-1. A bare-foot test was made from the casing shoe at 7068 ft to the cement plug at 7557 ft.

Test No	1
Date	September 14th 1967
Interval Tested	7,068' - 7,557'
Pressure Recorder	7,014'
Casing	Open Hole
Mud Weight	11.6
Packer Set	7,023'
Cushion	1,000' Water
Choke	5/8"
Init. Mud Press	4,217

Period 1	Flow Period	6 Minutes
	Closed	30 Minutes
	Init. Flow Press	2,442
	Final Flow Press	2,640
	Cip	3,074
Period 2	Flow Period	31 Minutes
	Closed	30 Minutes
	Init. Flow Press	2,626
	Final Flow Press	3,082
	Cip	3,092
	Final Mud Press	4,183

Remarks : Recovered 12.3bbls slightly muddy water (cushion), 4.5bbls mud, and 67.3bbls HGC, gas and slightly OCM.

Fig. 4. Drill stem tests from well 22/11-1.

J-Block

Nelson was not the only field to be undiscovered before Arbroath. The J-Block area was undiscovered by well 30/12-1. This well is situated near the prolific J-Block 30/7 (Fig. 1). The J-Block contains multiple oil reservoirs in Palaeocene sands, the Chalk and the pre-Cretaceous. 30/12-1 contains oil in several intervals (Fig. 5; DTI Released Well Records). A total of five drill stem tests were performed and Fig. 6 lists the results of DSTs 3 to 5. Once again there seems to have been a lack of enthusiasm for the testing. In particular there was very little attempt to measure the reservoir pressures. The well was drilled with heavy mud weights which suggests that the lower reservoirs are overpressured. Overpressuring is one of the reasons why so many North Sea reservoirs are so commercial because overpressuring 'preserves' good rock properties at depth and provides extra drive energy for production.

The testing also did not take fluid samples with the care and thoroughness that they are taken today. From the well reports lodged with the Department of Energy, it is unclear whether the sampled reservoirs were dry gas, wet gas, volatile oil or 'black' oil. Gas, even wet gas with a lot of condensate, is not as valuable as oil. However, many chalk fields have such poor reservoir qualities that oil cannot be produced at commercial rates. Volatile oil contains a high amount of gas in solution and when the reservoir pressure is dropped the gas comes out of solution and drives the oil towards the wells. This solution-gas-drive enables some chalk fields to produce at commercial rates.

The North Sea is one of the few areas in the world where chalk reservoirs are commercial. It is the combination of

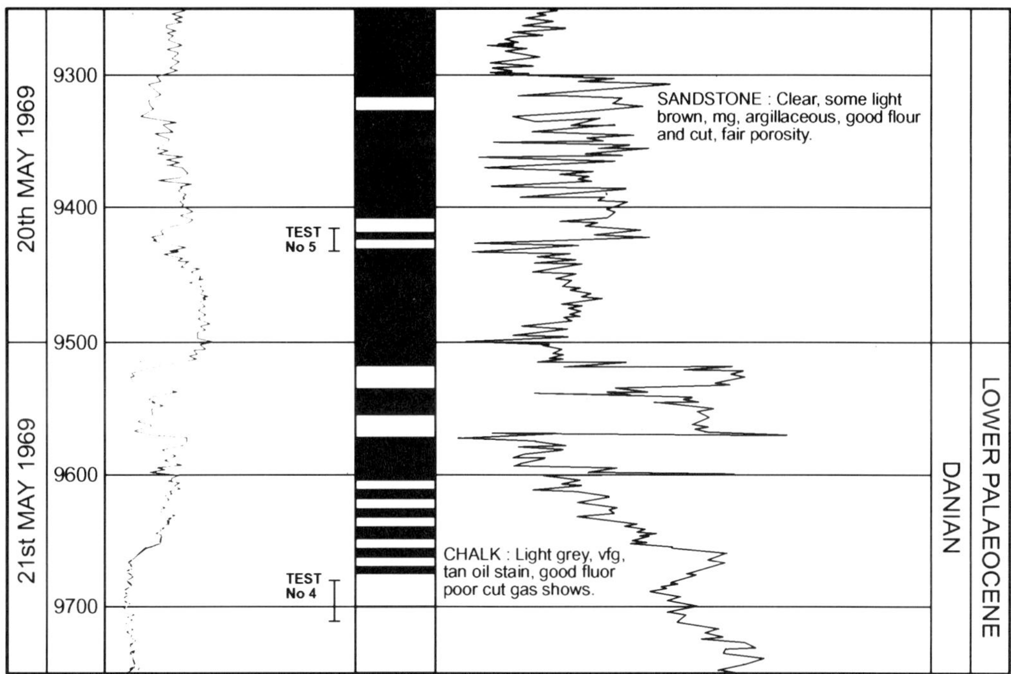

Fig. 5. Section of the operator's composite log from well 30/12-1.

Test No 3 — Perf. 9,965' - 9,980'
IF 30 - ISI 60 - FF 240 - FSI 60.
Recovered : Gas to surface in 85 min. 6 psi on 1/4 bubble hose, 3 gal 1 pint oil (41° A.P.I.)
Reversed Out : 55 bbls water cushion, 3 bbls gas cut mud.

Test No 4 — Perf. 9,680' - 9,710'
IF 30 - ISI 60 - FF 240 - FSI 120.
Recovered : Gas to surface - too small to measure, 5 1/2 gal oil (41° / 43.2° A.P.I.)
Reversed Out : 56 bbls water cushion, 2 bbl gas and oil cut mud.

Test No 5 — Perf. 9,416' - 9,431'
IF 15 - ISI 30 - FF 120 - FSI 60.
Recovered : Very weak blow, 56 bbl, water cushion, 3 bbl mud and formation water.

Fig. 6. Drill stem tests from well 30/12-1.

overpressured reservoirs and volatile oil that help make these reservoirs viable in the North Sea. Unfortunately these factors seem to have been missed during the testing of 30/12-1 and the well was abandoned as a dry hole.

Scott

Another very large and important oil field is the Scott Field, (Fig. 1). Scott came on stream in 1993 with the development starting soon after well 15/21a-15 revealed the potential size of the field. 15/21a-15 was drilled in July 1987. The well encountered some 400 feet of oil-bearing sands. Figure 7 shows the early appraisal wells. An additional well was drilled much earlier, in 1976, right in the middle of the field. This well was 15/22-3, the Scott Undiscovery well.

The main reservoir objective in this area is the extensive shallow-marine Piper sands and the hydrocarbons are often trapped in tilted fault blocks. The first two wells in block 15/22 were drilled just south of Scott on a shallower structure now called Telford. 15/22-1 found hydrocarbons but the well was not tested because of the high costs and difficulties of testing in winter. This well had only a thin sand interval but 15/22-2 drilled downdip found much thicker sands. 15/22-3 was then drilled on the crest of the Scott structure. The composite log from this well is shown in Fig. 8. It is interesting to compare this well with Fig. 9 which shows the typical Jurassic lithostratigraphy in the area. Note that the sands are absent in 15/22-3 with the well going straight from the high gamma-ray Kimmeridge Clay into the lower gamma-ray 'Basal Shale' at a depth of approximately 11 525 ft.

When it is suspected that a section is missing from a well then the dipmeter sometimes provides supporting evidence. If there is enough confidence that a section is missing in a well then occasionally the well is side-tracked in an attempt

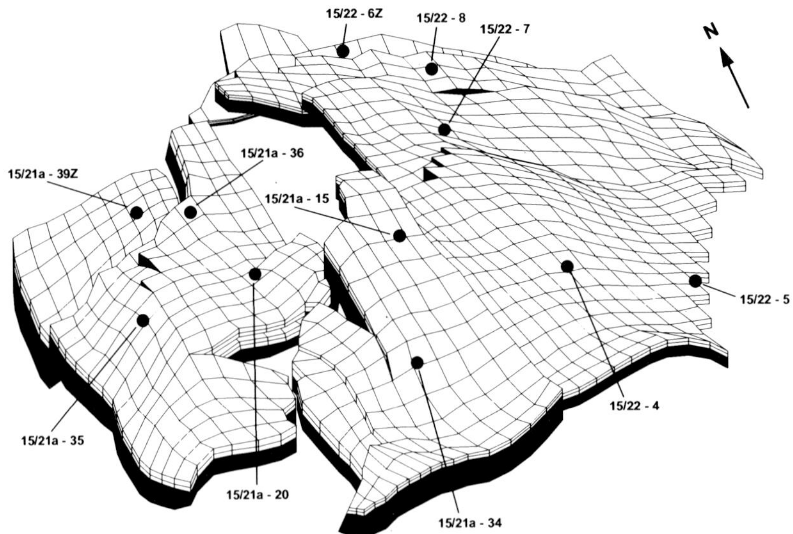

Fig. 7. Scott reservoir simulation grid 3D display. Undiscovery well 15/22- 3 was drilled through the 'hole' at the central crest of the field.

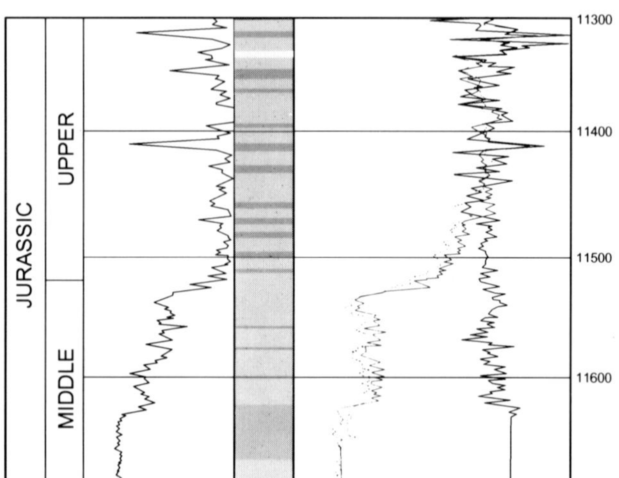

Fig. 8. Section of the operator's composite log from well 15/22-3. There are no significant sands in this section of the well. The interval below 11 620 ft is volcanics. The section from 11 530–11620 ft is the basal shale of Fig. 9. Above 11 530 ft, the high gamma ray (left-hand curve) indicates the Kimmeridge Shale.

to drill the complete sequence. Figure 10 shows the dipmeter from 15/22-3. An event, an unconformity or a fault, can be seen at 11 525 ft. Dipmeter presentation and interpretation are nearly always obscure and ambiguous. It is rare for a well to be side-tracked on dipmeter evidence. This well was no exception.

If 15/22-3 had been side-tracked, the Scott Field would have been discovered, but instead the field lay undiscovered for several more years.

Wytch Farm

The largest onshore oil field in Western Europe is said to be the Wytch Farm oil field in Dorset. It is odd that Wytch Farm or one of its satellite fields was not discovered earlier because nearby there are some famous oil seeps at Kimmeridge, (Fig. 1). There are also large fault throws likely to create trapping structures and thick reservoir quality sands are present at Bridport (Fig. 11).

Geochemical work led to the theory that the area was unlikely to be sourced by oil derived from marine kerogen and for this reason there was little drilling. However, Vic Colter of British Gas was more enthusiastic about the prospectivity of the area and was much encouraged when a well on the Isle of Wight discovered traces of marine kerogen. There was now no reason for not drilling and a well was drilled near Wytch Farm. The well was Wareham 1. The well was abandoned because it was thought to have no significant reservoir sands. There was then little enthusiasm for further exploration.

Figure 12 shows the interpreted structure of the Wytch Farm oil field. The larger reservoir is now the Sherwood Sandstone but initially it was the shallower Bridport Sandstone that was developed. This sand is overlain by the thin Inferior Oolite. The Bridport Sands themselves are part of the Lias and overlay the Mercia Mudstone.

Figure 13 shows the wireline logs from Wareham 1. The original stratigraphic interpretation is shown on the left and shows only very thin sands. With this interpretation, the only section of interest was the Inferior Oolite which was considered too thin to be a commercial reservoir.

Vic Colter disagreed with the initial interpretation and his interpretation is shown on the right of Fig. 13. The SP log shows thick permeable intervals. Vic Colter argued that the spikes on the microlog are caused by hardbands similar to those seen on the cliffs at Bridport, (Fig. 11), and so thick Bridport sands are present below the Inferior Oolite in the well. As a result of Vic Colter's re-interpretation of Wareham 1, another well was drilled, this time at Wytch Farm. The discovery of Wytch Farm owes a great deal to the work of Vic Colter.

It is reassuring that an individual can make such a significant impact on the fortunes of a company. Without the work of Vic Colter, Wytch Farm may well have never been discovered. The importance of the work of John Bains, a colleague of Vic Colter's, is illustrated in the final example of this paper.

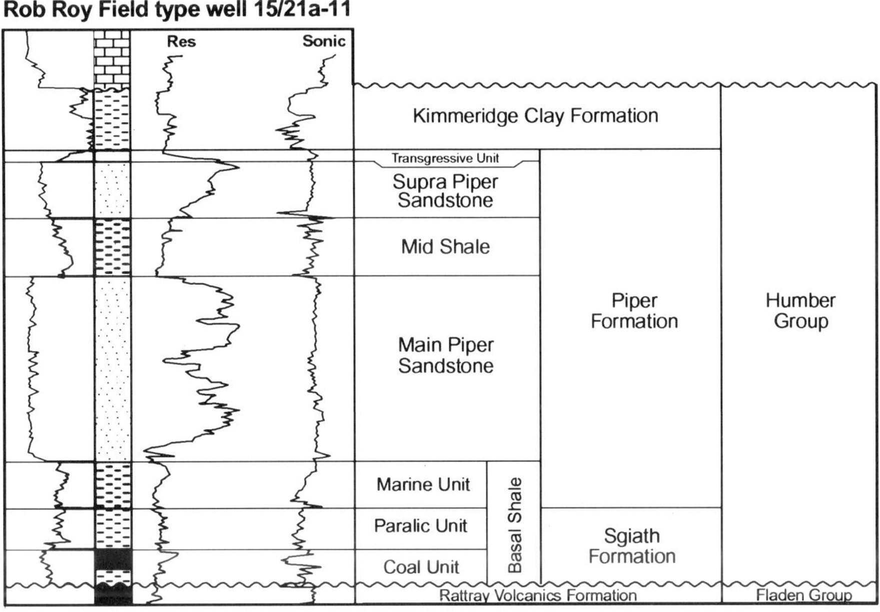

Fig. 9. Jurassic lithostratigraphy of Block 15/21. A comparison of well 15/21a-11 with 15/22-3 shows that the Piper Sand section is missing from the latter.

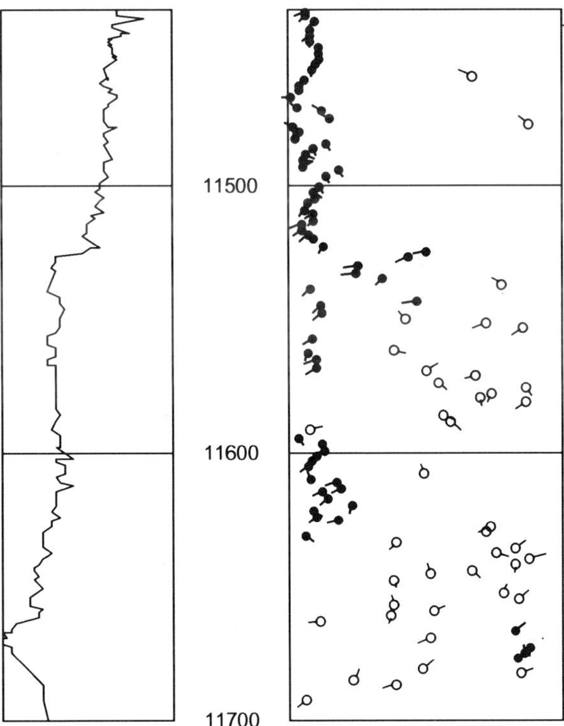

Fig. 10. A standard 'tadpole' presentation of the dipmeter log from well 15/22-3. The dipmeter measures the bedding 'dips' and how they change when a fault or unconformity is drilled. Unfortunately the results of the log are seldom available in time to enable a well to be side-tracked when the reservoir is missed, as in this well.

Fig. 11. The thick reservoir quality sands at Bridport.

South Morecambe

Perhaps the most spectacular of all the 'Undiscovery' wells is the Morecambe Undiscovery well, 110/8-2X. The Morecambe Field is situated in the Manx Basin. It is a huge gas field with reserves estimated at 4.5×10^{12} scf (Colter & Barr 1975; Colter 1978). Figure 1 shows the location of Morecambe Field. The dominant acreage position was held in 1978 by Hydrocarbons UK (later to be British Gas). Much of the acreage was originally licensed to Gulf Oil who drilled two wells. The second well, drilled in 1969, was 110/8-2X and was drilled into the Morecambe Field.

The objectives of the well were to drill the thick Permo-Triassic sediments in a large north–south trending, fault controlled structure having 13 000 acres of closure and 1200 ft of relief. The End of Well report (DTI Released Well Records) listed the following results:

> The Triassic section is thicker than expected, with approximately 7700 ft of section consisting of 2950 ft of salt, siltstone and shale overlying 4750 ft of predominated tight and hard Bunter sandstone. The upper 900 ft of sandstone is generally tight but does contain some beds of coarse, friable, porous sand, and the remaining 3850 ft is characterised by hard sandstone with abundant laminations of shale and clay.
>
> The only significant show was in the upper 20 ft of the Bunter sandstone immediately underlying the salt section. A maximum gas show of 230 000 ppm methane was recorded from 3272–92 ft.

The gas 'show' of 230 000 ppm is so large it would be more correct to call it a gas kick. The composite log of this section is shown in Fig. 14. Note the days on the left side show that it took four days to control the well. The gas kick was caused by unexpectedly high pressures. The high pressures were the result of drilling into the top of a very large gas column. Just how large the gas column is can be seen from the resistivity logs in Fig. 15. The laterolog was recorded on a linear scale. The resistivity is very low in the interval 3900–4000 ft but gradually increases going up the well. From this log, the approximate gas–water contact seems to be at 3870 ft. With the top of the sand at 3270 ft, this gives a 600 ft gas column. This is a huge column for a 'dry hole'. There are very few discovery wells with so much hydrocarbon.

The opinion of the operator was very different, as is shown from the following extract taken from the End of Well Report.

Zones of Interest

> 3275–3305′ (30′)
>
> Good shows of gas with 20′ drilling break. Porosity varies from 15–23%, being lower in the basal part of the sand.
>
> Water saturations vary from 30% average in the upper 25 feet to above 70% below 3300′.

So instead of 600 ft of gas sands, the well was thought to contain only 30 ft. Figure 16 (DTI Released Well Records) is a copy of the spreadsheet used for the log interpretation. The water resistivity, Rw, is not listed but can be derived by dividing the 'F–Rw' column by the 'F' column. At 4130 ft a water resistivity of 0.011 ohm m has been used. This is a very low value of water resistivity. However, salt-saturated formation waters do approach this resistivity. Figure 14 shows that the salt forms the cap rock for the gas. so it is highly likely that the sand in such close contact with the salt will have a salt-saturated formation water. Oddly, the other intervals analysed in the spreadsheet have used a much higher water resistivity of 0.11 ohm m which is a typical resistivity of seawater-based drilling mud. The result of using this high water resistivity is that the water saturations that are calculated are too pessimistic (the 'FRw/Rt' column on the extreme right of the spreadsheet). Instead of calculating low

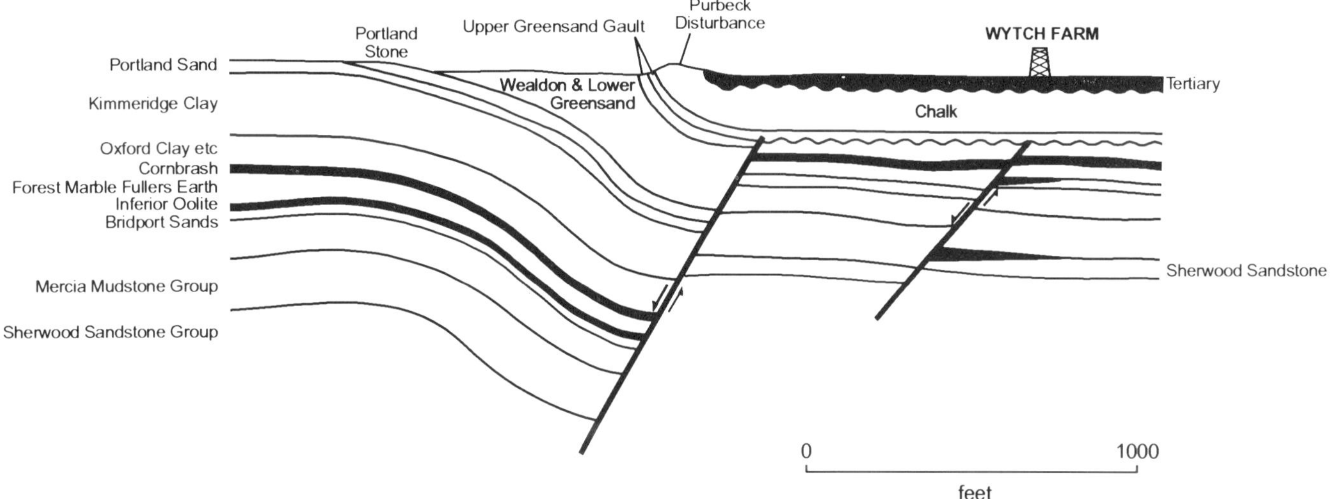

Fig. 12. Wytch Farm stratigraphic section.

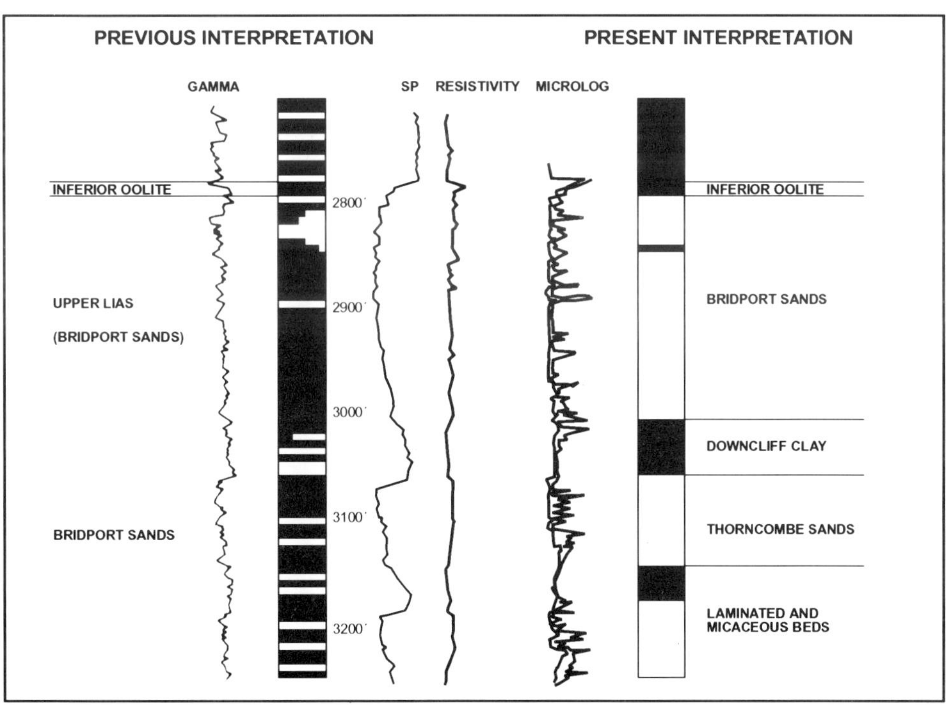

Fig. 13. The two interpretations of the wireline log from Wareham 1.

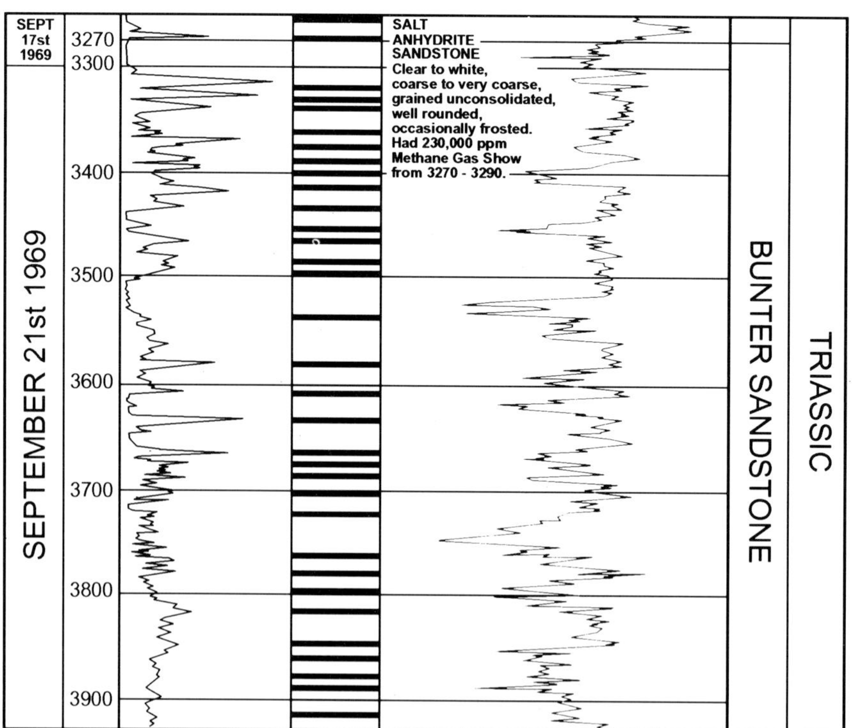

Fig. 14. The composite log for the reservoir section of the Morecambe Undiscovery well (110/8-2x). The mud was increased in weight after 3290 ft, which suppressed the gas shows and gave the impression that only the top 20 ft contained gas.

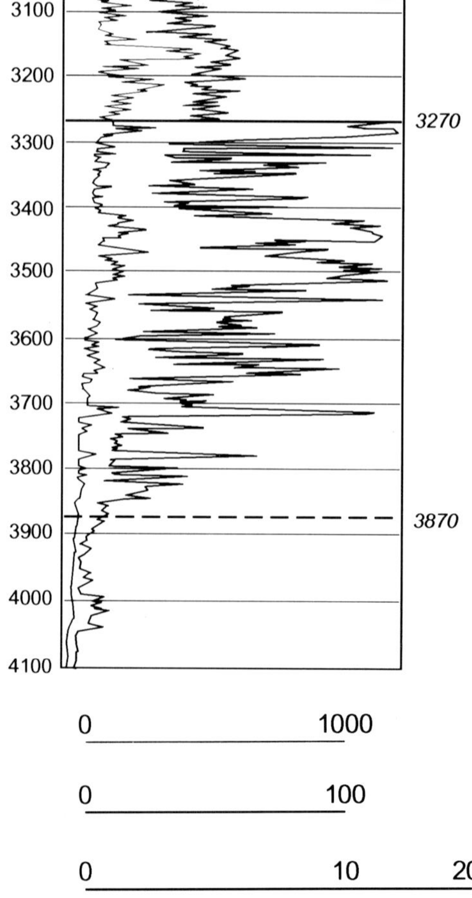

Fig. 15. The laterolog from well 110/8-2x, indicating that the gas column extends from 3270–3870 ft.

GOPC Log Interpretation — Well : Gulf / NCB 110 / 8 - 2X — Formation : Bunter Sandstone

Depth	Rll	Rxo Rmll	B (c)	ØFD	F	Ro F -Rw	Sw	Δt	Øs	FRW / RT
3275	19	4	2.44	14.7	39	4.20	48	70	12.0	.23
77	20	5	2.45	14.0	43	4.65	48	71	13.0	.23
79	48	10	2.36	20.2	19	2.05	23	77	17.5	.043
81	21	6	2.37	19.5	21	2.26	33	76	17.0	.107
84	39	6	2.31	23.5	14	1.50	19.8	80	20.0	.039
88	20	5	2.34	21.5	17	1.84	30	78	18.5	.092
92	15	7	2.32	23.0	15	1.62	33	95	32	.11
96	10.5	9	2.40	17.5	26	2.80	52	80	20	.267
3300	7.4	6	2.45	14.0	43	7.65	79	81	21	.63
04	6	2	2.43	15.3	34	3.68	78	75	16	.61
08	12	X	2.60	3.5	800	80	100	65	8	71.0
4130	.25		2.35	19.5	21	.227	97	85	22	.94

Fig. 16. A copy of the spreadsheet submitted to the Department of Energy to support the abandoning of well 110/8-2x. The formation water resistivity can be found by dividing the 'Ro column by the 'F' column. At 4130 ft the water resistivity is 0.011 ohm m, but at 3308 ft is 0.1 ohm m.

Fig. 17. Vic Colter (left), A. J. Bains (centre) and Martin Ford (right), all contemporaries at British Gas and now retired.

water saturations which are typically found at the top of a large gas column, the log interpretation gave high water saturations that are typically found near a gas–water contact.

After drilling 110/8-2X, Gulf decided to relinquish their acreage. Having no further interest in the area, Gulf gave a copy of the logs to Hydrocarbons GB (now British Gas) who had acquired adjacent licences in 1972. The logs were given to the log analyst, A.J. (John) Bains. John Bains saw that the logs showed that 110/8-2X was a major gas well and convinced his superiors. Sportingly, British Gas told Gulf that they thought there was a lot of gas in the well. Gulf, in response, re-analysed the well, this time using a computer. They came up with the same results as their initial interpretation and informed British Gas.

John Bains was asked to present his work to the Department of Energy. The Department of Energy were impressed and insisted that John Bains and a Gulf representative should be put together to discuss and agree an interpretation of the well. The outcome was that Gulf

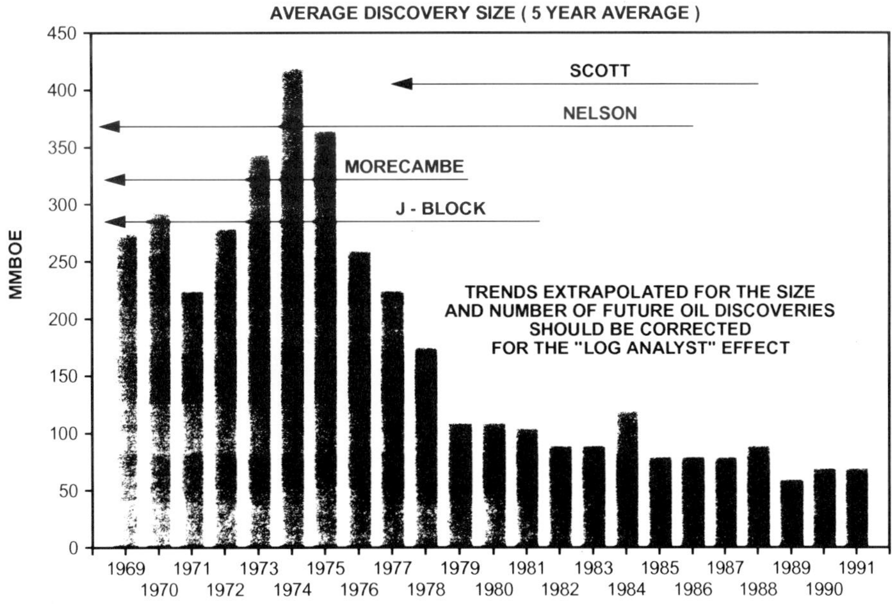

Fig. 18. Plots such as this are used to predict trends for the future of UKCS exploration. These plots can be misleading if they are not corrected for the Undiscovery effect.

realized their error and requested the Department of Energy for permission to withdraw their notice of relinquishment of the block. The Department of Energy refused. Instead the licences were awarded to British Gas who drilled the discovery well, 110/2-1, in 1974. Figure 17 is a photograph of John Bains flanked by Vic Colter (left) and Martin Ford (right).

This paper has shown that Undiscovery wells have affected past and present hydrocarbon exploration. Undiscovery wells also distort our perception of the future as illustrated in Fig. 18. We frequently use trends from plots such as Fig. 18 to predict future exploration success. If these plots are to be reliable they should be corrected for the Undiscovery effect. Thus, Undiscovery wells affect not only the past and the present but also our perception of the future of oil exploration.

Without the work of John Bains and others in rediscovering Undiscovery wells, the UK oil and gas industry would be very different and the UK would be poorer. The examples of Undiscovery wells that have been used are all of old public domain wells. In this author's opinion, Undiscovery wells are still being drilled today, especially in new basins or new plays.

REFERENCES

ABBOTS, I. L. (ed.) 1991. *United Kingdom Oil and Gas Fields 25 years Commemorative Volume*, Geological Society, London, Memoir.

BILSLAND, M. *et al.* 1989. *Predicting Hydrocarbon Saturation in Thin Beds.* SPWLA.

COLTER, V. S. 1978. Exploration for gas in the Irish Sea. *Geologie en Mijnbouw*, 57, 503–516.

——— & BARR, K. W. 1975. Recent developments in the geology of the Irish Sea and Cheshire Basins. *In: Petroleum Geology and the Continental Shelf of North-West Europe*, (Vol. 1). Applied Science Publishers, London, 61–75.

DTI. Department of Trade and Industry, Microfiche of Released Well Records.

Part Three: State of the Art and the Way Forward

Basin modelling history and predictions

James E. Iliffe and Martin R. Dawson[1]

PGS Tigress, PSTI Technology Centre, Exploration Drive, Aberdeen AB23 8GX, UK

[1] *Present address: Flomerics Ltd, 81 Bridge Rd, Hampton Court, Surrey KT8 9HH, UK*

ABSTRACT: Hydrocarbon potential assessment involves predictions of source and reservoir rock distribution and type, burial and temperature histories, oil and gas generation, expulsion and migration, accumulation and retention. Basin models have been developed to assist explorationists quantify hydrocarbon potential.

The earliest basin models predicted the maturity of rocks today based on non-decompacted burial history calculations with a constant geothermal gradient assumed throughout the geological history. Observed maturity parameters were ranked into indices. Predictive models of maturity were calibrated from observed maturities. The wholly deterministic and empirical modelling was often carried out in isolation from the other basin analysis activities.

Over the past 15 years basin modelling has been revolutionized due to scientific breakthroughs, mathematical innovation, dynamic advances in IT and improved 2D and 3D seismic data acquisition, processing and interpretation techniques. More recently, analyses of parameter sensitivities, rigorous testing and thorough investigation into techniques and assumptions has documented the limitations of models. This has led to levels of accepted practices and techniques, and more measured statements of potential, complete with error estimates and ranges of results.

The questions that are being answered at present by basin modellers and technology are not only where are the hydrocarbons now, but also how much is there, how did it get there and when, what type of hydrocarbons are there, and how certain are these predictions?

The challenges facing the basin modelling research and development community are in providing the 3D geometrical framework for maturity and generation modelling and in reducing the uncertainty in the sensitive parameters. There is a desperate need for more reliable and better quantified calibrants. Expulsion, sublimation of gas from kerogen, and pressure effects on hydrocarbon generation are very important. As more sophisticated basin models are developed the historical geological problem arises of the reliability and scale of extrapolation and interpolation of data for the framework. Model development must proceed hand-in-hand with efforts to ascertain more geological characteristics and lithology distribution from the seismic response and stochastic and statistical methods. Underpinning all models must be databases of knowledge for testing model veracity.

KEYWORDS: *hydrocarbon potential, generation, migration, thermal conductivity*

INTRODUCTION

Exploration has always been an investigative process where the rock record provided in the remotely sensed seismic and down-hole media provides the clues. The geoscientist uses various accumulated knowledge and techniques to recognize trends, interpolate, extrapolate, visualize and reconstruct a geological history of a basin that is consistent with the data. In the past, this task has been wholly manual, but the advent of the computer age has revolutionized data acquisition, processing, interpretation and mapping. Basin models are the numerical tools for simulating the physical processes that occurred during basin formation, they provide the framework with which geological, geophysical and geochemical observations and data must be consistent.

The principal objective of basin modelling is to reduce exploration risk.

The distribution, timing and quality of potential source, reservoir and seal rocks, the structural development, hydrocarbon generation, expulsion and migration are all integrated to rank sub-basins and prospects. Identification of the most sensitive parameters for the prospectivity assessment of the region is a critical part of the process. The multiple question and parameter variation approach requires that basin modelling results be quoted as a range under the specific model conditions. The actual process of attempting to honour the data in a basin through modelling often provides more insight into the problem than might appear in the final analysis. The success of the modelling is significant, but the journey taken

From K. Glennie & A. Hurst (eds), 1996, *AD1995: NW Europe's Hydrocarbon Industry*, Geological Society, London, pp. 83–105

to achieve the solution is often invaluable. In considering model results, it is important not only to consider how right the answer is, but also, how wrong could it be. A useful conclusion might not always be the tentative location of hydrocarbons, but also perhaps confident assessment of where they are very likely absent.

Basin modelling techniques provide the historical view of the basin development, which is key to the understanding and exploration effectiveness.

Several excellent basin modelling review papers have been published recently. An overview of basin processes and modelling is provided by Allen & Allen (1990). Hermanrud (1993) gives a broad and deep review of basin modelling techniques. He discusses compaction, thermal processes, fluid flow and overpressuring, hydrocarbon generation, primary migration, secondary migration and leakage, general mathematical development and modes of application. Lerche (1993), in the same excellent volume, sets out the theoretical aspects in basin modelling, including forward, inverse and pseudo-inverse models. Lerche also addresses some of the thornier problems in modelling, like faults and halokinetic effects, and sets out some clear areas of concern in basin modelling which should be addressed. A very good review of maturity and source rock issues past and present, and with respect to the North Sea, is set out by Cornford (1990).

This paper will summarize the historical development of basin modelling, address some of the more elusive practical problems like transient effects, unconformities, lithological parameters, thermal conductivities and sensitivities, and offer an opinion as to the directions the science will take in the future. The interested reader will be directed to more detailed discussions of subjects where appropriate.

PRINCIPLES

Physics

Gravity, tectonics and radioactive decay are the basic physical driving forces to a dynamic basin system, providing the kinetic and thermal energy. Gravity causes much of the deformation in a basin. The geometry of the basins are often determined by a combination of tectonic forces, and inherent crustal weaknesses. The actual causes and effects of various tectonic forces are still a matter of debate beyond the scope of this paper. Thermal energy is provided from several sources; by the deep-seated base of the lithosphere; a contribution from the crust/lithosphere interface through underplating; and a contribution from within the crust from the radioactive decay of basement rocks. Further, usually localized heating may be caused by the catastrophic emplacement of intrusions, and minor amounts of heat are emitted from formations within the sedimentary pile that are rich in uranium, potassium and thorium. Turcotte & Schubert (1982) elegantly relate fundamental physical processes to geological phenomena.

Rock properties govern the response of the basin system to the forces. The rock properties' behaviour also varies with the pressure and temperature of the system, and probably also with geological age (a difficult hypothesis to test). Heat energy is distributed through a basin mainly by phonal conduction; that is the 'knock on' vibration of adjacent molecules within the minerals/fluids comprising the bulk rock. Heat may also be exchanged and carried (advected) by fluids flowing through a system. The distribution of the heat energy is therefore contingent on the thermal conductivity and the specific-heat capacity of the media. Kappelmeyer & Haenel (1974) provide valuable information on geothermics.

Rock compressibility, porosity, permeability and fluid viscosity are just some of the important properties which affect fluid flow. A collation of rock properties is contained in Clark (1966). A more up-to-date collation of pertinent records would be of great help, as most information is currently rather disseminated.

Fluids within the basin system are also dynamic. The solubility of gas into liquid hydrocarbons varies with pressure and temperature amongst other factors (e.g. Glasø 1980).

Chemistry

Basin chemistry is known to have profound effects on the geology of a basin, but the thermodynamics of the open fluid–rock diagenetic system is very poorly understood. Diagenesis is voraciously studied in reservoir formations but the diagenetic linkage between the shale system and the reservoir system is still tentative. The chemical breakdown of organic material under varying temperature conditions is better understood, and forms one of the cornerstones of petroleum basin modelling (see Ungerer 1993 for recent review). However, even the use of organic geochemistry to predict hydrocarbon generation has its limitations (see Lerche 1993).

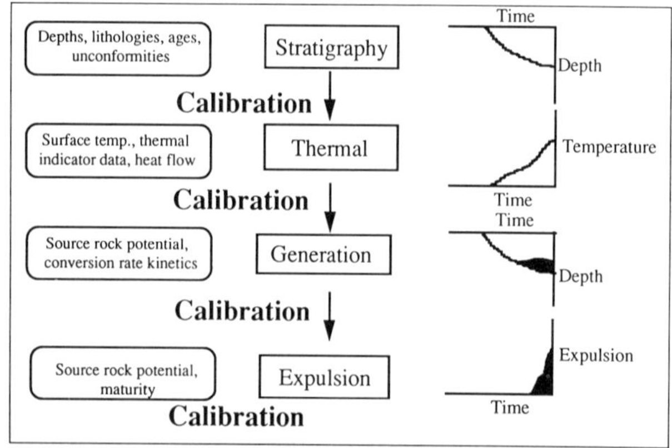

Fig. 1. The principles behind basin simulation to predict hydrocarbon charge.

The basin modelling process

A typical basin modelling study begins with the interpretation of the seismic and well data. The basin modeller will then want to locate the areas of mature source rock. Ideally, the modelling and interpretation should be closely linked so that the sensitive parameters, determined by modelling, can be concentrated upon at the interpretation stage. An initial phase of modelling will ascertain the problems to be addressed.

The burial depth history of the rocks in the well is determined by backstripping and sediment decompaction (Fig. 1). The process continues with an assessment of the present thermal state of the well, its geothermal gradient or heat-flow condition. The present day is typically used as the known pivotal point from which some reasoned estimate of the thermal history is constructed. A thermal history is the heat-flow history or geothermal-gradient history, and the corresponding surface temperature variation through geological time. The temperatures of the rocks at every geological

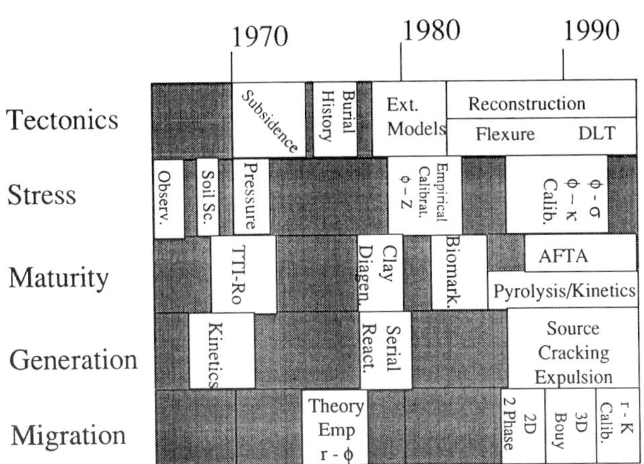

Fig. 2. The paragenesis of ideas that formed the science of basin modelling.

age can be calculated, using steady state or transient solutions to generate a temperature history. Once a temperature history has been determined for all the formations, the histories can be used in models that predict the maturity of the rocks. Maturity refers to a prediction of the relative transformation of any organic material in the section to oil and gas. Maturity may be measured and described using several criteria (see for example Cornford 1990). In the instances where actual source-rock horizons are known, they may be modelled specifically using refined forward kinetic models of source-rock transformation (e.g. Ungerer 1993). In addition, the process of primary expulsion, that is the release of hydrocarbons from the source rock into the carrier systems, may be simulated using a variety of mechanisms, none of which alone sufficiently simulate reality (see below).

Thus, basin modelling has provided a time framework for the key temperature and maturity, generation and expulsion events in a basin development–culminating in the 'charge' history. Assessment of the secondary migration, volumetric accumulation and preservation of the hydrocarbons, requires additional knowledge of the basin geometrical development, and hindrances to flow.

At every point in the modelling, all the available measured data must be used to verify the modelled results. The iterative process of calibration is central to the veracity of basin modelling studies. The calibration process is more fundamental than the application of the modelling, since it is the selection of the physical model itself that must be consistent with the observed analytical data, which in many cases should be viewed sceptically. This is a crucial area where basin modelling could be soundly improved.

BASIN MODELLING: THE BEGINNING

Basin modelling is computationally intensive. The development of basin modelling has, therefore, largely reflected developments in IT, both in hardware and software.

The historical development of the basin modelling techniques, which are typically combined in petroleum software today, occurred in parallel from many different subject domains. The paragenesis of ideas is set out in summary in Fig. 2.

Basin modelling developed from several independent basic observations. Sleep (1971) recognized that the seafloor depth appeared to increase with the age of the crust from the Mid Atlantic Ridge to the US continental shelf, which germinated the concept of thermal subsidence. Breakthroughs in the development of mathematical models that simulated the breakdown of organic matter to liquid and gaseous products were taking place at the same time in the geochemistry laboratories (Tissot 1969; Lopatin 1971; Tissot & Espitalie 1975; Waples 1980). The geohistory approach of van Hinte (1978), which reconstructs depth histories, and backstripping–the sequential removal of sediment back in time with the restoration of the historical sediment thicknesses and depths (Perrier & Quiblier 1974; Watts & Ryan 1976)–together with the implementation of empirical compactional models (Magara, 1976) to restore porosity into unloaded rocks, laid the foundation of the modern basin modelling approaches.

The first computer programs to integrate these processes appeared in the late 1970s to early 1980s (e.g. Tissot & Welte 1978; Welte & Yukler 1981; Guidish *et al.* 1984), and oil companies realized the enormous potential that predicting maturity and generation had for reducing exploration risk. These early models were essentially limited by the hardware, software and output facilities available at the time, commonly running in batch mode on mainframe computers with rudimentary interfaces and graphic output. However, since the late 1980s, commercial versions of basin modelling software have become available, which run on desktop computing facilities. In a rapidly evolving area of geological modelling, these packages offer not only increasingly sophisticated techniques, but also highly interactive graphical user interfaces and high quality printed output, a reflection of the rapid changes in information technology.

Basin modelling software is now routinely used in commercial exploration. Whilst many of the early developments in basin modelling software were driven directly by advances in the science or the computer technology, many of the future developments in basin modelling software will be driven by the need to more closely integrate basin-modelling techniques into exploration companies' information-systems framework.

CRUSTAL EXTENSION, SUBSIDENCE AND STRUCTURAL RECONSTRUCTION

At the same time as geochemical basin models were developing, crustal subsidence was becoming an important subject.

A quest for the definitive extensional model of subsidence seems to have been instigated by McKenzie (1978). Since McKenzie's seminal paper, the relationship between crustal extension, Airy subsidence and heat flow has been a matter for considerable debate. Numerous authors with models with various numbers of crustal layers with various plausible properties have been developed that are all consistent with some observations (Falvey 1974; Jarvis & McKenzie 1980; Royden *et al.* 1980; Vierbuchen *et al.* 1982; Hellinger & Sclater 1983). The development of flexural models (Watts & Ryan 1976; Watts *et al.* 1982; Kusznir & Karner 1985; Kusznir *et al.* 1991) have provided a valuable insight into the complexity of the real processes, but there is still some debate as to the calibration, dynamics and properties of the crust and lithosphere as basins develop. The subject of inheritance and isostatic, flexural cycles have not yet been addressed. A recent innovative approach has been to use earthquake dislocation theory to predict subsidence (Kusznir *et al.* 1995).

The early links between lithospheric subsidence and extension and maturity coincided with renewed interest in brittle deformation geometries in the upper crust (MacKenzie & McKenzie 1983). It was recognized that the upper crust undergoes brittle deformation, and that rocks become more ductile with depth. Measurements of extension became a benchmark for basin comparison. However, it quickly became obvious that crustal and lithospheric extension did not give corroborative extension amounts (β factors) (Hellinger & Sclater 1983). The link between β factors, faulting episodes and plate-tectonic movement with thermal events has yet to be clarified. Three basic models of lithospheric extension have been proposed; pure-shear models (with a horizontal detachment) with both uniform (McKenzie 1978) and non-uniform (Hellinger & Sclater 1983) versions; an intrusion model (Royden *et al.* 1980), where extended crust is replaced by igneous intrusive material and; simple-shear (with dipping detachment) models which include large-scale crustal shear zones (e.g. Wernicke & Burchfiel 1982; Wernicke, 1985) and anastomosing type shear zones (e.g. Kligfield *et al.* 1986)

Improved technology, such as deeper and better resolution seismic data and visualization aids, has provided the geologist with a better insight into extensional structures and their geometries.

The numerical and analogue modelling of such structures has also deepened our understanding of these sometimes subtle structures (e.g. Waltham 1989; McClay 1995).

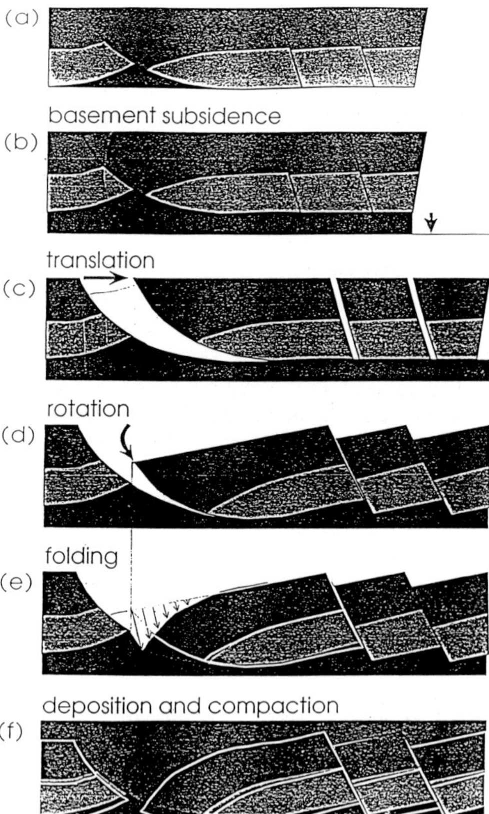

Fig. 3. The steps in palinspastic reconstruction in section modelling (after Schultz-Ela 1991).

Geometrical reconstruction

As geochemical basin modelling developed, and predictions of maturity and the hydrocarbons generated were being addressed, questions arose as to where the generated hydrocarbons went next, and the migration became more important. This involved the modelling of fluid flow (see later) and geometrical reconstruction. Models that perform all or part of these processes have been constructed in 2D (Kligfield *et al.* 1982; Moretti & Larrere 1989; Fjeldskaar *et al.* 1990; Kusznir *et al.* 1991) and, more recently, in 3D.

The reconstruction and balancing of cross-sections are very useful methods for testing the viability of a geological, geometrical interpretation, and can also provide estimates of extension. Full section balancing restoration involves the processes of: backstripping and decompaction; structural reconstruction and geodynamical restoration. The sequence for reconstruction is shown in Fig. 3 from Schultz-Ela (1991).

Current deformation restoration techniques are not realistic deformation mechanisms but are geometrical approximations that work in certain instances. These techniques include rotation; translation; simple shear (either vertical, Chevron, or constant-heave method or modified Chevron method), or inclined synthetic or antithetic; slip-line reconstruction; flexural slip (bed-length); palaeovertical method (Nunns, 1991), or non-specific (Moretti & Larrere 1989).

Each of these methods has its own basic assumptions, but two-dimensional geometrical restoration has the following general limitations in modelling:

1. deformation of the regional level for the reconstruction;
2. effects of 'unseen' faulting not accounted for in the restoration (Higgs *et al.*, 1991);
3. salt effects (pressure solution, piercement);
4. shale diapirism effects;
5. relative timing of fault development;
6. antithetic fault development;
7. problems establishing the correct (meaning one that works) deformation technique.

Since the major driving force of secondary migration is buoyancy, the restoration of the geometry of seals and carrier beds is important to basin modellers. In particular, reconstructions for the times of peak hydrocarbon generation and expulsion for the carrier-system hydrology to be modelled. At the prospect scale, reconstructions can constrain geometries for volumetrics in 3D, and may be used to investigate fault seals and spillages. The power of the geometrical reconstruction is through its visualization of the structure.

FLUID FLOW, COMPACTION AND OVERPRESSURING

Compaction

The earliest basin models ignored the effects of compaction when a burial history was constructed. Since shale can loose up to 60% of its volume, and sands 35%, due to compaction through burial to 4 km depth, compaction severely affects depth reconstruction back in time. Quantification of the observation of compaction phenomena (fitting of a curve to data) began as early as 1930 by Athy. Gardener *et al.* (1974) observed that the stratigraphic age of sediments had an impact on the change in seismic velocity and so presumably on porosity in a formation. Standard compaction models such as Sclater and Christie (1980), Baldwin & Butler (1985), Falvey & Middleton (1981) and Middleton (1980, 1983) do not consider the age criterion. Compaction was therefore included in basin models at an early stage (Magara 1976), the algorithms based on some earlier observations. Compaction may be defined as the mechanical loss of porosity due to increase of load stress. Porosity changes may be caused by several other mechanisms as set out by Waples & Kamata (1993), the most important of these being precipitation/dissolution phenomena. More recently, the effect of temperature has been implied to have an effect on compaction (Giles *et al.* in prep).

Fluid flow and overpressure

Most basin models simulate compaction in a forward sense by calculating a rate of sedimentation and determining the volume of fluid escape out of the deposited formations. The rate of escape is limited by the permeability of the formation. If some fluid is trapped within the pores, then the trapped fluid will act to support part of the overburden framework of rock pressing down on it. The fluid in the pores then becomes overpressured relative to the hydrostatic state.

Many workers have attempted to model fluid flow in 1, 2 and 3D (Yukler *et al.* 1978; Smith 1971; Bethke 1985 1986; Cao 1985; Glezen & Lerche 1985; Dutta 1986; Nakayama 1987; Ungerer *et al.* 1987; Audet & McConnell, 1992) and all have made some assumptions of how frame pressure or effective stress is related to porosity, and how porosity is related to permeability.

Since a fluid flow model works in a forward time sense, there is no guarantee that the results of the complex input

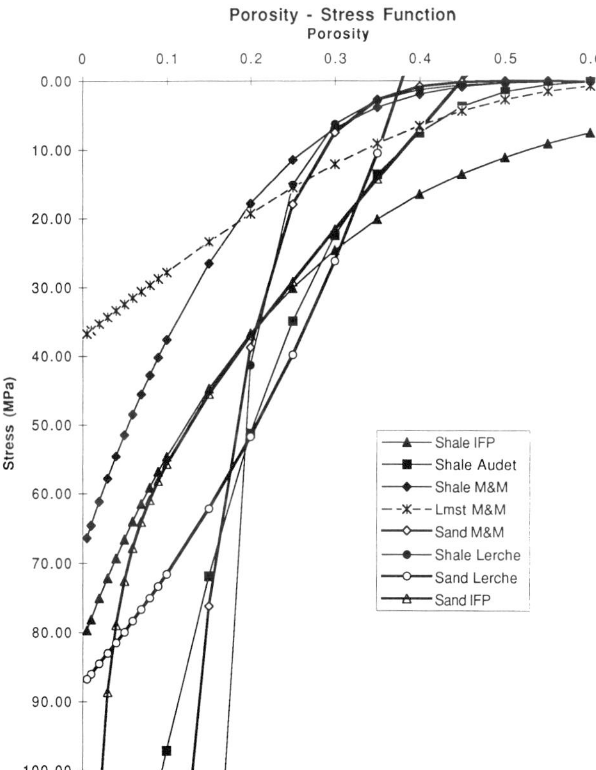

Fig. 4. Stress–porosity curves for basic lithologies published in the literature.

will be consistent with the data the user would like to comply with, in terms of pressure–depth, porosity–depth, and of layer thicknesses. Therefore, re-adjustment routines are used, which alter compactional parameters or masses in the modelling to minimize the difference between the modelled resulting section and present-day measured data.

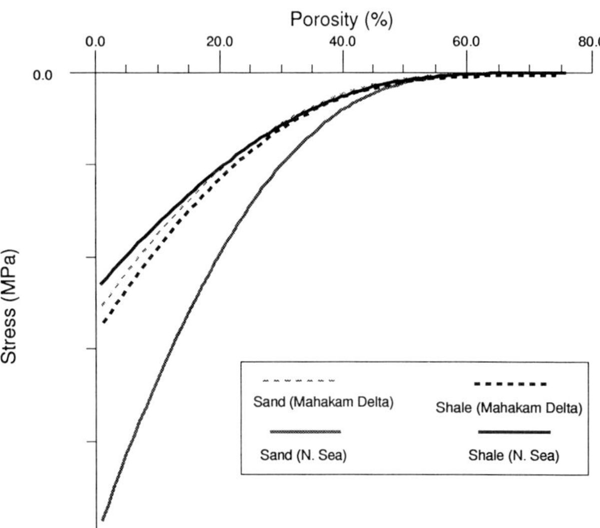

Fig. 5. Comparison of sand and shale porosity–stress relationships from the North Sea and Mahakam Delta (modified from Schneider *et al.* 1993).

Porosity–stress relationships and soil mechanics

Effective stress–porosity curves in the literature are quite varied (Fig. 4). The change in porosity with rock stress is a function of the rock's compressibility, a relationship that has been widely used in basin models.

The effective stress–porosity curve for different lithologies may be calculated from measurements of pore pressure and porosity for formations at various pressures from various wells (e.g. Schneider *et al.* 1993). Schneider *et al.* illustrate further problems, in that what is described as sand behaviour from the North Sea (Brent Formation), is almost identical to shale behaviour from the Mahakam Delta (Fig. 5). Perhaps this is a feature of clay chemistry and colloidal behaviour (Burland 1990) or an age effect.

Although the geotechnical sciences have studied void ratio and porosity effects at stresses typically less than 4 MPa (see Burland 1990 for a review), geological research at higher pressures and temperatures is urgently needed to provide a solid analytical basis for our model calibrations.

Porosity–permeability relationships

Permeability has units of area. The amount of fluid that may move through a formation is controlled by its permeability. Shales have very low permeabilities, sandstones are highly permeable. However, permeabilities act on different timescales (see Ungerer *et al.* 1987), and rocks that seem impermeable in an engineering sense may be relatively permeable in the geological basin modelling sense.

Permeability measurement is difficult in shales so reference data are very sparse. Several relationships have been published but there is a difference of opinion as to the actions of shales at low porosities (see Olsen 1960, and Ungerer *et al.* 1987). The actual microscopic-scale phenomena have not yet been elucidated. Theoretical relationships are likely to be valid only down to a certain depth, below which, other phenomena take over (Schneider *et al.* 1993).

The general forms of the porosity–permeability curves used in basin models are shown in Fig. 6. The Olsen (1960) behaviour curve tends to maintain a near-constant permeability at low porosities (Fig. 6). This is contradicted by other shaped curves, such as the Kozeny-Carman (e.g. Ungerer *et*

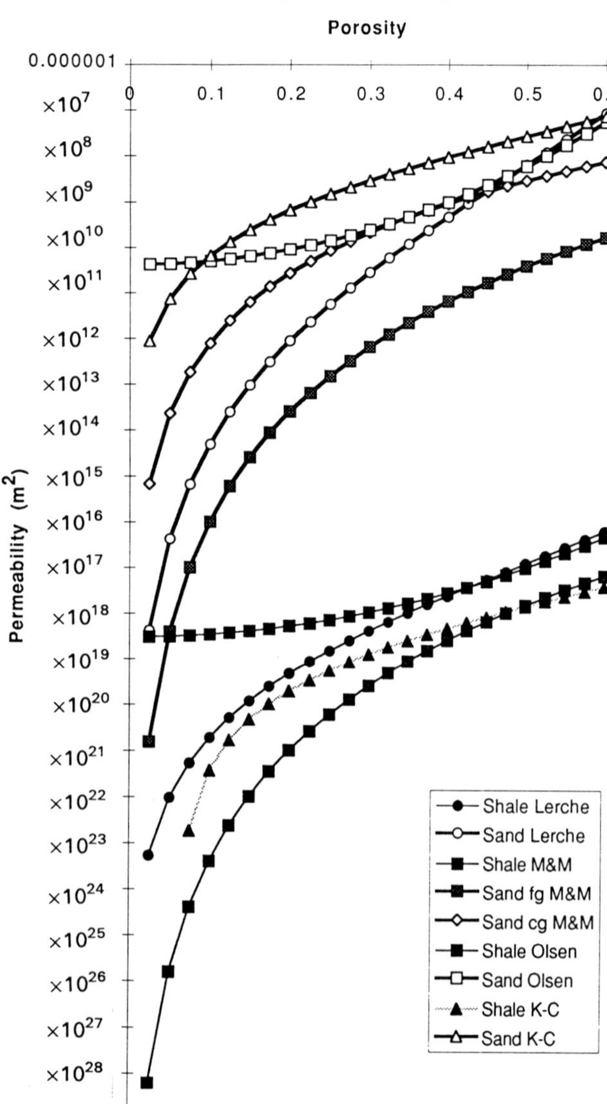

Fig. 6. Porosity–permeability relationships for basic lithologies published in the literature.

al. 1990) or power type curve (e.g. Mann & Mackenzie 1990; Lerche 1990*a*). An intermediate situation is an exponential approach, which becomes a straight line on a logarithmic plot (Fig. 6).

The flow through a system is controlled by the permeability of the finest-grained rocks, the clays, shales and silts, and also salts and cherts. Calibrations of porosity and permeability for these lithologies certainly requires convincing documentation. If more complex systems are to be modelled, with more detail of lateral facies changes for example, then the properties of fine-grained rocks will need to be much better understood, once again, through thorough analytical work.

Variability within published data such as Lambe & Whitman (1979) and Reike & Chilingarian (1974) show a wide range of hydraulic conductivities. Audet & McConnell (1992) in their review, suggest that clay mineralogy and pore-fluid chemistry, along with porosity, are key criteria. Recently, further advances towards permeability prediction, using a geometric model, have been advanced by Bryant *et al.* (1993) and Bryant & Blunt (1992).

There is a physical limit of internal pore pressure that a rock can withstand before it breaks or bursts (see Nakayama 1987). There are two ways that this may be modelled, by inputting a Poissons ratio (g) for each rock and letting the maximum pressure the rock can withstand be equal to

$$P_{\max} = S \times \frac{g}{1-g}. \tag{1}$$

Alternatively, the fracturing stress may be set at a fraction of the lithostatic stress (S), which for shales (which are the most commonly overpressured horizons) is equal to:

$$P_{\max} = 0.85 \times S \tag{2}$$

When the fracturing pressure is attained by the fluid-flow modelling, then the permeability is increased to a larger permeability, similar to sandstones, to simulate the opening-up of the pores during fracturing. (e.g. see Nakayama 1987). This, albeit dated, approach has recently been provided with some analytical basis from the work of Capuano (1993) on the Frio overpressured shales of the Northern Gulf Coast of Texas.

Overpressuring

The causes of overpressuring is another topic of voluminous literature. In addition to the process of compaction disequilibrium already described, some areas may become overpressured as part of a pressured cell (Mann & Mackenzie 1990). Within such cells, some areas export pressure, so sandstones are under pressured with respect to their seals. Other areas up-dip are overpressured relative to their seals due to the lateral and up-gradient import of pressure to the lower-pressure areas. Additional causes of overpressuring are aquathermal pressuring (Barker 1972), mineral transformations, such as smectite to illite (Dutta, 1986), and hydrocarbon generation (Nakayama 1987).

THERMAL MODELLING

Present thermal state

Present-day temperatures in wells are the important pivotal point for reconstructing the past thermal history. We should

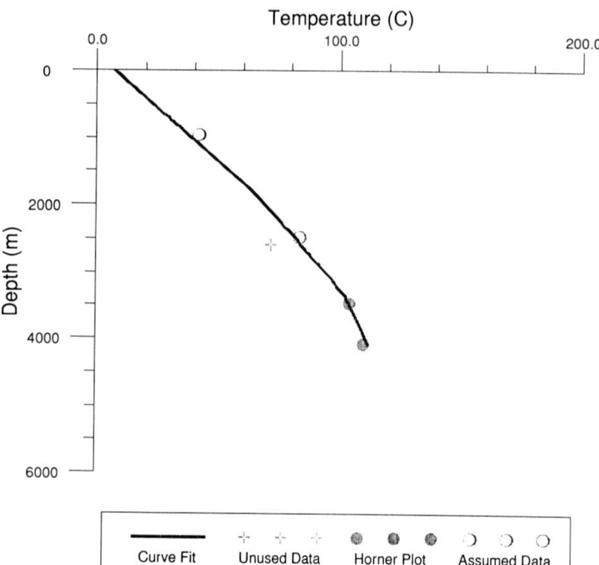

Fig. 7. Multiple down-hole temperature optimization.

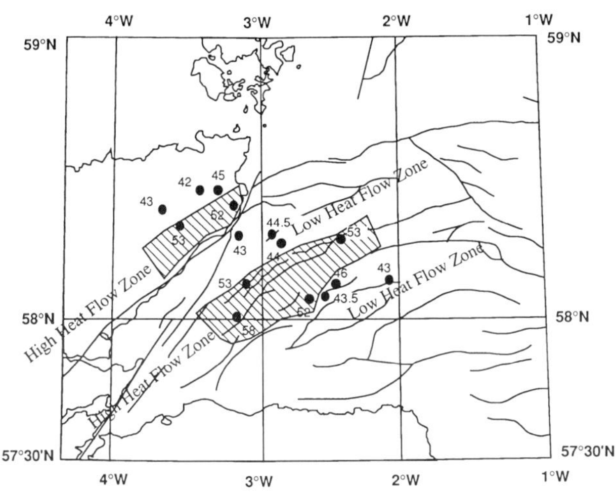

Fig. 8. Inner Moray Firth heat-flow map.

know more about the temperatures today than back in time, but accurate temperatures are very difficult to determine other than by drill stem tests, which are not so common (e.g. Hermanrud *et al.* 1988).

Several algorithms exist for correcting bottom-hole temperatures (temperatures measured by logging tools), for example the Horner Method (Fertl & Wichmann 1977). The Horner method requires three parameters to be recorded; depth, time since circulation and the observed temperature, for at least two times so that a sensible bottom-hole correction may be obtained. A heat-flow calculation that optimizes on several temperature measurements to find the best value will provide a better approximation.

The objective is to determine a regional heat flow that may be used in the undrilled kitchen areas of a basin. Any significant tilt in strata, underlying salt or core of basement core rocks, or hydrodynamic flow, may render the perfectly calculated heat flow invalid for use in the kitchen regions. Just such a problem may be observed in the Haltenbanken region of Norway (Viki & Hermanrud 1990; Jensen & Doré 1993) and in the Inner Moray Firth region (Thompson *et al.* in prep).

Thermal history

To determine the temperature history of the sediments, estimates of the surface-temperature history boundary condition must be made, and combined with either a geothermal-gradient history or heat-flow history.

Determining temperature histories using a heat-flow history requires knowledge of not only the depths, but also the matrix thermal conductivity and porosity of the sediments.

The variation of surface temperatures through time is a function of palaeolatitude, palaeoclimate, and palaeobathymetry. The transients caused by small time-scale surface events have geologically negligible effects (see Kappelmeyer & Haenel 1974 for a review).

A two- or three-dimensional heat-flow model should be used in regions of strongly tilted beds where heat flow will focus towards the highs and away from the lows. This is possibly observed on a heat-flow map of the Inner Moray Firth (Thompson *et al.* in prep) where the South Bank High is higher than the surrounding grabens (Fig. 8). A higher heat flow has also been observed in the Haltenbanken, offshore Mid Norway (Jensen & Doré, 1993; Vik & Hermanrud 1990). The Haltenbanken appears to have been a long-lived structural high and the focus of basement heat flux.

So when considering using a heat flow from a calibration well for modelling the deeper parts of a basin, it is crucial to know the well's structural setting, whether it is on a high or in a basin, and the geometrics and composition of the strata beneath (this is also true when using extensional models of subsidence to ascertain heat-flow histories). The heat-flow history of a basin centre is likely to be significantly less than observed on the structures.

Table 1. Data used for calibrating thermal history

Occurrence	Indicator	
Woody matter	Vitrinite reflectance	
Shales OM	Hopane isomerization	
Shales OM	Sterane isomerization	
Shales OM	Steroid aromatization	
Apatites – mainly sandstones	AFTA	
Limited to model assumptions	Tectonic subsidence	
Clays	Clay transformation	
OM > 1% TOC	Rock Eval Data	HI
		PI
		T_{max}
		TR observed
Source horizons	Mass balance	
Source horizons	Source rock kinetics	

Maturity

Illustrating the maturity of specific source rocks in terms of transformation ratio is now standard practice. Maps with vitrinite reflectance in % Ro equivalent for oil and gas windows are becoming less common, and only of use in areas with no proven source rocks of any description. Maturity indicators, such as vitrinite reflectance and biomarkers, have been fully described by other authors (e.g. Cornford 1993).

It is of immense value to have many different recordings of the thermal history of a basin through the use of multiple thermal indicators. Some indicators in Table 1 are available throughout the geological section, others are confined to the source rock. Each of the indicators in Table 1 is sensitive over a specific temperature range (Fig. 9). By combining several indicators that occur in different lithologies from different analytical techniques, material and, most importantly, stratigraphic levels, the geologist can get very useful clues as to the different histories of the individual strata. An example of this is in the Inner Moray Firth, where biomarker data from the Kimmeridge Clay Formation have been widely used to indicate uplift (e.g. Pearson & Duncan 1996), whereas the vitrinite reflectance is relatively insensitive in the same temperature range (Fig. 10). However, more recent measurements of Ro from the Devonian strata (e.g. Downie pers. comm., 1994) suggest that Devonian rocks have undergone some significant earlier heating event, not experienced by the Jurassic strata. The importance of such an event in this example is that it would serve to cook the Devonian lacustrine source rocks significantly, resulting in a loss of initial source potential for the later Late Jurassic to Cretaceous hydrocarbon generation phase at this source level (Thompson *et al.* in press). The power of using several techniques to validate an assumption cannot be overstressed,

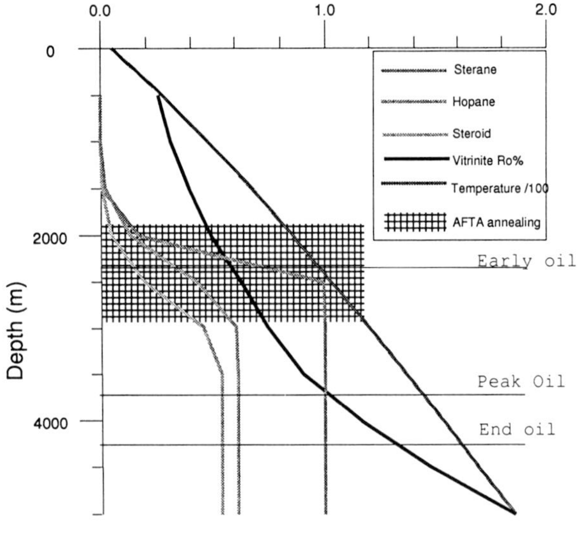

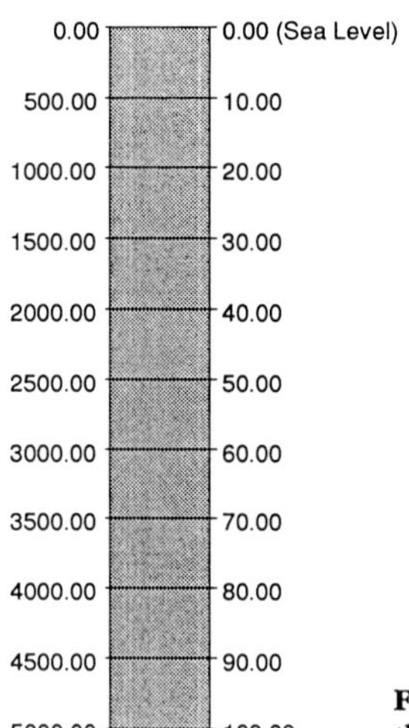

Fig. 9. Temperature range for various thermal indicators.

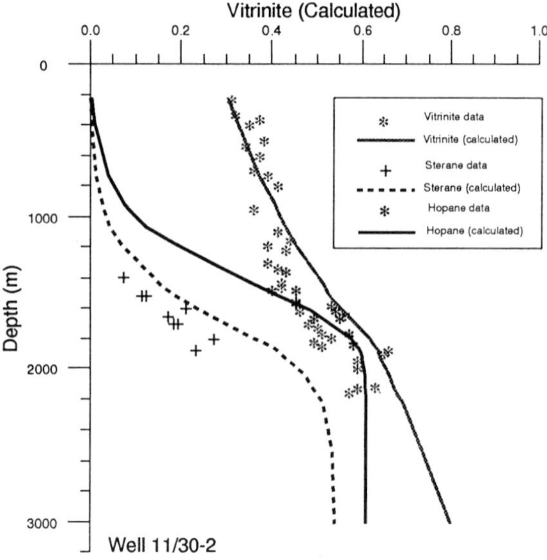

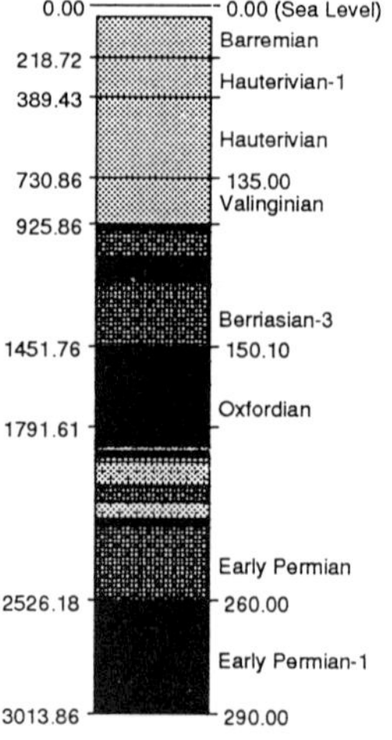

Fig. 10. Thermal indicator data from the Inner Moray Firth.

particularly in complexly inverted basins such as the West of Shetlands, Inner Moray Firth and Cardigan Bay areas of the UKCS.

Recent improvements in the past few years in assessing palaeotemperatures have been made with the use of apatite fission track data (Green *et al.* 1986). The apatite fission track data provide key information on timing of maximum temperatures. Several kinetic schemes for using apatite fission tracks to predict temperature histories exist (Laslett *et al.* 1987; Carlson 1990; Crowley 1993), and the calibration of a forward kinetic model that satisfactorily accounts for variability in apatite chlorine composition is developing (Corrigan 1993).

The use of fluid inclusions to determine palaeo-fluid content and provenance has proved valuable to the petrologist. The use of homogenization temperatures corrected for salinity can also provide palaeotemperatures, and if hydrocarbons exist within the inclusions, then palaeopressures may also be assessed (Swarbrick 1994). This could have major implications for modelling, as inclusions are as ubiquitous as cements. Apatites, however, are very scarce in some rock types.

Basin models, with perhaps the exception of Bethke's (1985), do not attempt to model diagenetic processes.

Clay mineralogy, as previously mentioned, is a very complex process. Illitization processes have been modelled by Dutta (1986) and more recently by Luo *et al.* (1993). The crucial temperature for this process is 100°C, which broadly coincides with the 3000 m level chemical-type seals described by Hunt (1990).

Calibration of thermal histories

Heat-flow history has been calibrated, or estimated, using a variety of techniques. Calibration usually involves matching maturity-indicator data to values predicted by a reliably calibrated model of corresponding maturity behaviour. So, for instance, vitrinite reflectance data are compared with predictions using one of several available VR models (e.g. Burnham & Sweeney 1989).

Automatic techniques for optimizing heat flow to match predicted and observed maturity data have used inverse techniques and tomography (e.g. Lerche 1990*a,b* and, more recently, Nielsen 1993, Gallagher & Morrow, in prep and Wei

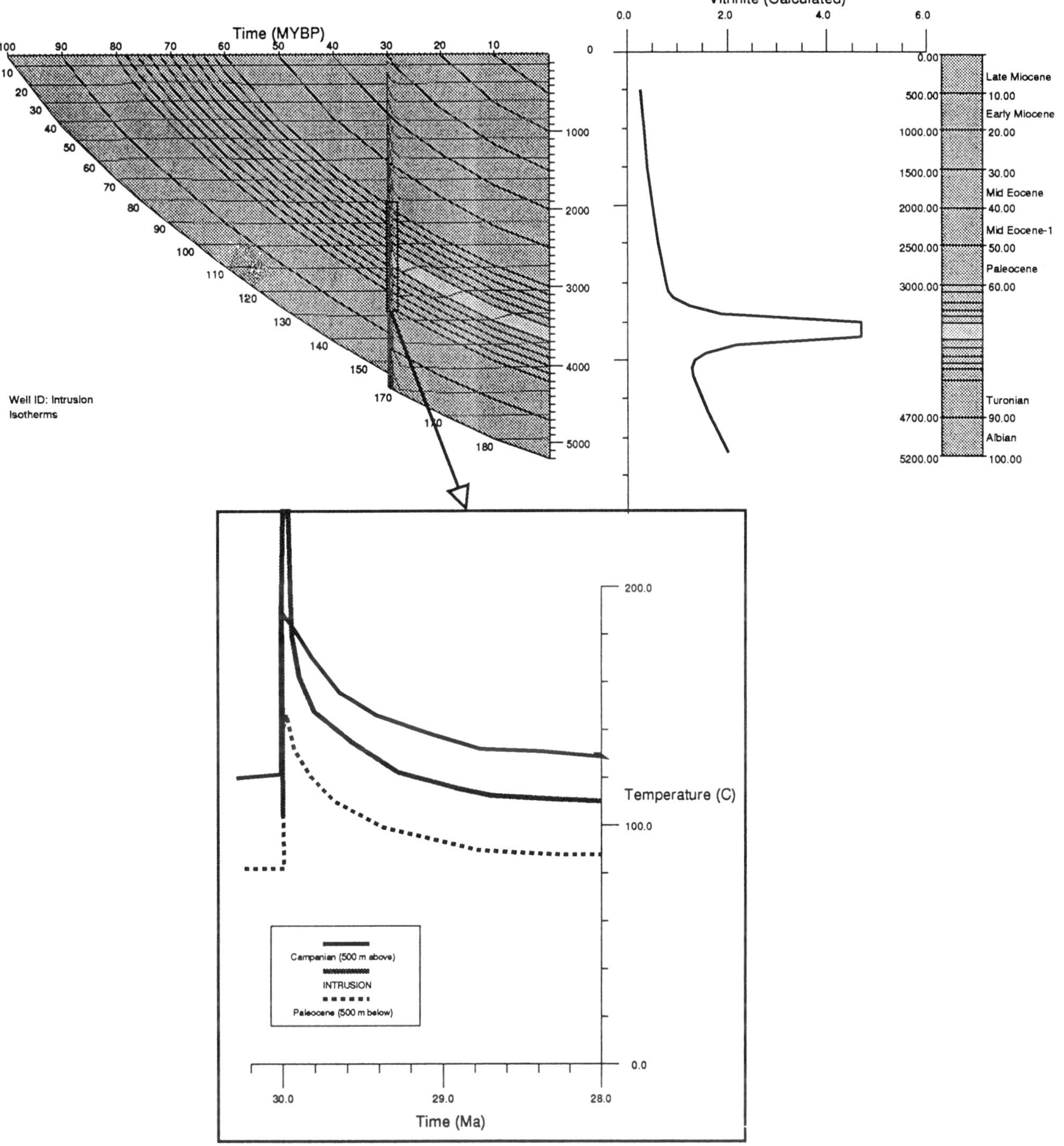

Fig. 11. Transient and maturity effects of intrusions.

1994). Such techniques, as in most applications, must be applied carefully in regions where the unaccounted for geology rather than heat flow may be the cause of maturity trends. All calibrations are non-unique, so geological cases should always be deterministically tested, so that an envelope of possible solutions may be presented. Quantifying this envelope has been the subject of research by Lerche (1993) and Thomsen (in press).

Tectonic subsidence has lasted as a classical technique for estimating heat-flow histories, particularly in virgin territory. Tectonic subsidence is the part of the down-dropping of the basin that is attributed to mechanisms other than sediment loading. These physical mechanisms of tectonic subsidence in extensional regimes are postulated to be initial rifting and thermal subsidence (McKenzie 1978). By taking the observed, so-called, 'tectonic' subsidence curve and comparing it with subsidence behaviour of rifts at different stages of extension (β factors) it is possible to determine the stretching factor (β) in the well. Using another equation, it is then possible to obtain estimates of heat-flow histories from models of how heat flow varies in rifts extended by different stretching factors using different physical stretching principles (e.g. McKenzie 1978; Royden *et al.* 1980). This technique has been used in the South Mozambique Graben (Iliffe *et al.* 1991) and Danish Basins (Nielsen & Balling 1990). Techniques for checking heat-flow histories and temperature

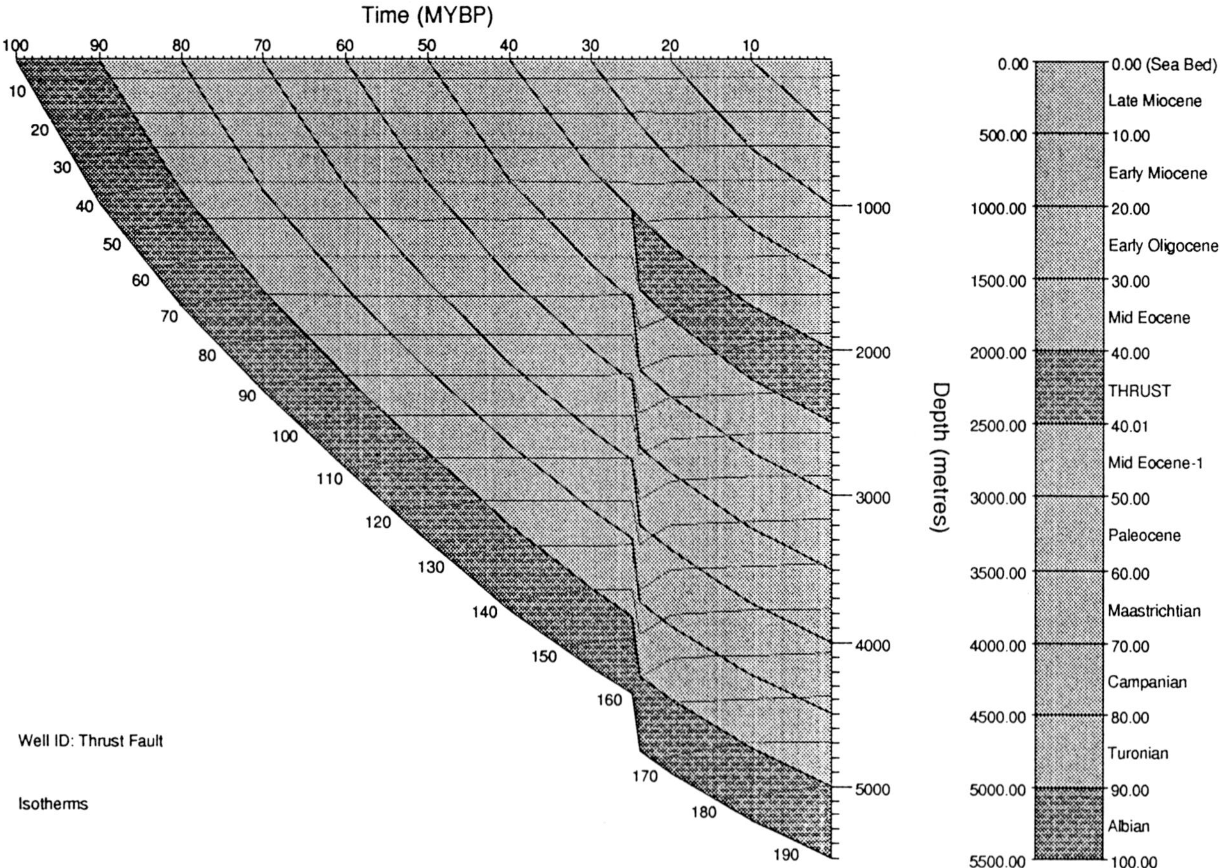

Fig. 12. Thermal effects of a thrust fault.

histories, or thermal calibrations, are shown in Table 1.

Thermal histories may be calibrated using source rock criteria from mass balance (Cooles *et al.* 1986) and observed TR versus predicted TR techniques. Pyrolysis techniques, be it by Rock Eval apparatus or Pyromat II, measure only the present-day residual potential. If one is confident of the source-rock kinetics (perhaps as a result of measurements nearby) then the temperature history can be checked by comparing observed TR (Forbes *et al.* 1991) and modelled TR from kinetic modelling of the same strata. This type of approach has been used by Dahl & Augustson (1993). Of course there is a circular argument here in that the kinetics may be unknown, in which case these could be estimated by adjusting the various kinetic parameters. This may be done by using a Gaussian activation energy distribution and adjusting either the mean or standard deviation to reduce the number of adjustable parameters.

A similar approach has been adopted by Nakayama & Cao (1993). Results can be checked when source material becomes available for analysis.

If source-rock kinetics are measured in the laboratory, then these may be used to constrain the upper temperature history limit of the source rock by testing for any kerogen conversion using the specific sample kinetic distribution with the predicted source-rock temperature history. If any conversion is predicted, then the temperature history is not consistent with the kinetic data. There should not be any transformation with a valid temperature history. This method is particularly useful when the kinetics are then run through wells which have a hotter source-rock temperature history, where source-rock transformation should be observed, the extent of which should also correspond to Rock Eval mass balance calculations.

Mass balance modelling is very sensitive to the particular immature source-rock analogue and again a range of likely or possible analogues should be used to investigate sensitivity and ranges of results.

Transient effects

Any geological event within the sediment pile, such as intrusions, sediment deposition, faulting or erosion, that happens very rapidly may have a transient thermal effect that usually lasts less than 0.5 Ma. The transient effect of crustal variation heat flow, however, lasts longer; for example, heat-flow fluctuations that occur at the base of the crust, as in some underplating, take about 2–3 Ma to reach the sediment pile and equilibrate. This effect is a function of thermal diffusivity. So the heat flow we are measuring in wells today is actually the thermal state 2–3 Ma ago (Nielsen 1993).

The effects of igneous intrusions have been discussed by Lerche (1990b). The insertion temperatures and thickness of intrusions make radical differences to the temperatures of the adjacent sediments. The overall effect of the intrusion is reasonably localized and certainly short-lived. However, if the incidence is indicative of a provincial trend, then this should be accommodated in a general thermal pulse. Figure 11 gives an indication of the duration of the transient thermal effect of the intrusion, and the impact it has on source-rock maturity in terms of vitrinite reflectance. As can be seen, the transient

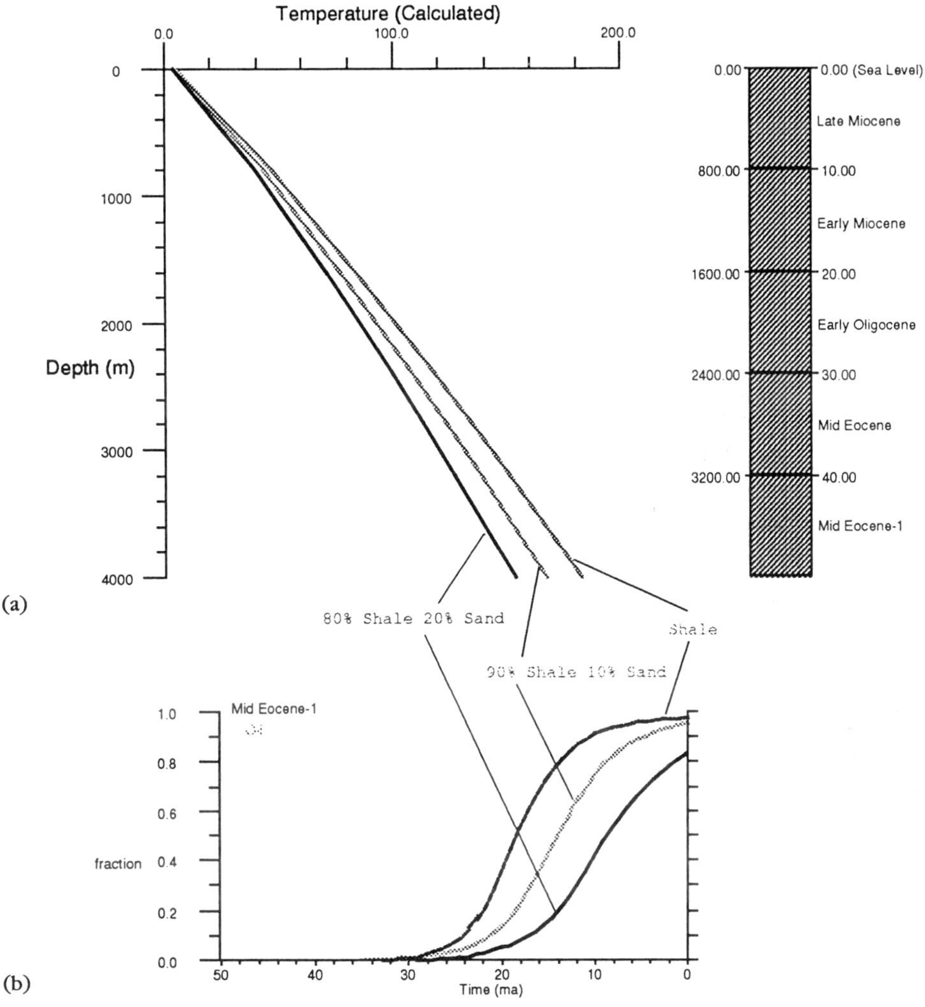

Fig. 13(a) Different temperature profiles produced by 1.1 in sections composed of 80:20 and 90:10 shale:sand ratios. **(b)** The hydrocarbon generation consequence of assuming a 10% increase in sand in pseudo-well stratigraphy illustrated as a transformation ratio change.

is short-lived, being about 1.5 Ma in duration, and through the maturity effect is very profound near to the intrusion, but a short distance from the intrusion, about 1.5 to 2 times its thickness, the rocks have a normal maturity trend.

Thrust faults have profound effects on the maturity of the section beneath and above the fault. The section beneath the thrust sheet has had hotter rocks placed on top of it. The juxtaposition of the thrust on top of colder rocks is not significant since the allochthon will equilibrate its temperature with the autochthon during emplacement. A more significant effect of the thrust intrusion is the rapid burial of sediments beneath the thrust fault, which have their geothermal gradient lowered for a transient period of time, but eventually heat up due to burial (Fig. 12).

Normal faulting in 1D has the effect of reducing the depth of burial that sediments would otherwise have.

The importance of thermal conductivities

Thermal conductivities are the key parameters affecting temperature distribution, which the geologist should be able to estimate more reliably. The analytical work outside volumes like Kappelmeyer & Haenel (1974), Somerton (1992) and Clark (1966) is fragmented, although the work of Brigaud & Vasseur (1989) and Demongodin *et al.* (1991) have provided more insight into factors influencing thermal conductivities such as anisotropy. End-member analysis easily illustrates the difference that thermal conductivity can make in a temperature calculation (e.g. Iliffe 1991), but it is often under-appreciated that this temperature calculation is cumulative; the effect of thermal conductivity change increases with the depth of the section; an important criterion in pseudo-well analysis. A more subtle example is a 4 km section of 80% shale and 20% sandstone, with a heat flux of 60 mW m^{-2} which has a temperature at 4 km of 154°C (assuming a constant 5°C at surface). A section of 90% shale and 10% sandstone has a base temperature at 4 km of 166°C (Fig. 13a). A 12°C difference caused by 10% difference in shale accumulated over 4 km. A difference in maximum temperature of 12°C is the difference of about 14% (TRS of 0.82 and 0.96 respectively) for a type II kerogen transformation history heated over a 20 Ma subsidence event (Figs 13a & b).

How well do we know our lithologies?

One approach to try and reduce the risk in estimating thermal conductivities has been to estimate them directly from the seismic (Leadholm *et al.* 1985; Ho *et al.* in prep). Thermal conductivities remain one of the most critical parameters controlling temperature history and so transfor-

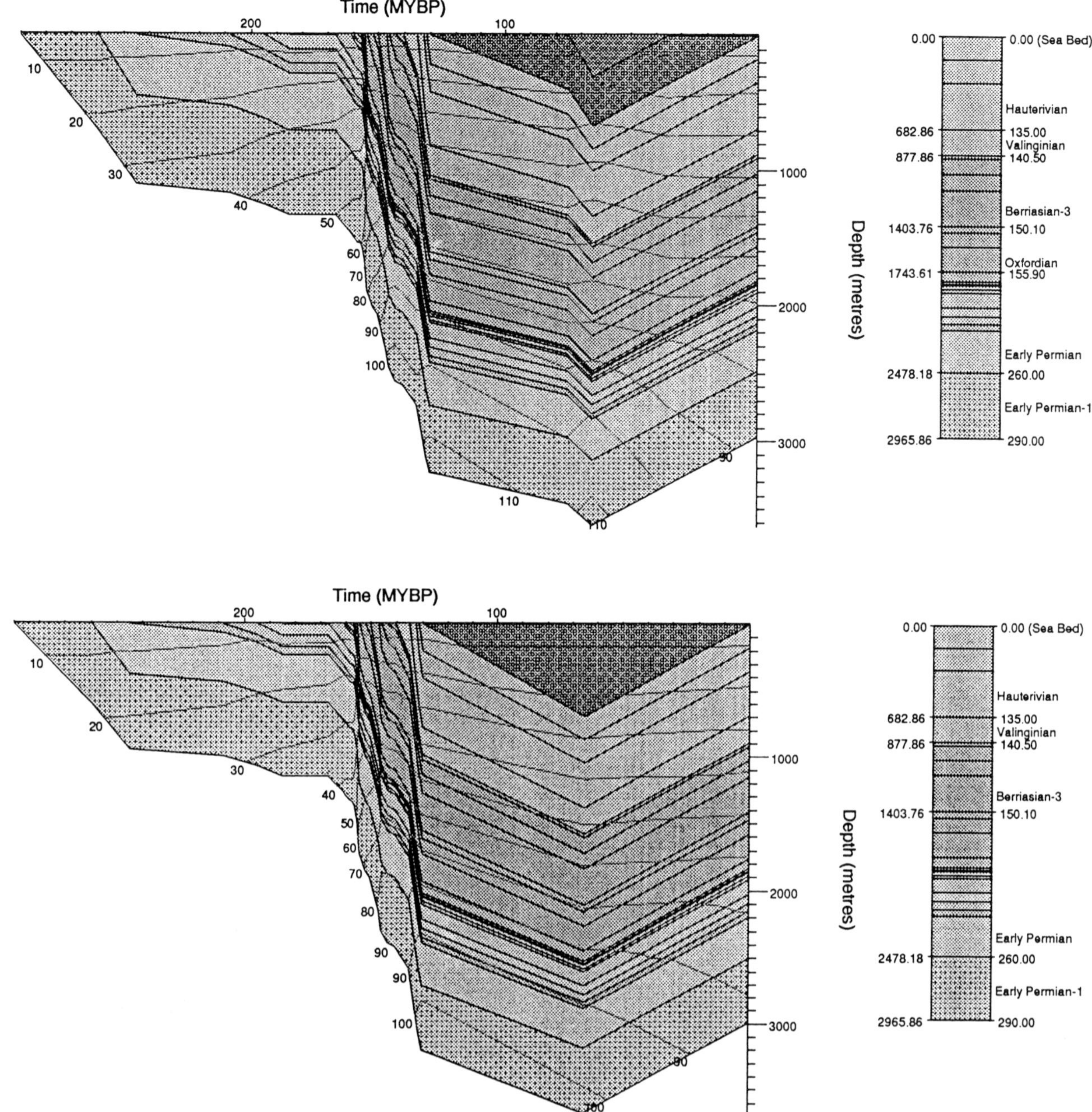

Fig. 14. Different interpretations of the Inner Moray Firth inversion event.

mation ratio. This is a subject area where a programme of analytical work could significantly reduce risk in economic basin modelling.

Calibrating inversion events

As stated above, the UKCS has many inverted basins. The timing of thermal events quite often hinges on establishing, not only the detailed chronostratigraphic framework, but also the complexity of unconformities.

A break in the rock record may be explained in many ways. A detailed regional geological understanding based on seismic and well data is needed to establish the complexities of unconformities, in terms of burial and erosion history, sediments that once existed but have been removed, and the burial depth (erosion magnitude) of the sediments represented by the unconformity. Figure 14 shows an example from the Inner Moray Firth with different interpretations of the same unconformity but with added detail. The techniques that may be used to determine unconformities are: VR data, biomarker data, AFTA, sonic slowness and seismic interpretation

It is very difficult to constrain the amount of sediment removed to better than a few hundred metres, mainly due to the fact that the temperature history is being modelled, and that this is a direct function of the thermal conductivity of the sediments removed, heat flow, burial depth and surface temperature history.

The consequences for the simple unconformity of changing lithology is shown in Fig. 15. AFTA data provide an added bonus of being able to date the period of maximum temperatures experienced by a rock beneath an unconfor-

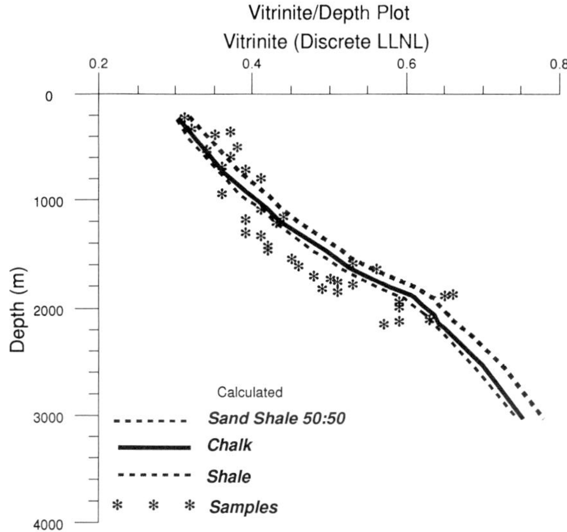

Fig. 15. The consequence of varying lithology within the Inner Moray Firth Tertiary inversion event.

mity which, if the geologist is provided with a good distribution of data down the hole, will identify the most important unconformity in a complex geohistory. AFTA data have been used to determine palaeo-heat flow using several techniques (Lerche 1990*a*). Examples of the use of sonic slowness are given by Thompson (1993).

SOURCE-ROCK POTENTIAL AND HYDROCARBON PRODUCTION

Pyrolysis methods and source-rock geochemistry

The source-rock geochemistry is measured using a Rock Eval apparatus. This is a pyrolysis machine connected to a Flame Ionisation Detector (FID). Results provide the amount of free liquid hydrocarbons in the pores of the rocks (S1) and the amount of hydrocarbon potential left in the rock (S2). The T_{max} value is the temperature at which the S2 fraction has reached its maximum production, and T_{max} is itself an indicator of maturity. Peters (1986) provides a useful insight into interpreting Rock Eval data.

Mass balancing

Mass balancing (Cooles *et al.* 1986) allows the comparison of down-hole geochemical Rock–Eval data for source rocks to be compared to the values for an immature equivalent of the source rock (an analogue). The petroleum generation index (the amount of hydrocarbons that have been generated from the source rock) and petroleum expulsion efficiency (the amount of the generated petroleum that has migrated out of the source rock into the carrier system) are calculated and may be used to determine volumes and masses of hydrocarbons generated, independently of any assumed thermal history. The mass balance technique is limited in its usefulness because wells are only rarely drilled into kitchen regions.

Hydrocarbon generation models

The model, in this context, is the mathematical formulae which describe the type of source-rock kinetic distribution. An analogue comprises the actual values and frequencies that fall under the specific distribution type (model).

There are four genres of models – discrete, compositional, gaussian and multiple reaction schemes – which are all based on the 1888 Arrhenius equation.

The discrete model consists of a series of parallel reactions (Fig. 16) with a single Arrhenius factor (pre-exponential factor) and a distribution of activation energies. Each activation energy has its own frequency.

The compositional model consists of a series of three or four discrete distribution reactions, each of which represents the reaction of a carbon fraction. There are two variations of this model. Ungerer, (1989) uses the carbon fractions $C_{15}+$, C_6–C_{14}, C_2–C_5, C_1. An example of an application of this model is given by Forbes *et al.* (1991). Braun & Burnham (1991) use C_5+, C_2–C_4, C_1, as their carbon fractions. Each fraction has its own activation energy distribution, initial petroleum potential (mg HC g^{-1} initial TOC) and pre-exponential factor.

A Gaussian oil generation model simply refers to the method of describing the distribution of activation energies (Ea). A Gaussian normal distribution is described by a mean activation energy and its standard deviation. The area under the distribution is the initial potential. Different source rocks will have different initial potentials, different mean Ea and standard deviations.

The Quigley *et al.* (1987) oil-generation model consists of two normal distributions of activation energies, each with its individual mean and standard deviations. The two distributions represent the labile and refractory parts of the kerogen. The lower mean activation energy distribution is for the labile component which converts to oil, the higher distribution corresponds with the refractory part of the kerogen which converts to gas.

Analogues for this model comprise a refractory: labile ratio. A review of the model is given in Mackenzie & Quigley (1988) and Quigley *et al.* (1987). Gas may be generated from previously generated oil or from the refractory part in the kerogen.

Gas-generation models are in the same form as oil-generation models: discrete, gaussian and compositional.

A compositional gas-generation model has a series of breakdown reactions for the various molecule chain lengths. The compositional gas-generation model can account for several cracking reactions occurring in parallel as set out by Ungerer (1989); Forbes *et al.* (1991), Behar *et al.* (1988), and also by Braun & Burnham (1991). In gas-generation models, a reactant gives rise to a set of products (which defines the model) in given proportions which are unique to a specific oil (therefore defines the analogue). In one multiple model there are a series of reactant-to-product reactions.

Two types of composite gas generation models are available. The IFP model takes the C_{15+} fraction and breaks it down into C_6–C_{14}, C_2–C_5, C_1 and coke in varying proportions (depending on the oil analogue). This reaction has a specific Arrhenius factor (A) and activation energy (Ea). The C_6–C_{14} fraction also breaks down to the C_2–C_5, C_1 and coke, again in varying proportions and at a specific A and Ea. The C_2–C_5 breaks down to C_1 and coke in proportions and with a unique reaction rate parameter (A and Ea). The C_1 degrades to coke also at a specific reaction rate determined by unique parameters. Analogues are created by varying the proportions of the products produced and possibly also the reaction parameters.

The LLNL model has two parts to the reaction: an oil to $CH_x + CH_4 +$ coke in varying proportions with specific rate parameters; and a reaction of CH_x to $CH_4 +$ coke. Analogues

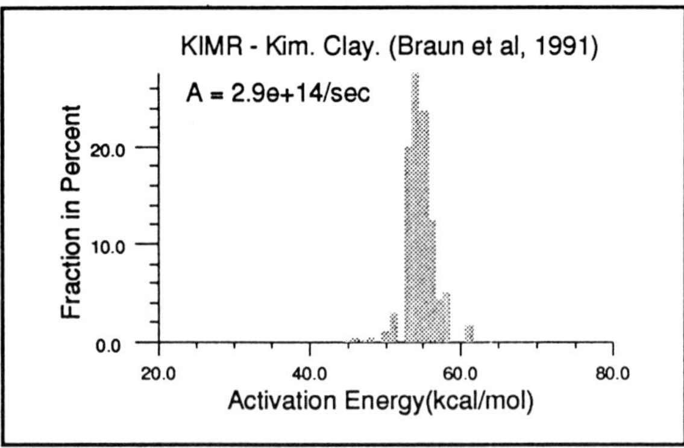

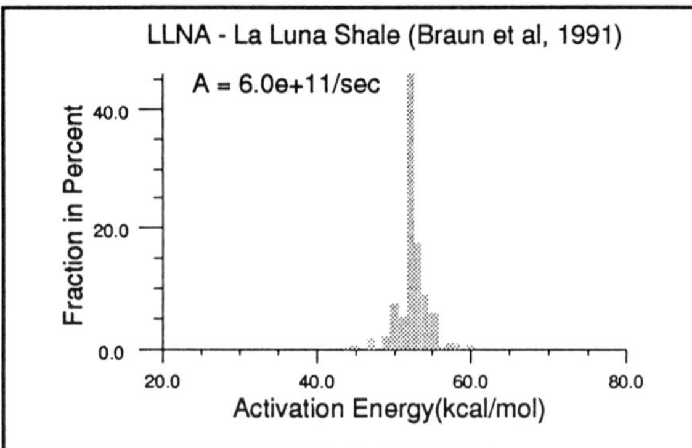

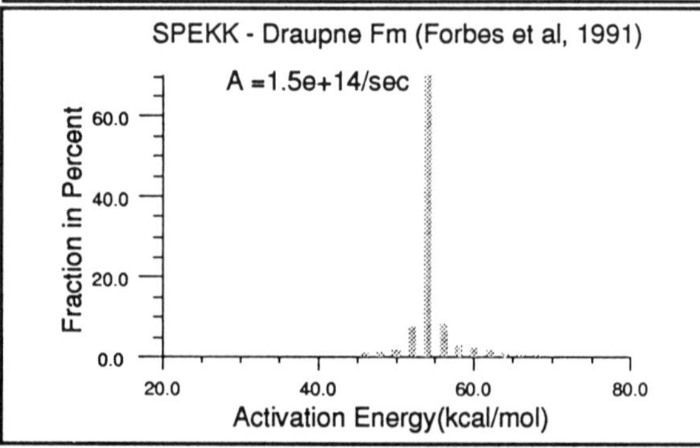

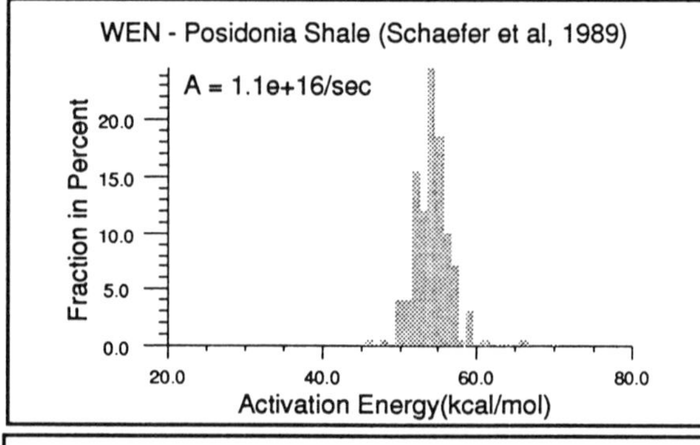

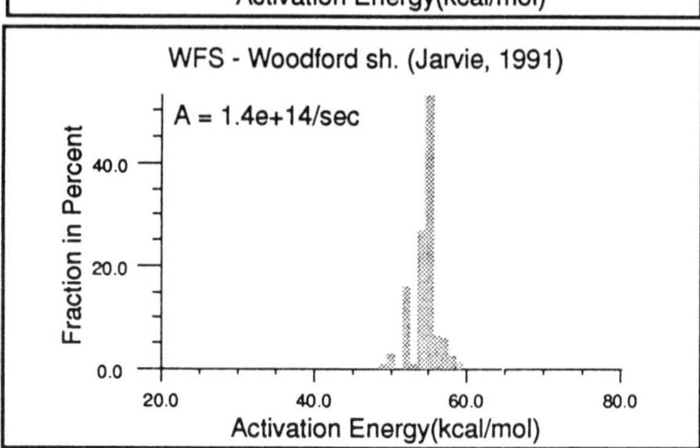

Fig. 16. Source-rock kinetic activation energy distribution for the discrete model.

are created by varying both the reaction parameters and the proportions of each product produced.

The multiple reaction scheme is a highly specific model of some temperature-dependent series of reactions and products (Braun & Burnham 1991). The reproducibility of the series of parameters, and the question of the generality of such unique and complex models with so many parameters to be determined, currently limit their routine use in practical exploration modelling.

Source-rock kinetics and transformation ratio

The use of source-rock kinetics has been one of the most important advances in basin modelling in the past few years. Source rocks appear to have very distinctive breakdown characteristics, which can be measured in the laboratory for simulation in models of geological time and temperature histories. An analogue will have a description containing information about source-rock and lab-determined parameters. If a discrete model is used, then a series of activation energies, each with a frequency, is assigned with a single pre-exponential number.

A sensitivity analysis of the published discrete activation distribution kinetics had the rather disturbing result of showing that there is as much variation within source rocks of the same traditional type as there are between the types themselves (Patel 1993). A comparison of the results of several type II kerogens that have all experienced the same temperature history is shown in Fig. 17. The implication of this result in any basin analysis is that it is not enough to suggest that the source rock is just, for example, type II. The approach should be to investigate several viable source-rock analogues and the sensitivity of the final result (the TR) noted.

The use of such analogues is greatly facilitated by maintaining libraries of these in a database, classified in a convenient cross-catalogue manner, such as geographical location, source-rock type and depositional setting.

The specific kinetic analogue used in any modelling will impact not only on the final maturity of the well or pseudo-well, but will also control the area of apparent maturity. This effect can be radical in bowl-shaped basins or reasonably flat-lying kitchen areas, with serious implications for the charge.

The kinetics of oil-cracking and gas-sublimation are very controversial. Figure 18 shows the results of using various published gas-cracking kinetic examples, and their impact on the resulting expelled gas:oil ratio (GOR kg/kg). Depending on the kinetics adopted, it is possible to get widely varying GOR values. This indication affects estimates of phase and volumetric charge calculations both for the kitchen and in the carrier trap systems (see below). Wherever gases are involved, the effects of pressure must affect any chemical processes. The kinetics of oil-cracking can have a significant input on economic predictions, and risk assessment.

Another interesting concept for kinetic modelling is the labile versus refractory kerogens, or the existence of gas-prone matter in the kerogen (Mackenzie & Quigley 1988). In measuring a kinetic analogue, and producing a kinetic distribution (albeit from a complex mathematical treatment) the total pyrolysis curves are taken into account. The sample is processed and what remains is just inert residue. During the total pyrolysis process, the labile, and the refractory and the oil-cracking kinetics, are all encapsulated in a single distribution of activation energies at a corresponding Arrhenius factor. The refractory amount may be estimated using pyrolysis-gas chromatography and taking the ratio of C_1

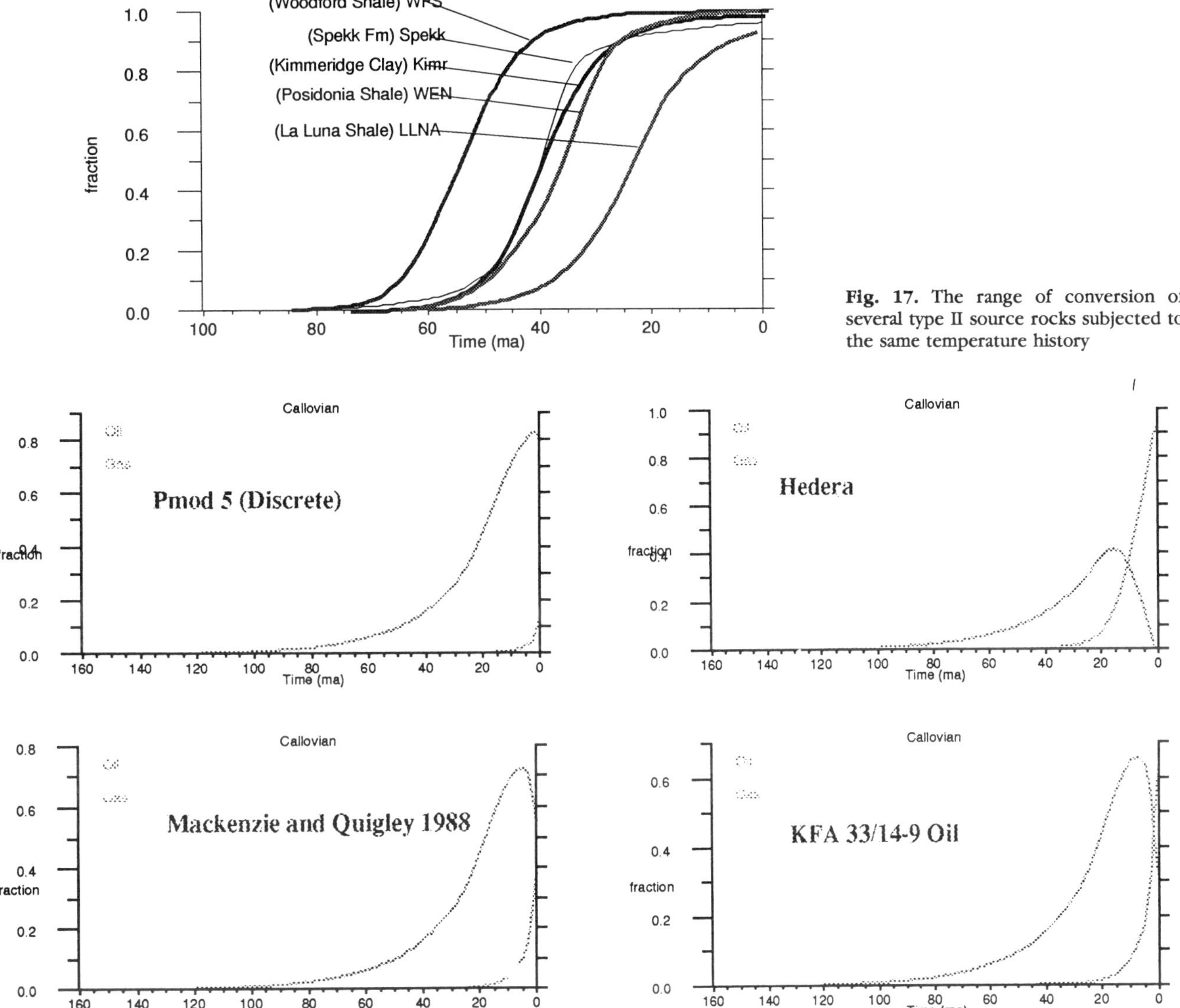

Fig. 17. The range of conversion of several type II source rocks subjected to the same temperature history

Fig. 18. The range of mass gas : oil ratio results for different oil cracking kinetic analogues. Each run has used the same temperature history and kerogen to oil generation kinetic analogue.

–C_5 molecules over the pyrolysis yield (Mackenzie & Quigley 1988), but it is difficult to compromise the Quigley *et al.* (1987) Gaussian labile and refractory kinetic schemes with the diverse measured kinetic signatures that characterize different source rocks. The answer probably resides in the actual mathematical optimization techniques for discrete kinetics, which do not favour more than one population of activation energies in a single distribution.

Expulsion

Expulsion (primary migration) from the source rock has been discussed and reviewed by several authors (Mackenzie & Quigley 1988; Duppenbecker & Welte 1991; Hermanrud 1993; Ungerer 1993). Microfracturing (e.g. Tissot & Pelet 1971; Meissner 1978), diffusion through networks of organic matter (Stainforth & Reinders 1989), pressure-driven flow through kerogen networks (McAuliffe 1979) and pressure drive controlled by Darcy's law for multiphase fluid flow (Ungerer *et al.* 1987) are the basic mechanisms presented.

However, most basin-modelling applications, use either saturation threshold methods or maturity-based calculations. Comments from Ungerer (1993) underlined the importance of further research into thermal dynamics, not only of liquid to gas, but also solid to liquid in source rocks. The excess pressure produced by hydrocarbon generation will add further drive-pressure to primary migration (Nakayama & Lerche 1987). This is not commonly incorporated into modelling schemes and has not yet been analytically quantified.

The saturation threshold (ST) is the amount of water-filled pore volume that must be displaced by oil to cause oil to leave the source rock and move to the nearest carriers. This ST value has been postulated to be typically 20% to 80%. It is possible to calibrate ST with Rock Eval S1 data (Ungerer 1993) provided that the average porosity and TOC of the layer is known. Expulsion takes place when the volume of oil generated fills the void spaces in a source layer to greater than the saturation threshold. The ST may vary with maturity and the volume of pore space varies with burial. The saturation threshold method is described in MacKenzie & Quigley (1988) and used in Forbes *et al.* (1991). The actual volume of liquid and gas is also controlled by the thermodynamics of the fluids (Dickey 1975; Welte & Yukler 1981; Durand *et al.* 1987; Leythaeuser *et al.* 1987); although this is not included in most basin models.

Expulsion efficiency is one of the principal controls on the gas:oil ratio of expelled products. Poor expulsion promotes the occurrence of high gas:oil ratios, since the oil that remains in the source rock will be cracked to gas at higher temperatures. A high expulsion efficiency reduces the oil left in the source rock for a late gas charge. There always appears to be some refractory gas potential in every source rock, which burns off at greater temperatures than those over which the cracking of oil is postulated to occur, thereby providing a late flushing of play fairways by gas.

Recently, it has been proposed that relative-permeability curves for source rocks are a function of both grain size and porosity (Okui & Waples 1993 : Okui *et al.* in prep). Relative-permeability curves derived from reservoir rocks may not be appropriate for the very fine-grained source rocks. Okui *et al.* (in prep.) suggest a dynamic saturation threshold mechanism, with initial expulsion occurring at very low saturations.

There is well documented evidence that relates expulsion efficiency empirically to source-rock initial potential and maturity (e.g. Cooles *et al.* 1986; Cornford 1990), but the processes are not fully understood. In the mean time, bulk methods and calibration must be the *modus operandi.*

SECONDARY MIGRATION

The principal driving forces for secondary migration through carrier systems are buoyancy, (the difference in densities between gas, oil and water) and pressure gradients. The resistance to these drives is capillary pressure, which is a function of pore-throat radius, interfacial surface tension and wettability of the petroleum–water–rock (e.g. Allen & Allen 1990). This system can be unbalanced by hydrodynamic forces (such as artesian fluid flow), which can promote or retard migration, depending on the force directions. The principles of secondary migration are set out in Hubbert (1953); Gussow (1954); Berg (1975); and Schowalter (1979).

Several authors have modelled secondary migration in two dimensions, (Bethke 1985; Nakayama & Lerche 1987; Welte, 1987; Ungerer *et al.* 1990; Person & Garven 1992). Three-dimensional approximations have also been attempted (Sylta 1991, 1993; Hermans *et al.* 1992; Taylor & Bandurski in prep).

Crucial parameters are discussed by Hermans *et al.* (1992), and include the carrier properties, seal properties and the retention factors (see Hermanrud 1993 for discussion). Seal properties have been a major source of risk in hydrocarbon exploration (Hermanrud 1993).

Hydrocarbons migrate orders of magnitude quicker (Hermanrud 1993) than does water produced by compaction. North (1985) indicated migration rates of 50–1200 m a^{-1}, and that distances can be immense (e.g. Dow 1977). Hermans *et al.* (1992) calculates hydrocarbon flow rates on the order of 0.1 m a^{-1} to 100 m a^{-1} depending on the carrier bed permeability.

The thermodynamics of the fluid properties during secondary migration are usually overlooked in models. The bubble point (the pressure and temperature at which gas exsolves from the petroleum fluid) expressed as a depth is

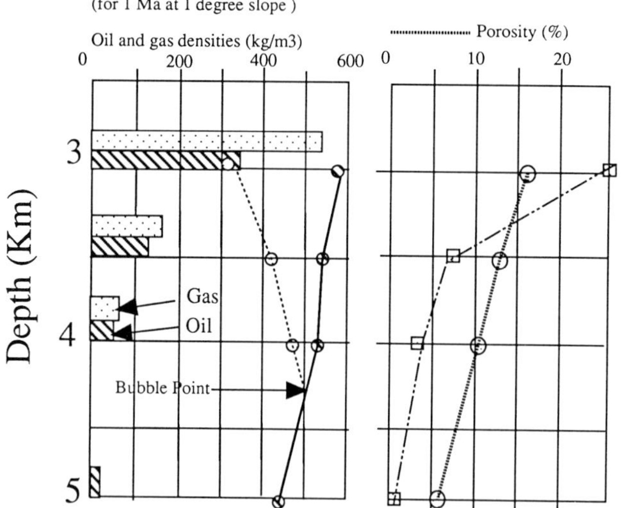

Fig. 19. Diagram showing the change in migration rate due to a change in dip at different depths and consequently different porosity and permeability values.

crucial to understanding the buoyancy pressure and the migration rate.

Most models assume a constant oil density of 800 kg m^{-3}. The *in-situ* density of oil as it migrates from source to carrier to trap, typically between 4500 m to 2500 m, is probably in the order of 400 to 600 kg m^{-3} . Depending on the generation and expulsion characteristics of the source rock and the pressure and temperature conditions, gas may not exist as a separate phase until 3000 m depth. The consequences of this are twofold. Firstly, a wider range of fine-grained rocks can perform as effective seals since the buoyancy pressure of the fluid is less than the equivalent gas column. Secondly, there will be an effective depth at which gas appears and results in an increase in buoyancy pressure on available seals, particularly in areas where columns can accumulate.

To simplify the above statements, migration in the deeper strata is likely to be more lateral along beds with effective seals, whereas above the bubble point depth, more vertical migration through seals is likely, particularly of the new gas phase.

The change in density of oil or gas and the bubble point line for a typical Norwegian Draugen Field hydrocarbon (40 API; Hermans *et al.* 1992) is shown in Fig. 19, derived using a geothermal gradient of 30°C km^{-1} and a hydrostatic pressure gradient of 11 MPa km^{-1}. The sensitivities of the bubble point curve to oil and gas densities and GOR is shown in Fig. 21.

Consideration of fluid thermodynamics has obvious applications to prospect appraisal and seal efficiency (Sales 1993). Two key criteria in such analyses of secondary migration and seal effectiveness are the prediction of porosity and permeability with depth, and the geometrical model used to relate porosity or permeability to pore-throat radius in the determination of capillary pressure. Secondary migration is very sensitive to lithological distribution and behaviour, which must come from further analysis of seismic data and depositional or stochastic predictive models.

The factors affecting the rate of migration are again porosity, fluid viscosity, permeability and the gradient of the strata. The particle-transport-rate equation given by Hermans *et al.* (1992) is, in part, evaluated in Fig. 19. This shows very

Table 2. Possible range of parameter values for equation (1).

Parameter	Min.	Max.	Unit	Determined	Constraints	Reducibility	Range	Relative importance
TR	0-	1	Fraction	Modelling	Model parameters	±0.2	40%	1
HI	0	700	kg/tonne	RE data	Statistics	±50	13%	6
TOC	1	10	% rock wt	RE data	Statistics	±1	20%	5
EFF_{th}	5	500	m	Seismic wells	Interpretation	±50	20%	4
ρBulk	2200	2750	$kg\,m^{-3}$	Modelled	Model parameters	±100	10%	7
A	1	*n*	km^2	Model mapping	Geometry + model parameters	±?	100%	2
EE	10	95	% HC generated	RE Data	Statistical analogues	±20	40%	3

conclusively that the rate is controlled primarily by the dip of the strata and secondarily by the porosity and permeability. At depths greater than 4 km (at low values of porosity and permeability depth using this specific behaviour) it can be seen that the rates are significantly curtailed. Bed dip in the range of 4 km to 2.5 km depth is crucial.

The migration rates for oil shown on Fig. 19 are at worst 8 km Ma^{-1} at 5 km depth at 1° dip and at best 1793 km Ma^{-1} at 3 km depth at 5° dip. Typical rates at 4 km to 3.5 km (in carriers adjacent to mature source rocks) are in the order of 50–100 km Ma^{-1}. In basin simulation time-scales this process is virtually instantaneous. The key criteria for effective secondary migration will therefore be the structural reconstruction at times of peak generation and expulsion, and the behaviour of seal, unconformities and faults. The charge amount, composition and timing will be a very important limiting factor.

MODELLING PARAMETER SENSITIVITY

Assessments of hydrocarbon potential will vary in their approaches and the questions being presented depending on the type of basin being evaluated, and the availability and type of data. In every assessment, the tenets of source, reservoir, seal, trap, maturity and charge, must be tested. The absence of a source horizon is critical. Similarly, a lack of reservoir strata will also seriously downgrade a region. Given the explorationist's imagination and the large array of plays to consider, it is unlikely that a lack of traps will be a determining factor for downgrading a region; this occurs at the prospect level. The question of maturity is often a key factor.

Source-rock maturity seriously affects the charge, either volumetrically or compositionally (e.g. Forbes *et al.* 1991). Charge into prospects is the result of the predicted masses generated to the migration pathways through time. Charge into prospects is important in that one requires enough to fill prospects and, with the fill/spill history, may control trap volumes. Sales (1993), however, presents a good discussion of the other criteria that affect prospect composition.

The process of carrying out a basin modelling study should be intrinsically tied to the interpretation work. Models often require certain geological events, such as unconformities to match the data. It is important that the details of these key events are not lost in the interpretation. For instance, it is the last few million years of basin subsidence that are exerting the greatest effect on maturity of source rocks. This key subsidence information is contained within the uppermost seismic reflectors and is often not explicitly part of the interpreters' normal mandate.

The interpreters' estimation of lithological distribution, both lateral and vertical, is also very important for the basin modeller, particularly in the delineation of aquifers and aquitards. This will control secondary migration and predicted pressure regimes, overpressuring and underpressuring (Mann & Mackenzie 1990).

Prospect delineation will also allow the models to concentrate on the key prospective migration routes. Models can then provide, not only the charge estimates, but also pressure, good estimates of porosity, permeability and temperature both within prospects and the sealing strata. With the provision of geometries and reservoir criteria like net:gross from detailed interpretation, it is possible to indicate a range of likely trap volumes and compositions, using PVT correlations (e.g. Glasø 1980) calibrated to any DST oils or to analogous source-rock oil data. All of these criteria can be used to risk the region or prospect.

The methodology and the assessment of the critical parameters that effect the overall charge of a region or prospect is discussed in more detail.

Determining masses of hydrocarbons generated and expelled

The calculation for estimating the amount of hydrocarbons that a kitchen region has generated and expelled is a simple multiplication (see Mackenzie & Quigley 1988):

$$ME = TR \times HI \times TOC \times \rho Bulk \times EFF_{th} \times A \times EE, \quad (1)$$

where TR = transformation ratio, HI = hydrogen index, TOC = total organic carbon, EFF_{th} = effective thickness of source rock, ρBulk = bulk density, A = area, EE = expulsion efficiency, and ME is the mass expelled.

The approximate possible ranges of values for all the parameters in this equation are shown in Table 2.

Although TR does not have the greatest range, it has the most critical effect since TR controls the mature area (A). Transformation ratio is derived from all the basin modelling assumptions and mathematical simulations, so the critical unknown criteria for mass calculations are those factors which affect transformation ratio. The source rock kinetics control TR, and kinetics are a function of time and temperature only. For accurate times, explorationists require the best possible stratigraphic ages.

The parameters that affect the temperature history of the source rocks are listed in Table 3. The first three parameters in Table 3 are lithological parameters and, along with the depth history, are simulated in the modelling. All these parameters may be constrained by data and appropriate calibration. Porosity, pressure, permeability, lithology and age data are used to calibrate the depth history and porosity.

Table 3. Factors affecting the temperature history of rocks

Factor	Min.	Max.	Units	Determined	Constraints
Thermal conductivity	1.2	6.0	$W m^{-1}$	Model	Data/model/analogues
Specific heat capacity				Model	Data/model/analogues
Porosity	1	70	%	Model	Log/core data
Heat flow history		?		Estimate	Calibrants
Surface temperature history	–25	40		Estimate	Palaeocene, water depth, latitude
Depth history	25	?		Model	Model assumptions
Present-day heat flow		200	$mW m^{-2}$	Model	Model parameters

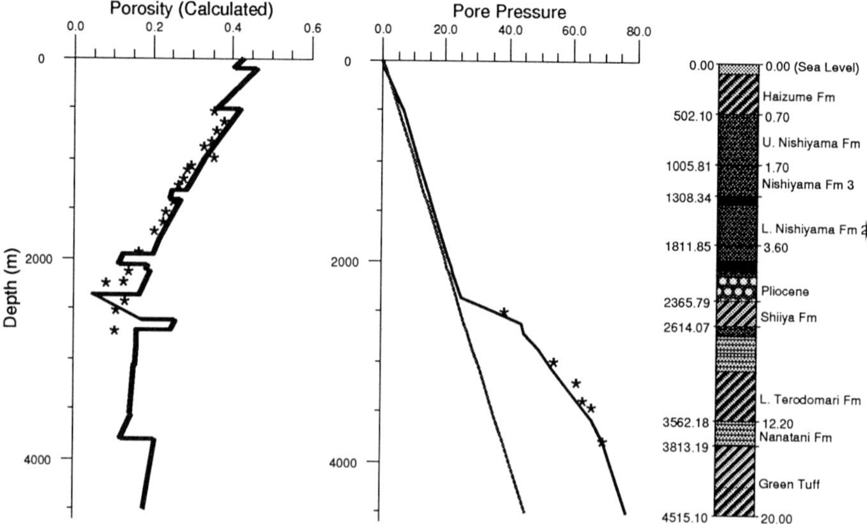

Fig. 20. Pressure and porosity calibration taken from well Iwafune Oki X (Nakayama 1987).

Figure 20 shows just such a calibration using one effective stress porosity behaviour (Audet & McConnell 1992) and permeability model (Olsen 1960), though these are by no means the only methods available.

Determining charge volume and phase

After the masses of hydrocarbons generated in a kitchen have been predicted through modelling and mapping, the questions posed are: is there enough charge to fill a prospect or lead; what is the likely hydrocarbon composition within the prospect?

Forbes *et al.* (1991) illustrated a good method for evaluating the first question using the Haltenbanken, Smørbukk Sør Field. Mackenzie & Quigley (1988), drawing on the work of Glasø (1980), England (1989) and Price *et al.* (1983), set out a method for determining the volumetrics at kitchen and at prospect.

The sensitive criteria in a volumetric charge calculation are investigated following the general Mackenzie & Quigley (1988) approach. The variables in a volume calculation are set out in Table 4.

The phase of hydrocarbon entering a trap will be contingent on the GOR, the API and gas gravity, for a specific pressure and temperature condition. These sensitivities are shown in Fig. 21. Obviously, the trap fullness is sensitive to the initial mass expelled towards the prospect, the trap capacity and any losses during migration.

Table 4. Parameters for volume calculations

	Sensitivity of phase
Kitchen pressure	4
Kitchen temperature	4
Kitchen gas : oil ratio (mass)	3
Prospect pressure	3
Prospect temperature	4
Prospect capacity	1
Oil density at surface ($kg m^{-3}$)	2
Gas density at surface ($kg m^{-3}$)	2
Mass expelled to prospect (kg)	1
Losses during migration (kg)	2

1 is Most Sensitive
4 is Least Sensitive

Uncertainties in the kinetics of the unproven and partially expired source rocks, and in the mechanism of expulsion make speculation of the GOR very subjective. Although possible, modelling GOR of the migrating fluids with a full equation-of-state approach has little practical meaning. A simpler approach that allows a wider ranging investigation of the results may be more practical.

Mackenzie & Quigley (1988), suggested that migration loss may be considered loosely as a function of carrier volume and residual saturation, 2% being the preferred residual saturation value. Forbes *et al.* (1991), suggested that the residual

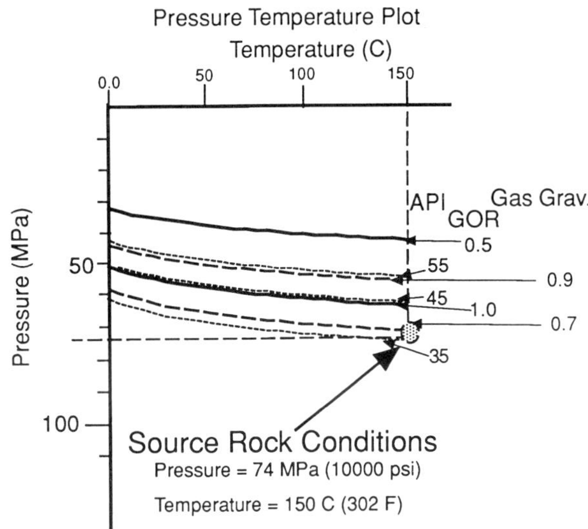

Fig. 21. Oil and gas density and GOR effects on the bubble point of petroleum at various pressures versus temperature conditions.

saturation lies between 1 and 3% and they had variations for each compositional hydrocarbon fraction. Both these groups of authors agree that residual saturation is a difficult parameter to constrain and calibrate, the danger being that it could be used as a useful free parameter itself for other 'calibration' purposes!

The phase of the hydrocarbons entering the trap is only part of the control on the composition in a prospect. The other criteria, which are a function of the trap character and geometry, are beyond the scope of this paper (see Sales 1993; Berg 1975; Downey 1984).

To summarize, estimating trap fullness and composition is very sensitive to several parameters that are poorly constrained. Critical parameters are those determining transformation ratio, namely source-rock kinetics, heat-flow history and thermal conductivities. Oil-to-gas cracking kinetics are critical to the determination of GOR, as is the expulsion criterion. Cracking kinetics seem to be a weak point in our understanding at present. Other parameters can be estimated or calibrated from nearby analogous situations, and have less of a bearing on the results.

BASIN MODELLING AND INFORMATION TECHNOLOGY

Despite shortcomings, basin modelling has proved to be of significant worth at various stages in exploration strategies, but there is significant room for improvement. However, rapid developments in both hardware and software engineering techniques implies that much of the present computer code will suffer from primitive legacy inertia.

Hardware developments

The hardware technology available to the pioneers of basin modelling science was limited in its capability to support anything more than the underlying calculations themselves. Typically, basin modelling programs ran on mainframe computers in batch mode, and output was in the form of post-processed hardcopy graphics. The appearance of desktop computers in the late 1970s and early 1980s, however, opened new opportunities in software development, allowing more widespread access to hardware. As a consequence, it facilitated the implementation of the simpler techniques of modelling outside of the mainframe environment, although the more numerically intense methods still required mainframe support.

Since the initial emergence of personal computers, the capabilities of these machines has dramatically increased, whilst in parallel there has been the development of desktop workstations that can support the multi-tasking and networking facilities previously found only on mainframe computers. These developments, in conjunction with evolving techniques of software development, enabled the delivery of the first commercial versions of basin modelling software for desktop computers in 1988. Most basin modelling software now operates either on personal computers or on their more powerful workstation counterparts, with the underlying hardware rarely being a limitation to the implementation of the numerical modelling.

At present, there is a practical separation between simpler, largely 1D models, implemented on personal computers, and the more numerically intense 2D and 3D models operating on workstation platforms. However, future developments in personal computing technology and the development of multi-tasking operating systems for such platforms is likely to result in the progressive erosion of this distinction.

Software developments

Of perhaps greater importance to the recent evolution of basin modelling software have been the changes in software design, development environments, and practice. Key developments in user interface technology, together with new paradigms of software development, have fundamentally altered the appearance and operation of modelling software.

Graphical user interfaces.

The most fundamental change in the appearance and operation of geological modelling software has been the development of graphical user interfaces. Initially developed for personal computers in the late 1970s and 1980s, such interfaces are now a *de facto* standard on all desktop computers and workstations. Usually designed to encourage user interaction with the software, using a pointing device as well as a keyboard, such interfaces provide methods for data input, operational control and visualization of the output as an integral part of the modelling operation.

This contrasts significantly with early modelling software which may now, in retrospect, best be described as programs whose function was to process pre-defined modelling parameters. Output in anything other than numerical form was generated by post-processor software, whilst input usually took the form of input files or command line input of parameter before processing commenced. Although this pre-processor/post-processor design of basin modelling software is still apparent in some packages, it is now more common for the software to operate interactively and incrementally using the graphical user interface.

A major recent advance in scientific data analysis has been the three-dimensional visualization of data. As yet not extensively utilized in basin modelling software, advanced computer graphics packages such as AVS provide facilities for the visualization, manipulation and animation of spatial data structures rendered in three dimensions. Future basin modelling packages will, undoubtedly, use such displays for the analysis of, for instance, maturity distributions and for the detailed analysis of migration pathways.

Software development practices

Software design is driven by user requirements and limited by hardware, budget and time.

Underlying the evolution of the appearance of software have been significant changes in the practice of software development and programming languages. Early modelling software was implemented as a sequence of numerical instructions and procedures, most commonly using Fortran.

Modern software is significantly more complex, and has required the use of a more structured approach to design and development.

A major recent evolution in software development practice has been the introduction of object-orientated techniques which have significantly extended the concepts of modular design and library re-use. Benefits of the object-orientated approach are better modularization, code re-use and rapid implementation.

Data management and exchange

The development of basin modelling software to date has largely taken place independently of other exploration software systems. However, the analysis undertaken using basin modelling software is dependent on the interpretations now routinely made using, for instance, seismic interpretation software. As a consequence there is an increasing need, particularly in exploration companies, to be able transfer data between systems.

A perceived solution to the data management problems has been to establish common data exchange formats and data models. Properly implemented, these define a common data structure and interfaces allowing information to be exchanged or shared between applications in a pre-defined way.

BASIN MODELLING – A VISION OF THE FUTURE

An integrated system supported by database technology

A vision of the future basin modelling system is as part of a fully integrated interpretation system supported by a fully functional common database. The system will encompass work-group computing and project-based basin assessments, with all raw data, interpretations, and versions thereof available to all domain specialists within the team. An example might be that basin modellers who require details of porosity–depth trends from logs can simply access the relevant computer-processed interpretations (CPIs) and place the data into the basin simulator for calibration. Feedback from well simulations at an early stage of interpretation may impact velocity models for time:depth conversion particularly in regions of inversion. The detailed sensing of the strata, which are acting as effective seals in a well's stratigraphy is crucial to understanding the development of overpressure and determination of migration surfaces just beneath these sealing formations. The level of detail available and used will approach that of field development. Modellers will be able to zoom quickly from operating at a basin-assessment scale into a prospect-charge scale using the same data, allowing focused interpretations and simulations.

Another aspect of the database function is to provide a central repository for electronic libraries of key data, be it source-rock kinetic analogues, reservoir analogues or thermal conductivities.

Yet another key domain for future basin modelling will be the coupling of seismic interpretation with forward seismic-sequence simulation. Building simulations that take a seismic interpretation, and given basic information about the ages of various units, will forward model the interplay of subsidence, uplift, deposition and sea-level. Basin modelling to assist the interpretation will be routine, just as structural reconstruction software presently constrains the interpretations of structural geologists. Detailed sequential deposition and palaeoenvironmental information will help to predict seismic attributes, and eventually may provide a basis for a more refined geological model for fluid-flow modelling. The approach based on horizontal sections will interface with a full 3D simulation, which will impinge on basin hinterlands.

The closer integration of interpretation and sequence simulation may breed the next generation of geostatistical models that predict the distribution of reservoir and seal properties over prospects.

Future functionality

It is immediately attainable to read CPIs and geological surface information into a basin simulator to create a data volume. Technology exists to allow the manipulation and visualization of the 3D volume, with wiremesh and various surface highlighting facilities, sectioning and full style and colour customizations. Animation of results in different views should be anticipated. The basin simulator data model must be able to simulate unconformities, honour complex geometries and their development through time, lateral facies variation, halokinesis, faulting and intrusions, perhaps pausing at meteoritic impact.

Scientific expectations and intangibles

The future in regional basin modelling lies in the continued strident research into seismic attribute analysis to determine rock thermal and mechanical properties. A rather dated technique using 1975 seismic data has recently been shown to have a good degree of success (Ho *et al.*, in press), provides some encouragement that the present-day quality of seismic source and reflection processing should reduce some of our unknowns.

What is not going to be available soon is a simulator that predicts diagenetic changes to rocks. Again the fundamental analytical work is ongoing and the chemical systems very complex.

As key supporting technologies develop in terms of seismic and well log attribute analyses and the incorporation of forward seismic stratigraphy simulation, basin modelling will become part of the interpretation phase, and predictions will be at a higher resolution, moving closer towards the field scale of operation. We will find that the voids in our understanding of fine-grained rocks and fluid–rock interactions are the fetters to development of effective more dynamic fluid flow simulators.

REFERENCES

ALLEN, P. A. & ALLEN, J. R. 1990. *Basin Analysis: Principles and Applications.* Blackwell Scientific Publications.

ATHY, L. F. 1930. Density, porosity, and compaction of sedimentary rocks: *AAPG Bulletin*, **14**, 84.

AUDET, D. M. & MCCONNELL, J. D. C. 1992. Forward modelling of porosity and pore pressure evolution in sedimentary basins. *Basin Research*, **4**, 147–162.

BALDWIN, B. L. & BUTLER, C. O. 1985. Compaction Curves. *AAPG Bulletin*, **69**, 622–626.

BARKER, C. 1972. Aquathermal pressuring role of temperature in development of abnormal-pressure zones. *AAPG Bulletin*, **56**, 2068–2071.

BEHAR, F. P., UNGERER, P., AUDIBERT, A & VILLALBA, M. 1988. Experimental study and kinetic modelling of crude oil pyrolysis in relation to the thermal recovery process. *Preprint of the Fourth UNITAR/UNDP Conference on Heavy Crude and Tar Sands*, **5**, paper 97, 1–14.

BERG, R. R. 1975. Capilliary Pressures in Stratigraphic Traps. *AAPG Bulletin*, **59**, 939–956.

BETHKE, C. M. 1985. A numerical model of compaction-driven groundwater flow and heat transfer and its application to paleohydrology of intracratonic sedimentary basins, *Journal of Geophysical Research*, **90**, 6817–6828.

——— 1986. Inverse hydrologic analysis of the distribution and origin of gulf coast-type geopressured zones. *Journal of Geophysical Research*, **91**, (B6), 6535–6545.

BRAUN, R. L. & BURNHAM, A. K. 1987. Analysis of chemical reaction kinetics using a distribution of activation energies and simpler models. *Energy & Fuels*, **1**, 153–161.

——— & ——— 1990. Mathematical model of oil generation, degradation, and explusion, *Energy & Fuels*,

——— & ——— 1991. 'PMOD: A Flexible Model of Oil and Gas Generation, Cracking, and Expulsion'. Paper presented at the 15th International Meeting on Organic Geochemistry Manchester, England, September 16–20, 1991.

——, ——, REYNOLDS, J. G. & CLARKSON, J. E. 1991. Pyrolysis kinetics for lacustrine and marine source rocks by programmed micropyrolysis. *Energy & Fuels*, **5**.

BRIGAUD, F. & VASSUER, G. 1989. Mineralogy, porosity and fluid control on thermal conductivity of sedimentary rocks. *Geophysical Journal*, **98**, 525–542.

BRYANT, S. & BLUNT, M. 1992. Prediction of relative permeability in simple porous media. *Physical Review*, **A46**, 2004–2011.

———, CADE, C. & MELLOR, D. 1993. Permeability prediction from geologic models. *AAPG Bulletin*, **77**, 1338–1350.

BURLAND, J. B. 1990. On the compressibility of and shear strength of natural clays. *Geotechnique*, 40. 329–378.

BURNHAM, K. & BRAUN, R. L. 1989. Development of a detailed model of petroleum formation, destruction, and expulsion from lacustrine and marine source rocks. *Advances in Organic Geochemistry*, **16**, (1–3), 27–39.

——— & SWEENEY, J. J. 1989. A chemical kinetic model of vitrinite maturation and reflectance. *Geochimica et Cosmochimica Acta*, **53**, 2649–2657.

CAO, S. 1985. *A quantitative dynamic model for Basin Analysis and its Application to the Northern North Sea Basin*. Masters thesis, University of South Carolina, USA.

——— & LERCHE, I. 1987. Geohistory, thermal history and hydrocarbon history of the northern North Sea Basin, *Energy Exploration & Exploitation*, **5**, 315–355.

CAPUANO, R. M. 1993. Evidence of fluid flow in microfractures in geopressured shales. *AAPG Bulletin*, **77**, 1305–1314.

CARLSON, W. D. 1990. Mechanisms and kinetics of apatite fission track annealing. *American Mineralogist*, **75**, 1120–1139.

CLARK, S. P. 1966. *Handbook of Physical Constants*, Geological Society of America, Memoir, **97**.

COOLES, G. P., MACKENZIE, A. S. & QUIGLEY, T. M. 1986. Calculation of petroleum masses generated and expelled from source rocks. *Advances in Organic Geochemistry*, **10**, 235–245.

CORNFORD, C. 1990. Source rocks and hydrocarbons of the North Sea. *In:* Glennie, K. W. (ed.) *Introduction to the Petroleum Geology of the North Sea*, third edn, Blackwell Scientific Publications, 294–361.

CORRIGAN, J. D. 1993. Apatite fission-track analysis of Oligocene strata in South Texas, USA: testing annealing models. *Chemical Geology* (Isotope Geoscience Section) **104**, 227–249.

CROWLEY, K. D. 1993. LENMODEL: A forward model for calculating length distributions and fission-track ages in apatite. *Computers and Geosciences*.

DAHL, B. & AUGUSTSON, J. H. 1993. The influence of Tertiary and Quaternary sedimentation and erosion on hydrocarbon generation in Norwegian offshore basins. In: Doré, A. G. *et al.* (eds). *Basin Modelling: Advances and Applications*. Norwegian Petroleum Society, Special Publication 3 Elsevier, Amsterdam, 419–431.

DEMONGODIN, L., PINOTEAU, B., VASSEUR, G. & GABLE, R. 1991. Thermal conductivity and well logs: a case study in the Paris basin. *Geophysical Journal International*, **105**, 675–69.

DICKEY, P. A. 1975. Possible migration of oil from source rocks in oil phase. *AAPG Bulletin*, **59**, 337–345.

DOW, W. G. 1977. Kerogen studies and geological interpetations. *Journal of Geochemical Exploration*, **7**, 9–99.

DOWNEY, M. W. 1984. Evaluating seals for hydrocarbon accumulations. *AAPG Bullein*, **68**(11), 1752–1763.

DUPPENBECKER, S. J. & WELTE, D. H. 1991. Petroleum expulsion from source rocks–insights from geology, geochemistry and computerized numerical modelling. *In: Proceedings of the Thirteenth World Petroleum Congress, Buenos Aires*.

DURAND, B., HUC, A. Y. & OUDIN, J. L.1987. Oil Saturation and Primary Migration: Observations in Shales and Coals from the Kerbau Wells Mahakam Delta, Indonesia. 173–194.

DUTTA, N. C. 1986. Shale compaction, burial diagenesis and geopressures: a dynamic model, solution, and some results. *In:* Burrus, J. (ed.) *Thermal Modelling in Sedimentary Basins*. Technip, Paris, 149–172.

ENGLAND, W. A. 1989. The organic geochemistry of petroleum reservoirs. *Advances in Organic Geochemistry*, **16**(1–3), 415–425.

FALVEY, D. 1974. The development of continental margins in plate tectonic theory. *APEA Journal*, **14**, 95–106.

——— & MIDDLETON, M. F. 1981. Passive continental margins; evidence for a pre-breakup deep crustal metamorphic subsidence mechanism. *Oceanologica Acta*, Colloque C3, Zbeme C.G.I., 103–104.

FERTL, W. H. & WICHMANN, P. A. 1977. How to determine static BHT from well log data. *World Oil*.

FJELDSKAAR, W., MYKKELTVEIT, J., JOHANSEN, H. *et al.* 1990. Interactive 2D Basin Modelling on Workstations. Presented at the SPE Conference June 25–28.

FORBES, P.L., UNGERER, P. M., KUHFUSS, A. B. RIIS, F. & EGEN, S. 1991. Compositional modeling of petroleum generation and expulsion: trial application to a local mass balance in the Smørbukk Sø Field, Haltenbanken aea, Norway. *AAPG Bulletin*, **75**, 873-893.

GALLAGHER, K. & MORROW, D. W. A novel approach for constraining heat flow histories in sedimentary basins. In: Duppenbecker, S. & Iliffe, J. E. (eds) *Basin Modelling*. Geological Society, Special Publications, in prep.

GARDENER, G. H. F., GARDENER, L. W. & GREGORY, A. R. 1974. Formation Velocity and Density - The Diagnostic Basics for Stratigraphic Traps, *Geophysics*, **39**(6), 770–780.

GILES, M. R., INDRELID, S. L. & JAMES, D. M. D. Compaction–The great unknown in basin modelling. *In:* Duppenbecker S. & Iliffe, J. E. (eds.) *Basin Modelling: practical applications*. Geological Society, London, Special Publication, in prep.

GLASØ, O. 1980. Generalized pressure–volume–temperature correlations. *Journal of Petroleum Technology*, **32**, 785–795.

GLEZEN W. H. & LERCHE, I. 1985. A model of regional fluid flow:sand concentration factors and effective lateral and vertical permeabilities. *Mathematical Geology*, **17**(3) 297-315.

GREEN, P. F., DUDDY, I. R., GLEADOW, A. J. W., TINGATE, P. R. & LASLETT, G. M. 1986. Thermal annealing of fission tracks in apatite. I. A ualitative description. *Chemical Geology*, **59**, 237–253.

GUIDISH, T. M., KENDALL, C. G. St. C., LERCHE, I., TOTH, D. J. & YARZAB, R. F. 1984. Basin evaluation using burial history calculations: an Overview: *AAPG Bulletin*, **69**, 92–105.

GUSSOW, W. C. 1954. Differential entrapment of oil and gas: a fundamental principle. *AAPG Bulletin*, **38**, 816–853.

HELLINGER, S. L. & SCLATER, J. G. 1983. Some comments on two-layer extensional models for the evolution of sedimentary basins. *Journal of Geophysical Research*, **88**, 8251–8269.

HERMANS, L., VAN KUYK, A. D., LEHNER, F. K. & FEATHERSTONE, P. S. 1992. Modelling secondary hydrocarbon migration in Haltenbanken, Norway. *In:* Larsen, R. M., Brekke, H., Talleraas, E. (eds.), *Structural and Tectonic Modelling and its application to Petroleum Geology*, Norwegian Petroleum Society, Special Publication **1**, Elsevier, Amsterdam, 305–323.

HERMANRUD, C. 1993. Basin modelling techniques – an overview. *In:* Doré, A. G. *et al.* (eds) *Basin Modelling: Advances and Applications*, Norwegian Petroleum Society, Special Publication **3**, Elsevier, Amsterdam, 1–34.

———, CAO, S. & LERCHE, I. 1990. Estimates of virgin rock temperatures derived from BHT measurements; bias and error. *Geophysic*, **55**, 924–931.

HIGGS, W. G., WILLIAMS, G. D. & POWELL, C. M. 1991. Evidence for flexural shear folding associated with extensional faults. *Geological Society of America Bulletin*, **103**, 710–717.

HO, T. Y., JENSEN, R. P., SAHAI, S. K., LEADHOLM, R. H. & SENNESETH, O., Comparative studies of pre- and post-drilling modelled thermal conductivity and maturity data with post drilling results: Implications to basin modelling and hydrocarbon generation. *In:* Duppenbecker S. & Iliffe, J. E. (Eds.) *Basin Modelling: Practical Applications*, Geological Society, London, Special Publication (in prep.).

HUBBERT, M. K. 1953. Entrapment of petroleum under hydrodynamic conditions. *AAPG Bulletin*, **37**, 1954–2026.

HUNT, J. H. 1990. Generation and migration of petroleum from abnormally pressured fluid compartments, *AAPG Bulletin*, **74**, 1–12.

ILIFFE, J. E. 1991. Basin analysis in extensional regimes. PhD thesis, University of South Carolina, USA.

———, LERCHE, I. & DEBUYL, M. 1991. Basin analysis and hydrocarbon modelling of the south mozambique graben using extensional models of heat flow: *Marine and Petroleum Geology*, **8**, 152–162 .

JACKSON, L. & MCKENZIE, D. 1983. The geometrical evolution of normal fault systems, *Journal of Structural Geology*, **5**(5) 471–482.

JARVIS, D. P. & McKENZIE, D. P. 1980. Sedimentary basin formation with finite extension rates. *Earth and Planetary Science Letters*, **48**, 42–52.

JENSEN, R. P. & DORÉ, A. G. 1993. A recent Norwegian Shelf heating event–fact or fantasy? In: Doré, A. G. *et al.* (eds) *Basin Modelling: Advances and Applications*, Norwegian Petroleum Society, Special Publication **3**, Elsevier,

Amsterdam, 85–107.

KAPPELMEYER, O. & HAENEL, R. 1974. *Geothermics with special reference to application.* Geoexploration Monographs, Series 1, No. 4, Gebruder Borntraeger, Berlin-Stuttgart.

KLIGFIELD, R. & CRESPI, J. 1984. Displacement and strain patterns of extensional orogens, *Tectonics*, 3(5) 577–609.

———, GEISER, P. & GEISER, J. 1986. Construction of Geologic Cross Sections Using Microcomputer Systems. *Geobyte*, 60–66.

KUSZNIR, N. & KARNER, G. 1985. Dependence of the flexural rigidity of the continental lithosphere on rheology and temperature. *Nature*, **316**.

———, MA, X. Q. & NORRIS, S. 1994. 2-D and 3-D forward and reverse structural modelling of rift basin formation: application to the Inner Moray Firth. *Abstracts of the Geological Society Conference on Basin Modelling Applications: What have we learnt.* London.

———, MARSDEN, G. & EGAN, S. S. 1991. A flexural-cantilever simple-shear/pure-shear model of continental lithosphere extension: application to the Jeanne d'Arc Basin, Grand Banks and Viking Graben, North Sea. *The Geometry of Normal Faults*, Geological Society Special Publication, **56**, 41–60.

——— & PARK, R. G. 1987. The extensional strength of the continental lithosphere: its dependence on geothermal gradient, and crustal composition and thickness. From: Coward, M.P., Dewey, J.F and Hancock, P.L (eds), *Continental Extensional Tectonics*, Geological Society Special Publication, **28**, 35–52.

LAMBE, T. W. & WHITMAN, R. V. 1979. *Soil Mechanics: SI version.* John Wiley & Sons, New York.

LARTER, S. 1988. Some pragmatic perspectives in source rock geochemistry, *Marine and Petroleum Geology*, **5**, 3.

LASLETT, G. M., GREEN, P. F., DUDDY, I. R. & GLEADOW, A. J. W. 1987. Thermal annealing of fission tracks in apatite, 2. Aquantative analysis. *Chemical Geology (Isotope Geoscience Section)*, **65**, 1–15.

LEADHOLM, R. H., HO, T. T. Y. & SAHAI, S. K. 1985. Heat flow, geothermal gradients and maturation modelling on the Norwegian continental shelf using computer methods *In: Petroleum Geochemistry in Exploration of the Norwegian Shelf*, Norwegian Petroleum Society, Graham & Trotman, 131–143.

LERCHE, I. 1990a. *Basin Analysis: Quantitative Methods, Volume 1.* Academic Press, Inc.

——— 1990b. *Basin Analysis: Quantitative Methods, Volume 2.* Academic Press, Inc.

——— 1993. Theoretical aspects of problems in basin modelling. *In:* Dore A. G. *et al.* (eds). *Basin Modelling: Advances and Applications.* Norwegian Petroleum Society, Special Publication, 3, Elsevier, Amsterdam, 35–65.

LEWIS, C. R. & ROSE, S. C. 1970. A theory relating high temperatures and overpressures. *Journal of Petroleum Technology*, **22**, 11–16.

LEYTHAEUSER, D., MULLER, P. J., RADKE, M. & SCHAEFER, R. G. 1987. Geochemistry can trace primary migration of petroleum: recognition and quantification of expulsion effects. 197–202.

LOPATIN, N. V. 1971. Temperature and geologic time as factors in coalification. *Izvestiya Akademii Nauk. SSSR*, (Ser. Geol.), **3**, 95–106.

LUO, X., BRIGAUD, F. & VASSEUR, G. 1993. Compaction coefficients of argillaceous sediments: their implications, signifcance and determination. *In:* Dore, A. G. *et al.* (eds). *Basin Modelling: Advances and Applications*, Norwegian Petroleum Society, Special Publication 3, Elsevier, Amsterdam, 321–3332.

MACKENZIE, A. S. & McKENZIE, D. 1983. Isomerization and aromatization of hydrocarbons in sedimentary basins formed by extension. Geological Magazine, **120**, 417–528.

——— & QUIGLEY, T. M. 1988. Principles of Geochemical Prospect Appraisal *AAPG Bulletin*, **72**, 399–415.

———, PRICE, I. LEYTHAEUSER, D., MULLER, P., RADKE, M. & SCHAEFER, R. G. 1987. The expulsion of petroleum from Kimmeridge clay source-rocks in the area of the Brae Oilfield, UK continental shelf. *In:* Brooks, J. & Glennie, K. (eds). *Petroleum Geology of North West Europe* (Vol 2), 865–877.

MAGARA, K. 1976. Water expulsion from elastic sediments during compaction-direction and volumes. *AAPG Bulletin*, **60**, 543–553. MANN, D. M. & MACKENZIE, A. S. 1990. Prediction of pore fluid pressures in sedimentary basins. *Marine and Petroleum Geology*, 7, 55–65.

McAULIFFE, C. D. 1979. Oil and gas migration-chemical and physical constraints. *AAPG Bulletin*, **65**, 761–781.

McCLAY, K. R. 1995. 2D and 3D analogue modelling of extensional fault structures: templates for seismic interpretation. *Petroleum Geoscience*, **1**, 163–178.

McKENZIE, D. 1978. Some remarks on the development of sedimentary basins. *Earth and Planetary Science Letters*, **40**, 25–32.

MEISSNER, F. F. 1978. Petroleum Geology of the Bakken Formation, Williston Basin, No. Dakota and Montana. *In: Economic Geology of the Williston Basin Symposium*, Montana Geological Society, Billings, 207–327.

MIDDLETON, M. F. 1980. A model of intracratonic basin formation, entailing deep crustal metamorphism, *Geophysical Journal of the Royal Astronomical Society*, **62**, 1–14.

——— 1983. Seismic geohistory analysis-A case history from the Canning Basin, Western Australia. *Geophysics*, **49**, 333–343.

MORETTI, I. & LARRERE, M. 1989. Locace: Computer-aided construction of balanced geological cross sections. *Geobyte*, 16–20.

NAKAYAMA, K. 1987. *A dynamic two-dimensional fluid flow model for Basin Analysis.* Phd Thesis, University of South Carolina, USA.

——— & CAO, S. 1993. An inverse method for determining kinetic parameters for kerogens from routein pyrolysis data, and application to the Peace River arch area, Alberta, Canada. *In:* Doré, A. G. *et al.* (eds). *Basin Modelling: Advances and Applications.* Norwegian Petroleum Society, Special Publication, **3**. Elsevier, Amsterdam, 419–431.

——— & LERCHE, I. 1987. Basin analysis by model simulation: Effects of geological parameters on 1D and 2D fluid flow systems with application to an oilfield. *Transactions of the Gulf Coast Association Geological Society.* **37**, 175–184.

NEILSEN, S. B. 1993. Fast Monte Carlo simulation of hydrocarbon generation. *In:* Dore A. G. *et al.* (eds). *Basin Modelling: Advances and Applications.* Norwegian Petroleum Society, Special Publication, **3**. Elsevier, Amsterdam, 1–34.

——— & BALLING N. 1990. Modelling subsidence, heat flow, and hydrocarbon generation in extensional basins. *First Break*, **8**, 23–31.

NORTH, F. K. 1985. *Petroleum Geology.* Allen and Unwin, Winchester.

NUNNS, A. G. 1991. Structural restoration of seismic and geological sections in extensional regimes, *AAPG Bulletin*, **2**, 278–297.

OKUI, A., SIEBERT, R. M. & MATSUBAYASHI, H. Simulation of oil expulsion by 1D and 2D basin modelling-Saturation threshold and relative permeabilities of source rocks In: Duppenbecker, S. and Iliffe, J. E. (eds). *Basin Modelling: Practical Applications.* Geological Society, London, Special Publication, in prep.

——— & WAPLES, D. 1993. Relative permeabilities and hydrocarbon expulsion from source rocks. *In:* Doré A. G. *et al.* (eds). *Basin Modelling: Advances and Applications*, Norwegian Petroleum Society, Special Publication, **3**, Elsevier, Amsterdam, 1–34.

OLSEN, H. W. 1960. Hydraulic flow through saturated clays. *Clays and Clay Minerals*, **19**, 151–158.

PATEL, S. 1993. *Effects of different kerogen types on the timing and duration of hydrocarbon generation; a sensitivity analysis using an activation model.* M.Sc. thesis, Aberdeen University.

PEARSON, M. J. & DUNCAN, A. D. 1996. Biomarker maturity profiles in the inner Moray Firth Basin and implications for inversion estimates. *In:* Hurst, A. *et al.* (eds) *Geology of the Humber Group: Central Graben and Moray Firth, UKCS.* Geological Society, London, Special Publications, **114**, 287–298.

PERRIER, R. & QUIBLIER, J. 1974. Thickness change in sedimentary layers during compaction history. *AAPG Bulletin*, **52**, 57–65.

PERSON, M. & GARVEN, G. 1992. Hydrogeologic constraints on petroleum generation within continental rift basins: theory and application to the rhine graben. *AAPG Bulletin*, **76**, 468–488.

PETERS, K. E. 1986. Guidelines for evaluation of petroleum source rock using programmed pyrolysis *AAPG Bulletin*, **70**, 318–329.

PRICE, L. C., WEGNER, L. M., GING, T. & BLOUNT, C. W. 1983. Solubility of crude oil in methane as a function of pressure and temperature. *Organe Geochemistry*, **4**, 201–221.

QUIGLEY, T. M., MACKENZIE, A. S. & CRAY, J. R. 1987. Kinetic theory of petroleum generation, *In:* Doligez, B. (ed.) *Migration of hydrocarbons in sedimentary basins*, Editions Technip, Paris, 649–666.

REIKE, H. H. III & Chilingarian, C. V. 1974. *Compaction of Argillaceous Sediments*, Elsevier, Amsterdam.

ROYDEN, L., SCLATER, J. G. & VON HERZEN, R. P. 1980. Continental Margin subsidence and heat flow: Important parameters in formation of petroleum hydrocarbons, *AAPG Bulletin*, **64**, 173–187.

SALES, J. K. 1993. Closure versus seal capacity-a fundamental control on the distribution of oil and gas. *In:* Doré, A. G. *et al.* (eds). *Basin Modelling: Advances and Applications.* Norwegian Petroleum Society, Special Publication, **3**, Elsevier, Amsterdam, 399–341.

SCHNEIDER, F. BURRUS, J. & WOLF, S. 1993. Modelling overpressures by effective-stress/porosity relationships in low-permeability rocks: empirical artifice or physical reality? *In:* Doré, A. G. *et al.* (eds). *Basin Modelling: Advances and Applications*, Norwegian Petroleum Society, Special Publication, **3**, Elsevier, Amsterdam, 333–341.

SCHOWALTER, T. T. 1979. Mechanics of secondary hydrocarbon migration and entrapment. *AAPG Bulletin*, **63**, 723–760.

SCHULTZ-ELA, D. D. 1991. Practical restoration of extensional cross sections, *Geobyte.*

SCLATER, J. G. & CHRISTIE, P. A. F. 1980. Continental stretrelizing: an explanation of the post-mid-Cretaceous subsidence of the Central North Sea Basin. *Journal of Geophysical Research*, **85**(B7), 3711–3739.

SLEEP, N. H. 1971. The thermal effects of the formation of Atlantic continental margins by continental breakup. *Geophysical Journal of the Royal Astronomical Society*, **24**, 325–350.

SMITH, J. E. 1971. The dynamics of shale compaction and evolution of pore-fluid pressures. *Mathematical Geology*, **3**, 239–263.

SOMERTON, W. H. 1992. Thermal properties and temperature-related behaviour of rock/fluids. *Developments in Petroleum Science*, **37**. Elsevier, Amsterdam.

STAINFORTH, J. G. & REINDERS, J. E. A. 1989. Primary migration of hydrocarbons by diffusion through organic matter networks, and its effects on oil and gas generation. *Advances in Organic Geochemistry*, **16**(1–3), 61–74.

STECKLER, M. S. & WATTS, A. B. 1978. Subsidence of the Atlantic-type continental margin off New York. *Earth and Planetary Science Letters*, **41**, 1–13.

SWARBRICK, R. E. 1994. Reservoir diagenesis and hydrocarbon migration under hydrostatic conditions. *Clay Minerals*, **29**, 463–474.

SWEENEY, J. J. 1990. BASINMAT FORTRAN program calculates oil and gas generation using a distribution of discrete activation energies. *Geobyte*.

SYLTA, O. 1991. Modelling of secondary migration and entrapment of a multicomponent hydrocarbon mixture using equaiton of state and ray-tracing modelling techniques. *In:* England, W. A. & Fleet, A. J. (eds). *Petroleum Migration*. Geological Society, Special publications, **59**, 111–112.

——— 1993. New techniques and their applications in the analysis of secondary migration. *In:* Doré, G. A. *et al.* (eds). *Basin Modelling: Advances and Applications*. Norwegian Petroleum Society, Special Publication, **3**. Elsevier, Amsterdam, 419–431.

TAYLOR, M. S. G. & BANDURSKI, E. L. Grid based hydrocarbon generation, migration and entrapment using a high-speed solution to the Arrhenius oil generation rate equation. *In:* Duppenbecker, S. & Iliffe, J. E. (eds). *Basin Modelling: Practical Applications*. Geological Society, London, Special Publication, in prep.

THOMPSON, K. 1993. *Tertiary tectonics and uplift of the Inner Moray firth and adjacent areas*. PhD Thesis. University of Edinburgh.

———, TURNER, J. D., ILIFFE, J. E. & PEARSON, M. Hydrocarbon generation and migration of the Inner Moray Firth: the influence of Cenozoic tectonics and implications for the filling history of the Beatrice Oilfield. *In:* Duppenbecker, S. & Iliffe, J. E. (eds.) *Basin Modelling: Practical Applications*, Geological Society, London, Special Publication, in prep.

THOMSEN, R. O. Aspects of applied basin modelling: sensitivity analysis and scientific risk. *In:* Duppenbecker S. & Iliffe, J. E. (eds). *Basin Modelling: Practical Applications*. Geological Society, London, Special Publication, in prep.

TISSOT, B. 1969. Premieres donnees sur le mecanismes et la cinetique de la formation du petrole dans les sediments: simulation d'une schema rationel sur ordinateur: *Revue Institut Francais Petrole*, **24**, 470–501.

——— & ESPITALIE, J. 1975. L'evolution thermique de la matiere organique des sediments: Applications d'une simulation mathematique: *Revue Institut Francais Petrole*, **30**, 743–777.

——— & PELET, R. 1971. Nouvelles donnsur les mecanismes de genese et de migration du petrole. Simuation mathematique et application a la prospection. *Proceedings of the 8th World Petroleum Congress*. Applied Science Publisher, 34–46.

———, ——— & UNGERER, P. H. 1987. Thermal history of sedimentary basins, maturation indices, and kinetics of oil and gas generation, *AAPG Bulletin*, **71**, 1445–1466.

——— & Welte, D. M. 1978. *Petroleum formation and occurrence*. Springer Verlag, New York, 500–521.

TURCOTTE, D. L. & SCHUBERT, G. 1982. *Geodynamics, Applications of continuum physics to geological problems*. J. Wiley, Sons, Chichester.

UNGERER, P. 1989. State of the art in kinetic modelling of oil formation and expulsion. *Advances in Organic Geochemistry*, **16**(1-3) 1–25.

———, 1993. Modelling of petroleum generation and expulsion, an update to recent reviews. *In:* Doré A. G. *et al.* (eds). *Basin Modelling: Advances and Applications*, Norwegian Petroleum Society, Special Publication **3**, Elsevier, Amsterdam, 219–232.

———, BURRUS, J., DOLIGEZ, B., CHENET, P. Y. & BESSIS, F. 1990. Basin evolution by integrated two-dimensional modeling of heat transfer, fluid flow, hydrocarbon generation, and migration. *AAPG Bulletin*, **74**, 309–335.

———, DOLIGEZ, B., CHENET, P.Y., BURRUS, J., BESSIS, F. *et al.* 1987. A 2-D model of basin scale petroleum migration by two-phase fluid flow application to some case studies. *In:* Doligez, B. (ed.) *Migration of Hydrocarbons in Sedimentary Basins*. Technip, Paris, 415–456.

VAN HINTE, J. E. 1978. Geohistory analysis – application of micropaleontology in exploration geology. *AAPG Bulletin*, **62**, 201-222.

VASSEUER, G. & VELDE, B. 1993. A kinetic interpretation of the smectite-to-illite transformation. *In:* Dore, A. G. *et al.* (eds). *Basin Modelling: Advances and Applications*, Norwegian Petroleum Society, Special Publication, **3**, Elsevier, Amsterdam, 173–184.

VIERBUCHEN, R. C., GEORGE, R. P. & VAIL, P. R. 1982. A thermo-mechanical model of rifting with implications for outer highs on passive continental margins. *AAPG Bulletin*, **66**, 765–778.

VIK, E. & HERMANRUD, C. 1993. Transient thermal effects of rapid subsidence in the Haltenbanken area. *In:* Doré, A. G. *et al.* (eds). *Basin Modelling: Advances and Applications*, Norwegian Petroleum Society, Special Publication **3**, Elsevier, Amsterdam, 107–118.

WALTHAM, D. 1989. Finite difference modelling of hanging wall deformation: *Journal of Structural Geology*, **11**, 433–437.

WAPLES, D. W. 1980. Time and temperature in petroleum formation – application of Lopatin's method to petroleum exploration. *AAPG Bulletin*, **64**, 916–926.

——— & KAMATA, H. 1993. Modeling porosity reduction as a series of chemical and physical processes. *In:* Doré, A. G. *et al.* (eds). *Basin Modelling: Advances and Applications*, Norwegian Petroleum Society, Special Publication **3**, Elsevier, Amsterdam, 303–320.

WATTS, A.B., KORNER, G. D. & STECKLER, M. S. 1982. Lithospheric flexure and evolution of sedimentary basins: *Philosophical Tranactions of the Royal Society of London*, **305**, 249–281.

——— & RYAN, W. B. F. 1976. Flexure of the lithosphere and continental margin basin. *Tectonophysics*, **36**, 25–44.

WEI, H. 1994. *Petroleum generation and migration in the North Sea, an experimental and numerical approach*. PhD thesis, Norwegian Institute of Technology, University of Trondheim, Norway.

WELTE, D. H. 1987. Migration of hydrocarbons, facts and theory. In: Doligez, B. (ed.) *Migration of Hydrocarbons in Sedimentary Basins*. Technip, Paris, 393–413.

——— & YUKLER, M. A. 1981. Petroleum origin and accumulation in basin evolution – a quantitative model: *AAPG Bulletin*, **65**, 1387–1396.

WERNICKE, B. 1985. Uniform-sense normal simple shear of continental lithosphere: *Canadian Journal of Earth Sciences*, **22**, 108-125.

——— & ÅBURCHFIEL B. C. 1982. Modes of extensional tectonics: *Journal of Structural Geology*, **4**, 105–115.

YUKLER, A., CORNFORD, C. & WELTE, D. 1978. One-dimensional model to simulate geologic, hydrodynamic and thermodynamic development of a sedimentary basin. *Geologische Rundschau*, **67**, 960–979.

The Central Graben: a dynamic overpressure system

Gordon M. Holm
Hydrocarbon Management International Limited, 93, Liberton Drive, Edinburgh EH16 6NR, UK

ABSTRACT: The Central Graben in the North Sea experienced extension during the Jurassic, with deposition of reservoirs in the Late Jurassic, overlain by the Kimmeridge Claystone, the principal source rock in the North Sea. A subsequent period of thermal subsidence commenced during the Early Cretaceous, and since then in excess of 4500 m of sediments have been deposited.

The Middle to Late Jurassic is overlain by claystone of the Late Jurassic and Early Cretaceous, and contains highly overpressured aquifers, with pore-pressures that approach the fracture gradient. Evidence is accumulating that the overpressure, while partly due to compactional effects, may principally be related to generation of hydrocarbons from the Kimmeridge Clay source rock and, in deeply buried reservoirs, the subsequent cracking of the previously generated hydrocarbons. In the deepest parts of the graben the Kimmeridge Clay is currently in the late stage of gas generation.

The formation of gas, directly from kerogen and from previously generated hydrocarbons, leads to a volume increase which, when isolated, increases the pore-pressure until it is equal to, or locally exceeds, the fracture gradient. The proposed model is that overpressure increases on an intermittent basis, with the pressure seal acting like a valve on a pressure cooker. The fluid volume (and pressure) increases until micro-fractures develop in the seal. With fluid escape, there is a proportional reduction of the fluid pressure in the aquifer. The pore-pressure therefore decreases to a level below the fracture gradient and the micro-fractures reseal. This process is intermittent and is likely to continue as long as the source rock is actively producing hydrocarbons and is connected to the reservoir.

This model has significant implications for the deep hydrocarbon prospectivity within the Central Graben. The most significant is considered to be that the classic Gussow model of hydrocarbon filling and subsequent spilling into adjacent traps, is supplemented by partial leaking through the seal. The typical history of a structure can be considered as early oil fill, and then subsequent displacement by gas as the source rock generates later stage hydrocarbons with the oil escaping at the spill point. However, in the proposed model, the gas will start to fill and displace the oil, but when the pore-pressure exceeds the fracture pressure, the gas can partially leak through the top seal. Thus oil may be preserved in structures which are connected to a late-stage gas generative kitchen, as the gas can bypass the oil by partial dissipation through the seal.

A corollary to this is that structures, where gas chimneys have been recognized, are not necessarily indicative of a presently breached seal with poor prospectivity. In the model described above, the gas chimney is probably indicative of intermittent gas release through the seal and the structure may therefore still contain hydrocarbons. It addition, it may provide migration routes from the deeply buried Jurassic source rock into the reservoirs of Cretaceous and Tertiary Age.

KEYWORDS: *UK Central Graben, Kimmeridge Clay Formation, overpressure*

INTRODUCTION

This study examined formation pressures within Quadrants 21–23 and the northern portions of Quadrants 28–29 (Fig. 1). The Central Graben of the North Sea is a prolific hydrocarbon province (Fig. 2) with oil, gas and condensate present in reservoirs of the Zechstein (Robson 1991; Trewin & Bramwell 1991), Triassic (Penge *et al.* 1993), Jurassic (Stevens & Wallis 1991; Glennie & Armstrong 1991), Cretaceous (Pekot & Gersib 1987) and Tertiary (Wills 1991; Armstrong *et al.* 1987) (Figs 3 & 4). The pre-Cretaceous reservoirs in the depocentre of the Central Graben are highly overpressured and contain large reserves of gas condensate.

From K. Glennie & A. Hurst (eds), 1996, *AD1995: NW Europe's Hydrocarbon Industry*, Geological Society, London, pp. 107–122

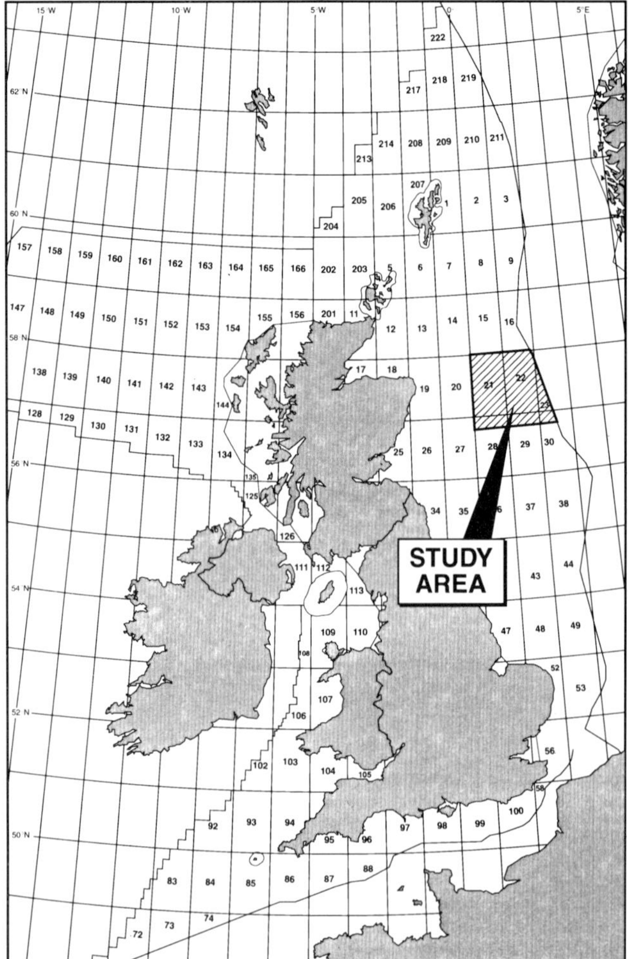

Fig. 1. Location map.

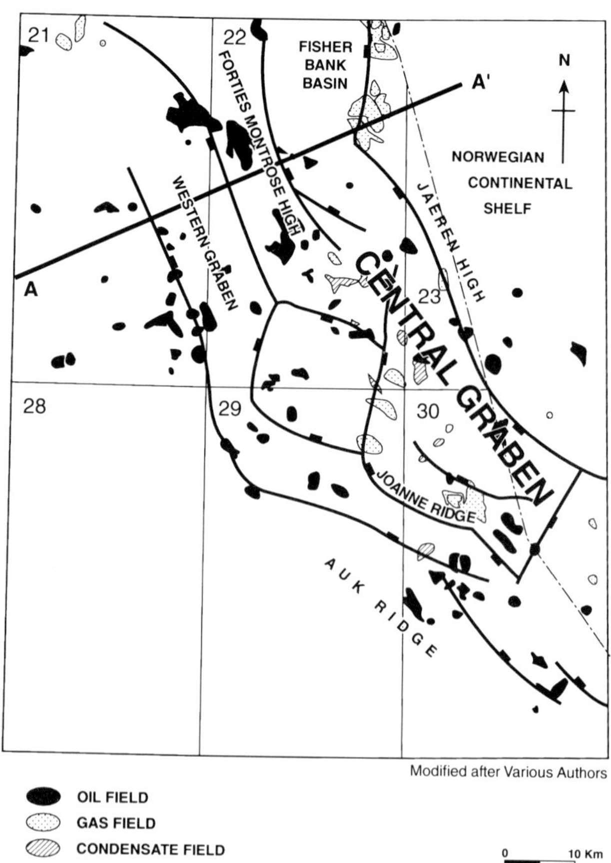

Fig. 2. Pre-Cretaceous tectonic elements and major oil and gas fields.

The deep condensate play has been the target of exploration during the last ten years and a recent press statement indicated that the first production from this play would commence in 1996 with the development of the Erskine Field, Block 23/26b.

Pressure gradients within these reservoirs exceed 0.9 psi ft^{-1} (20.34 kPa m^{-1}) and locally appear to have exceeded the fracture gradient, so causing intermittent or permanent loss of the hydrocarbon fill. A typical pore-pressure profile through the deep Central Graben indicates a multilayered system (Gaarenstroom 1993). There is normal pressure within the Palaeocene sandstone and the underlying Chalk Group. The claystones and marls of the Early Cretaceous Cromer Knoll Group act as the transition zone, and there is a rapid increase in the pore-pressure gradient to greater than 0.9 psi ft^{-1} (20.34 kPa m^{-1}) at the top of the Jurassic. The underlying reservoirs of the Fulmar, Pentland and Cormorant Formations, have pressure gradients similar to or slightly lower than those encountered in the Kimmeridge Claystone Formation (see Fig. 9).

Generation of overpressure can be explained by a number of processes (Mouchet & Mitchell 1989). The effect of undercompaction is well documented (Magara 1978). Overpressure generated by organic matter transformation has been discussed in detail for the intra-continental basins of the Rockies (Meissner 1978; Spencer 1987). Cayley (1987) relates the deep overpressure in the Central Graben to the current generative state of the Kimmeridge Claystone, and this has been further developed (Leonard, 1993; Gaarenstroom *et al.* 1993). This study will examine evidence that overpressure in the deep Central Graben is the result of ongoing transformation of organic matter.

The author proposes a model of a Dynamic Overpressure System where active hydrocarbon generation produces pore-pressures which may equal the fracture gradient. A scenario is proposed, where there is intermittent fracturing of the seal, with fluid and pressure reduction. With subsequent annealing, the pore-pressure will start to build up again. This process may be cyclical, and would continue as long as the reservoir is linked to an active kitchen (Ortoleva 1994). Thus in a Dynamic Overpressure System, it is not necessary that a trap would be filled with oil to spill and then be displaced by later generated gas. Instead, the hydrocarbon contacts are dictated by the phase in the cycle of charge from the source rock and pressure loss through the seal (Sales 1993). In special cases where the capillary pressure is too great to allow leakage of the oil, then gas alone may leak through the seal fractures. In this latter case, we may see gas bypass with no displacement of earlier generated oil (Hodgson *et al.* 1992). These ideas are of great importance for the prospectivity of the Central Graben and similar geological provinces.

TECTONO-STRATIGRAPHIC HISTORY OF THE CENTRAL GRABEN

The Central Graben is the southern arm of an aulocogen with the Viking and Witch Ground Grabens forming the northern and western arms of this failed rift system. The Central

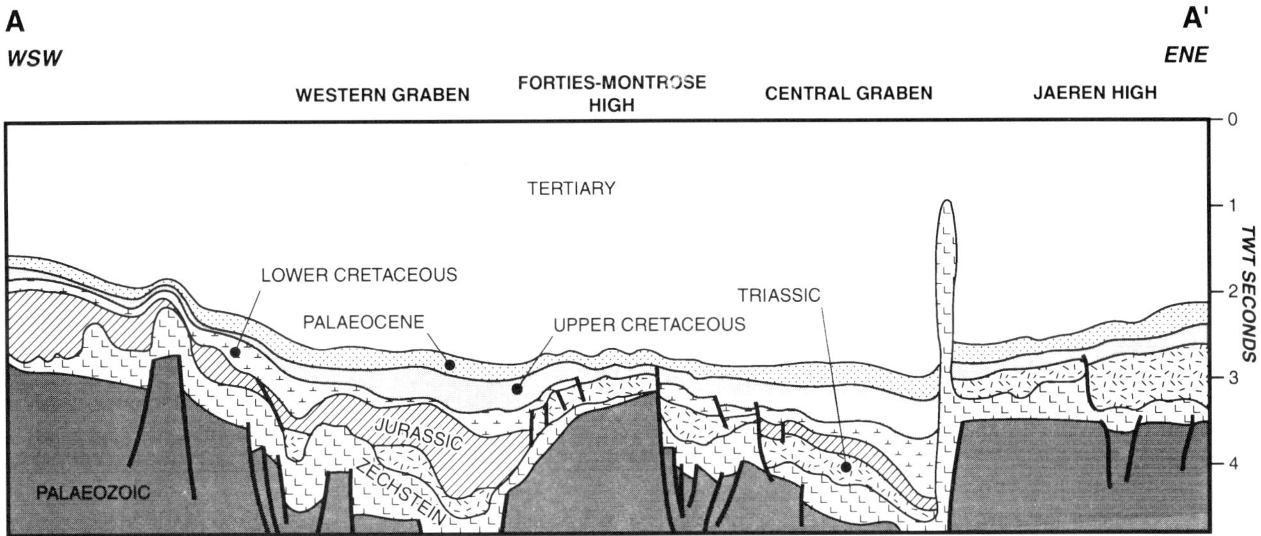

Fig. 3. Cross-section through the Central Graben (modified after Sears *et al.* 1993).

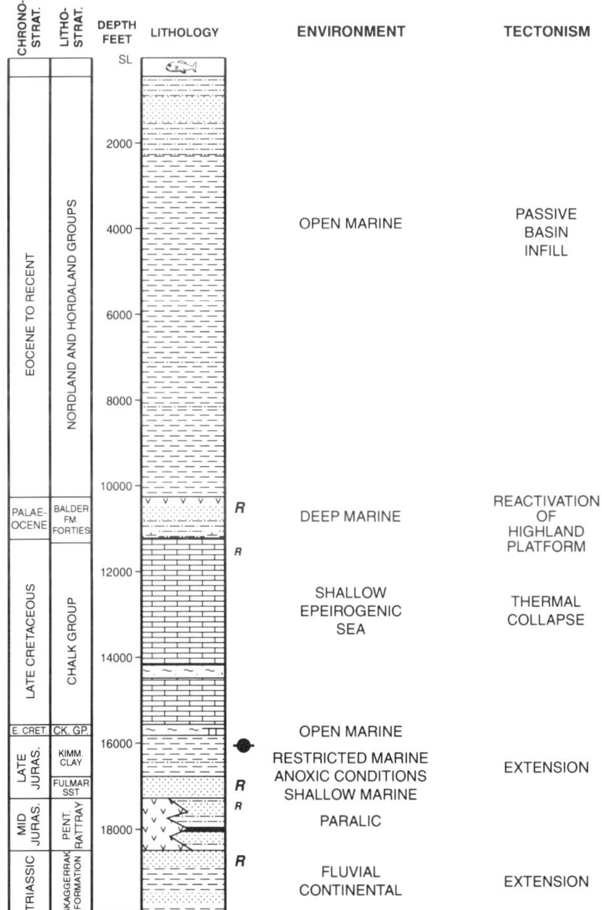

Fig. 4. Central Graben stratigraphic column.

Graben trends to the south-southeast, initially with two arms, the Western Graben and the Fisher Bank Basin, separated by the stable block of the Forties-Montrose High. The two arms combine as they pass into the Norwegian and Danish Sectors (Fig. 2).

During the Late Permian in Western Europe, there was a major intracontinental sea, stretching from the Northern North Sea to Eastern Poland. There were five major periods of desiccation which were associated with the deposition of the Zechstein evaporites. The Stassfurt Salt (Z2) has the greatest thickness of halite and is the significant factor in subsequent halokinesis within the Central Graben. Rifting commenced during the Triassic, with continental conditions prevailing. The sediments were principally red brown siltstones and claystones deposited in ephemeral lakes. Intermittent rivers flowed through the rift valley, created by the proto-graben, and these formed large braidplains. Halokinesis commenced during the Triassic (Fisher & Mudge 1990). At the end of the Triassic the climate became more pluvial in nature, and deposition of the Late Triassic Skagerrak Sandstone occurred in a series of braidplains.

Early Jurassic sediments are absent in the Central Graben, either due to non-deposition or to erosion associated with thermal doming. In the Middle Jurassic there were outpourings of an extensive suite of volcanic deposits centred around the Forties-Montrose High. Coeval with the Rattray Volcanics is the Pentland Formation which consists of a series of paralic sediments (Fig. 5).

Rifting in the Upper Jurassic resulted in large fault blocks, which show evidence of active rotation during sediment deposition. The structural picture is further complicated by halokinesis, thus the sediments of the Upper Jurassic may show considerable local variations in thickness (Fig. 6). This coincided with a world-wide period of sea-level rise. In the Central North Sea, the Late Jurassic sea advanced from the southeast, and deposition of the Fulmar Sandstone took place along the shoreface during this major transgression (Fig. 5). This culminated in the development of a major flooding event with sea-bottom conditions becoming anoxic, thereby allow-

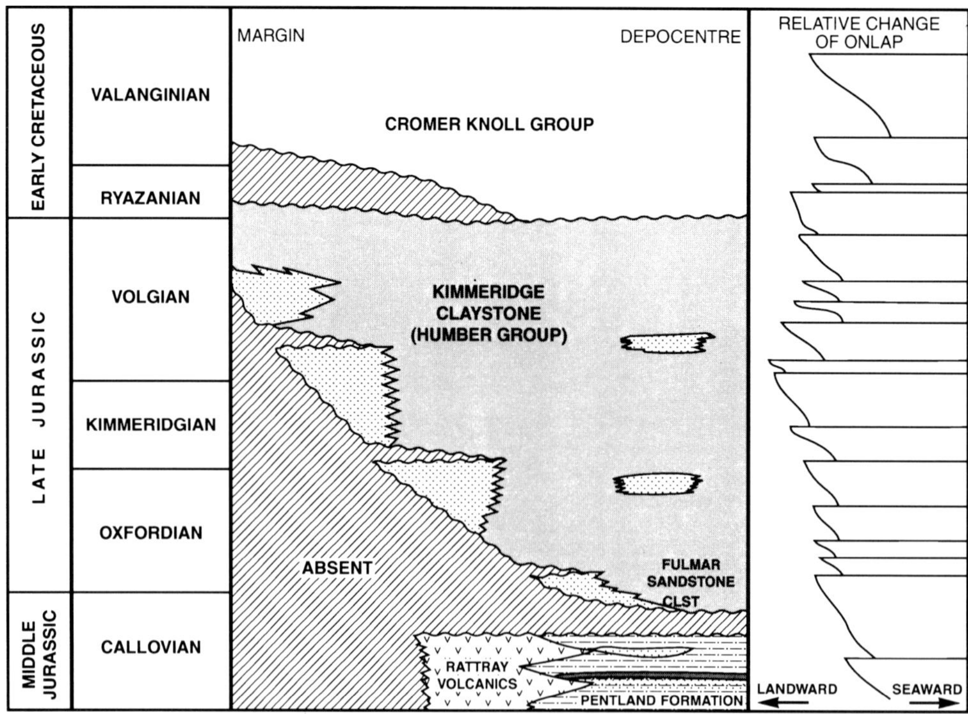

Fig. 5. Central Graben Mesozoic chronostratigraphic diagram.

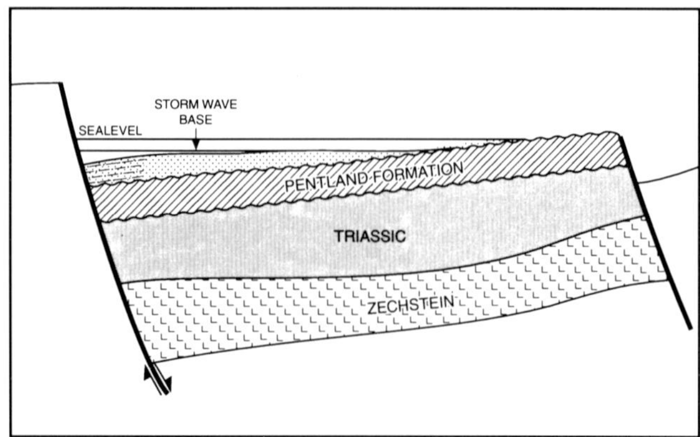

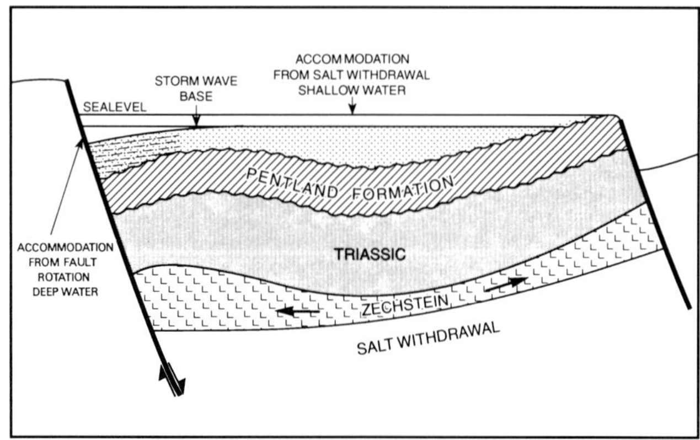

Fig. 6. Fulmar Sandstone depositional model.

ing deposition of the Kimmeridge Claystone source rock. This is a world-class source rock responsible for most of the oil found in the North Sea.

Deposition of claystones continued in the Early Cretaceous, but by now aerobic conditions had been established and there was also an increasing content of carbonate resulting in marls. Extensional faulting was no longer active, and the graben moved into a period of thermal subsidence. This allowed accommodation space to be made available for basin infill. These tectonic conditions continued into the Late Cretaceous with deposition of the Chalk Group. Thermal collapse continued through the Palaeocene, but uplift to the west in the Scottish Highlands and the Shetland Platform resulted in a major influx of coarse clastics, which were fed into the Central Graben by a system of turbidity currents. The Palaeocene sandstones are significant reservoirs throughout most of the North Sea. The Eocene to Recent sediments are principally claystone and siltstone, with gravels, tills and sandstones in the Pleistocene.

SOURCE ROCK AND MATURATION

The Late Jurassic Kimmeridge Claystone is the major source rock in the North Sea. It was deposited during a period of maximum flooding when sea-bottom conditions changed from dysaerobic, during the deposition of the Heather Formation siltstone, to anoxic, during deposition of the Kimmeridge Claystone Formation (KCF). The anoxic conditions provided a perfect environment for preservation of the amorphous kerogen, and TOC values of up to 15%, with an average of 8%, are found. The oil yield is 50 bbl/acre/%TOC, and this is capable of supplying significantly more oil than can be trapped in the known reservoirs in the North Sea (Cayley 1987).

The burial history of the KCF consists of almost continual

Fig. 7. Central Graben Kimmeridge Claystone maturity.

subsidence, with the exception of local erosion and hiati in the Early Cretaceous. Thus the maturity of the KCF is dependent on its current depth of burial and on the geothermal gradient (Fig. 7). Cayley (1987) modelled the maturity of the Kimmeridge Claystone, and his results indicate that the KCF entered the oil window at approximately 9500 ft (2896 m), and the windows for wet gas/ condensate at 12500 ft (3810 m) and dry gas at 14500 ft (4420 m). The modelling used in this study does not significantly disagree with those conclusions (Fig. 8) and, furthermore, indicates that at 17500 ft (5334 m) the KCF is post-mature for dry gas generation.

RESERVOIRS

The Upper Jurassic Fulmar Sandstone

The Fulmar Sandstone is the principal Pre-Cretaceous reservoir in the Central Graben and forms the reservoir in the Fulmar, Clyde, Gannet and Kittiwake fields. Though regionally extensive, it is variable in thickness and often found in isolated lenses. The variability of the thickness of the Fulmar Sandstone can be explained by a combination of the primary deposition and the available accommodation space (Fig. 6). The thickest deposits often exhibit a lozenge shape, which may reflect active salt withdrawal during deposition. Therefore we have two types of tectonism, fault-block rotation and salt withdrawal, coeval with the deposition of the Fulmar Sandstone. For the fault-block rotation model, one would anticipate a thicker deposit of sand juxtaposed against the hanging-wall, while a thin sand or bald crest would be present near the hinge point. Where salt withdrawal was active, if the sand supply was in equilibrium with the accommodation space, great local thicknesses could form.

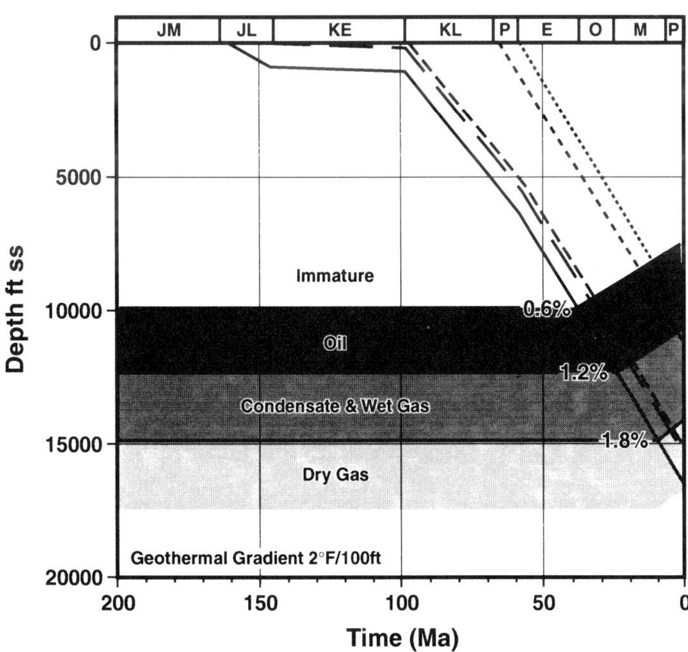

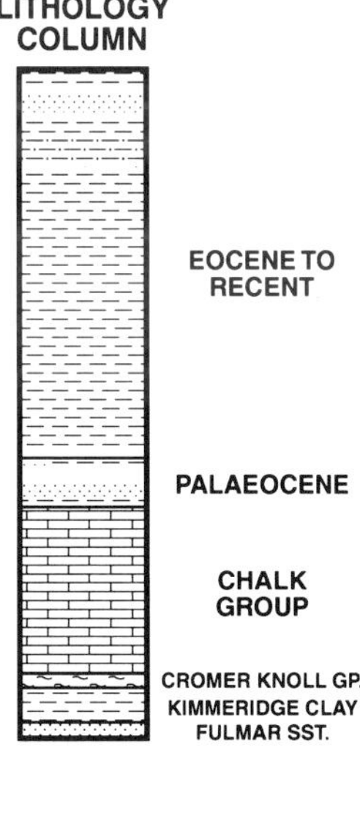

Fig. 8. Central Graben burial curve.

The lateral discontinuity of this reservoir may have been important in creating separate pressure cells.

The Fulmar Sandstone is a general term for a number of sandstones, which vary in age from Oxfordian to Volgian, and were deposited unconformably at the frontal edge of an advancing sea-level rise. The older deposits are found in the centre and the north of the graben while younger deposits are present on the graben margins and in the south.

The porosity of the Fulmar Sandstone typically varies from 15–25% (Stevens & Wallis 1991; Stockbridge & Gray 1991), and locally may be higher especially when preserved by overpressure. The thickness of the Fulmar Sandstone is highly variable, from a few feet as a basal lag to in excess of 1300 ft (396 m) (Price *et al.* 1993). Donovan *et al.* (1993) believe this to be the result of primary deposition of a series of wave-dominated bars with subsequent truncation by intra-Kimmeridgian unconformities. Additionally, the sands were deposited below wave base by intermittent storm events, but remained in aerobic conditions; this allowed extensive bioturbation.

SECONDARY RESERVOIRS

Triassic Skagerrak Formation

The Late Triassic Skagerrak Formation was deposited intermittently in a series of braidplain channels and fans on a broad floodplain (Fisher & Mudge 1990). The channel sands are not always in direct communication with each other, which decreases the productivity of these reservoirs. The porosity varies between less than 8% to 15% and locally 20%. The Skagerrak Formation forms the reservoir in the Marnock gas condensate field.

The Middle–Upper Jurassic Pentland Formation

The Middle Jurassic Pentland Sandstones were deposited in a deltaic environment in a series of fluvial channels. These sandstones are interbedded with claystone, siltstones and coals (Knox & Cordey 1993*b*). The sandstones typically have poor poroperm properties, and although they may contain hydrocarbons, they are rarely highly productive on test. Reservoir pressures from this formation tend to be inaccurate as the tight nature of the sands give poor results from RFT pressure tests, which have either not fully built-up or were supercharged. Where there are valid measurements, these sandstones appear to be in pressure communication with the overlying Fulmar Sandstone.

The Upper Jurassic Kimmeridge/Heather Sandstones

A number of wells have encountered thin sandstones within the Upper Jurassic claystones and siltstones. These vary from a few feet to more than 50 ft (15.2 m) and are now called the Ribble Sandstone, if within the Kimmeridge Clay and the Freshney if within the Heather Siltstone (Knox & Cordey 1993*b*). The sandstones are very clean and have a 'box car' motif on wireline logs, may have good poroperm characteristics and are often hydrocarbon bearing. The sandstones are found in wells within this area only occasionally and it is considered that they will have poor connectivity and will likely form stratigraphic traps, completely enclosed by the claystones and siltstones. Pressures within these sands are often slightly higher than the older Fulmar Sandstone, presumably indicating their limited aquifer volume. Pressures within these sandstones may be a more accurate measure of the pressure of the encapsulating Kimmeridge Claystone.

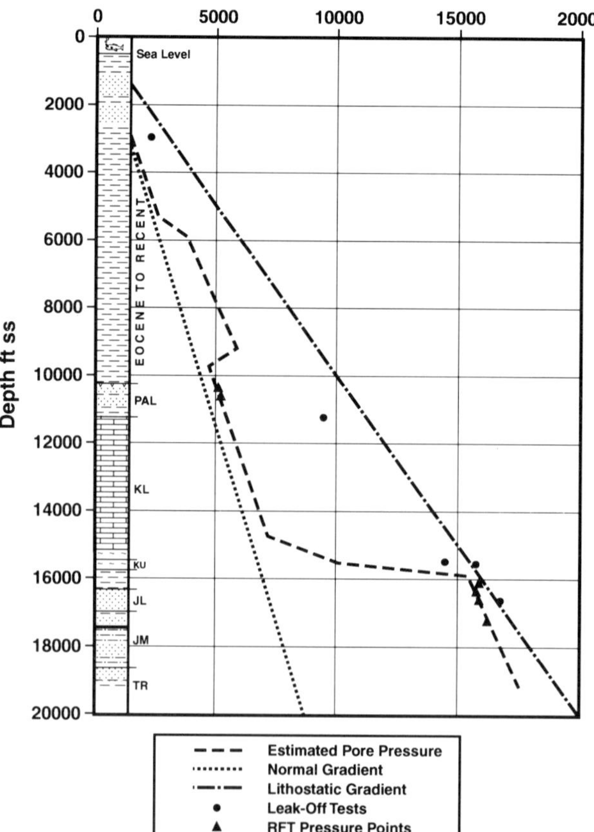

Fig. 9. Example of typical Central Graben pressure–depth plot.

VERTICAL DISTRIBUTION OF PRESSURE

The pressures exhibited in a vertical column in this area indicate a multi-layered pressure environment (Fig. 9). The pressure is normal at the surface and then builds up through the Oligocene sequence to approximately 0.6 psi ft^{-1} (13.56 kPa m^{-1}), with this pressure gradient continuing down through the Eocene. This apparent overpressure is encountered at the base of a large accumulation of Tertiary claystones and siltstones which were deposited rapidly (185 ft (56 m) Ma^{-1}) (Fig. 12) and is probably associated with compactional disequilibrium. Pressure data over this interval are derived indirectly from 'D'-exponents and the records of the mud weights used during drilling. There are no direct measurements of pressure in this lithological unit.

The underlying Palaeocene sandstones are a major aquifer, which extends north from the Central Graben into the Witch Ground and Viking Grabens. The sandstones are interconnected and, at present, deepest in the Central Graben, rising to the west until they are close to sea floor in the Moray Firth. These sandstones are typically normally pressured and act as a significant pressure sink.

The underlying Chalk Group is in pressure continuity with the Palaeocene sandstones and likewise is normally pressured in the study area. In the south of the Central Graben the turbidite systems are no longer sand prone due to their distal location, and the Chalk Group is sealed by claystones of Palaeocene age. In these localities the Chalk Group is overpressured to between 0.6 and 0.7 psi ft^{-1} (13.56–15.82 kPa m^{-1}), similar to the value of 0.676 psi ft^{-1} (15.28 kPa m^{-1}), quoted for Ekofisk (Pekot & Gersib 1987).

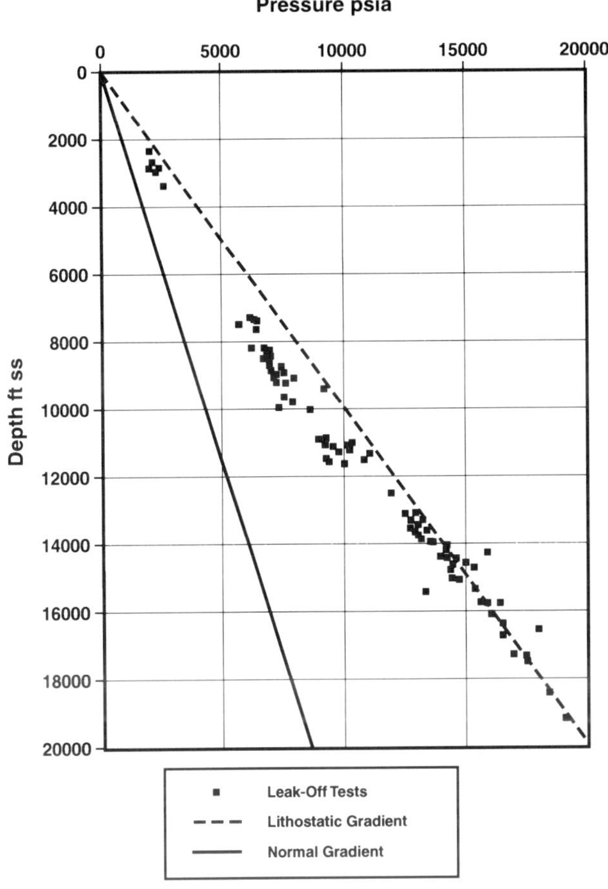

Fig. 10. Leak-off test values (of wells penetrating the pre-Cretaceous).

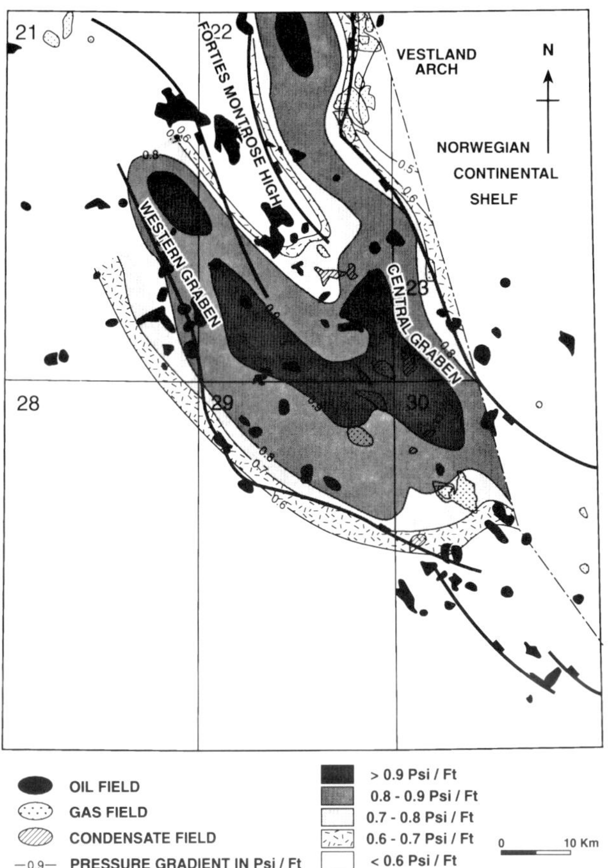

Fig. 11. Central Graben distribution of pre-Cretaceous overpressure.

The Lower Cretaceous consists of marls and claystones of the Cromer Knoll Group, and these act as both a seal and transition zone to the extreme pressures found in the Jurassic.

The pressure gradient to the top of the Jurassic varies between 0.6 to 0.97 psi ft^{-1} (13.56–21.92 kPa m^{-1}), depending on the position of the well in relation to the regional pressure distribution (Figure 11). The pressure in the Kimmeridge Claystone is measured indirectly using connection gas, mud weight and directly from RFT measurements from thin sands occasionally present within this unit. The Kimmeridge Claystone is typically the reservoir seal to the underlying formations, and the organic matter transformation within this formation may be a cause of the overpressure.

The underlying sandstone reservoirs of the Fulmar, Pentland and the Skagerrak formations have pressures marginally less than those recorded for sandstones within the overlying Kimmeridge Claystone. RFT evidence appears to indicate that where the Fulmar Sandstone overlies sandstones in the Pentland or Skagerrak formations, then they are in approximate pressure continuity.

REGIONAL DISTRIBUTION OF PRE-CRETACEOUS OVERPRESSURE

The distribution of the pre-Cretaceous overpressure has been plotted using the pressure gradient calculated back to sea-level (Fig. 11) rather than a potentiometric surface. The pre-Cretaceous reservoirs are not in pressure continuity but form discrete cells (Gaarenstroom *et al.* 1993) and it is the lateral seals separating these cells that maintain the overpressure. The contouring takes into account the megafeatures of the overpressure distribution but does not show the boundaries of the cells. These appear to be defined by reservoir sandstone morphology and faults, some of which are not readily recognized on seismic. Therefore Figure 11 shows an overview of the pressure distribution but has not attempted to define the limits of the individual pressure cells.

The pattern of distribution of the overpressure shows two arms in the deepest parts of the Central Graben and Western Graben where the overpressure gradient exceeds 0.9psi ft^{-1} (20.34 kPa m^{-1}). These reflect the deepest parts of the basin where both reservoirs and the Kimmeridge Clay source rock are presently most deeply buried. The pressure gradients diminish on either side of the basin, probably reflecting the migration of fluids up deep-seated faults to the younger sediments. This is considered to be one route of migration of hydrocarbons from the 'source kitchen' into the shallower formations (Cayley 1987). The pressure differential, between the extremely high 'overpressures' in the basin centre compared with the basin margins and the overlying Palaeocene reservoirs, acts as the driving mechanism for primary migration of hydrocarbons.

ORIGINS OF THE OVERPRESSURE

The origin of overpressure has received much attention over the last 40 years (Dickinson 1953; Magara 1978; Hedberg 1974; Fertl 1976; Chapman 1980; Meissner 1985; Mouchet &

Mitchell 1989; Hall 1993) The principal methods for generating overpressure are summarized below:

undercompaction of shales;
organic matter transformation;
aquathermal expansion;
smectite to illite conversion;
tectonic uplift.

Undercompaction of shales

Undercompaction of claystone is well documented throughout the world (Dickinson 1953; Magara 1978; Bishop 1979; Plumley 1980; Hall 1993) particularly in rapidly subsiding basins. The combination of large supply of sediment and poor permeability in the clays allows overpressure to build up due to a failure of the dewatering process to reach equilibrium. This process is recognized to be in association with excess claystone porosity (Mouchet & Mitchell 1989), therefore the majority of the overpressure detection methods are based on discriminating between a normal claystone compaction curve and one where there is evidence of a trend of enhanced porosity. Claystone compaction curves have been measured in a number of areas throughout the world (Athey 1930; Burst 1969; Magara 1978) and these indicate that, following subsidence under normal compaction conditions, a claystone will achieve a minimum porosity of between 3 and 10% at approximately 11 000 ft (3352 m).

A basin where the overpressure has been derived from compactional disequilibrium can have the following major characteristics.

- present in young, active, rapidly subsiding basins, rarely older than Tertiary.
- the maximum pressure is present near the base of the compacting unit, with a long transition zone upwards until normally compacted sediments are encountered.
- it can be recognized by the use of 'D' Exponent, and other methods of claystone porosity measurements.
- the maximum depth for completion of water expulsion is *c.* 11 000 (3352 m); therefore this may define the active limit of this overpressure generating mechanism, unless exceedingly high depositional rates (> 1000 ft (305 m) Ma^{-1} years) are prevalent.
- the pattern of effective stress will follow a straight line curve with the ratio of effective to total stress being constant with increasing depths.
- the pore-pressure will only exceed the fracture gradient at shallow depths where the bearing capacity is low compared to the fluid pressure (Bowers 1994).

Organic matter transformation

Organic matter transformation was initially proposed as the reason for overpressure in the Western Basins of the USA (Meisner 1978, 1985; Law & Dickinson 1985; Spencer 1987). Subsequent work in the North Sea (Cayley 1987; Gaarenstroom *et al.* 1993) and the Gulf of Mexico (Hunt *et al.* 1994) indicated that the initial pressures produced by oil generation were considerably enhanced when the hydrocarbon system became gas generative or oil was subsequently cracked to gas. The extreme pore-pressure produced by gas generation approaches, and may breach, the fracture gradient, with subsequent fluid loss through the seal.

Organic matter transformation has only recently been considered to be significant as a process in overpressure generation. The process of hydrocarbon generation from kerogen has been increasingly investigated over the last twenty-five years, both theoretically and in laboratory experiments (Lopatin 1971; Tissot & Welte 1984). The kinetics of the alteration of kerogen to hydrocarbons are a function of both time and temperature (Lopatin 1971), and typically commences in the region of 90°C with the generative process proceeding exponentially with a doubling for every 10°C increase in temperature (Waples 1981). The process involves the initial generation of oil with a smaller amount of associated gas. This continues with oil generation reaching a maximum and then declining as the kerogen becomes gas generative. The kerogen eventually becomes post-mature for gas and the transformation of the organic matter is complete. In addition to the direct production of gas from kerogen, gas may also be produced by catagenesis of earlier produced oil into gas (Hunt, 1979).

With the transformation of kerogen to oil there is considered to be a 10–20% volume increase (Oskaya 1988). Meissner (pers comm.) considers that there is an initial volume loss on oil generation as the kerogen contracts with production of a fluid phase. There is no further contraction of the kerogen, which forms a skeletal structure, and with continued production of oil and gas there is a volumetric increase of 10%. The oil generation occurs between approximately 8000 ft (2438 m) and 11 500ft (3505 m), and is a function of time, temperature and burial history. The volumetric increase will lead to an increase in pore-pressure within a closed system, or a system where the leakage rate is slower than the volumetric increase produced by continuing hydrocarbon generation.

When oil converts to gas by catagenesis, one volume of oil produces 534 volumes of gas at standard temperature and pressure (Barker 1990), and it is assumed that these volumes will be similar for gas produced directly from kerogen. At down-hole conditions, the volume increase associated with gas generation from kerogen will be a function of the gas expansion factor (GEF). Therefore, in an example with a GEF of 300 volume/volume (typical for deep Central Graben), there will be a 78% volumetric increase during the transformation of kerogen (and earlier generated oil) to gas. The volume expansion will decrease if the GEF is greater, and alternatively increase if the GEF is less than 300. The volumetric increase during oil generation is relatively small compared to gas generation, and as this process is concurrent with claystone compaction it may be difficult to distinguish the overpressure contribution from other processes, which may be active within a water-wet system.

Source rocks tend to proceed from oil to gas generation at approximately 11 500 ft (3505 m) depth of burial, although this will vary depending on the time/temperature conditions within the basin. At this point, we would expect to see a large increase in the fluid volumes generated and, subsequently, a large jump in the pore-pressure where confining conditions prevail. This has been observed in the Gulf of Mexico (Plumley 1980) and the North Sea (Cayley 1987) and many other basins throughout the world (Rehm 1972). Hunt (1990) observed this phenomenon to be on a global scale, and considered that this may be due to a diagenetic seal which could be universally present at these depths. Local evidence indicated inconsistencies in the global seal theory for the North Sea, but an explanation is still required for the large increase in overpressure which is often found at depths of *c.* 11 500 ft (3505 m). As discussed above, the increase in volume associated with generation of oil from kerogen, and gas from kerogen are 10–20% and 50–75% respectively. Therefore, if the sealing/leaking capacity of the system

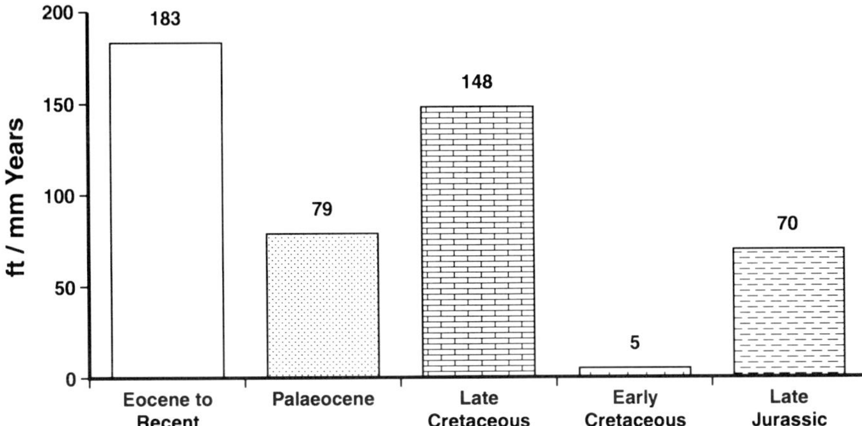

Fig. 12. Depositional rates in the Central Graben.

remains constant, on passing through the generative change from oil to gas there will be a large build up of fluids, which is being generated faster than it can be dissipated. The dissipation through the porosity/permeability of the sealing rocks would not be large enough to match the fluid volumes generated; therefore the pore-pressure will approach the fracture gradient and episodic breaching will occur. The mechanism of overpressure generation from the gas phase of organic matter transformation may explain the many observations of zones of high overpressures consistently present at these depths, in basins throughout the world.

Aquathermal expansion

Aquathermal expansion was proposed by Barker (1972) as a mechanism for overpressure generation. As basin subsidence occurs, water trapped interstitially or within an aquifer will heat up directly relative to the geothermal gradient, with a resultant expansion of water according to the coefficient of thermal expansion. He recognized that for overpressure to be generated and maintained, there would have to be a perfect seal, and he calculated that it would only require leakage of 5% of the volume for the pressure of the system to return to normal gradient. Daines (1982) demonstrated that claystones act as an aquitard, therefore there would be fluid leakage of any incremental volume; he considered it unlikely that aquathermal expansion was a valid mechanism for overpressure generation.

Smectite to illite conversion

Smectite to illite conversion has been proposed as another method of generating overpressure (Burst 1969). Perry & Hower (1972) indicated that an initial mixture of 75% smectite and 25% illite would alter to 80% illite and 20% smectite with the release of 15.5% water. It is this latter volume which supplies the excess volume necessary for overpressure generation. This process is believed to occur at 90–100°C (194–212°F) (Mouchet & Mitchell 1989) corresponding to depths between 6500 ft (1981 m) and 9750 ft (2972 m). Colten-Bradley (1987) argued that overpressure could not be produced by the smectite to illite transformation as this reaction can only occur in a drained system, whereas a closed system would be required for the pressure to increase during release of the bound water. She considered that the coincidence of overpressure with the depths for smectite to illite transformation may be caused by the considerable by-products associated with this process, and these could produce diagenetic seals. Mouchet & Mitchell (1989) concluded that although clay diagenesis may be a contributory factor to overpressure, it is unlikely to be significant on its own.

Tectonic uplift

Tectonic uplift is not an overpressure generating mechanism but changes the physical conditions of a trapped volume of fluid, whilst maintaining sealing conditions. The most common example is that of a raft overlying a salt diapir. Here, the aquifer has been isolated and uplifted to a new depth. By retaining the original fluid pressure, the reservoir is now overpressured with respect to the new depth.

EXPLANATION FOR OVERPRESSURE IN THE CENTRAL GRABEN

A number of significant observations can be made regarding the overpressures of the Central Graben, which may suggest that the fluid expansion and increased volume associated with hydrocarbon generation or conversion of oil to gas, is the most significant element in the generation of the extreme overpressures encountered in the graben depocentre. These observations are listed below:

1. The pressures at depths below 11 500 ft (3505 m) in the pre-Cretaceous of the Central Graben are extremely high and approach the lithostatic gradient (Fig. 9) at the top of the Jurassic.
2. There is no evidence of undercompaction in the Kimmeridge Claystone or in the claystones and marls of the Early Cretaceous Cromer Knoll Group.
3. The sediments of the Early Cretaceous Cromer Knoll Group overlying the Kimmeridge Clay were deposited slowly (Fig. 12) with a number of hiati and local unconformities on the rotated fault blocks. This is atypical of overpressure generated from compactional disequilibrium, where high deposition rates are usually present.
4. The extreme overpressures are coincident with areas where the Kimmeridge Clay has a vitrinite reflectance greater than 1.2% and is currently mature for, and actively generating, wet and dry gas (Figs 7 & 11). These

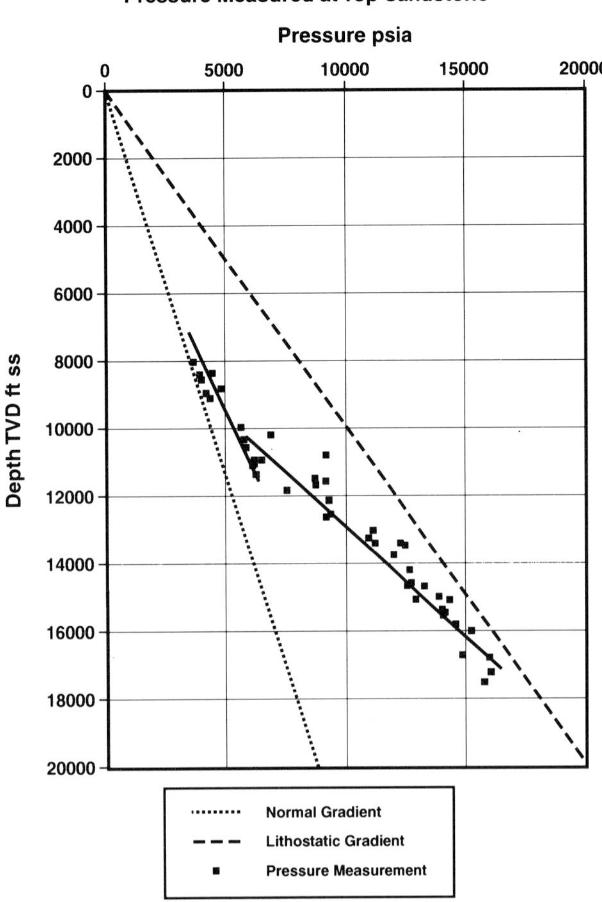

Fig. 13. Pre-Cretaceous pressures in the Central Graben.

zones were also the areas of most rapid sedimentation during the Tertiary.

5. The principal hydrocarbon type found in the deep reservoirs is gas condensate, sometimes with residual asphaltene. This combination may indicate cracking of earlier generated oil to condensate.
6. The distribution of pressures within the pre-Cretaceous of the graben indicates two populations (Fig. 13), the first above 11 500 ft (3505 m) and close to a Normal Gradient, with a second below 11 500 ft (3505 m) which, at greater depths, approaches the Fracture Gradient. The apparent division of these populations close to 11 500 ft (3505 m) is similar to the depths where the kerogen starts to generate wet gas, with a large increase in volume. This is deeper than the optimum depth for ending the transformation of smectite to illite, 9750 ft (2972 m) (Mouchet & Mitchell, 1989) and the depth at which claystone compaction would be complete in a normally compacting basin.
7. The Cromer Knoll seal to the overpressure is locally very thin, < 100 ft (31 m). If the Jurassic sediments had excess pressures built by compactional disequilibrium during the Tertiary, these excess pressures are likely to have been rapidly dissipated by upward fluid loss obeying Darcy's law (Deming 1994).

Based on the above circumstantial evidence, the principal factor in creating the extreme overpressures observed in the Central Graben is the generation of gas from the Kimmeridge

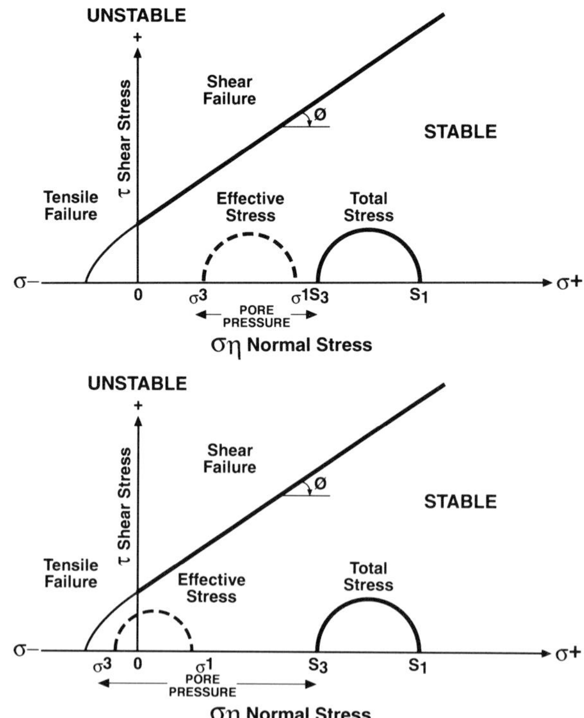

Fig. 14. Mohr-Coulomb stress failure envelope.

Clay source rock, and conversion of earlier generated oil to gas within a closed system. This generation is continuing today. A dynamic charging system is responsible for the extreme overpressures, the fracture of the seals and the primary expulsion of hydrocarbons through a system of vertical fractures and faults.

BREACH OF PRESSURE SEAL

The idea of primary migration through microfractures in the source rock was proposed some years ago (Du Rouchet 1981; Gaarenstroom *et al.* 1993; Leonard 1993) and similarly it has long been recognized that in cases where the pore-pressure exceeds the fracture pressure, the seal can be breached with subsequent loss of hydrocarbon charge. In order to understand the generation of fractures, it is easier to use effective stress as defined by Terzaghi's Law (Terzaghi & Peck 1948), which states that:

$$P = S - \sigma v$$

where S is the overburden load, σv is the effective vertical stress of the rock type and is related to the porosity, and P is the pore-pressure. Du Rouchet (1981) undertook a detailed analysis of the effects of fracture propogation on fluid migration, and noted that S3 (the effective horizontal stress) is related to S1 the effective vertical stress by a relationship:

$$S3 = f.S1$$

where f is the confining stress coefficient. Typically below 500 m (1640 ft) S3 is always less than S1.

For a fracture to open, the pore-pressure must be greater than S3, and requires additional pressure greater than the capillary pressure, Pc, to allow the fluid to leak off into a more permeable zone. Du Rouchet (1981) defined the pressure required to initiate a fracture with subsequent escape of fluids, as the effective stress plus the capillary pressure of the neighbouring permeable zone:

$$Pp \geq S3 + Pc$$

He quotes values for the capillary pressure of between 10 and 50 bars (150 psi and 450 psi). At a depth of 14 500 ft (4420 m) therefore, the increase in pressure gradient, to dissipate fluid through the seal into a permeable zone, varies between 0.01 and 0.03 psi ft^{-1} (0.23–0.68 kPa m^{-1}). Thus, the additional effect of the capillary pressure can be considered negligible at these depths.

To initiate fractures, the pore-pressure must increase to enable the effective stress to approach the fracture pressure (Fig. 14). In the deep Central Graben, the fracture gradient (Fp) approximates to the lithostatic gradient of 1 psi ft^{-1} (22.6 kPa m^{-1}) (Fig. 10). Gaarenstroom *et al.* (1993) assume that the minimum of the leak-off pressure measurements is a more accurate value for anticipating the initiation of fracturing. As the pore-pressure is in excess of 0.9 psi ft^{-1} (20.34 kPa m^{-1}), any small increase in pressure could induce fracturing. Du Rouchet (1981) points out that effective stress does not become negative. It is therefore likely that fractures which opened when the pore-pressure equalled S3 will close after fluid loss and subsequent reduction of pore-pressure. In the event that pore-pressure again builds up to exceed S3, then the process is repeated, and this can continue as long as the system is supplied by a continuing influx of hydrocarbons from the source rock.

Chen *et al.* (1990) modelled the effect of oscillatory fluid release from overpressured compartments and concluded that during subsidence, rocks may be present in three zones, the unfractured (or healed) zone, the oscillatory zone and the fractured zone (Fig. 15a). They concluded that oscillatory zones are often quite extensive and will continue in this state of disequilibrium as long as the switching events are active. Ortoleva (1994) noted that with subsidence and increasing fluid pressure, the basin is driven from a state of equilibrium, and pore-pressure builds up until hydrofracturing is induced. His modelling indicated that seal breach and pressure build-up is episodic (Fig. 15b). This conclusion of rupture and resealing of fractures is very significant, as it indicates that seals will typically leak intermittently and the permanent rupture of a seal is likely to be an unusual event.

The pore-pressure gradients in these deep reservoirs are very close to the fracture gradient, and it is highly likely a small increase in the pressure may be sufficient to cause microfracturing of the seal (Fig. 9). This increase in pressure could be derived from an increase in the height of the hydrocarbon column, an increase in the volume of fluids generated from kerogen, conversion of oil to gas, or possibly a change in the tectonic stress regime. The evidence for a leaking seal comes from the presence of vertical migration routes in the Central Graben (Sears *et al.* 1993), the observation of gas chimneys (Leonard 1993), and the phenomenon of 'gains and losses' which occur when drilling through the Upper Jurassic in these wells. A possible example may have occurred in well 23/26b-4. While drilling through the claystones of the Cromer Knoll Group this well experienced an influx of gas condensate. When the kick was controlled and drilling resumed, no evidence of permeable horizons was observed in the cuttings samples or the logs. An explanation for this kick may be that this well penetrated a fracture, bleeding off condensate from the underlying Kimmeridge Claystone and Upper Jurassic reservoirs of the Erskine Field.

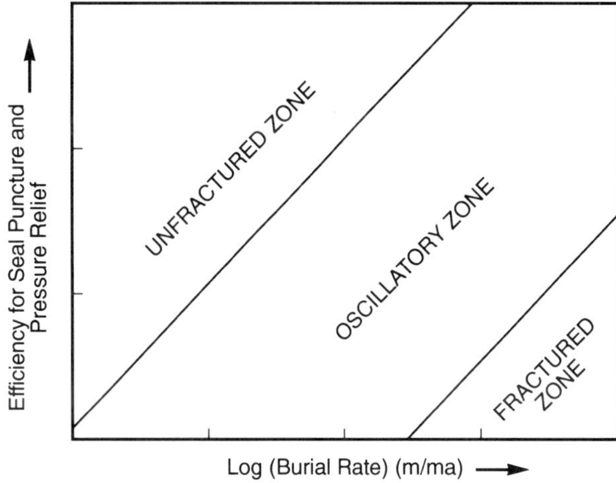

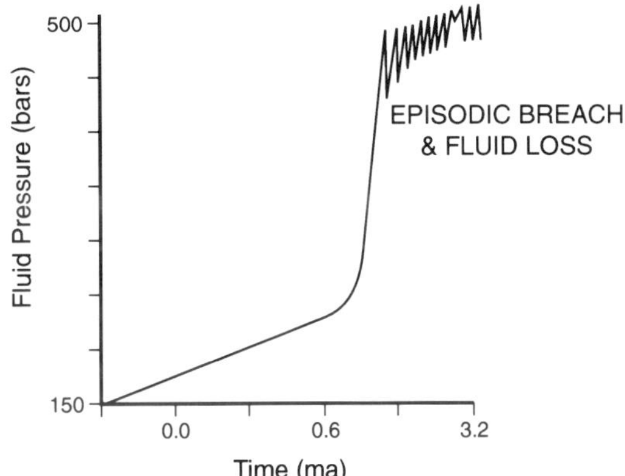

Fig. 15. Episodic pressure build-up and release. (a) Modelling of the efficiency rate for pressure seal puncture and burial rate (modified after Chen *et al.* 1990). (b) Modelling of pressure changes during basin development and subsidence (after Ortoleva 1994).

The model proposed is one of cyclical, intermittent breaching of the pressure seal as the effective stress increases, and then subsequent annealling (resealing) of the microfractures, after partial fluid escape and the subsequent loss of pressure (Fig. 16). The reservoir would then commence recharging as it is in direct contact with the source rock. This model is similar to the method discussed by Leonard (1993) and the author proposes that such a system be called a **Dynamic Overpressure System.** It is recognized that fluid leakage may occur intermittently through capillary seal failure (Schowalter 1979), but the propinquity of the pore-pressure and fracture gradients make it highly probable that the intermittent leakage occurs through episodic hydrofracturing and annealling of the seal. This is the explanation given by Wensaas *et al.* (1994) for leakage through the seal in the Gullfaks South Field.

The timing for this type of recharge cycle is not known, but basin modelling (Ortoleva 1994) indicates that these cycles repeat in less than 100 000 years. Roberts & Nunn (1995), in modelling expulsive events through faults, considered that the expulsive event took place over 100 years whilst it required 10 000–500 000 years before the pressure builds up sufficiently to initiate the next cycle of seal breaching.

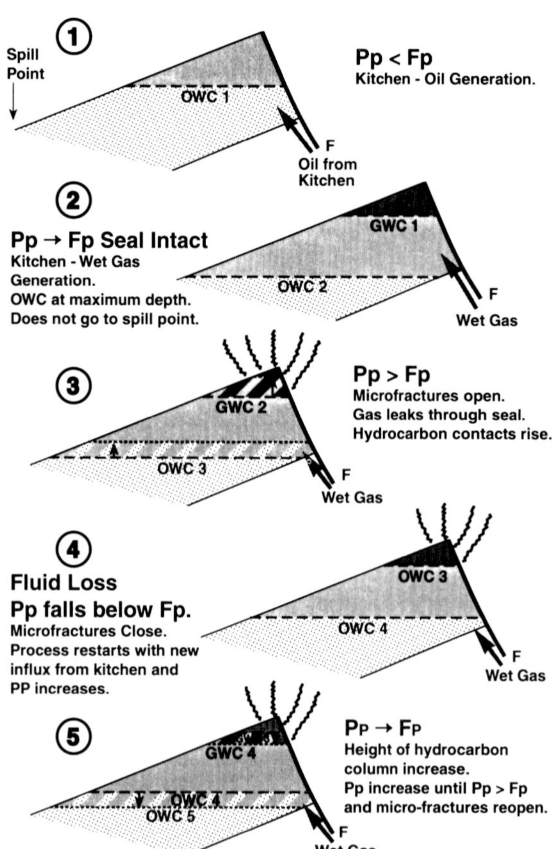

Fig. 16. Filling history in a Dynamic Overpressure System.

IMPLICATIONS FOR PORE-PRESSURE PREDICTION

In the Dynamic Overpressure System, with overpressure derived from hydrocarbon generation in addition to undercompaction, it is important to recognize which pore-pressure prediction methods can still be applied with confidence.

The methods which cannot be confidently used

D Exponent and its derivatives

These techniques rely on estimating the normal compactional curve and then recognizing the transition zone over an overpressured system. This method is unlikely to work as claystone undercompaction is not a factor in overpressure produced from organic matter transformation, and any transition zone may be too small for a trend to be recognized.

Shale factor

This is a measure of the smectite to illite transformation, and as this process is mostly complete by 9750 ft (2972 m), it is not considered relevant for overpressure associated with depths in excess of 12 500 ft (3810 m).

Methods which may still be effective

ROP

It has been recognized that overpressured zones are often associated with an increase in the rate of penetration following drilling through a seal. Therefore this 'drill break' can be associated with an overpressured zone, but may also indicate a change in lithology as porous sandstone drills considerably faster than claystone.

Temperature

Monitoring of the temperature of the mud coming out of the flow line may indicate a potential overpressure zone. It should be noted that this again relies on establishing trend lines, and there is a delay from the mud heating up at the bit (presumed hottest point in the hole) and the measurement when the mud has been circulated out at surface. The more porous overpressured zone acts as an insulator, therefore there is an increase in the temperature recorded following penetration of an overpressured zone (Mouchet & Mitchell 1989).

Normalized gases

The gases are measured from the mud; therefore there is a time lag similar to temperature readings. A common feature of deep wells in the Central Graben is that the quantity and types of gas increase on approaching the overpressured zone. If there is leakage through fractures and micro-fractures, then it can be expected that methane, as the lightest and most mobile of the alkanes, will be detected first, and on approaching the overpressure zone there will be a sequential increase in C_2–C_5 gases. Normalization techniques may be used to try to correlate between wells the increases of drilled gases.

Seismic velocity methods

There is often a velocity contrast between the seal and the overpressured zone. This may sometimes be detected prior to drilling by modern sophisticated methods of analysing stack velocities. In the event that the transition zone is known and has a velocity impedance, then a 'Look-Ahead' VSP may be employed during drilling, to refine the prognosis of the top of the overpressured zone.

PPFG from Sperry Sun

This is a proprietary method, which can utilize log data from offset wells, or MWD data in real time, to calculate effective stress. The effective stress will increase constantly with increasing depth in the case of undercompacted shales, but where there has been additional fluid influx there will be a reduction of effective stress, referred to as unloading (Ward *et al.* 1994) and recognition of the change of effective stress will indicate the overpressured zones.

It is very important to recognize the limitations of the overpressure prediction methods. Therefore prior to drilling a well, a detailed analysis of the geology and the potential overpressure generating systems should be undertaken. Then the most appropriate detection methods should be incorporated into the well monitoring. In dealing with any overpressure system it is important to remember that most detection methods are *post priori*, therefore by the time an overpressured zone is recognized, the drill bit will already have penetrated this zone and there may be a danger of fluid influx.

IMPLICATIONS FOR EXPLORATION

The realization that we are dealing with Dynamic Overpressure Systems has a number of important implications for exploration:

- oil and gas can leak intermittently through the seal as well as at the spill point;
- hydrocarbon contacts may rise and fall and are a function of the state of discharge and recharge within a reservoir controlled by this type of system;
- gas by-pass of oil with gas loss through the seal; thus oil may remain in a structure despite being connected to a gas-generative source rock;
- gas chimneys over structures are not necessarily indicative of breached seals but of the presence of a deeper Dynamic Overpressure System;
- there may be potential for production from actively generating source rocks, particularly where permeability is enhanced by fractures. This would be similar to the fields in California where oil is produced from the fractured Monterey source rock.

In the Gussow model of oil and gas emplacement (Gussow 1954), the reservoirs would initially fill with oil and then spill into the next structure on the migration route, after the first structure was full. If the structure remained in communication with the source rock when it became gas generative, then the gas would enter the structure and displace the oil. In the Dynamic Overpressure System, the hydrocarbon contacts are a function of the position in the cycle of charge, displacement and leakage through the seal. Therefore initially the structure would fill with oil followed by a gas charge. If the excess pressure was sufficient to open the microfractures in the seal, then subsequent loss of fluid volume and pressure would occur. Thus the fluid contacts would rise within the structure until the pressure drop was sufficient for the fractures to close, and at this point the process would begin again (Fig. 16). This is similar to the Class 3 Trap proposed by Sales (1993), where leakage of both oil and gas can occur. Sales (1993) considered that the hydrocarbon contacts would remain at a constant level as there would be equilibrium between the losses through the seal and the continuing influx. Hydrocarbon contacts will change as a function of stage of discharge/recharge. This is supported by Roberts & Nunn (1995), who modelled episodic discharge through a seal. They noted that the expulsive cycle was extremely rapid whilst the build-up cycle typically lasts 50–2500 times the length of the expulsive event. Therefore, following the expulsive event, there would be a very rapid rise in the hydrocarbon contacts, and these would slowly lower with continuing fluid recharging. Thus the specific height of gas and oil/water contacts within a structure is determined by the current stage in the Dynamic Overpressure System.

A model similar to the above is described by Hodgson *et al.* (1992), for the Mungo Field (Block 23/16a). This is a Tertiary reservoir which has a large hydrocarbon column associated with a salt diapir. In this case the trap is filled with oil despite being in direct contact with a gas generative source-rock kitchen. Hodgson *et al.* (1992) propose that there was early emplacement of oil in the structure, and then gas entered it. The added pressure from the buoyancy effect of the gas was sufficient to open the microfractures in the seal. This allowed gas to leak, but the pore entry pressure (capillary pressure) was only sufficient to allow loss of gas. The fractures did not open sufficiently to allow leakage of the more viscous oil. Thus there was gas bypass, leaving oil in a structure which lies in a gas-prone fairway. This process appears to be ongoing as a gas chimney is visible on seismic section over the structure.

Gas chimneys observed on seismic have traditionally been taken to indicate a structure with a breached seal, with loss of hydrocarbons which migrate directly through it into the overlying stratigraphic sequence. In the Dynamic Overpressure Model, it is predicted that the presence of a gas chimney would overlie intact structures as the gas is a function of the intermittent release of pressure through the seal. These have been observed overlying the oil fields of the Ekofisk complex (Pekot & Gersib 1987; Fig. 18) and overlying the Gullfaks Field in the Viking Graben (Miles 1990). As both oil and gas may leak through the seal, it follows that in addition to the conventional migration routes, the explorationist should investigate hydrocarbon migration routes overlying the seal (Fig. 17).

It should be noted that in the Central Graben, gas chimneys are not observed in the Chalk Group overlying the Jurassic but only in the Tertiary claystones. This probably results from the varying permeability of these formations. Although the Chalk has low permeability, there are discrete migration routes, either through more permeable layers or via fractures. Such migration routes are rarely present in the Tertiary claystones whose permeability is typically 10 000 times less than the chalk. Therefore gas migrates slowly through claystones and is probably disseminated within the available porosity. This has the effect of scattering seismic energy, thus the phenomenon of a gas chimney is observed on seismic sections.

The Kimmeridge Clay is currently generating wet gas below 12 500 ft (3810 m) and large volumes of drilled gas are released into the well bore during drilling. In the Central Graben, the source rock has not been an exploration target to date. Given the large volumes of gas which may be generated directly within the Kimmeridge Clay there may be exploration potential in areas where gas production would be enhanced by fracture permeability.

CONCLUSIONS

The Central Graben of the North Sea is a multi-layered hydrostratigraphic system consisting of a deep highly overpressured system in the pre-Cretaceous, overlain by Palaeocene sandstones. These sandstones form a normally pressured regional aquifer which act as a pressure sink. The distribution of the pre-Cretaceous overpressure is characterized by a number of pressure cells, probably separated by both lateral facies changes within the permeable horizons and by sealing faults

There have been a number of theories proposed for the origin of overpressure, with undercompaction of shales and organic matter transformation considered to be the most prominent in different basins throughout the world. The pre-Cretaceous overpressure in the Central Graben is probably the result of a combination of claystone undercompaction and hydrocarbon generation. The extreme overpressure present below 12 500 ft (3810 m) approaches the lithostatic gradient and is probably due to organic matter transformation, principally gas generation.

The pore-pressure may locally and intermittently exceed the fracture gradient. When this occurs, microfractures in the seal will open. There will be a loss of fluid, followed by a subsequent fall in pore-pressure. At this point the pore-pressure will fall below the fracture gradient and the fractures will reseal. The active gas generation will then start to recharge the system with increasing pore pressure until it breaches the fracture gradient again. This process will work

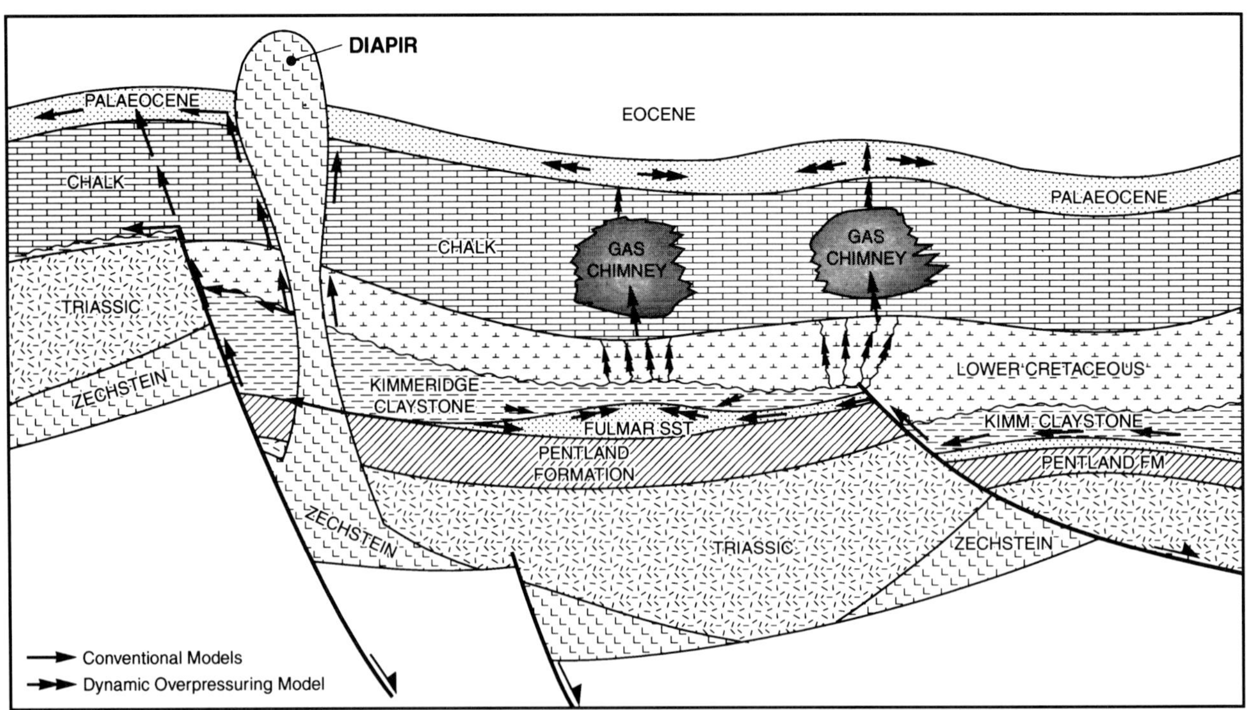

Fig. 17. Central Graben migration route models.

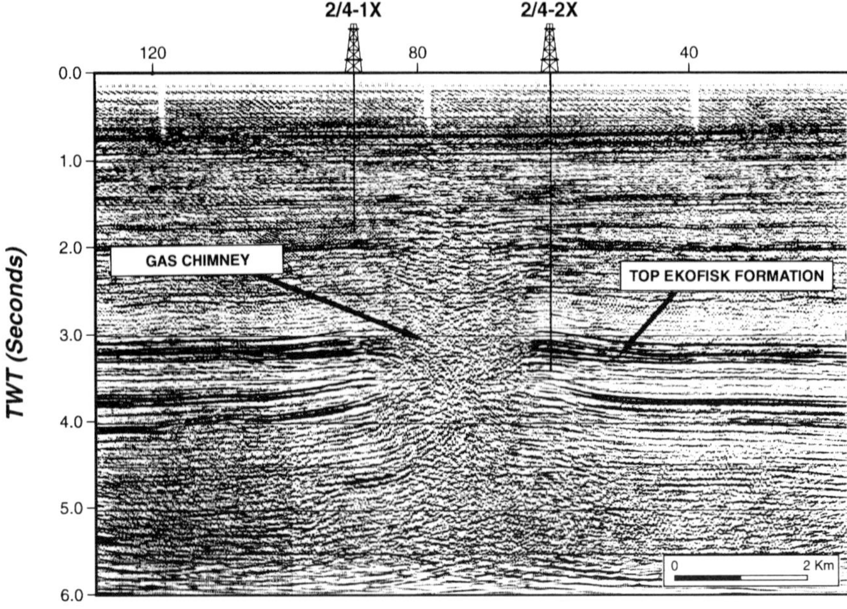

Fig. 18. Gas cloud on a seismic section from the Ekofisk Field (reproduced from Pekot & Gersib, 1987 with permission from the NPF).

on an episodic basis, with the fluid loss occurring rapidly and the recharge process taking up to 5000 times longer than the expulsive events. The process of episodic discharge and recharge will continue as long as the system is charged from the actively generating kitchen.

The author has called this process a **Dynamic Overpressure System.** It consists of a combination of overpressure charging up a system, followed by episodic discharge through the seal when the pore-pressure exceeds the fracture gradient. Recognition of a Dynamic Overpressure System is important both for overpressure recognition and in hydrocarbon exploration. In this system, the classical methods of overpressure detection cannot be used with confidence, and it is important for drilling safely to recognize that one is dealing with a Dynamic Overpressure System.

The Dynamic Overpressure System has a number of important implications for exploration. The effect of intermittent discharge through the seal implies that there may be unconventional migration routes. Gas bypass may occur, without displacement of the oil at the spill-point and gas chimneys overlying a structure are not indicative of a completely breached trap, but may indicate an active Dynamic Overpressure System. Therefore the knowledge that a basin contains a Dynamic Overpressure System can lead to the recognition of additional prospectivity.

I would like to thank Phil Newman and Colin Oswald of Elf Caledonia for reviewing various versions of this paper, the drawing office of Elf Caledonia for preparation of the diagrams, Dave Derby and Richard Swarbrick for stimulating discussions regarding overpressure problems and two anonymous reviewers whose comments and additional references have helped to improve the quality of this paper.

REFERENCES

ARMSTRONG, L., TEN HAVE, A. & JOHNSON, H. D. 1987. The geology of the Gannet Fields, Central North Sea, UK Sector, *In:* Brooks, J. & Glennie, K. (eds). *Petroleum Geology of the Northwest Europe: Proceedings of the 3rd Conference.* Geological Society, London, 533–548.

ATHEY, L. F. 1930. Density, porosity and compaction of sedimentary rocks, *American Association of Petroleum Geologists Bulletin*, **14**, 1–24.

BARKER, C. 1972. Aquathermal pressuring: Role of temperature in development of abnormal pressure zones. *American Association of Petroleum Geologists Bulletin*, **56**, 2068–2071.

——— 1990. Calculated volumes and pressure changes during the thermal cracking of oil to gas in reservoirs, *American Association of Petroleum Geologists Bulletin*, 74, 1254–1261.

BISHOP, R. S. 1979. Calculated compaction states of thick abnormally pressured shales. *American Association of Petroleum Geologists Bulletin*, **6**, 918–933.

BOWERS, G. L. 1994. Pore-pressure Estimation from Velocity Data: Accounting Mechanisms besides Undercompaction. Paper SPE 27488, presented at SPE Meeting, Dallas, February 15–18.

BURST, 1969. Diagenesis of Gulf Coast clayey sediments and its possible relation to petroleum migration. *American Association of Petroleum Geologists Bulletin*, **53**, 73–93.

CAYLEY, G. T. 1987. Hydrocarbon migration in the Central North Sea, *In* Brooks, J. & Glennie, K. (eds). *Petroleum Geology of the North-west Europe.* Graham & Trotman, London, 549–556.

CHAPMAN, R. E. 1980. Mechanical versus thermal causes of abnormally high pore-pressures in shales, *American Association of Petroleum Geologists Bulletin*, **64**, 2179–2183.

CHEN, W., GHAITH, A., PARK, A., & ORTOLEVA, P. 1990. Diagenesis through coupled processes: Modeling approach, self-organisation and implications for exploration. *In:* Meshri. I. D. & Ortoleva, P. J. (eds) *Prediction of Reservoir Quality through Chemical Modeling.* AAPG Memoir, **49**, 103–130.

COLTEN-BRADLEY, V. A. 1987. Role of pressure in smectite dehydration – Effects on geopressure and smectite-to-illite transformation. *American Association of Petroleum Geologists Bulletin*, **71**, 1414–1427.

DAINES, S. R. 1982. Aquathermal pressuring and geopressure evaluation, *American Association of Petroleum Geologists Bulletin*, **66**, 931–939.

DEMING, D. 1994. Factors necessary to define a pressure seal. *American Association of Petroleum Geologists Bulletin*, **78**, 1005–1009.

DICKINSON, G. 1953. Geological aspects of abnormal reservoir pressures in Gulf Coast Louisiana. *American Association of Petroleum Geologists Bulletin*, **37**, 410–432.

DONOVAN, A. D., DJAKIC, A. W., IONNIDES, N. S., GARFIELD, T. R. & JONES, C. R. 1993. Sequence stratigraphic control on Middle and Upper Jurassic reservoir distribution within the UK Central North Sea, *In:* Parker, J. R. (ed.) *Petroleum Geology of the Northwest Europe: Proceedings of the 4th Conference.* Geological Society, London, 251–270.

DU ROUCHET, J. 1981. Stress fields, a key to oil migration, *American Association of Petroleum Geologists Bulletin*, **65**, 74–85.

FERTL, W. H. 1976. Developments in Petroleum Geoscience, (vol 2), *Abnormal Formation Pressures*, Elsivier, Amsterdam.FISHER, M. J. & MUDGE, D. C. 1990. The Triassic, *In:* Glennie, K. W. (ed.) *Introduction to the Petroleum Geology of the North Sea.* Blackwell Scientific Publications, Oxford, 191–218.

GAARENSTROOM, L., TROMP, R. A. J., DE JONG, M. C. AND BRANDENBURG, A. M. 1993. Overpressures in the Central North Sea: Implications for trap integrity and drilling safety. *In:* Parker, J. R. (ed.) *Petroleum Geology of the Northwest Europe: Proceedings of the 4th Conference.* Geological Society, London, 1305–1313.

GLENNIE, K. W. & ARMSTRONG, L. A. 1991. The Kittiwake Field, Block 21/18, UK North Sea, *In:* Abbots, I. L. (ed.) *The United Kingdom Oil and Gas Fields, 25 Years Commemorative Volume*, The Geological Society London, 339–346.

GUSSOW, W. C. 1954. Differential entrapment of oil and gas: a fundamental principle, *American Association of Petroleum Geologists Bulletin*, **38**, 816–853.

HALL, P. L. 1993. Mechanisms of overpressurring: an overview, *In:* Manning, D. A. Hughes, C. P. L. & Hughes, C. R. (eds) *Geochemistry of Clay-Pore Fluid Interactions*, Chapman & Hall, London.

HEDBERG, H. 1974. Relationship of methane generation to undercompacted shale, shale diapirism and mud volcanoes. *American Association of Petroleum Geologists Bulletin*, **58**, 661–673.

HODGSON, N. A., FARNSWORTH, J. & FRASER, A. J. 1992. Salt-related Tectonics, sedimentation and hydrocarbon plays in the Central Graben, North Sea, UKCS, In: Hardman, R. F. P. (ed.), *Exploration Britain*, Geological Society, London, Special Publication **67**, 31–63.

HUNT, J. M. 1979. *Petroleum Geochemistry and Geology.* W. H. Freeman and Co., San Francisco.

——— 1990. Generation and Migration of Petroleum from Abnormally Pressured Fluid Compartments, *American Association of Petroleum Geologists Bulletin*, **74**, 1–12.

———, WHELAN, J. K., EGLINTON, L. B. & CATHLES, L. M. 1994. Gas generation – a major cause of deep gulf overpressures. *Oil and Gas Journal*, July 18, 59–63.

KNOX, R. W. O'B, & CORDEY, W. G. 1993a, *Lithostratigraphic nomenclature of the UK North Sea, 2. Cretaceous of the Central and Northern North Sea.* UKOOA.

——— & ——— 1993b. *Lithostratigraphic nomenclature of the UK North Sea, 3. Jurassic of the Central and Northern North Sea.* UKOOA.

LAW, B. E., & DICKINSON, W. W. 1985. Conceptual model for origin of overpressured gas accumulations in low permeability reservoirs, *American Association of Petroleum Geologists Bulletin*, **69**, 1295–1304.

LEONARD, R. C. 1993. Distribution of sub-surface pressure in the Norwegian Central Graben and applications for exploration. *In:* Parker, J. R. (ed) *Petroleum Geology of the Northwest Europe: Proceedings of the 4th Conference.* Geological Society, London, 1295–1303.

LOPATIN, N. V. 1971. Temperature and geological time as factors in coalification (in Russian). *Izv. Akad. Nauk SSSR, Seriya geologicheskaya*, 3, 95–106.

MAGARA, K. 1978. Developments in Petroleum Science vol. 9, *Compaction and Fluid Migration*, Elsevier, New York.

MEISSNER, F. F. 1978. Petroleum geology of the Bakken Formation, Williston Basin, North Dakota, *In: 24th Annual Conference, Williston basin Symposium, Montana Geological Society*, pp. 207–227.

——— 1985. Regional hydrocarbon generation, migration and accumulation pattern of Cretaceous strata, Powder River Basin, *American Association of Petroleum Geologists Bulletin*, **69**, 856.

MILES, J. A. 1990. Secondary migration paths in the Brent Sandstone of the Viking Graben and East Shetland Basin. Evidence from oil residues and subsurface data. *American Association of Petroleum Geologists Bulletin*, 74, 1718–1735.

MOUCHET, J. P. & MITCHELL, A. 1989. *Abnormal pressures while drilling: Origins–Predictions–Detection– Evaluation.* Manuels technique Elf Aquitaine No. 2. Elf Aquitaine Editions, Boussens.

ORTOLEVA, P. J. 1994. Basin compartmentation: Definitions and Mechanisms, *In:* Ortoleva, P. J. (ed.) *Basin Compartment and Seals.* AAPG Memoir, **61**, 39–51.

OSKAYA, I, 1988. A simple analysis of oil-induced fracturing in sedimentary rocks. *Marine and Petroleum Geology*, **5**, 293–297.

PEKOT, L. J. & GERSIB, G. A. 1987. Ekofisk, *In:* Spencer *et al.* (eds) *Geology of the Norwegian Oil and Gas Fields.* Graham & Trotman, London.

PENGE, J., TAYLOR, B., HUCKERBY, J. A. & MUNNS, J. W. 1993. Extension and salt tectonics in the East Central Graben, *In:* Parker, J. R. (ed.) *Petroleum Geology of Northwest Europe: Proceedings of the 4th Conference.* Geological Society,London, 1197–1209.

PERRY, E. & HOWER J. 1972. Burial diagenesis in Gulf Coast pelitic sediments. *Clays Clay Mineral*, **18**, 165–177.

PLUMLEY, W. J. 1980. Abnormally high fluid pressures: Survey of some basic principles. *American Association of Petroleum Geologists Bulletin*, **64**, 414–430.

PRICE, J., DYER, R., GOODAL, I., MCKIE, T., WATSON, P., & WILLIAMS, G. 1993. Effective stratigraphic subdivision of the Humber Group and the late Jurassic evolution of the UK Central Graben, *In:* Parker, J. R. (ed.) *Petroleum Geology of Northwest Europe: Proceedings of the 4th Conference.* Geological Society, London, 443–458.

REHM, W. 1972. Worldwide occurrences of abnormal pressures, Part II. Paper SPE 3845, presented at SPE Meeting, Baton Rouge, May 15–16.

ROBERTS, S. J. & NUNN, J. A. 1995. Episodic fluid expulsion from geopressured sediments. *Marine and Petroleum Geology*, **12**, 195–204.

ROBSON, D. 1991. The Argyll, Duncan and Innes Fields, Blocks 30/24, 30/25a UK North Sea, *In:* Abbots, I. L. (ed.) *The United Kingdom Oil and Gas Fields 25 Years Commemorative Volume.* Geological Society, London,

219–226.

SALES, J. K. 1993. Closure vs seal capacity – a fundamental control on the distribution of oil and gas, *In:* Dore. A. G. *et al.* (eds), ***Basin Modelling: Advances and Applications***, NPF Special Publication **3**, Elsevier, Amsterdam, 399–414.

SCHOWALTER, T. (1979). Mechanisms of secondary hydrocarbon migration and entrapment, ***American Association of Petroleum Geologists Bulletin***, **63**, 723–760.

SEARS, R. A., HARBURY, A. R., PROTOY, A. J. G. & STEWART, D. J. 1993. Structural styles from the Central Graben in the UK and Norway, *In:* PARKER, J. R. (ed.) ***Petroleum Geology of the Northwest Europe: Proceedings of the 4th Conference.*** Geological Society, London, 1231–1244.

SPENCER, C. 1987. Hydrocarbon generation as a mechanism for overpressuring in the Rocky Mountain Region, ***American Association of Petroleum Geologists Bulletin***, **71**, 368–388.

STEVENS, D. A. & WALLIS, R. J. 1991. The Clyde Field, Block 30/17b, *In:* Abbots, I. L. (ed.) ***The United Kingdom Oil and Gas Fields, 25 Years Commemorative Volume***, Geological Society, London, 279–286.

STOCKBRIDGE, C. P. & GRAY, D. I. 1991. The Fulmar Field, Block 30/16 & 30/11b, UK North Sea, *In:* Abbots, I. L. (ed.) ***The United Kingdom Oil and Gas*** Fields, 25 Years Commemorative Volume, Geological Society, London, 309–316.

TERZAGHI, K & PECK, R. B. 1948. ***Soil Mechanics in Engineering Practice.***, J. Wiley, New York.

TISSOT, B. PP., & WELTE, D. H. 1984. ***Petroleum Formation and Occurrence.*** Springer-Verlag, Berlin.

TREWIN, N. H. & BRAMWELL, M. G. 1991. The Auk Field, Block 30/16, UK North Sea. *In:* Abbots, I. L. (ed.) ***The United Kingdom Oil and Gas Fields, 25 Years Commemorative Volume.*** Geological Society, London, 227–236.

WAPLES, D, 1981. ***Organic geochemistry for exploration geologists.***, Burgess Pub. Co.

WARD, C. D., COGHILL, K. & BROUSSARD, M. D. 1994. The application of petrphysical data to improve pore and fracture pressure determination in the North Sea Central Graben HPHT wells. Paper SPE 28297, presented at SPE Meeting, New Orleans, September 25–28.

WENSAAS, E. T., SHAW, H. F., GIBBONS, K., AAGAARD, PP. & DYPVIK, H. 1994. Nature and Causes of overpressure in the mudrocks of the Gullfaks Area, North Sea. *Clay Minerals*, **29**, 439–449.

WILLS, J. M. 1991. The Forties Field, Blocks 21/10, 22/6a, UK North Sea, *In:* Abbots, I. L. (ed.) ***The United Kingdom Oil and Gas Fields, 25 Years Commemorative Volume.*** Geological Society London, 301–308.

Using 3D seismic to understand the structural evolution of the UK Central North Sea

J.W. Eggink, D.E. Riegstra and P. Suzanne

Shell UK Exploration and Production, Shell-Mex House, Strand, London WC2R 0DX, UK.

ABSTRACT: Interpretation of regional 3D seismic data suggests that the structural evolution of the Central North Sea occurred in three successive structural regimes. Two successive extensional tectonic regimes from Late Palaeozoic to Early Cretaceous were followed by a predominantly compressive tectonic regime from the Late Cretaceous into the Tertiary.

Tectonic movements were controlled by the relative orientation of the stress fields with respect to the fault strikes of older major Variscan fault trends. The amount and direction of strike-slip movement was controlled by a gradual clockwise rotation of the minimum effective stress in the horizontal plane from approximately NE–SW to E–W in the first two phases. Within this framework, halokinesis is only of local importance, and serves to amplify the tectonically controlled structuration.

The structural model developed here explains the observed distribution of fields and structures in the Central Graben, as well as enabling prediction of structural development in its less well explored portions.

KEYWORDS: *geology, structure, strike-slip, stress, UK Central North Sea*

INTRODUCTION

The structural evolution of the Central North Sea Graben (Fig. 1) has been the subject of an increasing number of studies, following the acquisition of 3D seismic datasets by various operators. Many of these studies have been documented in the *Petroleum Geology of North-West Europe: Proceedings of the 4th Conference* (Parker 1993). A review of the range of structural models which has been proposed for the North Sea over the last decade, is given by Williams (1993). The two end-members comprise models invoking, on the one hand, simple normal (listric or planar) faulting, and on the other, more complicated models related to wrench tectonics (strike/oblique slip).

The structural complexity of the Central North Sea (CNS) has been enhanced because of the presence of (Early Palaeozoic) basement lineaments, attributed to the Caledonian and Variscan orogenies and the presence of the Trans European Fault Zone (Coward 1990; Bartholomew *et al.* 1993). These lineaments, many of which trend NW–SE to NNW–SSE in the CNS, have controlled the subsequent Mesozoic and Cenozoic evolution of the Central North Sea Basin.

Salt withdrawal has played a role in the structuration of the area. However, it may be argued that the movement of salt was, in many cases, triggered by reactivation of fault zones and, as such, has been of secondary importance to the structural development. Examples of the relationship between faulting and halokinesis in the CNS are given by Erratt (1993).

First published in *Petroleum Geoscience*, Vol. **2**, 1996, pp. 83–96.

The role of oblique slip in the structural evolution of the North Sea has been documented by Bartholomew *et al.* (1993). Studies over the Central Graben of the UK and Norway, applying oblique-slip tectonics, have been described by Sears *et al.* (1993). Both models describe rifting in the North Sea as a response to an approximate E–W extensional system from the Late Jurassic onwards.

A structural study by Thomson & Underhill (1993) in the Inner Moray Firth Basin indicates that most structural styles appear to have developed as a result of dip-slip extension and thermal subsidence. Two extensional phases are recognized in the Inner Moray Firth during the Permo-Triassic and Late Jurassic, followed by regional uplift and local inversion as a result of compression during the Tertiary.

The model proposed for the Central Graben in this paper agrees with an overall extensional setting from Late Palaeozoic to Early Cretaceous as suggested by Thomson & Underhill (1993) and Bartholomew *et al.* (1993). The minimum effective stress is suggested to have rotated over this period from NE–SW to ENE–WSW. In the Late Cretaceous to Tertiary, the maximum effective stress moved into the horizontal plane in an approximate N–S direction, with the minimum effective stress only then orientated E–W (Fig. 2), where it broadly remains today.

DATABASE AND METHODOLOGY

The database used for this structural evaluation comprises an extensive regional 2D seismic grid, together with some fifteen 3D seismic datasets (Fig. 3). The total areal coverage of the 3D surveys amounts to 5000 km^2. Five key horizons, which were mapped on the 2D and 3D data, were selected for the compilation of regional maps over the entire Central Graben :

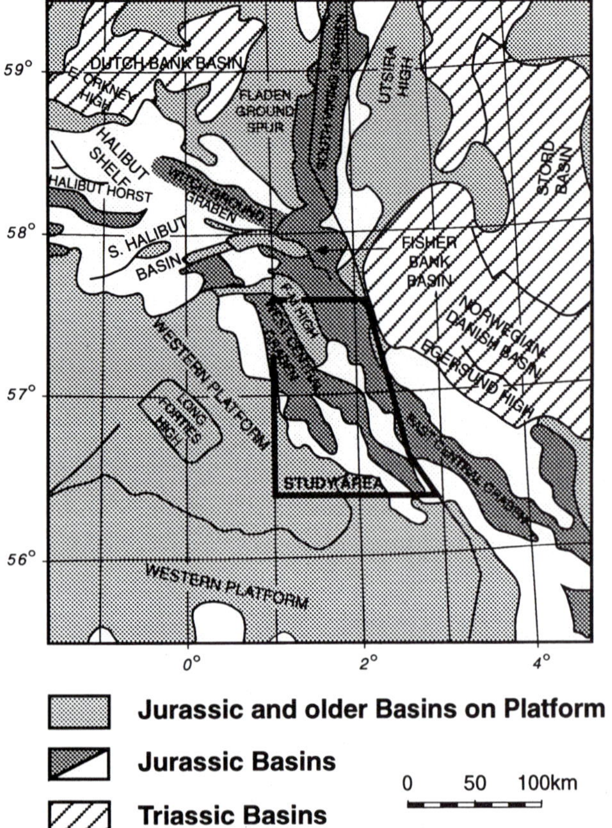

Fig. 1. Location map of study area. Note overall NW–SE trend of UK Central Graben, a relict of old NNW–SSE basement lineaments, which were subject to later rifting and compression.

Top Rotliegend, Base Cretaceous, Base Chalk, Top Chalk and Top Balder (Fig. 4).

Illumination techniques have been applied to these regional maps to enhance the expressions of structural geometry (e.g. Fig. 6). Dedicated software is used to simulate the effect of a light source shining onto a surface. The maps are illuminated from an azimuth which is approximately at right angles to the prevailing structural grain in order to highlight fault trends in an optimal way. The angle of inclination of the illuminating source is chosen so as to produce areas of light and shadow which enhance the expression of highs, lows, and fault trends. In some areas the effect of interpolated 2D data can be observed on the illumination displays.

THE CENTRAL GRABEN

The Central Graben area of the North Sea is a NW–SE trending rift basin bounded by large faults which are offset at intervals by W–E trending faults (Fig. 1). Within the basin the Western and Eastern Central grabens can be identified, separated by a series of intra-graben highs such as the Forties–Montrose High. The northern end of the basin is defined by the 'triple junction', where the Witch Ground Graben and South Viking Graben meet the northern extension of the Eastern Central Graben. In the southeast, the Eastern Central Graben extends into offshore Norway and Denmark.

PROPOSED MODEL

An E–W cross-section over the Central Graben shows the predominantly extensional setting of this rift basin with major normal displacements along the main boundary faults (Fig. 5). Faulting has not been restricted to dip-slip movements in an extensional setting. The presence of old structural lineaments has influenced structuration in the Central North Sea, and both

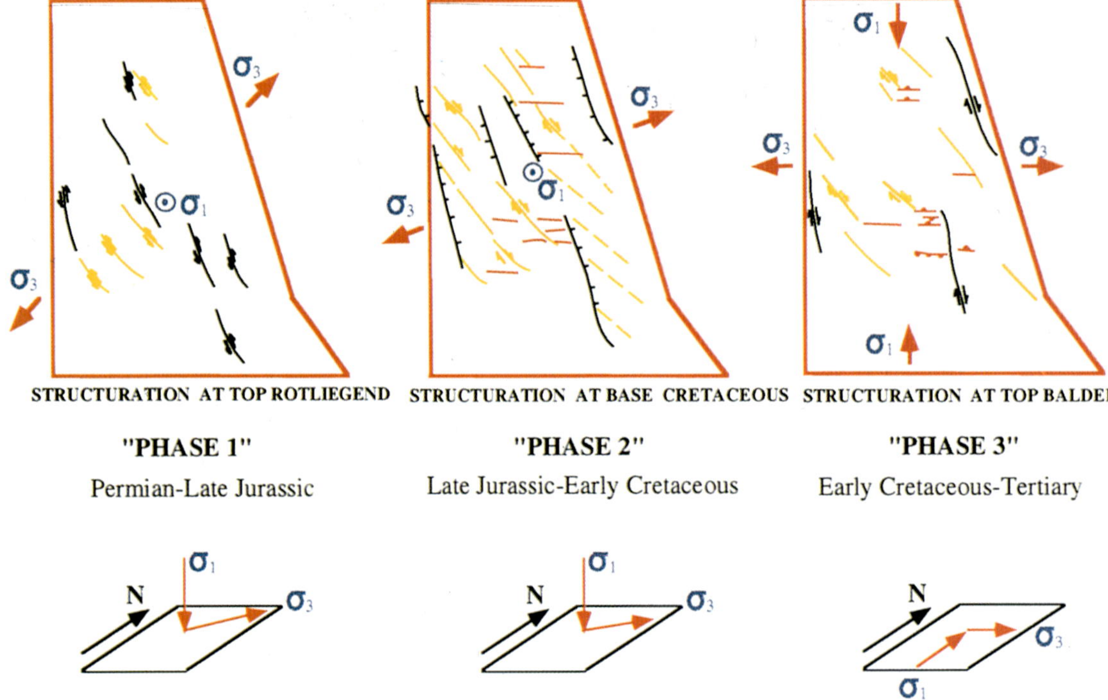

Fig. 2. Conceptual overview of the evolution of stress regimes. The major change over time was from an overall extensional regime from the Permian to the Early Cretaceous (Phases 1 and 2) to a compressional regime from the Early Cretaceous to the present (Phase 3). Continuous rotation of the minimum effective stress (σ_3, indicated with red arrows) from NE–SW to E–W during the course of Phases 1and 2 controlled movements along the main pre-Permian faults (indicated in black), and their associated riedels (indicated in yellow). Movements changed from sinistral oblique slip in Phase 1 to pure extension and dextral oblique slip in Phase 2. During these movements E–W extensional accommodation faults (shown in red) developed between the main faults. In Phase 3, continued clockwise rotation of the minimum effective stress, combined with the now N–S orientated maximum effective stress, produced dextral transpressional movements and inversions of the E–W accommodation faults.

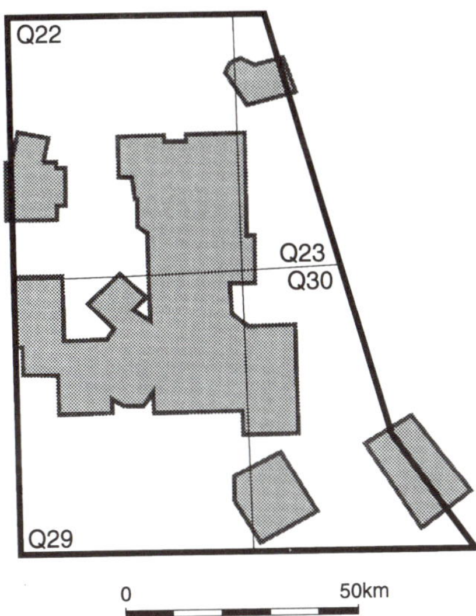

Fig. 3. 3D seismic coverage over study area. Outside the indicated areas, extensive 2D regional seismic data were used.

oblique-slip and strike-slip faulting have occurred in both transtensional and transpressional settings.

The geometry of the fault pattern observed at key horizons suggests that a simple three-phase tectonic model can be applied over the Central Graben. While the first two phases are interpreted to be extensional in nature (from Late Palaeozoic to Early Cretaceous), the last phase (from Late Cretaceous to Tertiary) is characterized by E–W orientated inversions as a result of N–S compression.

Phase 1 (Late Palaeozoic–Early Mesozoic)

An illuminated display of the Top Rotliegend isochron map highlights three main fault trends (Fig. 6):

- the main boundary faults of the Central Graben and its central highs essentially trend NNW–SSE;

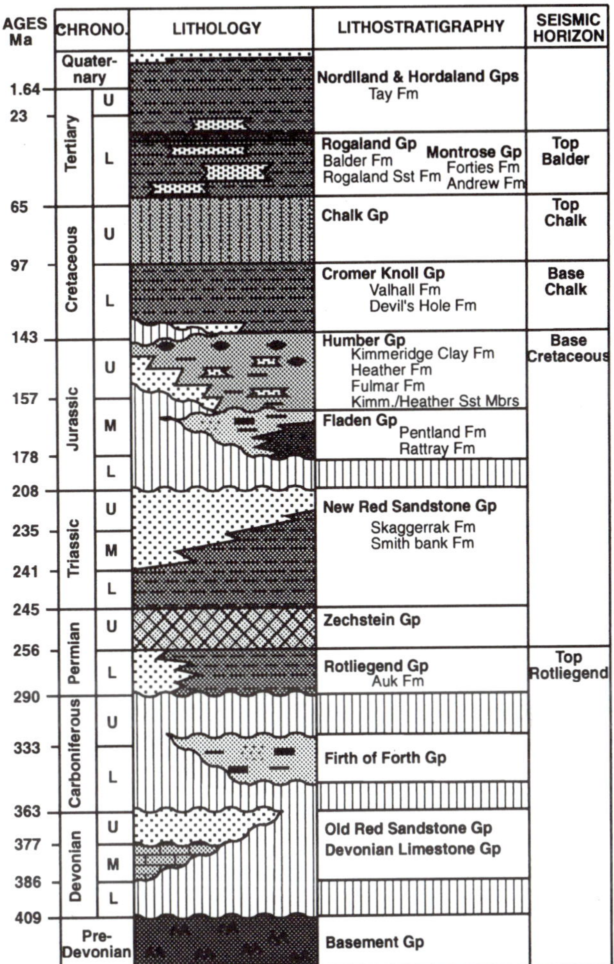

Fig. 4. Central North Sea stratigraphy and mapped seismic horizons.

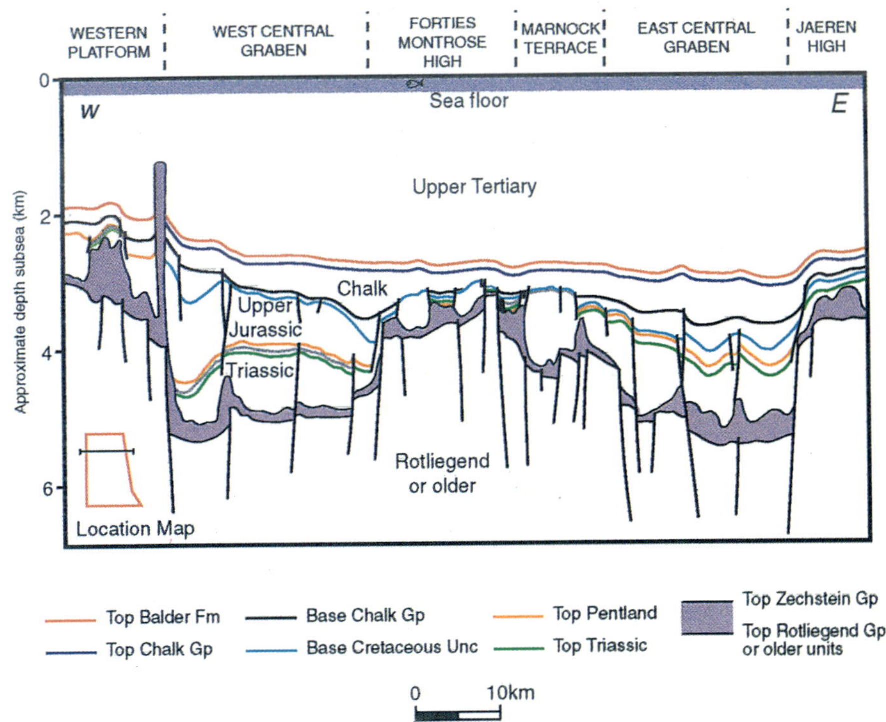

Fig. 5. Schematic east-west cross-section of the Central Graben. The overall extensional character of the Central Graben is evidenced by the observed major normal displacements along the main boundary faults.

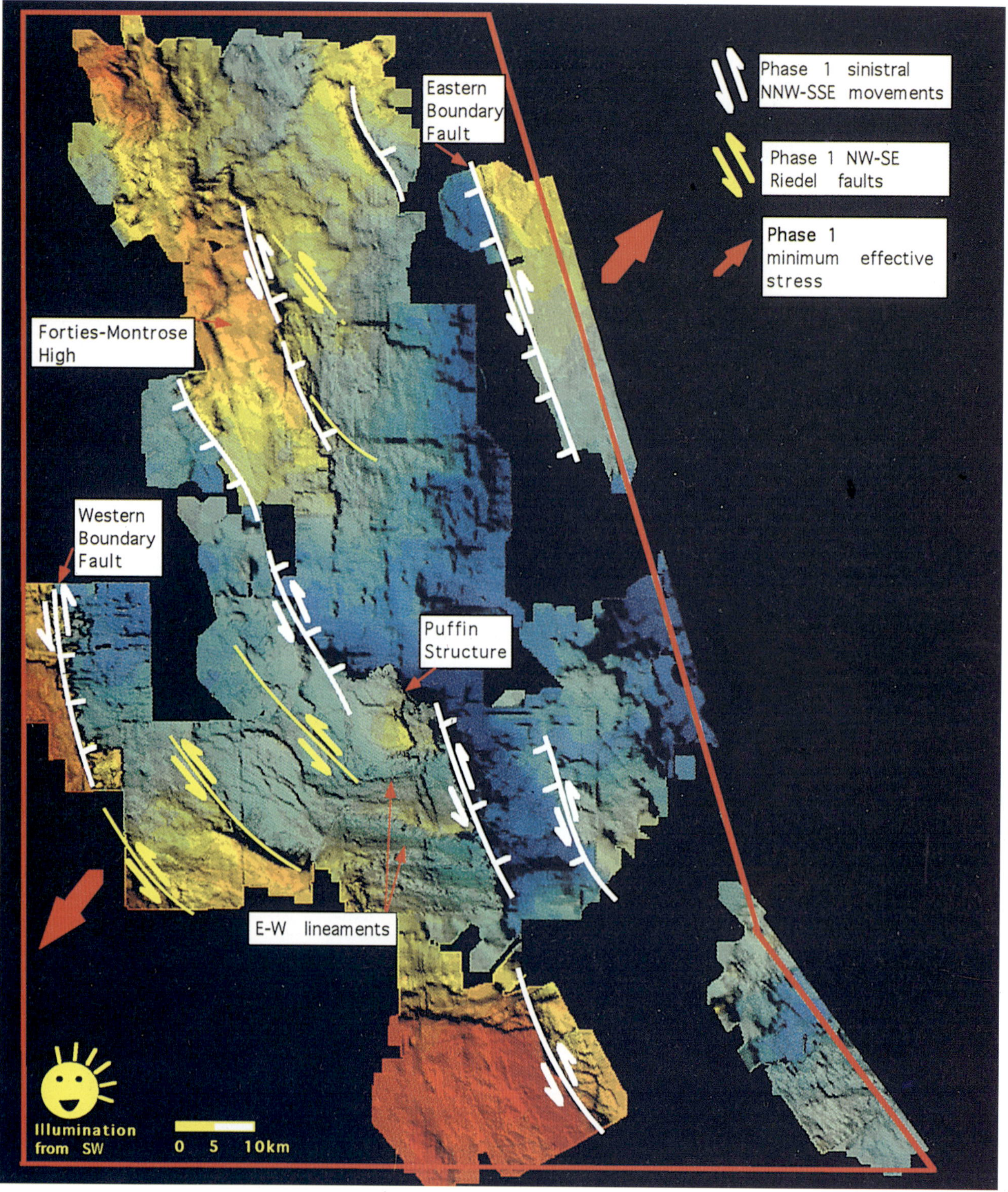

Fig. 6. Illumination map at top Rotliegend level. The map is created by simulating the effect of shining light (in this case from a source to the SW, at an elevation of 45 degrees) onto the mapped surface.

- a series of NW–SE trending faults connect the main boundary faults;
- E–W normal faults bound the isolated highs in the graben.

There is a debate as to when the main boundary faults were formed. They are probably at least as old as the Triassic (deduced from the observed thickness difference between the Triassic on the platforms and in the graben, Smith *et al.* 1993). The boundary faults are more likely of Late Permian or older age, as the southern and western boundaries of the North Zechstein Basin were delimited by the same fault trends.

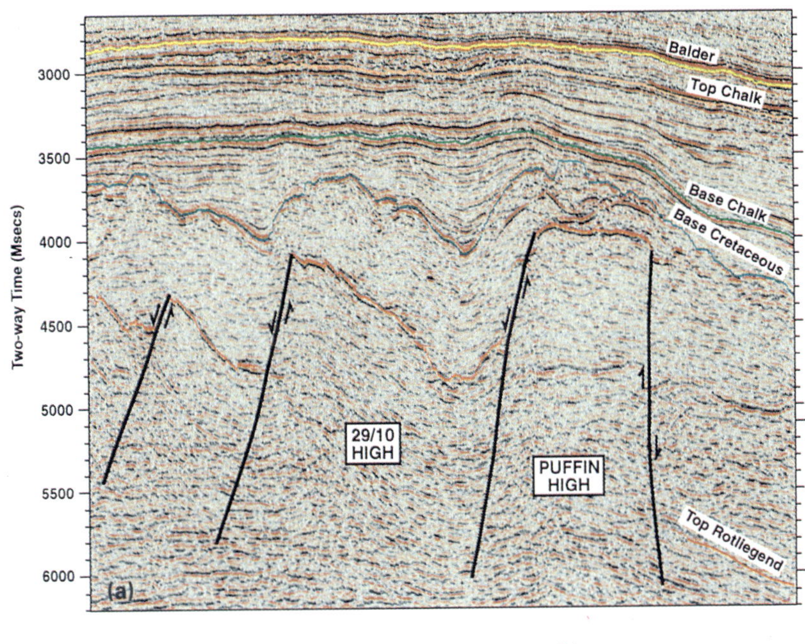

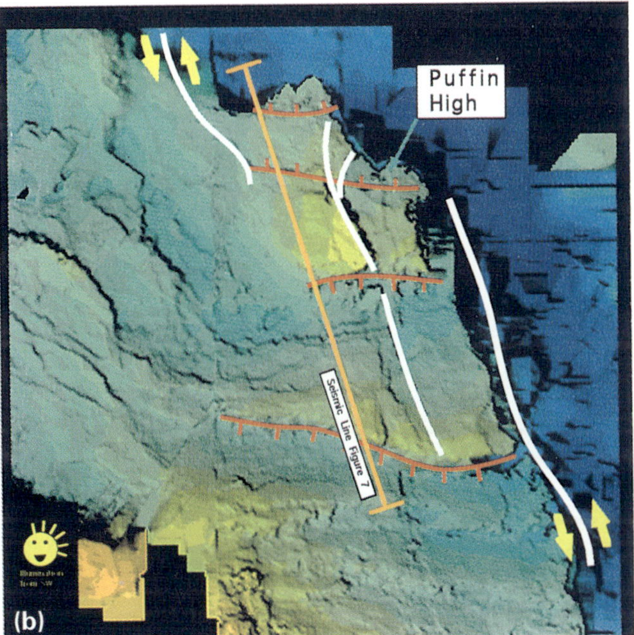

Fig. 7. (**a**) A 3D seismic line over Block 29/10 and Puffin High. Note the large normal displacements along faults at Top Rotliegend level. These normal faults are interpreted to have been created in the Late Permian when sinistral oblique slip along two major NNW–SSE trending faults created a local low (pull-down) in the Puffin area, bounded to the N and S by E–W trending normal faults. (**b**) Location map of seismic line (overlies Rotliegend illumination).

Detailed seismic interpretation, carried out over selected areas in the Central Graben (the Puffin Structure (blocks 29/4,5,9,10) and the Western and Eastern Boundary faults), has resulted in a better understanding of the observed structural geometry.

The Puffin Structure is located between two *en echelon* NNW–SSE trending faults and is crossed by a number of E–W faults. A seismic cross-section over the Puffin High indicates that these E–W orientated faults have normal displacements (Fig. 7). This observation suggests that the Puffin horst block resulted from sinistral oblique slip along the two *en echelon* NNW–SSE trending basement faults.

Indications of sinistral oblique slip in the earlier stages of the basin history have also been observed along the boundary faults of the Central Graben. Detailed interpretation of the Western Boundary Fault in Block 21/25, suggests the development of a series of right-stepping faults above the Western Boundary Fault (Fig. 8). In the northeastern area of the Central Graben similar right-stepping faults have been tentatively interpreted.

The occurrence of a series of right-stepping faults above a basement fault has been described before (for the CNS by, e.g. Bartholomew *et al.* 1993). Such an arrangement has been modelled in sandbox experiments such as described by Koopman *et al.* (1987). Figure 9 shows the result of such a sandbox experiment. Sinistral movement applied along an underlying fault results in the development of a series of right-stepping faults in the overburden.

During this tectonic phase the major NW–SE intra-basinal faults developed. They are interpreted as (mega) riedel faults and are trending at an angle of 10–30 °C with respect to the major NNW–SSE boundary faults. The mega-riedel faults are best expressed during the younger tectonic phases (see below) when reversal from a sinistral movement to a dextral movement along the faults took place, enhancing structural relief.

An isochore map constructed between Top Rotliegend and Base Cretaceous (Fig. 10) shows that a considerable thickness of sediments was deposited in a series of major depocentres, adjoining the NNW–SSE trending faults. These deep basins are interpreted to have been developed in a transtensional (oblique-slip) setting, with basinal axes trending NNW–SSE to N–S. An E–W seismic line, situated some ten kilometres southeast of Puffin, shows that thick Triassic and Jurassic sediments were deposited in these deep basins (Fig. 11).

The observation of a sinistral horizontal component in this extensional setting (mainly observed along NNW–SSE pre-existing trends) is interpreted to indicate an approximate NE–SW minimum effective stress (σ_3) direction (Figs 2 & 6).

The timing of the first structural phase is envisaged to be approximately Late Palaeozoic to Early Mesozoic, probably Permian to Late Jurassic. Permo-Triassic sequences are

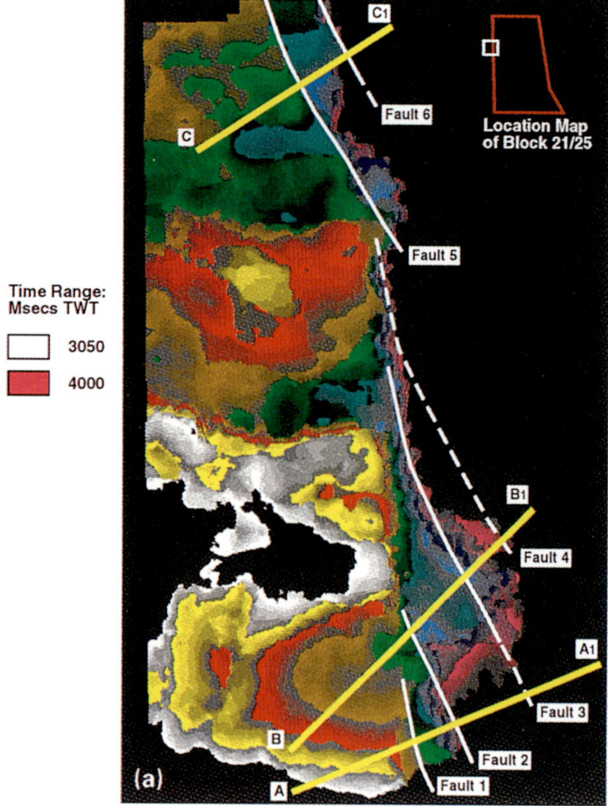

Jurassic is interpreted to be the result of this doming event (Fig. 11).

Phase 2 (Late Jurassic–Early Cretaceous)

During this time interval, normal displacements were still active along the main boundary faults of the Central Graben, indicating a dominantly extensional regime (Fig. 5). The illumination display generated at Base Cretaceous level (Fig. 12) shows again the three main fault trends which were observed at Top Rotliegend level. The NW–SE orientated (mega) riedel faults are even better expressed at this level. Along these faults pull-apart and pop-up structures have been developed.

An illumination map at Top Triassic level in the Greater Curlew area (block 29/8) shows an example of these pull-apart and pop-up structures (Fig. 13). The NW–SE (mega) riedel fault is delineated by a break of slope. It is composed of a series of right-stepping *en echelon* faults, evidencing earlier (Phase 1) sinistral movement. North of this fault a series of kilometre wide pull-apart basins and pop-up structures has been generated by (Phase 2) right lateral displacements along (secondary) riedel faults.

Fig. 14 shows the distribution over the Central Graben of pre-Cretaceous hydrocarbon discoveries. These appear to be dominantly aligned along the NW–SE faults. The location of these structures (together with the salt diapirs) along the NW–SE faults is envisaged to be the result of the reversal

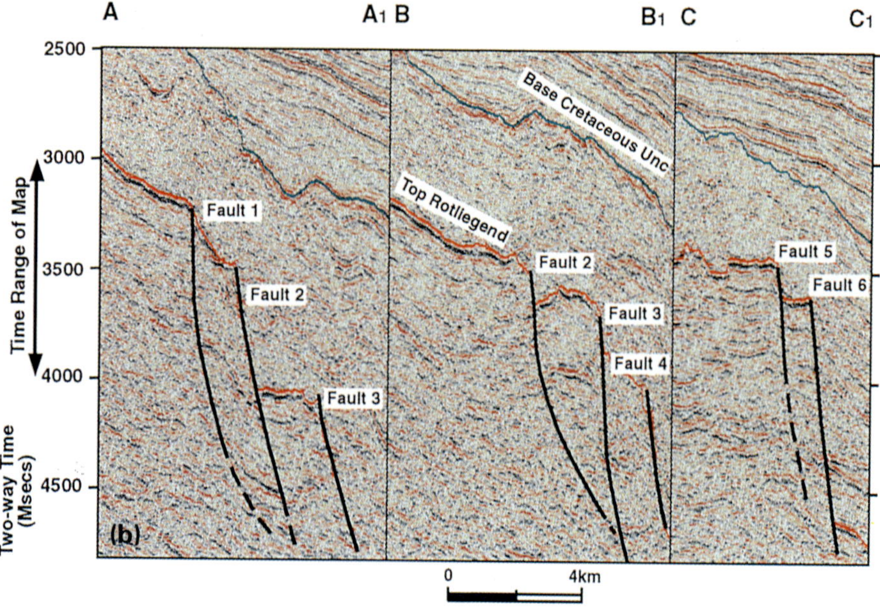

Fig. 8. The Western Boundary Fault, an example of right-stepping faults in block 21/25: (**a**) Top Rotliegend 3D isochrome map; (**b**) seismic lines. Six individual right-stepping faults can be traced along this part of the Western Boundary Fault at Top Rotliegend level. At depth these faults are interpreted to converge into the deeper-seated main fault.

interpreted to onlap the Rotliegend, suggesting that movement along the faults started as early as Late Palaeozoic. Thickness variations of the Triassic (Skagerrak) sequence could be indicative of a continuation of this phase into the early part of the Mesozoic (e.g. Smith *et al.* 1993). Dating of the phase, however, is difficult for various reasons: the effect of halokinesis on the deposition of the post-salt sequences in the Central Graben; the decoupling effect of salt on the Mesozoic sequences which overly the Palaeozoic; the difficulty in seismic interpretation of the Base Triassic, even on 3D seismic data; and the effect of Middle Jurassic doming in the Central North Sea. A shift in depocentres between the Triassic and

of movement along these (mega-riedel) faults: releasing bends were changed into restraining bends resulting in structuration.

Along the major NNW–SSE trending faults no clear indications of oblique shear (reversal of movement) have been observed, only normal vertical dip-slip occurred.

A (time) isochore constructed over the interval Base Cretaceous to Base Chalk confirms that depocentres developed along the main NNW–SSE faults, as well as in NW-SE trending basins along the riedel faults (Fig. 10). Basin axis polarity has shifted somewhat to the west as compared to the first phase, where mainly NNW–SSE to N–S trending basins were observed.

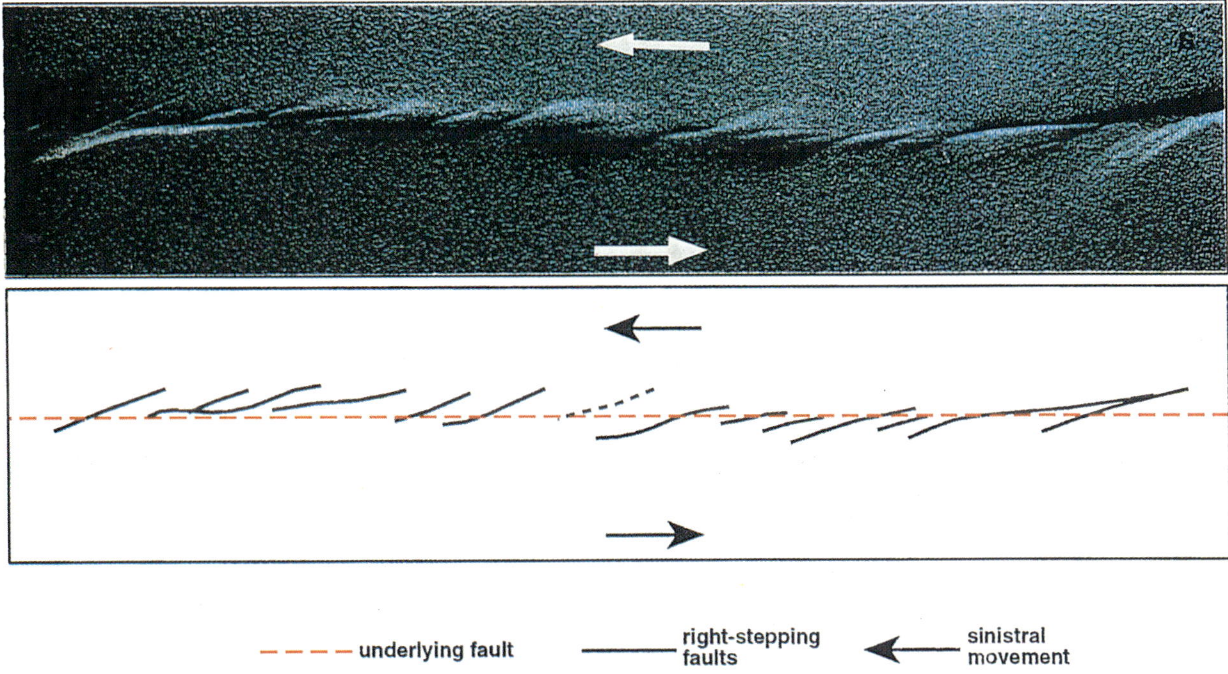

Fig. 9. A sandbox experiment. Sinistral motion along a deep-seated fault produces a right-stepping fault pattern in the sedimentary overburden. This pattern is seen at Rotliegend level along the Western and Eastern boundary faults of the Central Graben (see Fig. 8).

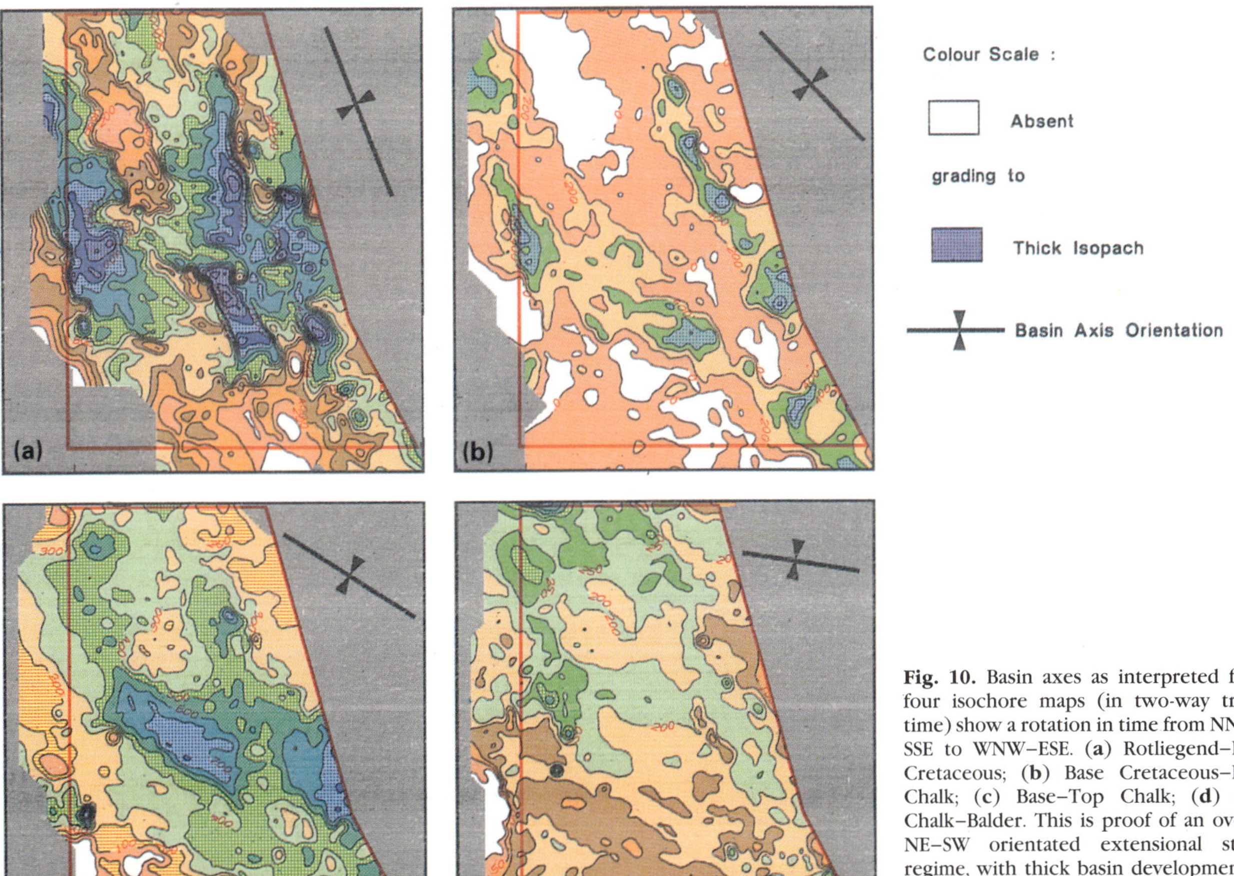

Fig. 10. Basin axes as interpreted from four isochore maps (in two-way travel time) show a rotation in time from NNW–SSE to WNW–ESE. (**a**) Rotliegend–Base Cretaceous; (**b**) Base Cretaceous–Base Chalk; (**c**) Base–Top Chalk; (**d**) Top Chalk–Balder. This is proof of an overall NE–SW orientated extensional stress regime, with thick basin development in the Mesozoic and the Cretaceous, grading to a broadly N–S compressional regime with little basin development in the Late Cretaceous/Tertiary.

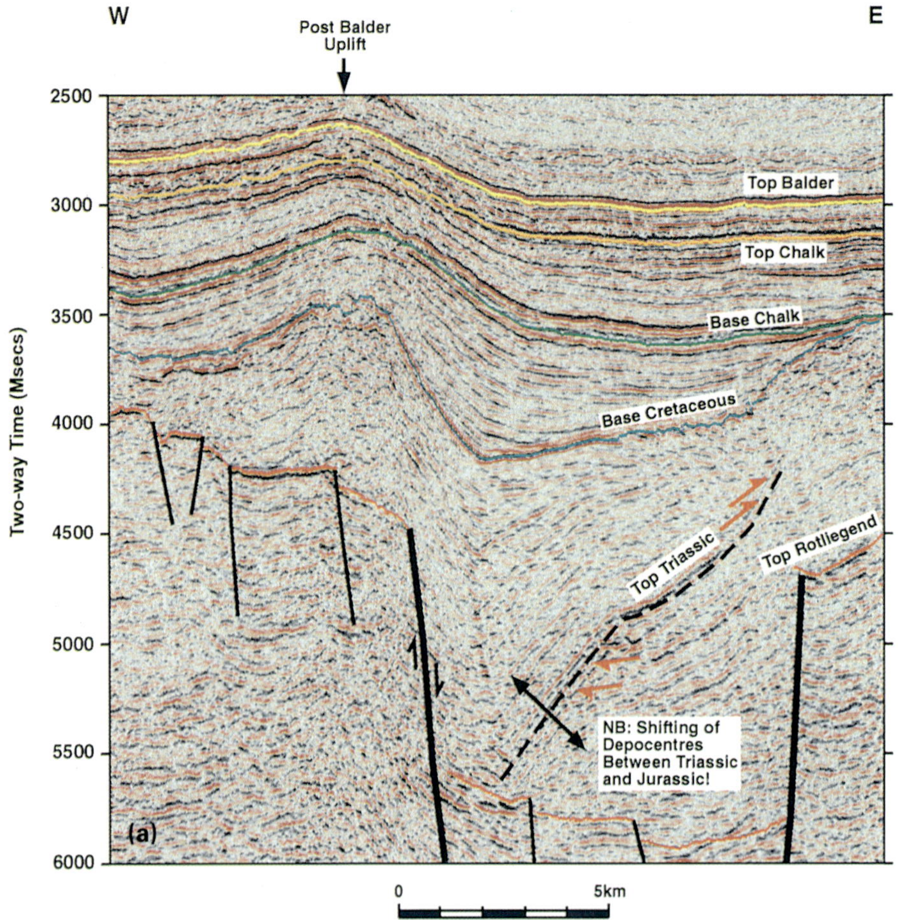

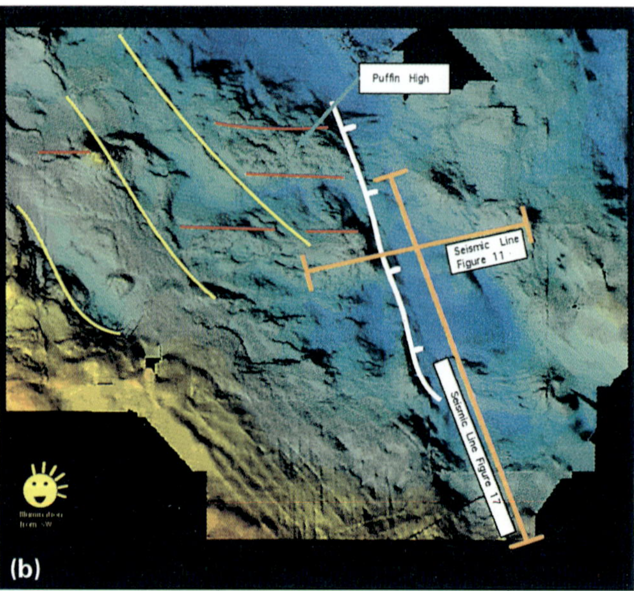

Fig. 11. (**a**) East–West 3D seismic line in the southern Central Graben showing thick Triassic, Jurassic and Cretaceous sediments. A major depocentre developed east of the main boundary fault as a result of extension during Phase 1 and (to a lesser extent) Phase 2. Later inversion during the Eocene and/or younger periods is thought to be the result of transpression related to the Alpine Orogeny (Phase 3). (**b**) Location map of seismic line (overlies Base Cretaceous illumination).

The observation of dextral lateral movements along the NW–SE riedel faults, together with normal displacements along NNW–SSE faults, suggests that the minimum effective stress in this extensional setting was now directed approximately ENE–WSW (Figs 2 & 12).

The first sediments that filled the basins which were created during this phase, belong to the Lower Cretaceous Cromer Knoll Group (Fig. 4). These strata show growth along the active faults, suggesting that this phase of dextral movement encompassed the Late Jurassic to Early Cretaceous periods. Of the three phases described in this paper, Phase 2 is interpreted to constitute the shortest and most intense.

The (time) isopach for the Chalk Group shows that the Chalk Basin was orientated WNW–ESE (Fig. 10). During the latter part of the second phase the major depocentre shifted more to the centre of the present-day Central Graben, probably as the result of thermal subsidence and continuing rotation of the minimum effective stress.

Phase 3 (Late cretaceous - Tertiary)

The third phase was dominated by the occurrence of major inversions and compressional features along the E–W orientated faults (Fig. 15). Compressional features are well expressed on a north–south seismic line over the Puffin High (Fig. 16). These features can be dated as being of Late Cretaceous to Tertiary age.

A major inversion occurred in the area southeast of Puffin, where a thick sequence of Jurassic, Cretaceous and Tertiary sediments was inverted. The inversion appears to have been episodic but continuous from Late Cretaceous to Miocene, with a sharp pulse occurring at Campanian time after deposition of the Hod Formation (Fig. 17).

The maximum effective stress apparently moved into the horizontal plane during this phase, probably in relation to the Alpine Orogeny (compression from south, beginning in the Campanian as the African continent commenced its

Fig. 12. Illumination map at Base Cretaceous level. See text and Fig. 6 for explanation of the illumination process.

northward drift). It is important to note that a major change occurred during the transition from the second tectonic phase to this third tectonic phase : a change from an overall oblique extension (transtension, related to Atlantic rifting) to strike-slip (transpression, related to the Alpine Orogeny).

The isopach constructed over the interval Top Chalk to Top Balder shows that no major depocentres developed during the Late Cretaceous to Tertiary (Fig. 10). The absence of such depocentres would be expected in a compressional rather than extensional setting. The isopach map suggests that only some (shallow) east–west orientated basins developed.

The inversions and compressional features which are

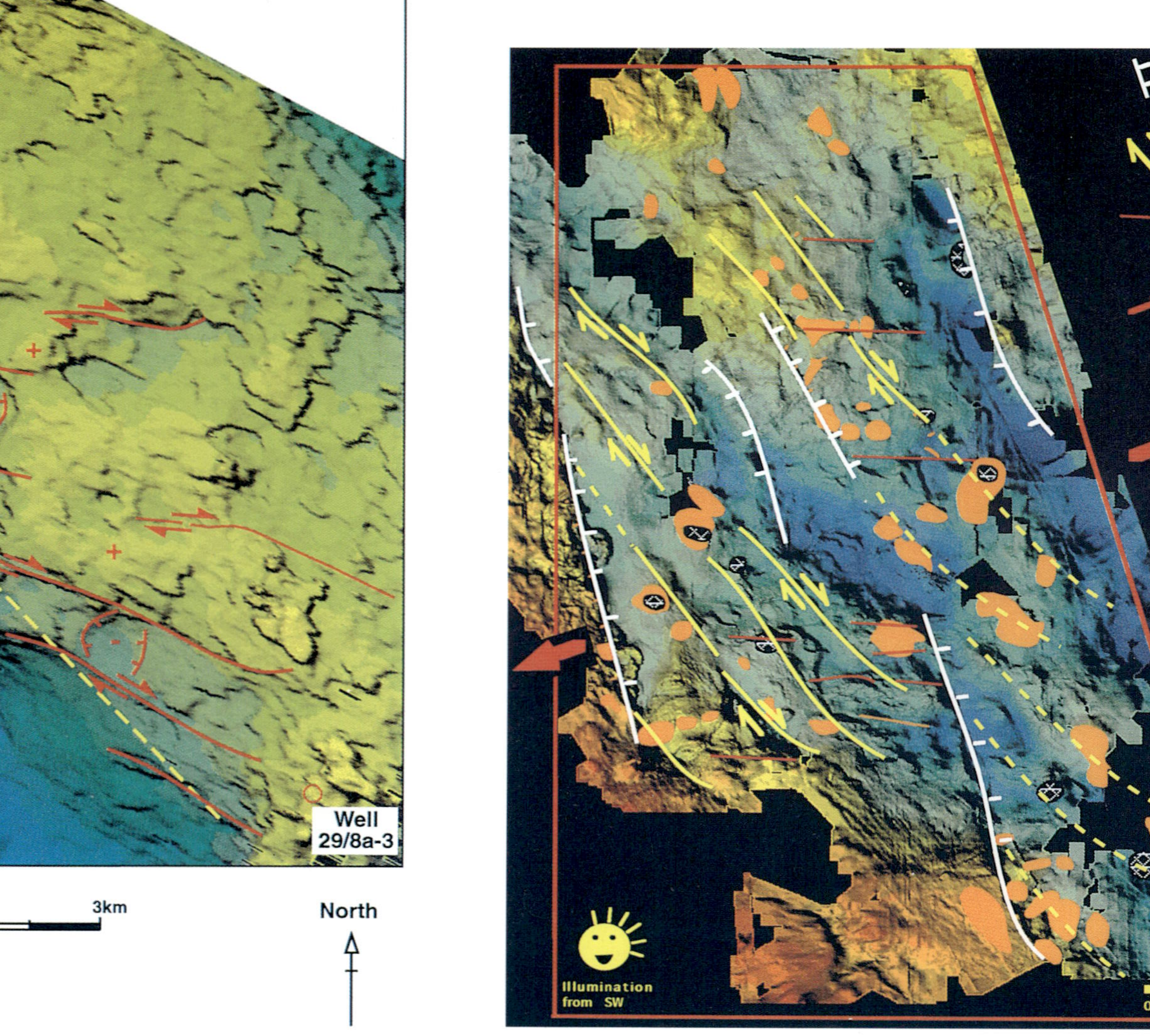

Fig. 13. Top Triassic illumination map with pull-aparts and pop-ups in the greater Curlew area. The main NW–SE fault is delineated by a break in slope. The red faults are riedel faults which are right-stepping and are rooted to the main NW–SE fault, evidencing earlier (Phase 1) extensional movements. The positions of kilometre-scale pull-apart basins and pop-ups are controlled by (Phase 2) right-dextral movement along the (red) riedel faults. Illumination of the top Triassic surface is from the north.

Fig. 14. Illumination map at Base Cretaceous level with distribution of pre-Cretaceous hydrocarbon discoveries and salt domes. Both types of features are dominantly aligned along the NW–SE faults.

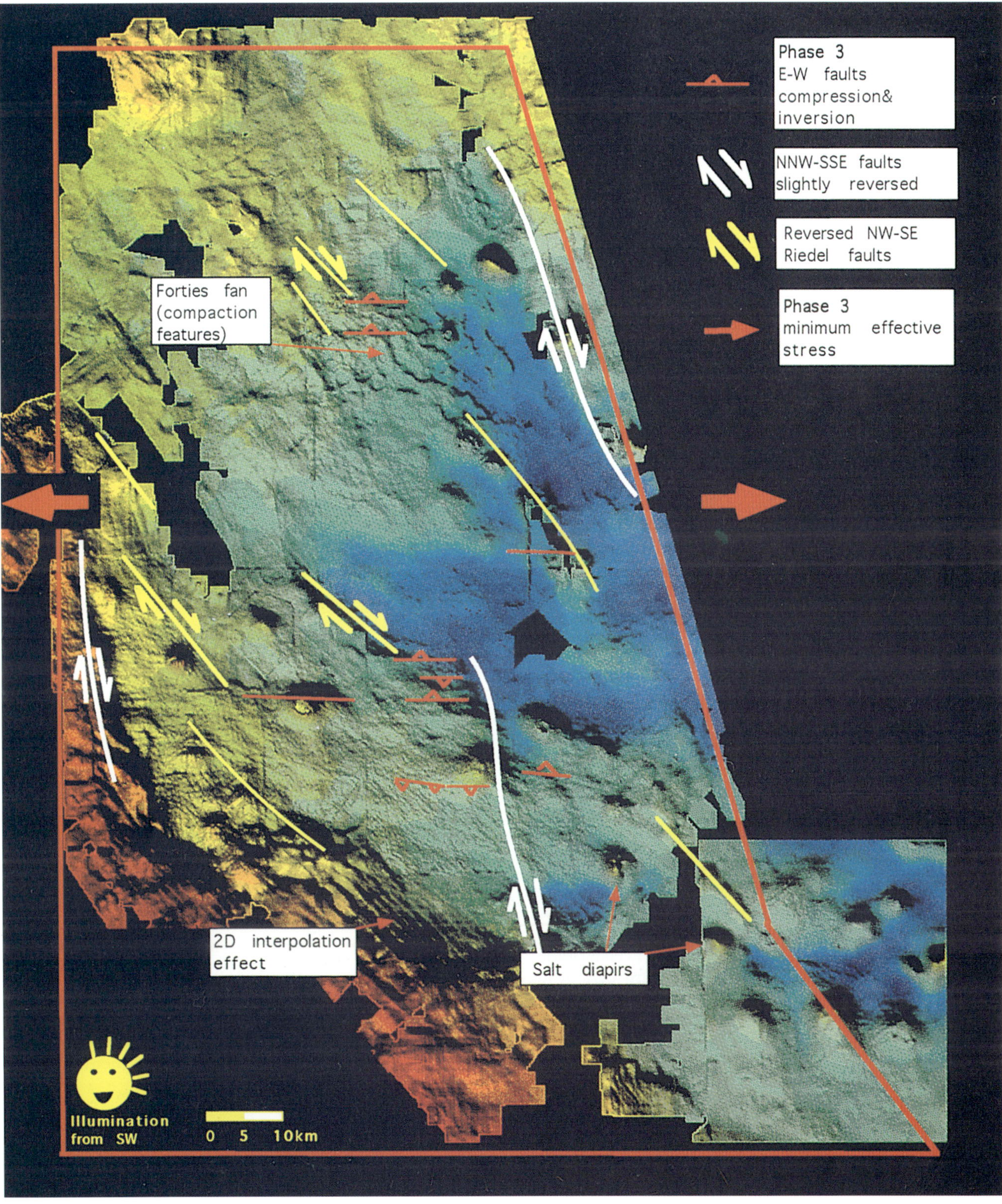

Fig. 15. Illumination map top Balder.

predominantly seen at E–W faults, together with the orientation of the shallow basins, suggest that the minimum effective stress field has (further) shifted in a clockwise mode to an approximate E–W direction. For completeness it should be noted that a later major subsidence phase took place during the Pliocene. This later susidence is not shown by the isochore maps as no shallower horizons have been mapped above the Balder Formation.

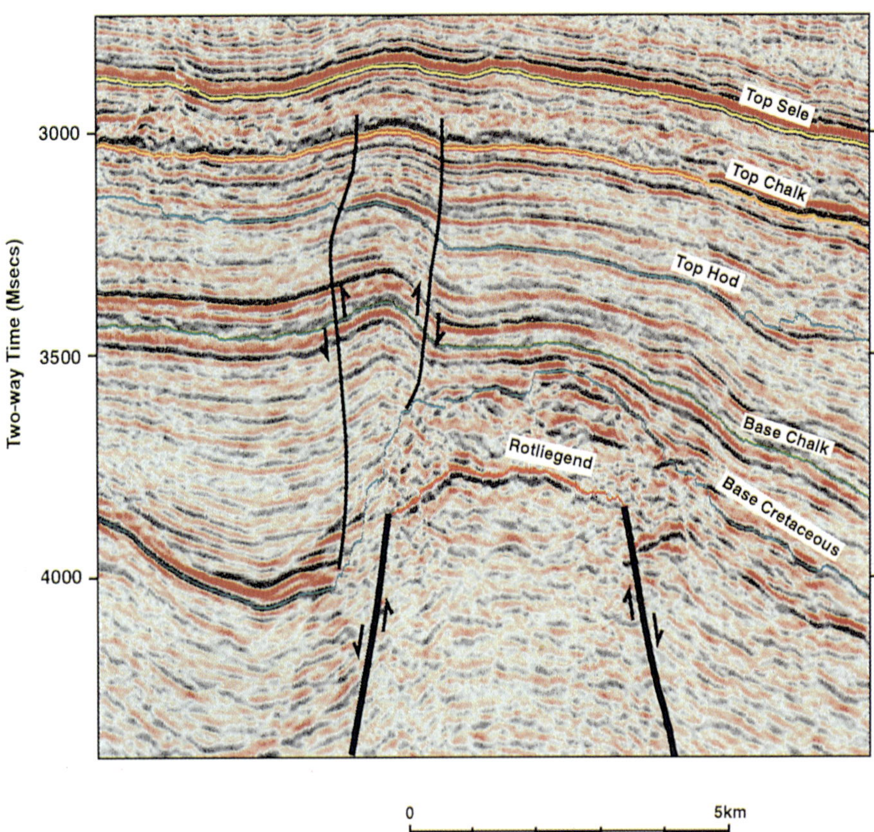

Fig. 16. 3D seismic line showing Tertiary compressional feature and movements in the Puffin area. The compressional feature observed at the southern side of the Rotliegend High (Puffin) is dated as post-Palaeocene. Location of this line is parallel to and 1 km east of the line shown in Fig. 7.

Halokinesis and fault tectonics

Structuration in the Central Graben has been locally enhanced by halokinesis of Zechstein salt. Salt swells, resulting in the distribution of pods and interpods (Skagerrak/Fulmar distribution, Smith *et al.* 1993) are located in areas where faults have been active during the first two phases.

Figure 18 demonstrates the interaction between faulting and the movement of salt in the southwestern part of the Central Graben. Initial sinistral movement along the NNW–SSE trending faults at Rotliegend level has controlled the initial deposition of thick Zechstein halite in the hangingwall. Differential loading, together with a reversal of movement along the riedel faults in the second tectonic phase, has resulted in the development of steep salt walls. The thicker walls typically trend NW–SE (along the reversed riedel faults) and occasionally pierce the overlying strata, forming diapirs. Continuous propagation of faults due to reactivation is observed in areas where salt has apparently been thin or absent. In the case of relatively thick salt, faults are offset towards the downthrown side due to salt decoupling.

The location of diapirs along the NW–SE riedel faults in the Central Graben seems to have been governed by fault tectonics. The diapirs are well expressed on the Top Balder illumination map (Fig. 15), and are mostly located at the intersection of E–W faults and the NW–SE orientated riedel faults. At these locations salt walls were locally inverted into diapirs during the third (compressive) tectonic phase.

Salt walls and diapirs are confined to zones of structural weakness (major fault zones). In the areas between major fault zones structuration has been governed by the tectonics described in this paper. This is well documented in the 29/8 area, where structuration north of the riedel fault, can be explained without invoking halokinesis (Fig. 13).

CONCLUSIONS

In the present paper a simple tectonic model is proposed, which explains the regional development of structures in the Central Graben. The model can be used to predict structuration in less well known areas, and as such will be of assistance in the ranking of Central Graben acreage for hydrocarbon prospectivity.

The model calls for an extensional tectonic setting during the Late Palaeozoic to Early Cretaceous period. From the Late Cretaceous onwards a compressional setting developed. The minimum effective stress rotated over this period from roughly NE–SW to E–W; this rotation explains the relative amounts and senses of dip and strike-slip observed on 3D seismic data. The three phases described in the model are presented as discrete phases of structuration (mainly driven by the limited number of interpreted horizons). It is likely that further studies will reveal a more transitional sequence of events, as is already suggested by the study of isopach maps.

Halokinesis is considered to be only of local significance, serving only to amplify the effects of tectonic structuration.

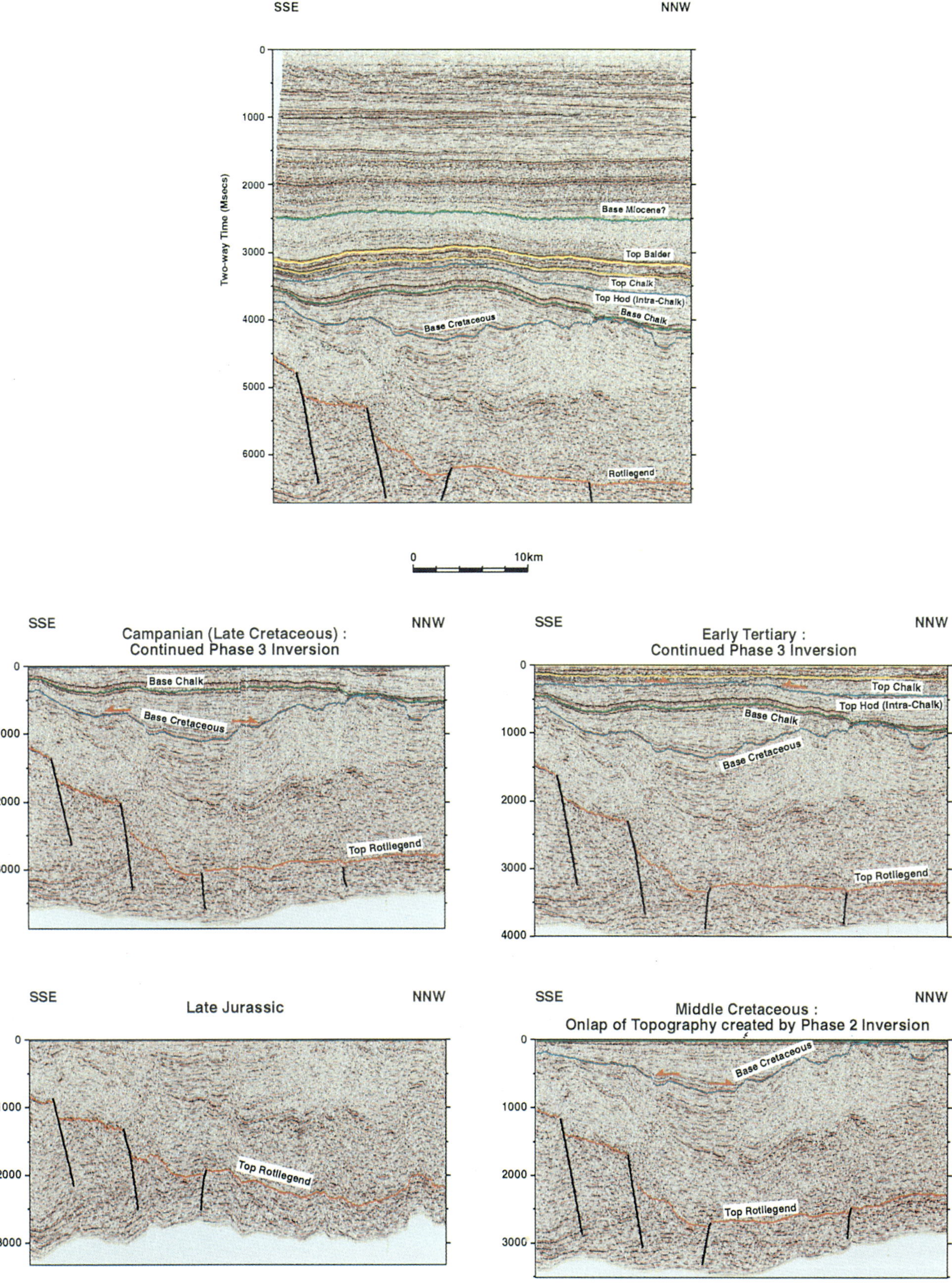

Fig. 17. 3D seismic line showing basin development and inversion in the Puffin area (for location, see Fig. 11). Starting from the bottom left, structural evolution of the basin is illustrated with 3D seismic sections flattened successively on the Base Cretaceous, Base Chalk, Top Hod (approximately top Campanian) and Top Balder seismic horizons, with the topmost section showing present-day structure. Late Jurassic and Early Cretaceous basin development is followed by inversion in Late Cretaceous and Tertiary times. The Top Hod event constitutes a marked onlap surface, probably related to the onset of the Alpine Orogeny.

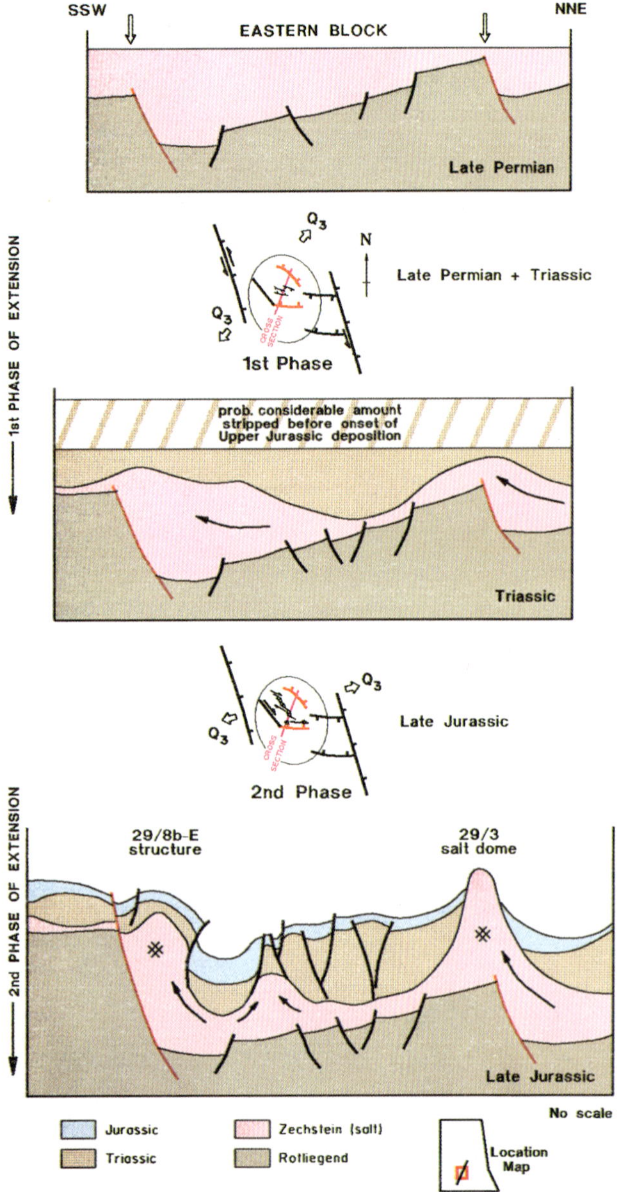

Fig. 18. Interaction of tectonics and halokinesis in the Curlew area . Initiation of salt pillows and salt walls is controlled by the location of large faults at Top Rotliegend level.

This paper is based on work performed by a large group of geologists and interpreters in the Central Graben Team of the Exploration Department of Shell UK. Special thanks are due to Jean Michele Larroque, whose (structural) interpretations were instrumental during the development of the described model, and Bruce Levell for his assistance in providing the paper with a well-balanced geological perspective. We wish to thank Shell UK Exploration and Production and Esso Exploration and Production UK Ltd for permission to publish this paper. In addition we would like to thank all partners of Shell/Esso in the CNS for their permission for publication.

REFERENCES

BARTHOLOMEW, I. D., PETERS, J. M. & POWELL, C. M. 1993. Regional structural evolution of the North Sea : oblique slip and the reactivation of basement lineaments. *In*: Parker, J.R. (ed.) *Petroleum Geology of Northwest Europe : Proceedings of the 4th Conference.* Geological Society, London, 1109–1122.

COWARD, M. P. 1990. The Precambrian, Caledonian and Variscan framework to NW Europe. *In*: Hardman, R. F. P & Brooks, J. (eds.) *Tectonic Events Responsible for Britain's Oil and Gas Reserves.* Geological Society. London, Special Publications **55**, 1–34.

ERRATT, D. 1993. Relationships between basement faulting, salt withdrawal and Late Jurassic rifting, UK Central North Sea. *In*: Parker, J. R. (ed.) *Petroleum Geology of Northwest Europe : Proceedings of the 4th Conference.* Geological Society, London, 1211–1219.

KOOPMAN, A., SPEKSNIJDER, A. and HORESFIELD, W. T. 1987. Sandbox studies of inversion tectonics. *Tectonophysics,* **137**, 379–388.

PARKER, J. R. (ed.) 1993. *Petroleum Geology of Northwest Europe : Proceedings of the 4th Conference.* Geological Society, London.

SEARS, R. A., HARBURY, A. R., PROTOY, A. J. G. & STEWART, D. J. 1993. Structural styles from the Central Graben in the UK and Norway. *In*: Parker, J. R. (ed.) *Petroleum Geology of Northwest Europe : Proceedings of the 4th Conference.* Geological Society, London, 1231–1243.

SMITH, R. I., HODGSON, N. & FULTON, M. 1993. Salt Control on Triassic reservoir distribution, UKCS Central North Sea. *In*: Parker, J. R. (ed.) *Petroleum Geology of Northwest Europe : Proceedings of the 4th Conference.* Geological Society, London, 1211–1219.

THOMSON, K. & UNDERHILL, J. R. 1993. Controls on the development and evolution of structural styles in the Inner Moray Firth Basin. *In*: Parker, J. R. (ed.) *Petroleum Geology of Northwest Europe : Proceedings of the 4th Conference.* Geological Society, London, 1167–1178.

WILLIAMS, G. D. 1993. Structural models for the evolution of the North Sea area. *In*: Parker, J. R. (ed.) *Petroleum Geology of Northwest Europe : Proceedings of the 4th Conference.* Geological Society, London, 1083–1093.

Secondary migration - visualizing the invisible – what can geochemistry potentially do?

Steve Larter[1], Paul Taylor[1], Mei Chen[1], Berni Bowler[1], Phil Ringrose[2] and Idar Horstad[3]

[1]*Fossil Fuels & Environmental Geochemistry (Postgraduate Institute): NRG, Drummond Building, University of Newcastle, Newcastle upon Tyne, UK*

[2]*Department of Petroleum Engineering, Heriot-Watt University, Research Park, Riccarton, Edinburgh, UK*

[3]*Saga Petroleum a.s., Kjørboveien 16, 1301 Sandvika, Norway.*

ABSTRACT: Lateral petroleum secondary migration takes place rapidly on geological time scales (km Ma^{-1} or greater) over ranges of up to several hundreds of kilometres in permeable (> 1 mD) carrier beds. The distribution of petroleum within carrier beds is controlled by petroleum buoyancy, hydrodynamics and capillary forces, and both theoretical and field studies of oil migration suggest that, in general, oil is transmitted through relatively small portions of carrier systems. The dimensions and residual oil saturations of actual carrier systems are difficult (if not impossible) to examine directly, but geochemical tracers within reservoired oil may provide us with important clues. We present in this paper a schematic illustration of how such technology may develop, and suggest that large oil fields may fill through very small net carrier volumes.

KEYWORDS: *petroleum migration, geochemistry, phenols, carbazoles*

INTRODUCTION

In this short review paper we summarize what we feel are key issues in the petroleum geology of secondary oil migration, and attempt to show that the geochemistry of oils and waters may give novel clues to the nature of the processes involved.

Definitions and primary migration

Figure 1 shows a schematic representation of petroleum transport within a generalized petroleum system. Petroleum migration involves the series of processes that transport petroleum from its site of generation in the source rock to the carrier system (**primary migration**), through the carrier system to the trap (**secondary migration**), and ultimately (in many cases) through the cap rock or reservoir seal (**tertiary migration**). In rich, oil-prone source rocks such as the Kimmeridge Clay Fm, a broad quantitative understanding of primary oil migration on scales up to *c.* 100 m has been achieved (England & Fleet 1991; Pepper 1991). The detailed mechanisms of petroleum expulsion are still under debate, but solid organic matter (kerogen plus adsorbed bitumen) is increasingly being considered to be important as a medium for petroleum sorption and transport within source rocks (McAuliffe 1978; Pepper 1991; Sandvik *et al.* 1992).

The source rock/carrier interface

The interface between an actively expelling source rock and the immediate carrier bed (C in Fig. 1) has not been extensively studied, and there are many critical issues relating to the development of high-oil-saturation stringers in the carrier from presumably more-spatially spread stringers in the source rock. This immediate interface between source rock and carrier may exert a profound control on the overall efficency of expulsion from the source unit.

Basic principles and rates of secondary migration in high-permeability (> 1 mD) lateral carriers

Hubbert (1953) showed that petroleum buoyancy, hydrodynamics and capillary forces were main factors controlling the distribution of petroleum in permeable carrier beds, and it is now generally agreed that in both laboratory and field settings the movement of petroleum in carrier beds occurs in high-petroleum-saturation 'rivers' (cf. England *et al.*, 1987).

Ranges of secondary petroleum migration distance vary greatly, even within basins such as those in the North Sea. In Upper Jurassic submarine fans interbedded with mature source rocks (e.g. the Brae Fm./Kimmeridge Clay Fm in the Miller Field, North Sea–Mackenzie *et al.* 1987), secondary migration distances of up to a few kilometres seem likely. Secondary migration distances from Jurassic source rocks to Jurassic and Triassic reservoirs in the Tampen Spur/East Shetland Basin province are probably on the order of a few tens of kilometres, and typically 20–30 km (Goff 1984). In the Troll Field petroleum system, migration distances from U. Jurassic sources to trap may be up to 100 km (Horstad & Larter 1995). More extreme migration ranges occur in other settings. In fractured shale source rock/reservoir systems, such as the Bakken Fm in the Williston Basin (USA, Canada) and the Monterey Fm in the Santa Maria Basin (USA), very short-range migration (<< 1 km?) is evident, and the distinction between primary and secondary migration becomes difficult. At the other extreme, in foreland basins such as the well-documented W. Canada Basin (Creaney & Allan 1990; Piggott & Lines 1991), migration distances from source to trap may reach many hundreds of kilometres. Secondary migration, therefore, appears to occur over distances ranging from metres to many hundreds of kilometres.

From K. Glennie & A. Hurst (eds), 1996, *AD1995: NW Europe's Hydrocarbon Industry*, Geological Society, London, pp. 137–143

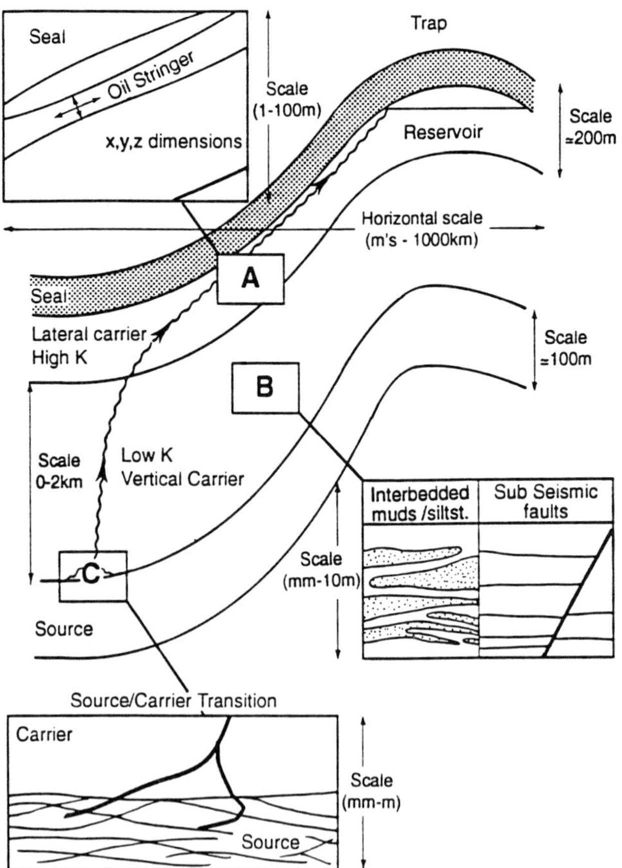

Fig. 1. Schematic representation of a petroleum system.

Reported rates of advancement of buoyancy-driven petroleum migration fronts or oil-stringers vary widely. Ringrose & Corbett (1994) have summarized the reported rates of petroleum migration in 'sandstone' carriers. In parametric studies of probable field rates, estimates of around 1 nanometre per second (10^{-9} m s^{-1}) were derived by England *et al.* (1987). In contrast, laboratory experiments involving very high-permeability bead and sand packs and small oil columns have produced buoyant flow rates of up to 10^{-4} m s^{-1} (Dembicki & Anderson 1989; Catalan *et al.* 1992). In more realistic experiments using gently dipping real rocks, Selle *et al.* (1993) produced observed petroleum migration rates as high as 10^{-2} m s^{-1}. More recently, Thomas & Clouse (1995) used a homogeneous sand pack with properties and petroleum flow rates scaled (using the Rappoport (1955) viscous/capillary-force scaling group) to be equivalent to a 100 mD permeability, 27 m block of sandstone dipping at 5°. Their scaled field bouyancy-driven petroleum flow rates were equivalent to 0.63×10^{-10} m s^{-1} (or 2 mm a^{-1}).

The reported flow rates for these experimental systems would give secondary migration rates in the range of km Ma^{-1}, with maximum rates up to thousands of km Ma^{-1}! However, Ringrose *et al.* (1995) caution that experiments involving homogeneous bead/sand packs or rocks may be most misleading when it comes to relating the results to petroleum migration in real systems. They point out that the experimental systems fail to consider the effects of sedimentary heterogeneity and architecture on fluid transport in sediments, and (perhaps more importantly) they do not take into account the retention of petroleum by the capillary-force trapping mechanisms involved when petroleum attempts to migrate through, for example, cross-bedded sandstones. Ringrose & Corbett (1994) suggested that in such circumstances, petroleum becomes trapped in high-permeability sections of the rock due to the higher capillary entry pressures in the low-permeability laminae. Using idealized model systems of graded-bed and cross-bedded sandstones, the authors show that at water (or oil) saturations of around 50%, the oil-relative permeabilities of the two systems differ by factors of two or more. Substantial differences also exist between the two systems in terms of residual oil saturation levels. Clearly, sedimentary architecture may have a substantial impact on migration rates and the degree of oil retention by the carrier.

England *et al.* (1987) considered that when continuous permeabilities of 1mD or greater are encountered in a potential carrier bed, lateral petroleum migration predominates over vertical migration (A in Fig.1). In many efficient petroleum systems, particularly those in which a high quality carrier/reservoir rock is directly in contact with mature petroleum source beds, it seems clear that petroleum flow occurs predominantly through buoyancy and hydrodynamic processes in those regions of the carrier bed characterized by high petroleum saturations (*c.* 50%) (Schowalter 1979; England *et al.* 1987; Catalan *et al.* 1992; Thomas & Clouse 1995), with perhaps only 1–10% of the porosity of the carrier system actually flowing petroleum (England *et al.* 1987; Larter & Horstad 1992).

Low-permeability vertical carriers

Very little is known about vertical petroleum transport through low-permeability siltstone/mudstone sequences, yet in certain oil-producing regions these may be the only available transport routes and may act as the rate-limiting portions of the migration path. Thus, in the North Sea, the mechanisms for the emplacement of liquid and gaseous petroleum within Lower Tertiary sandstones in the Viking Graben region (e.g. Frigg Field area) from Upper Jurassic source rocks separated by over 2 km of Cretaceous and Lower Tertiary mudstones, remain elusive (cf. Goff 1984).

In Fig. 1, inset B summarizes the two main processes thought to contribute to petroleum flow through mudstone/siltstone sequences. In such situations, migration through faults, including those seismically unresolvable (sub-seismic faults), is often advocated on the basis that 'mudrock sequences are impermeable'. However, recent work on the characterization of siltstones and mudstones in terms of pore-throat radii distributions and probable permeabilities indicates that variation in pore-throat radius in siltstones and mudstones in the North Sea in the depth range 2 to 3.5 km ranges from 80 to 21000 angstroms (Clayton & Hay, 1994), suggesting that 'fine grained' sequences may indeed contain viable carrier pathways. Clayton & Hay (1994) further suggest that the seal leak-off pressures in many North Sea oil and gas fields, which show seismically defined gas chimneys above them, indicate that the seals are failing by capillary failure (i.e. flow through pore systems) rather than fracture failure (flow through fractures). This intriguing observation suggests the possibility that on a large scale, siltstone/mudrock sequences can perhaps transmit petroleum through interconnected silty mudstone or even siltstone horizons (B in Fig 1). A radical notion is that the effective large-scale permeability of thick, fine-grained rock sequences actually increases with the scale of the measuring box, as continuous small stringers of coarser-grained sediments eventually link opposite sides of the box! As indicated by geophysical logs, mudstone sequences are known to be very heterogeneous (MacQuaker & Gawthorpe 1993; Aplin *et al.* 1995) but, as yet, very little has been done to evaluate and quantify the extent of the heterogeneity in natural systems.

A definition of the probable mechanisms involved in the transport of petroleum through thick, fine-grained rock sequences, including seals, is clearly very important because such mechanisms will greatly affect petroleum charge volumetrics. The role of organic matter in controlling migration rates through fine-grained sediments has been discussed by Whelan *et al.* (1984) and becomes especially relevant when we consider the geochemical tracer approaches outlined below. Although diffusion may contribute to the migration of light hydrocarbons through mudrocks (Leythaeuser *et al.*, 1982), secondary oil migration requires Darcy flow. The migration of oil phases through mudrock pore systems is likely to result in far greater residual oil retention in the carrier compared with fracture-dominated migration pathways. These issues relate to the understanding and prediction of intra-reservoir petroleum migration and seal-failure mechanisms, and provide an area for future research.

Secondary migration - the invisible rivers!

Potential carrier beds may be defined using seismic methods. The inferred small size of the oil-transmitting 'rivers' or conduits within the carrier beds suggests that away from local accumulations, these will be very difficult (or even impossible) to resolve seismically, especially if the oil-saturated zones have a generally tubular or digitate, rather than sheet-form, geometry. Similarly, the concentration of flowing petroleum within quite small regions of the carrier beds, and the low probability of encountering these features in exploration wells, means that they are rarely, if ever, drilled and effectively sampled. Indeed, when oil stain is encountered in 'dry holes' it is usually not clear if the stain is related to a carrier system or to a failed accumulation with perhaps quite different fluid saturation and composition properties (cf. Miles 1990). As most oil companies prefer/aim to drill their wells in the low fluid-potential zones of structural or stratigraphic traps rather than carrier beds, it is probable that few carrier systems have been successfully sampled, and even fewer recognized and studied.

Oil stains in exposed settings (e.g. oil seeps, oil-stained cliff exposures or other near-vertical outcrop sections) are obvious field examples of carrier systems to study. However, compared to 'real', deep subsurface carriers, they may be in inappropriate hydrogeological or structural positions, and may have petroleum-saturation distributions unlike an active carrier system. Present-day petroleum-saturation distributions may be the result of local water flow subsequent to petroleum emplacement, with all the consequences relating to fluid redistribution and chemical alteration of the oil. It is usually either difficult or impossible to validate these field examples of carrier beds unless the observed distributions of fluids are compared with fluid-flow models carried out at appropriate scales. Thus, while in-depth investigation of oil-stained potential carriers from a coupled flow modelling-geochemical standpoint is desirable, worthy and encouraged (and we are currently engaged with others in such studies) it is by no means certain that this will lead to significant advances in our understanding of secondary migration!

England *et al.* (1987) and Larter & Horstad (1992) have discussed the importance of quantifying the oil retained in carriers with regard to oil charge calculations for prospect analysis. It is clear that the inaccuracies involved in estimating carrier bed petroleum losses render geochemical volumetric prospect analysis using current methods a very doubtful pursuit. Changes of only a few per cent in the average bulk residual oil saturation in carriers can convert giant fields to dry holes (Larter & Horstad 1992). Such losses of petroleum to the carrier system may be particularly significant in fine-grained siltstone/mudstone carrier systems if petroleum flow occurs through the pore network.

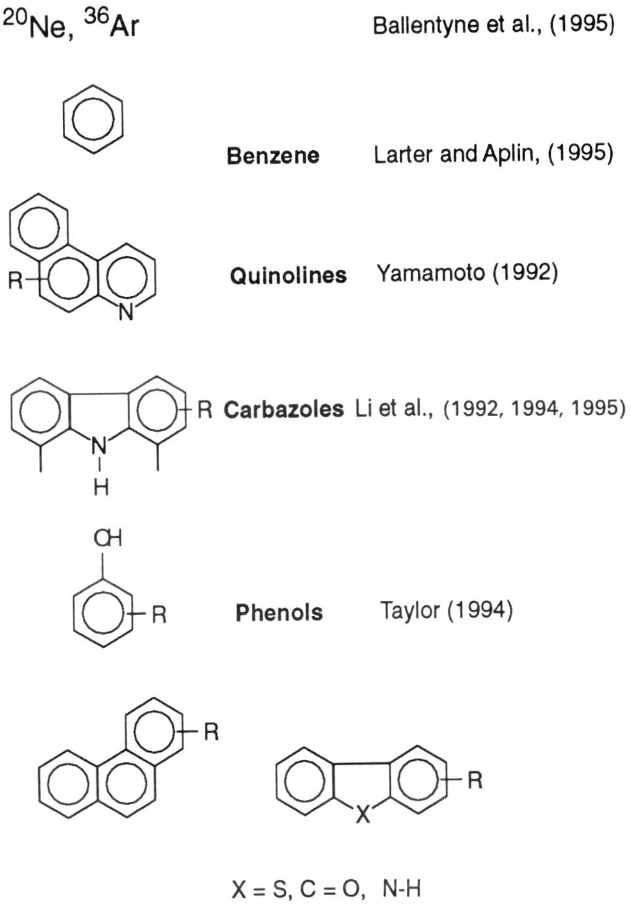

Fig. 2. Typical geochemical tracers in the petroleum system.

CAN GEOCHEMISTRY HELP?

Filling histories

Conventional oil–source rock and oil–oil correlations have proven utility as tools for the elucidation of general, large-scale relationships between source rocks and petroleum in traps. Recently, observations of maturity gradients in petroleum columns within fields (cf. England 1990; Horstad *et al.* 1990) and of oil family relationships between oil pools (Horstad *et al.* 1995) have been used to infer more-detailed charging scenarios for several North Sea oil fields. Horstad *et al.* (1990), for example, suggested that the Brent Group reservoir in the Gullfaks Field was charged locally from the west, whereas petroleum in the Cook and Statfjord fms was charged from the north and east of the field. Regional oil–oil correlation studies have subsequently strengthened this view and have additionally suggested that Upper Jurassic sands in the Tampen Spur area to the west of Gullfaks are critical parts of the regional carrier system, with the potential to be exploration targets in their own right (Horstad *et al.* 1995).

Geochemical tracers (geotracers) – assessing migration range

Another approach to understanding the secondary migration of oil has been the use of tracers present in the petroleum system. Tracers are chemical components in the petroleum system (occurring in the petroleum phase or in the water or in both) which distribute themselves systematically and hence predictably by fluid/fluid partition or fluid/rock adsorption processes. A knowledge of the partition/sorption processes and their efficiencies may, in theory, tell us something about relative volumes of the phases. Some of the geochemical tracers used to date are shown in Fig. 2. Both Yamamoto (1992) and Li *et al.* (1992, 1994, 1995) have used nitrogen compounds in petroleum as internal tracers to qualitatively assess migration range in migrated oils by studying the partition of nitrogen compounds from crude oils into carrier-bed waters and their sorption to carrier-bed mineral and organic matter.

Larter & Aplin (1995) developed some general systematics concerning the use of tracers for petroleum systems for the elucidation of petroleum migration range, and for an initial attempt at assessing oil/carrier interaction. Using the partition of the water-soluble hydrocarbon benzene from oil into formation water, they concluded that for a North Sea oil field, the oil charge had equilibrated with approximately 100 times its own volume of water during migration. Ballentyne *et al.* (1995) studied the partition of atmosphere-derived ^{20}Ne and ^{36}Ar noble gases from carrier-bed waters into petroleum in the Magnus Field petroleum system, and concluded that the Magnus petroleum had equilibrated with *c.* 110 times its volume of water.

Larter & Aplin (1995) pointed out that crude oil is a complex mixture of compounds, many of which are very water soluble and/or sorb strongly to solid phases in rocks. Their persistence in migrated oils suggests that during migration petroleum interacts only weakly and inefficiently with the carrier medium. Potentially, these interactions may provide a means by which we can measure the dimensions of the actual carrier systems. The partition of alkylphenols between rock, oil and water may be an optimal system for migration studies in that the distribution properties of phenols seem ideally suited to the migration problem (Taylor 1994; Larter *et al.* 1994)

Tracer systematics

An ideal secondary migration tracer component would initially exist naturally in only one part of the system (i.e. oil or water or rock). The tracer would be totally conserved within the system, being neither produced nor consumed by chemical or nuclear reaction, and would distribute itself predictably between the components of the system to an easily determined extent. In this regard, the specific ^{36}Ar and ^{20}Ne noble gas tracers of Ballentyne *et al.* (1991, 1995) are ideal, being inert, solely atmosphere derived, and existing in basin waters at concentrations controlled by atmospheric equilibration of the pore-waters at the time of deposition. During migration, ^{20}Ne and ^{36}Ar partition into petroleum from the water surrounding the petroleum in the carrier bed. The method is quite involved in terms of sampling and analysis but it seems to have great potential for assessing the general dimensions of petroleum carrier systems.

Alkylphenols (Fig. 2) are another potential tracer of petroleum migration in the subsurface. Alhough the systematics of their occurrence in petroleums are as yet incompletely understood, and there are questions concerning their origin and fate in sedimentary basins which remain to be resolved (Taylor 1994), alkylphenols seem to have ideal characteristics to serve as migration tracers. Macleod *et al.* (1993) have shown that alkylphenols in petroleum systematically partition into basin waters. Taylor (1994) showed data from a fill-spill sequence of oils from the Tampen Spur (North Sea) in which phenol concentrations decreased systematically with increasing distance of secondary migration over a range of *c.* 25 km. Despite systematic decreases in alkylphenol concentration in the petroleums, the relative abundances of alkylphenols substituted with from 0–3 alkylcarbons remained largely unchanged.

Taylor (1994) suggested that alkylphenols were removed from migrating petroleums by a dual partition process involving the partition of phenols from oil into pore-waters and then from waters onto solid phases in the carrier. Taylor (1994) and Larter *et al.* (1994) suggested that solid organic matter in the carrier was an important element in the partition process. The alkylphenols which partition most easily into the carrier waters are those that subsequently partition least efficiently from waters into solid organic matter. It is probable that local equilibria between the phases are achieved rapidly on a geological time-scale (Larter & Aplin 1995). The consequence of this dual oil→water→solid partition is that the distributions of alkylphenols (i.e. the relative proportions of phenols with 0–3 alkylcarbons) in oils remain relatively uniform, while the concentrations of alkylphenols decrease with increased exposure of petroleum to the carrier bed. Such a partitioning system, if validated, can potentially provide much useful information on the dimensions of petroleum systems.

Dimensions of a petroleum system – a preliminary assessment

What follows is an illustration of how petroleum geochemistry may be able to scale migration systems. It is presented so as to give an idea of the potential (rather than a real) application of geochemistry as we have insufficient data for a reliable assessment at this stage, and the petroleums studied are from two, albeit somewhat similar, systems. However, the general principles can be described and broad conclusions about migration can still be drawn.

Figure 3 shows a crossplot of relative migration distance versus C_0–C_3 alkylphenol concentration using data obtained by analysis of reservoired crude oils from the North Sea. Four of the data points are from Taylor (1994) and represent oils on a spill-fill sequence in the Tampen Spur area of the North Sea. The remaining data are from Chen (1995) and represent DST oils sampled across the Miller Field. Taylor's (1994) data have been normalized such that the sample with the highest phenol concentration has a similar relative concentration to the sample in the Miller set with highest phenol concentrations, i.e., the plot shows relative concentration relationships versus distance for the combined data. The abscissca of the plot represents an estimate of the relative petroleum migration distance, i.e. the distance (km) from a nominal reference point, not from the source. In the case of the Tampen Spur samples, the reference point is the sampled reservoir nearest the start of the carrier system. For the Miller oils, the distance represents an in-field distance from the major fill point of the field in the SE of Miller measured

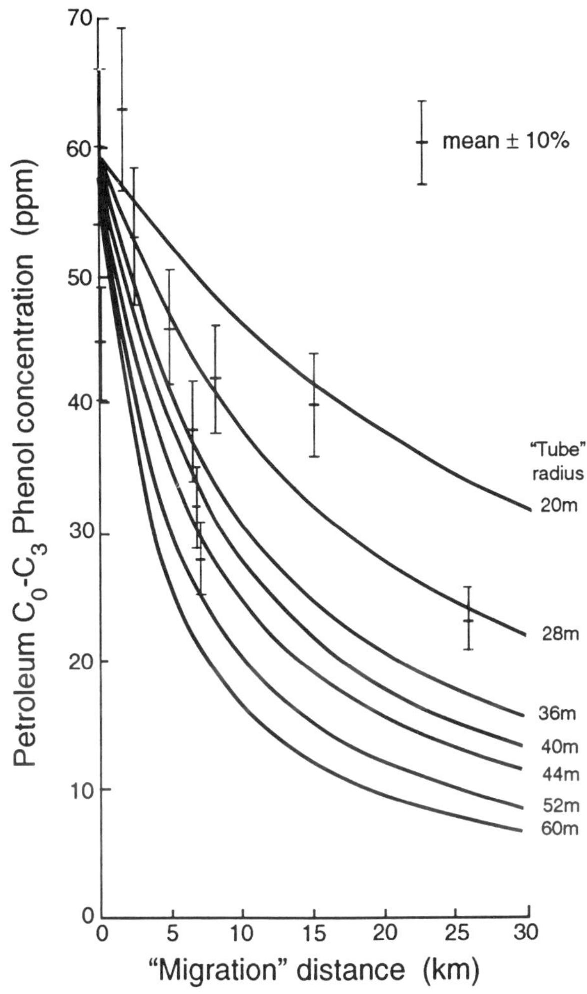

Fig. 3. C_0–C_3 Alkylphenol concentrations versus relative migration distance for petroleums from the Tampen Spur system and from the Miller Field, North Sea. The series of solid line curves represent the variation in the theoretical concentration of C_0–C_3 alkylphenols with increasing migration distance through hypothetical, cylindrical, oil-saturated carrier zones ('tubes') of different radii.

towards NW Miller. The plot therefore compares the variation in relative alkylphenol concentration of some North Sea oils in different petroleum systems with relative secondary migration distance. While there is considerable scatter (error bars are 10%) a general trend of decreasing phenol concentration with increasing migration distance is evident. It should be noted that the most-migrated Tampen Spur oil (at *c.* 25 km) has suffered mild biodegradation; the observed depletion of alkylphenols in this sample may therefore reflect losses due to biodegradation or to migration processes or both.

Larter & Aplin (1995) presented a simplified partition model of the behaviour of geochemical tracers in sedimentary basins. Their equation derived for the variation of tracer concentration with the relative concentrations of equilibrated petroleum/water/rock can be generalized to account for the presence of both minerals and organic matter on the solid phase. If diffusion processes during migration are ignored, then the concentration of phenols in a petroleum (Coil) equilibrated once and directly with a water-wet carrier bed containing minerals and solid organic matter can be described by the equation shown in Table 1.

If the oil-saturated portion of the carrier is assumed to have an arbitrary cylindrical geometry, V_{rock} can be described in terms of migration distance and the radius of the oil-saturated carrier. The assumption of a cylindrical geometry for the migration channel is purely one of convenience, enabling us to illustrate the general approach to the problem. In reality, the oil would probably migrate as stringers or 'braided streams', many of which may not reach the trap (these would comprise part of the residual oil of the carrier system).

The parameters and values used in the illustration are shown in Table 1.

Table 1. Equation and parameters used in illustration

Parameter	Value/units
C_{oil} is the measured concentration of phenols in the migrated oils	ppm
$C_{initial}$ is initial phenol content of the oils	59ppm
V_o is the volume of oil equilibrated	$1.59 \times 10^8 m^3$
P is the average oil/water partition coefficient for C_0–C_3 phenols	15
Porosity is carrier porosity	0.25
F_{org} is the volume fraction of organic matter in the carrier	0.01
$K_{organic}$ is the organic matter/water distribution coefficient	$0.6 m^3.kg^{-1}$
K_{min} is the mineral/water distribution coefficient	$0.025 m^3.kg^{-1}$
$Dens_{org}$ is the density of the organic matter	$1200 kg.m^{-3}$
$Dens_{min}$ is the density of the minerals in the carrier	$2600 kg.m^{-3}$
S is the oil saturation in the carrier	0.85
V_{rock} is the volume of rock equilibrated with the oil charge.	m^3

$C_{oil} = (C_{initial}.\ V_o.P)/(V_o.P + V_{rock}.porosity.\ (1\text{-}S).K_{mineral}.V_{rock}.Dens_{min}\text{-}K_{min}.V_{rock}.Dens_{min}.F_{org}\text{-}K_{min}.V_{rock}.Dens_{min}.porosity + K_{min}.V_{rock}.\ Dens_{min}.\ porosity.F_{org} + K_{organic}.V_{rock}.F_{org}.Dens_{org} - K_{organic}.V_{rock}.\ F_{org}.\ Dens_{org}.porosity)$

As discussed earlier, Fig. 3 shows the observed (C_0-C_3) alkylphenol concentrations obtained in two North Sea case studies plotted against approximate relative migration distance. Also plotted on the diagram are curves for theoretical concentration versus distance for hypothetical, cylindrical, oil-saturated carrier zones of different radii. While this model is unrealistic and is presented solely as an illustration of the potential of the method, it is interesting to note that it appears that the observed variation in phenols concentration is broadly equivalent to a $\times 10^9$ barrel oil charge equilibrating with a single conduit of only 30 m in radius. Such small subsurface features are unlikely to be visible on seismic records or to be penetrated by drilling.

Clearly, before this technology can be applied to oil exploration a more detailed understanding of the sources and sinks of phenols (and other tracer molecules) in petroleum systems (especially biodegraded oil systems) is needed, as are distribution parameters (partition and distribution coefficients) for tracers under subsurface pressure and temperature conditions and the process involved in removal of tracer compounds. Such information will appear during the next few years from both field studies, case histories and laboratory-based experiments, and it is possible that the approach which we have illustrated schematically may provide petroleum geoscientists with tools for the direct evaluation of carrier-bed dimensions from the analysis of reservoired petroleum suites.

CONCLUSIONS

Lateral secondary migration of petroleum takes place rapidly on geological time scales over distances of up to many hundreds of kilometres in permeable (> 1mD) carrier beds. The distribution of petroleum within carrier beds is controlled by petroleum buoyancy, hydrodynamics and capillary forces. Theoretical and field studies of oil migration indicate that, in general, only small portions (1–10%) of carrier systems are actually involved in the flow of oil. The dimensions and residual oil saturations of actual carrier systems are difficult, if not impossible, to directly examine in the field, but potentially, petroleum geochemical tracers may allow petroleum geoscientists to measure migration range and dimensions through the geochemical analysis of reservoired petroleums. First indications are that totally oil-saturated carrier volumes associated with large oil fields may be very small, and thus local structure and sedimentary architecture are perhaps the most critical factors controlling the location of the actual keyholes or filling points through which a reservoir is filled.

Note added in proof

Recent work has indicated that for some tracers, such as the benzocarbazoles, the mechanisms of tracer removal involve non-equilibrium processes, so the example shown in this paper is highly schematic.

We thank the EU Joule II and Thermie programs for support of our secondary oil migration research program and acknowledge L. Schwark, P. Corbett, D. Carruthers, D. Leythaeuser, D. Stoddart, N.Telnaes, E. Elewaut and the SOM group for stimulating and useful discussions. At Newcastle, Andy Aplin, Maowen Li, Gordon Macleod, Barry Bennett and Martin Jones contributed greatly to discussions of the concepts presented here. Yvonne Hall, Barbara Brown and Christine Jeans prepared the manuscript and graphics. The manuscript has been greatly improved by reviews from Andy Aplin, Ken Glennie and Chris Cornford. PT acknowledges the support of a NERC studentship (GT4/90/GS/100).

REFERENCES

APLIN, A. C., YANG, Y. & HANSEN, S. 1995. Assessment of BETA, the compression coefficient and its relationship to detailed lithology. *Marine and Petroleum Geology* **12**, 955–963.

BALLENTYNE, C.J., O'NIONS, R.K., OXBURGH, E.R., HORVATH, F. & DEAK, J. 1991. Rare gas constraints on hydrocarbon accumulation, crustal degassing and groundwater flow in the Pannonian Basin. *Earth Planetary Science Letters*, **105**, 229–46.

——, —— & COLEMAN, M. L. 1995. A Magnus Opus: He, Ne and Ar isotopes in a N. Sea oil field. *Geochimica et Cosmochimica Acta*, **60**, 831–849.

CATALAN, L., XIAOWEN, F., CHATZIS, I. & DULLIEN, F. A. L. 1992. An experimental study of secondary oil migration. *American Association of Petroleum Geologists Bulletin*, **76**, 638–650.

CHEN, M. 1995. *Response of Pyrrolic and Phenolic Compounds to Petroleum Migration in Reservoir Processes.* PhD thesis, University of Newcastle Upon Tyne, UK.

CLAYTON, C. J. & HAY, S. J. 1994. Gas migration mechanisms from accumulation to surface. *Bulletin Geological Society of Denmark*, **41**, 12–33.

CREANEY, S. & ALLAN, J. 1990. Hydrocarbon generation and migration in the W. Canada sedimentary basin. *In:* Brooks. J. (ed.) *Classic Petroleum Provinces.* Geological Society, London, Special Publication **50**, 189–202.

DEMBICKI, H. Jr, & ANDERSON, M. J. 1989. Secondary migration of oil: Experiments supporting efficient movement of separate, buoyant oil phase along limited conduits. *American Association of Petroleum Geologists Bulletin*, **73**, 1018–1021.

ENGLAND, W. A. 1990. The organic geochemistry of petroleum reservoirs. *Organic Geochemistry*, **16**, 415–425.

—— & FLEET, A. J. 1991. (eds) *Petroleum Migration.* Geological Society, London, Special Publication **59**. 1–6.

——, MACKENZIE, A. S., MANN, D. M. & QUIGLEY, T. M. 1987. The movement and entrapment of petroleum fluids in the subsurface. *Journal of the Geological Society, London*, **144**, 327–347.

GOFF, J. C. 1984. Hydrocarbon generation and migration from Jurassic source rocks in the East Shetland basin and Viking Graben of the Northern N. Sea. *In:* Demaison, G. & Murris, R. J. (eds) *Petroleum Geochemistry and Basin Evaluation.* American Association of Petroleum Geologists, Memoir, **35**, 273–303.

HORSTAD, I. & LARTER, S. R. 1995. Petroleum migration, alteration and remigration within the Troll Field, Norwegian North Sea. *American Association of Petroleum Geologists Bulletin* (in press).

——, ——, DYPVIK, H., AAGAARD, P., BJØRNVIK, A.M., JOHANSEN, P.E. & ERIKSEN, S. 1990. Degradation and maturity controls on oil field petroleum column heterogeneity in the Gullfaks Field, Norwegian North Sea. *Organic Geochemistry*, **16**, 497–510.

——, —— & MILLS, N. 1992. A quantitative model of biological petroleum degradation within the Brent Group reservoir in the Gullfaks Field, Norwegian North Sea. *Organic Geochemistry*, **19**, 107–117.

——, —— & —— 1995. Migration of hydrocarbons in the Tampen Spur area, Norwegian North Sea–A reservoir geochemical evaluation. *In:* England, W.A. & Cubitt, J. (eds) *The Geochemistry of Reservoirs.* Geological Society, London, Special Publication, **86**, 159–183.

HUBBERT, M.K. 1953. Entrapment of petroleum under hydrodynamic conditions. *American Association of Petroleum Geologists Bulletin*, **37**, 1954–2026.

LARTER, S. R. & APLIN, A. C. 1995. Reservoir geochemistry: methods, applications and opportunities. *In:* England, W. A. & Cibitt, J. (eds) *The Geochemistry of Reservoirs.* Geological Society, London, Special Publication, **86**, 103–123.

——, ——, CORBETT, P., EMENTON, N., CHEN, M.& TAYLOR, P. 1994. Reservoir geochemistry: A link between reservoir geology and engineering? Society of Petroleum Engineers, Paper No. 28849.

—— & HORSTAD, I. 1992. Migration of hydrocarbons into Brent Group Reservoirs – some observations from the Gullfaks Field, Tampen Spur Area North Sea. *In:* Morton, A. C., Haszeldine, A. C., Giles, R. S. & Brown, S. (eds) *Geology of the Brent Group.* Geological Society, London, Special Publication, **61**, 441–452.

LEYTHAEUSER, D., SCHAEFER, R.G. & YUKLER, A. 1982. Role of diffusion in primary migration of hydrocarbons. *American Association of Petroleum Geologists Bulletin*, **66**, 408–429.

LI, M., LARTER, S. R. & FROLOV, Y. B. 1994. Adsorptive interactions between petroleum nitrogen compounds and organic/mineral phases in subsurface rocks as models for compositional fractionation of pyrrolic nitrogen compounds in petroleum during petroleum migration. *Journal of High Resolution Chromatography*, **17**, 230–236.

——, ——, STODDART, D. & BJORØY, M. 1992. Practical liquid chromatographic separation schemes for pyrrolic and pyridinic nitrogen aromatic heterocycle fractions from crude oils suitable for rapid characterisation of geochemical samples. *Analytical Chemistry*, **64**, 1337–1344.

——, ——, —— & —— 1995. Fractionation of pyrrolic nitrogen compounds in petroleum during migration: derivation of migration-related geochemical parameters. *In:* England, W. A. & Cubitt, J. (eds) *The Geochemistry of Reservoirs.* Geological Society, London, Special Publication, **86**, 5–32.

McAULIFFE, C. D. 1978. Oil and gas migration – chemical and physical constraints. *American Association of Petroleum Geologists, Bulletin*, **63**, 761–781.

MACGREGOR, D. S. & MACKENZIE, A. S. 1986. Quantification of oil generation and migration in the Malacca Strait Region. *Proceedings of the 15th Annual Convention of the Indonesian Petroleum Association, 7–9 October 1986, Jakarta.*

MACKENZIE, A. S., PRICE, I., LEYTHAEUSER, D., MULLER, P., RADKE, M., & SCHAEFFER, R. G. 1987. The expulsion of petroleum from Kimmeridge Clay source-rocks in the area of the Brae oilfield, UK continental shelf. *In:* Brooks, J. & Glennie, K. (eds) *Petroleum Geology of NW Europe.* Graham & Trotman, London, 865–877.

MACLEOD, G., TAYLOR, P. N., LARTER, S. R. & APLIN, A. C. 1993. Dissolved organics in formation waters: insights into water-oil-rock ratios in petroleum systems. *In:* Parnell, J., Ruffell, A. & Moles, N. (eds) *Geofluids 93: Contributions to an international conference on fluid evolution, migration and interaction in rocks.* Geological Society, London, pp. 18–20.

MACQUAKER, J.H.S & GAWTHORPE, R.L. 1993. Mudstone lithofacies in the Kimmeridge Clay Formation, Wessex Basin, southern England: Implications for the origin and controls of the distribution of mudstones. *Journal of Sedimentary Petrology*, **63**, 1129–1143.

MILES, J.A. 1990. Secondary migration routes in the Brent sandstones of the Viking Graben and East Shetland basin: Evidence from oil residues and subsurface pressure data. *American Association of Petroleum Geologists Bulletin*, **74**, 1718–1735.

PEPPER, A. S. 1991. Estimating the petroleum expulsion behaviour of source rocks – a novel quantitative approach. *In:* England, W. A. & Fleet, A. J. (eds), *Petroleum Migration.* Geological Society, London, Special Publication, **59**, 9–32.

PIGGOTT, N. & LINES, M. D. 1991. A case study of migration from the W.Canada basin. *In:* England, W. A. & Fleet, A. J. (eds), *Petroleum Migration*, Geological Society, London, Special Publication, **59**, 207–226.

RAPPOPORT, L. A. 1955. Scaling laws for use in design and operation of water-oil flow models. *Transactions of the American Institute of Mechanical Engineers*, **204**, 143.

RINGROSE, P. S. & CORBETT, P. W. M. 1994. Controls on two phase fluid flow in heterogeneous sandstones. *In:* Parnell, J. (ed.) *Geofluids: Origin and migration of fluids in sedimentary basins.* Geological Society, London, Special Publication, **78**, 141–150.

———, LARTER, S. R., CORBETT, P. C. & CARRUTHERS, D. 1995. Discussion of 'Scaled physical model of secondary oil migration' by M. M. Thomas and J. A. Clouse', *American Association of Petroleum Geologists, Bulletin*, **79**, 19–29.

SANDVIK, E. J., YOUNG, W. A. & CURRY, D. J. 1992. Expulsion from hydrocarbon sources: the role of organic absorption. *Organic Geochemistry*, **19**, 77–87.

SCHOWALTER, T. T. 1979. Mechanics of secondary hydrocarbon migration and entrapment. *American Association of Petroleum Geologists Bulletin*, **63**, 723–760.

SELLE, O. M., JENSEN, J. I., SYLTA, O., ANDERSON, T., NYLAND, B. & BROKS, T. M. 1993. Experimental verification of low dip, low rate, two phase secondary migration by means of gamma ray absorption (extended abstract). *In:* Parnell, J. (ed.) *Geofluids: Origin and migration of fluids in sedimentary basins.* Geological Society, London, Special Publication, **78**.

TAYLOR, P. N. 1994. *Controls on the Occurrence of Phenols in Petroleums and Waters.* PhD Thesis, University of Newcastle upon Tyne, UK.

THOMAS, M. M. & CLOUSE, A. 1995. Scaled physical models of secondary oil migration. *American Association of Petroleum Geologists Bulletin*, **79**, 19–29.

WHELAN, J. K., HUNT, J. M., JASPER, J. & HUC, A. C. 1984 Migration of C1–C8 hydrocarbons in marine sediments. *Organic Geochemistry*, **6**, 683–694.

YAMAMOTO, M. 1992. Fractionation of azaarenes during oil migration. *Organic Geochemistry*, **19**, 389–402.

Recent developments in reservoir engineering and their impact on oil and gas field development

R. H. Davies and H. Niko
Shell Internationale Petroleum Maatschappij, P.O. Box 162, The Hague, The Netherlands

ABSTRACT: With much of the reservoir engineering development activities prior to 1986 being directed to new processes such as EOR, reservoir engineering of today has, like the other petroleum engineering disciplines, become part of an integrated effort to extract the maximum amount of oil from a reservoir. We will discuss some of the new developments in reservoir engineering which had a real impact on oil field operations in Shell and on the working practices of the individual reservoir engineer. Examples of recent advances in reservoir engineering are: (1) progress in the field of measuring residual oil saturations to water under representative conditions which will enable a more realistic assessment of trapped/bypassed oil in water floods such as those in large North Sea fields; (2) improved understanding of the production behaviour of horizontal wells based on analytical and numerical modelling which led to successful applications in Gabon and Oman; (3) advances in our understanding of production in naturally fractured reservoirs which provided the basis for a unique field experiment in the Natih Field in Oman; (4) understanding of the mechanism of fracturing in water injection wells, a process which has large cost-saving potential. The one factor largely responsible for the change in working practices of individual reservoir engineers is the availability of modern integrated IT technology. Moreover, working with uncertainty in a structured way has become routine in a reservoir engineer's work.

KEYWORDS: *Reservoir engineering, residual oil saturations to water, horizontal hole development, naturally fractured reservoirs, fractured water injection wells, uncertainty.*

INTRODUCTION

The intention of the present paper is to discuss examples of recent developments in reservoir engineering. Although there is a natural bias towards those that have taken place within the Shell Group in the past five years, it is probably true to say that, industry-wide, very similar developments have taken place. While reservoir engineering was very process oriented up to the early 1980s (in the hay-days of EOR), it is very much aware today that it must operate in a multi-disciplinary petroleum engineering environment, enabling the extraction of as much oil as possible from a reservoir in an efficient manner. The areas in which reservoir engineering in the Shell Group has made significant strides towards understanding very complex phenomena and simultaneously contributed to significant improvement in field operations are: realistic assessment of residual oil saturations, horizontal hole development, production from naturally fractured reservoirs, and the understanding of the significance of fracturing in water injection wells. Apart from progress in specific areas, the working practices of reservoir engineers have changed significantly. First, the vast number of user-friendly computer tools available today creates very efficient working conditions but also demands an extremely high degree of computer literacy. Moreover, working with uncertainty has become part of the daily routine for every reservoir engineer.

In the following, each of the areas mentioned above will be described in more detail, including not only the innovations as they originated typically from research but also examples of field cases where they had a direct and measurable impact.

First published in *Petroleum Geoscience,* Vol. **1**, 1995, pp. 193–203.

REALISTIC ASSESSMENT OF RESIDUAL OIL SATURATIONS TO WATER

Until the mid-1980s, the determination of residual oil saturations to water was mainly driven by EOR. Now it is almost exclusively directed towards optimizing the performance of water floods. Residual oil saturation can be determined both *in situ* by techniques such as logging or single well tracer testing (Wellington & Richardson 1994*a,b*) and by special core analysis. We will concentrate here on special core analysis, because this is the area in which some major breakthroughs have been achieved. Special core analysis is hardly ever carried out for the sole purpose of determining residual oil saturations. One usually measures either relative permeability or capillary pressure, both essential input parameters in reservoir simulation, and obtains the residual oil saturation automatically as a result of these measurements.

The two major complications in special core analysis are the

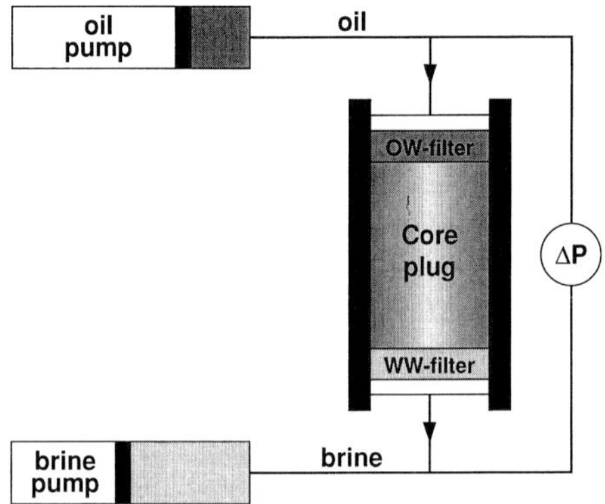

Fig. 1. Schematic set-up of CAPRICI.

unknown state of sample homogeneity and wettability. It is only since the CAT scanner was introduced in EP laboratories that reasonable control over core sample homogeneity could be exercised, and cases where up to 90% of the core plugs suggested for analysis had to be rejected because of small-scale heterogeneity are well known in Shell. Wettability is known to have a large effect on residual oil saturation. Since methods for measuring wettability *in situ* are still at an experimental stage, special core analysis is still largely used to determine the state of wettability. There is one school of thought which favours preserving core samples in their native state as much as possible and then proceeding with flooding experiments. Within Shell, the method of restoring the wettability prior to carrying out flood experiments is preferred. 'Restoring' includes thorough cleaning of the core sample, thus making it water-wet, subsequently 'ageing' it by leaving the rock in contact with crude oil and formation brine for a couple of weeks, and then carrying out the flood experiment. 'Ageing' and core flood are sometimes carried out at true reservoir pressure and temperature, but there is sufficient evidence that working at reservoir temperature and with tank oil may suffice.

Besides the two main complications mentioned above, there is the difficult choice of measuring technique. In the past, the steady state, displacement (= unsteady state) and centrifuge techniques were available, each of which was claimed to yield residual oil saturation and relative permeability and/or capillary pressure. Without wanting to enter into a discussion on the pros and cons of these three techniques, it must be said that experimental artefacts often make the interpretation of such a special core analysis experiment highly unreliable. We should note here that more and more core floods after 'ageing' indicate that the rock is mixed wet rather than water wet as was hitherto believed. This means that a 'capillary end effect' is present during the laboratory water flood which cannot be squeezed away, except at extremely high rates or after excessively long flooding times, resulting in artificially high 'residual' (= remaining) oil saturations. Thus, the capillary pressure is interfering with a relative permeability measurement in the steady state or displacement apparatus. Conversely, we can say that relative permeability effects interfere with a capillary pressure measurement in the centrifuge. The centrifuge does, on the other hand, yield the true value of residual oil saturation. Very recently, Kokkedee & Boutkan (1993) have added a new measuring technique, called CAPRICI, to our portfolio, enabling the simultaneous determination of capillary pressure, relative permeability and resistivity (Fig. 1). An experiment in the new apparatus proceeds at very low flow velocities such that the pressure differences are equal to the capillary pressure. In contrast to the other experimental techniques, CAPRICI allows uninterrupted experimentation, i.e. proceeding from first drainage via ageing to imbibition, then to second drainage, etc. An additional advantage over the centrifuge technique is that, in the case of water wet rock, capillary pressures over both spontaneous and forced imbibition regions can be recorded.

In order to demonstrate the problem often faced in displacement experiments, results of a reservoir condition core flood are shown in Fig. 2. It was first believed that this core flood gave the correct value of residual oil saturation, but a subsequent centrifuge test using tank oil indicated a much lower value. These apparently irreconcilable results could, however, be brought in line by (numerically) simulating the core flood using data obtained by centrifuge such as capillary pressure and relative permeability. Figure 2 indicates that in this mixed wet core it would have required a very long time to reach the true residual oil saturation in the core flood. These findings may have important consequences for the interpretation of currently ongoing water floods, most of which were designed under the assumption that rock is generally water-wet. If the rock is really mixed wet rather than water-wet, meaning that the residual oil saturation is 20% or less

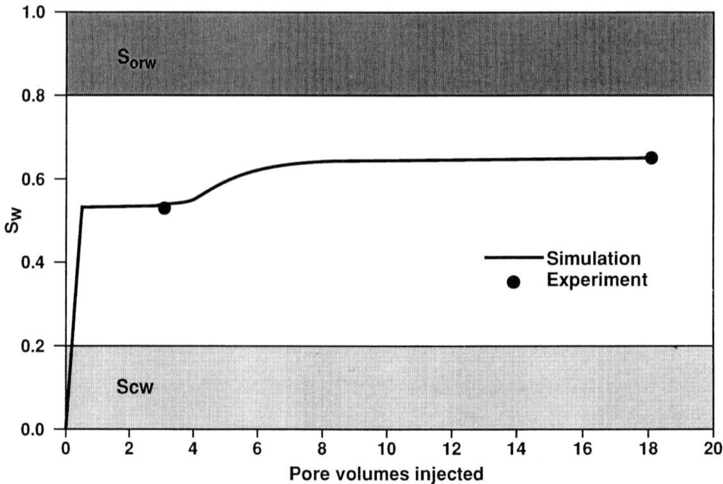

Fig. 2. Example of a reconciliation between centrifuge and displacement experiment.

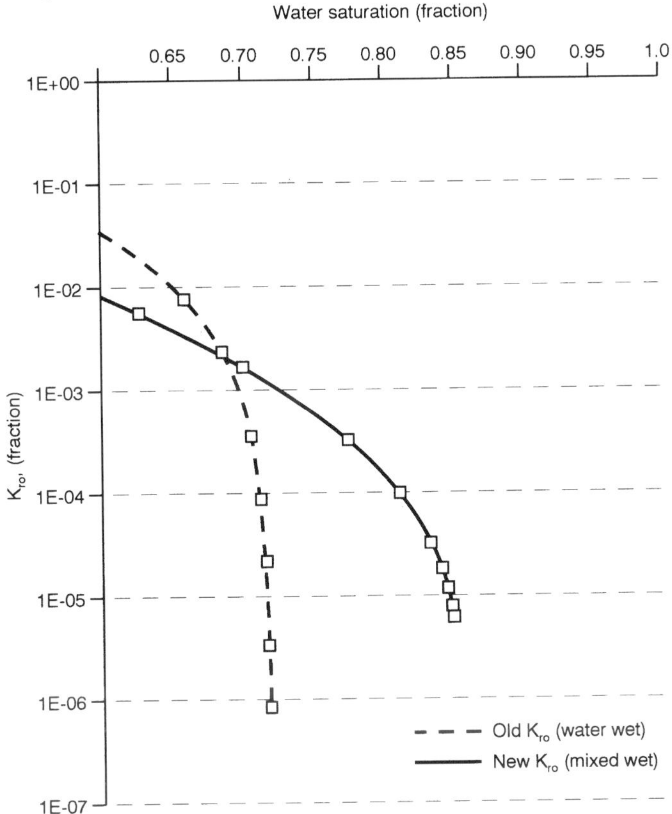

Fig. 3. Comparison between new and old relative permeability curves (tail end).

compared to 30% assumed so far, by-passing of movable oil is taking place on a much larger scale than was believed so far. This provides even less scope for EOR and more scope for infill drilling. We would also like to mention that (numerically) simulating small-scale core floods as a means of quality control is rapidly becoming a routine within Shell.

In order to verify this concept in the field, a study on a small element of the Brent Field was carried out. The study comprised a fine grid simulation of a water flood in a cross section comprising the highly permeable Etive and the Rannoch formations. There is full vertical communication between the formations and water is assumed to be injected into the Rannoch. The simulations were carried out using both an 'old' set of relative permeabilities (water-wet cores, $S_{orw} = 0.28$) and a 'new' one (mixed wet cores, $S_{orw} = 0.15$). Figure 3 shows the characteristic shape of the two relperm curves. A detailed comparison of water saturation profiles over the cross section at different times with those obtained from electrical logs is shown on Fig. 4a for the mixed wet case and Fig. 4b for the water-wet case. The comparison suggests that the relperms relating to the mixed wet condition give a much better agreement with the saturation profiles derived from the logs, particularly near the boundary between Etive and Rannoch. Note also the strong time (or throughput) dependence of the remaining oil saturations in the case of the mixed wet conditions which is due to extremely low values of oil relative permeability in the low oil saturation range (Fig. 3).

HORIZONTAL HOLE DEVELOPMENT

The design of reservoir engineering tools and techniques for horizontal hole development within Shell was a considerable effort stretching over some three years, part of which has been published in the open literature (Niko 1992). Noteworthy is the work of Dikken (1989) that introduced a novel concept of critical rates, whereby pressure losses in the horizontal portion of the borehole were incorporated. Next, the work of Konieczek (1990) presenting a method of arriving at production forecasts for horizontal wells in highly permeable oil reservoirs completed below gas caps, deserves mention. Critical conditions were assumed with the well being beaned back continuously after gas breakthrough. Tiefenthal (1992) extended the work of Konieczek (1990), covering also supercritical conditions. Although there are many areas of application for horizontal hole development in Shell's operations, such as improving the productivity in tight reservoirs, trying to connect with heterogeneous sand bodies or natural fracture systems, oil rims are by far the most important targets of horizontal hole drilling. For instance, there are literally billions of barrels of oil contained in thin oil columns in Shell's acreage in Nigeria, which are so far considered unproducible by vertical wells. We will thus centre our discussion on tools developed for predicting the production performance in these types of reservoir.

An oil rim is defined here as a laterally continuous oil column. Some of the typical situations encountered in practice are illustrated (Fig. 5), showing an oil rim overlain by gas, another one underlain by water and a third one sandwiched between a gas cap and bottom water. The tools available for predicting the performance of horizontal wells in situations such as shown in Fig. 5, range from simple analytical ones to complex reservoir simulation. When using these tools, the engineer must realise their respective strengths and weaknesses. The analytical tools relate mainly to the process,

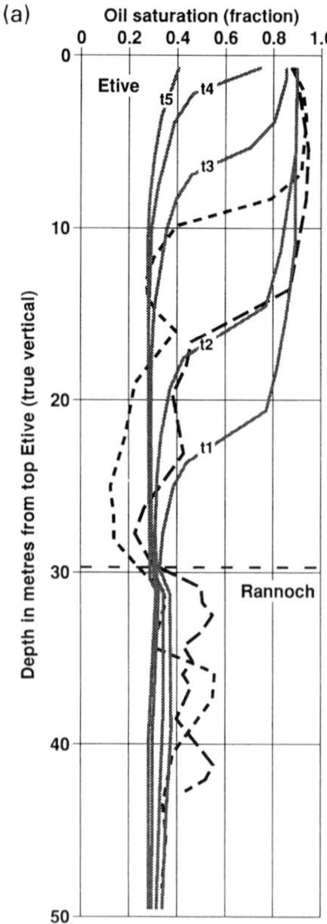

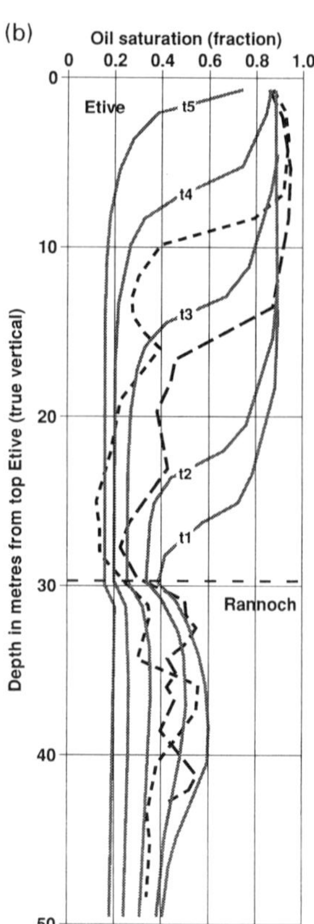

Fig. 4. (a) Simulations with the old relative permeability model, compared to logs. (b) Simulations with the new relative permeability model, compared to logs.

Fig. 5. Oil rim reservoirs.

but are weak in reservoir description, usually only describing to a pancake shaped reservoir. They are easy to use and are ideal for screening, e.g. for determining the optimum vertical position of a well, well spacing etc. After a screening stage, reservoir simulation, incorporating a realistic geological model, is nearly always used. Present simulation tools, including local grid refinement near the ends of the horizontal well, are well suited for both reservoir element studies and full field simulation. A typical reservoir element simulation study, might start with the optimum well position found in the screening stage, trying to answer questions like: is this position still safe given the uncertainty in the geological model? or how much free gas can be produced before significant amounts of oil resaturate the gas cap?

At this stage, we would like to show typical results from some of those screening models discussed earlier. Shown in Fig. 6 are dimensionless production forecasts for a bounded reservoir element (Konieczek 1990; Tiefenthal 1992). The definitions of the dimensionless quantities are also given in Fig. 6 (for supercritical conditions; for critical conditions, Δp should be replaced by $\Delta \rho g h_0$). The top two graphs relate to critical conditions while the two at the bottom relate to supercritical conditions. In both cases we are dealing with an oil rim overlain by a gas cap. Concentrating on the critical conditions first, we see a typical oil rate vs. time graph on the right and a 'type curve' plot, i.e. the product rate times cumulative oil production vs. cumulative oil production, on the left. The horizontal line on this plot relates to the period in which the deformation of the gas/oil contact has not yet reached the outer boundary of the reservoir element. Note that this so-called 'transient' period is considerably longer than in the case of a vertical well and is the source of a large amount of oil production. In the case of supercritical production, i.e. when free gas production is allowed, there is a strong influence of the width of the reservoir element (indicated by the dimensionless parameter inside the plot) on the oil rate decline. Apart from dimensionless oil rate decline graphs, similar graphs describing the dimensionless rise in GOR are available for this case (not shown). We have also extended the screening model described above to three-phase coning in oil rims located between gas caps and bottom water. When translated to typical field conditions, critical rates in horizontal wells can be extremely high. For example, the first horizontal oil well tested in the western part of the Troll Field in Norway (Seines *et al.* 1992) produced more than 20 000 b/d from a 65 feet oil rim in a highly permeable reservoir sandwiched between gas cap and water.

As an illustration of the application of the screening model described above to oil rims overlain by gas caps, Fig. 7 shows the sensitivity of cumulative oil production to an assumed economic limit of 100 b/d for a 48 feet oil rim under critical conditions, to *in situ* oil mobility and well spacing. The length of the horizontal well is assumed at 1600 feet and the well is placed at a distance from the base equal to 1/3 of the oil

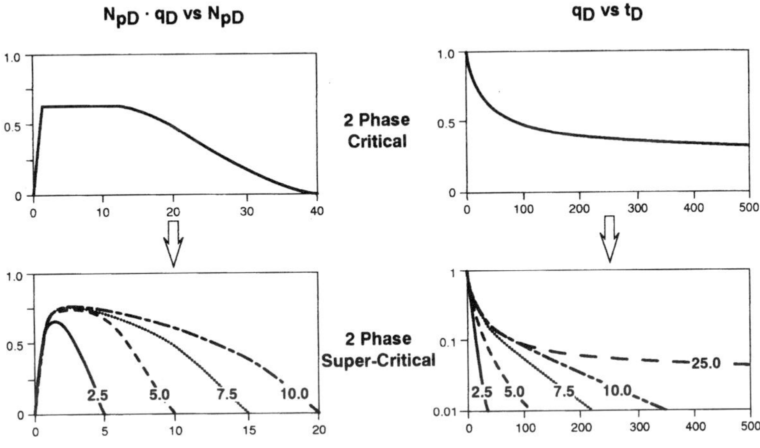

Fig. 6. Production behaviour under critical and super-critical coning (two-phase systems). © SPE 25048.

column height. Figure 7 shows that, depending on the well spacing, oil recoveries of $2–7 \times 10^6$ barrels are obtainable for this horizontal well when the *in situ* oil mobility lies between 1 and 10 D/cP.

Some 560 horizontal wells had been drilled by the Shell Group by end 1994 world-wide, with the number still increasing. It is noteworthy that 47% of these wells are targeted at oil rims, 14% have been drilled for sweep efficiency improvement, 17% for reservoir connectivity improvement, 12% for PI increase and the rest for drainage of attic oil and improved gravity drainage in naturally fractured reservoirs.

We will devote a short discussion here to the application of horizontal wells in the development of the Rabi Field in Gabon. The Rabi Field, located onshore in Gabon, contains a STOIIP of some 1.75 bstb, contained in an oil rim of 148 feet, sandwiched between a gas cap and an active aquifer. A typical cross section through the crest of the field, together with a horizontal hole trajectory, is shown in Fig. 8. Note that typical horizontal hole lengths in Rabi are 2500 feet. The production performance of 12 selected horizontal wells over a three-year period is given in Fig. 9, including a comparison with results of reservoir simulations carried out in the late 1980s for a representative element of the field. It appears that the results obtained by reservoir simulation are largely confirmed by reality. It must be stressed that some of the earlier wells have been tubing constrained, more recently drilled wells have come in with initial well rates close to 15 000 b/d. The rate decline in Rabi is partly caused by an increase in water cut. Moreover, wells are beaned back once the free gas enters the borehole. Although not shown in Fig. 9, horizontal wells in Rabi outperform their vertical counterparts typically by a factor of three, with the corresponding cost ratio being considerably less than that.

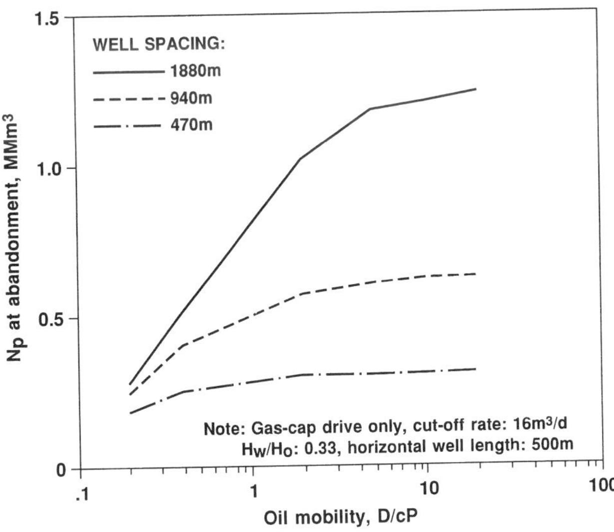

Fig. 7. Oil recovery at abandonment for 15m oil column (no excess gas production).

PRODUCTION FROM NATURALLY FRACTURED RESERVOIRS

Production mechanisms in fractured reservoirs differ strongly from those in non-fractured ones. The fracture network delineates matrix blocks in which most of the oil is located. The fracture system which sometimes contains less than 1% of total pore space will contribute most to fluid flow conductivity. Shell has had a long tradition of working with fractured reservoirs, including a 25-year long involvement in the large fractured fields in Iran. While a significant understanding of the basic reservoir mechanisms in naturally fractured reservoirs had been developed in this period, inclusion of these ideas into an in-house state-of-the-art fractured reservoir simulator took place only very recently (Boerrigter *et al*, 1993). We will briefly describe how this was done and subsequently give an example of a major fractured field simulation. We will highlight how it has led to a complete field redevelopment programme in a major field and show preliminary field responses from this programme.

The most important recovery mechanism in a densely fractured reservoir with high relief is gas/oil gravity drainage. It takes place in the secondary gas cap, where the matrix filled with oil is surrounded by gas-filled fractures. Oil from a matrix block is then produced into the fractures by gravity forces at a rate mainly governed by the density difference between oil and gas, matrix permeability, oil viscosity, oil relative permeability and the height of the gas/oil capillary transition zone relative to the height of the draining oil column in the matrix. Popta *et al.* (1990) recently published results of *in situ* gas saturation measurements in the rock matrix located within the secondary gas cap of the Natih Field in Oman, using the

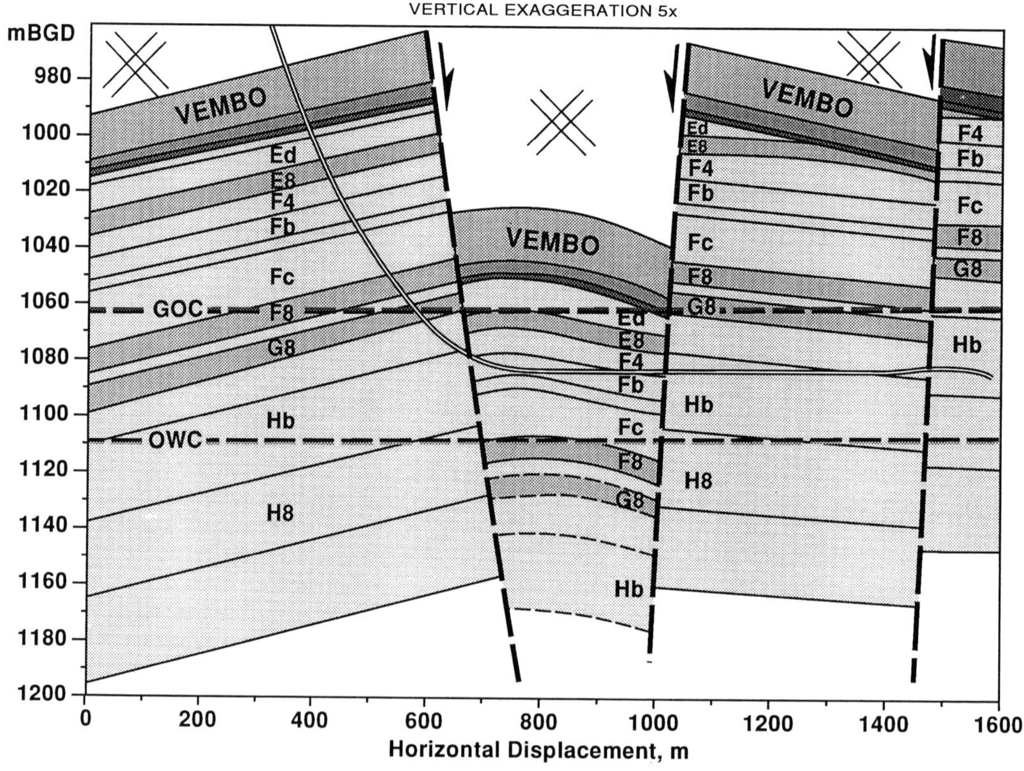

Fig. 8. Well trajectory Rabi 66.

borehole gravimetric tool, which pointed to a very efficient gas/oil gravity drainage process. When modelling gas/oil gravity drainage in a stack of matrix blocks surrounded by fractures, several hundred feet high, two important aspects of the process and the reservoir model must be fully accounted for. The first one is capillary continuity between matrix blocks. When present, the gas/oil capillary transition zone is relatively short compared to total stack height, and oil can drain down to its residual saturation, except close to the very base of the reservoir. In addition, there is the so-called block-to-block interaction. It has the effect that oil draining out from one matrix block will be absorbed by the underlying matrix blocks, whenever this is possible, through gravity and capillary forces. Shell's in-house fractured reservoir simulator allows both the modelling of capillary continuity in the matrix (dual permeability option) and block-to-block interaction. The latter is

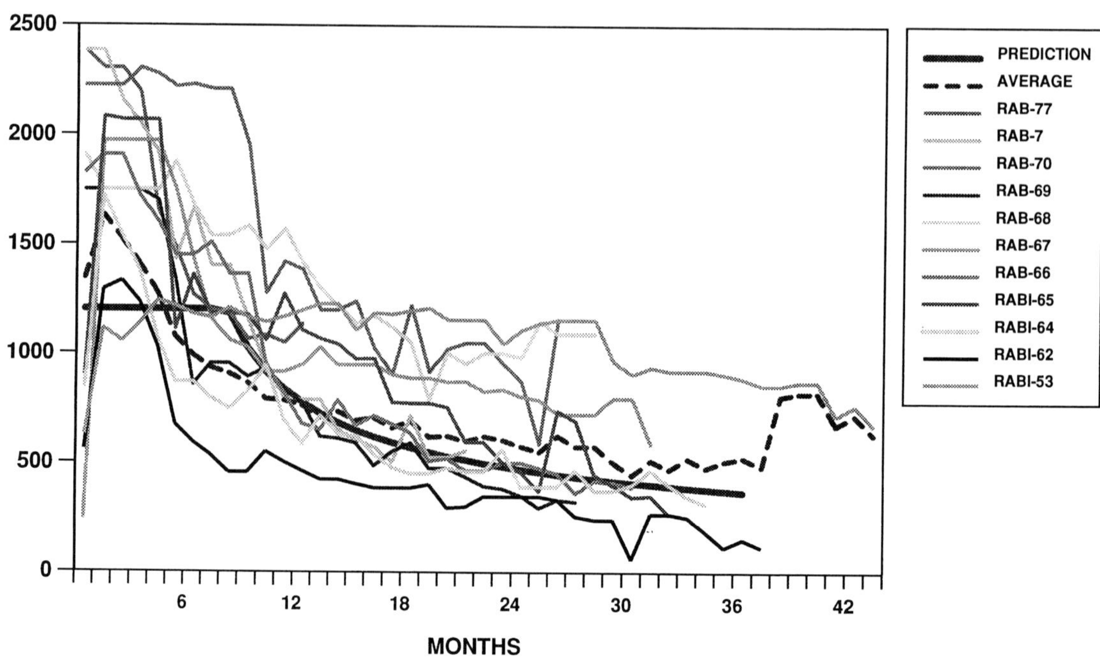

Fig. 9. Rabi horizontal well performance.

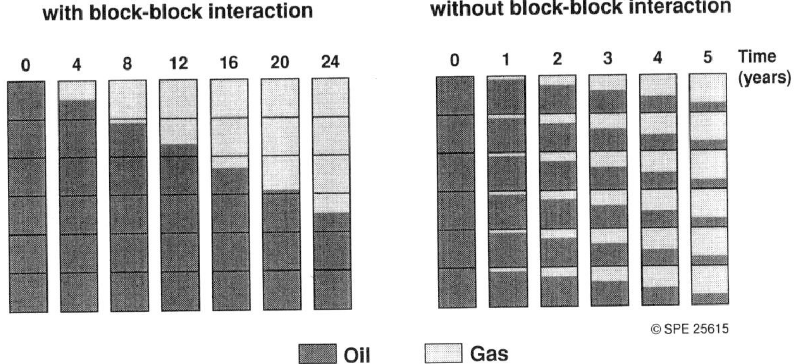

Fig. 10. Gravity drainage in homogeneous, fractured stack.

achieved by assigning special relative permeability functions for the fracture–matrix and fracture–fracture connections.

In order to show how important it is to include block-to-block interaction and capillary continuity in the simulation of fractured reservoirs, the drainage of a fractured stack both without and with block-to-block interaction is shown in Fig. 10. In the former case, blocks drain independently, resulting in relatively fast depletion of the stack. In the latter, drainage of the stack proceeds considerably slower. Figure 11 shows the production performance of the stack for cases with or without capillary continuity, combined with the presence or absence of block-to-block interaction. In addition, a barrier to vertical flow was assumed to be either present or absent in the middle of the stack. Note that capillary continuity has a large influence on the ultimate recovery while block-to-block interaction largely affects the timing of when this recovery is obtained. Although the left part of Fig. 11 suggests that block-to-block interaction can be mimicked by assuming capillary continuity in case of no barrier, the right part shows that this is definitely not true for the more likely case where flow barriers are present.

The efficiency of water drive in naturally fractured reservoirs depends to a large extent on the wettability of the matrix. In case of oil wetness, we are left with water/oil gravity drainage which can be shown to be less efficient than gas/oil gravity drainage. In case of water wetness, water/oil gravity drainage is reinforced by imbibition. Depending on the rate of advance of the water/oil level in the fracture system, we may have counter current imbibition in which matrix blocks are 'immersed' behind a rapidly advancing fracture water/oil level or concurrent imbibition in which the water movement in the fracture lags behind the advance of the water in the matrix. A combination of imbibition and water/oil gravity drainage can, in principle, also be a very efficient recovery process.

The recognition of the potential of gas/oil gravity drainage together with the application of state-of-the-art fractured reservoir simulation technology, is starting to have its impact on some of Shell's major fractured fields. A prominent example is the Natih Field in Oman (Fig. 12). The internal reservoir architecture consists of layers with variable properties and of an extensive natural fracture system, particularly near the crest. The STOIIP of Natih amounts to some 3 billion barrels. The field was brought onto production in 1967, producing first by primary depletion. Several secondary recovery processes were tried in the field in the 1970s and 1980s. Water injection was tried first, but proved unsuccessful. Simultaneous injection of water at the flanks and gas at the crest was carried out next, until the decision was taken during the late 1980s to go for gas injection only. The factors contributing towards this decision were: (1) direct field evidence of the efficiency of the gas/oil gravity drainage mechanism (Popta *et al.* 1990), (2) the recognition that the rock matrix exhibited mixed wettability (which makes water drive unattractive), and (3) improved capability of both geological modelling and simulating naturally fractured reservoirs. The greatest technical challenge of going for gas injection in such a late stage of the field life was that, in order to achieve maximum benefit from gas/oil gravity drainage, the remaining fracture oil rim (see Fig. 12) had to be driven down dip into the previously watered-out area. There is little field

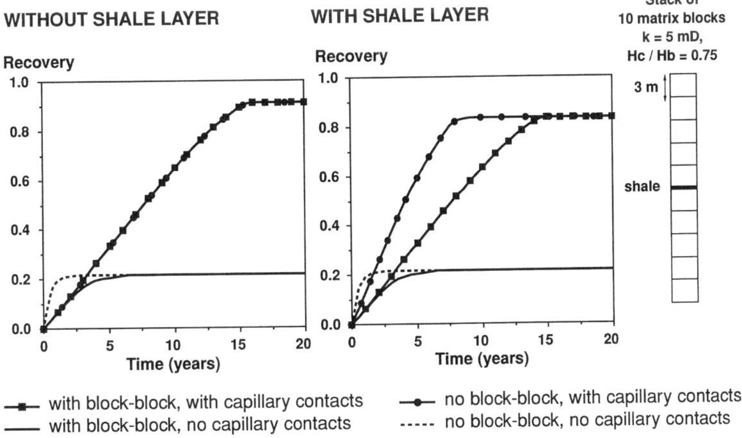

Fig. 11. Effect of block-to-block interaction and capillary contacts on the gas/oil gravity drainage process. © SPE 25615.

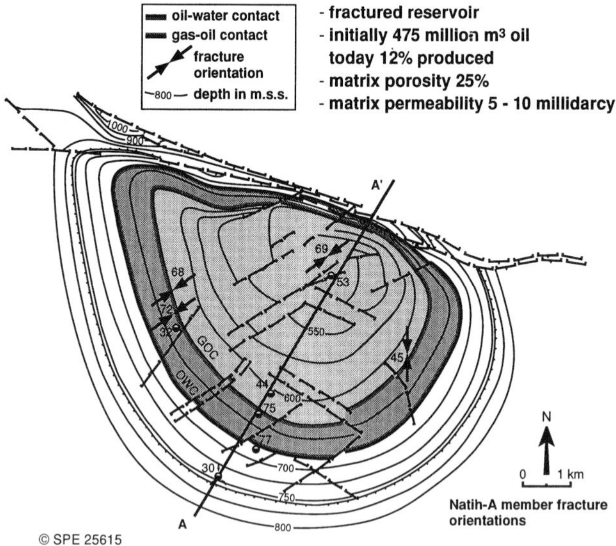

Fig. 12. Natih Field, Oman, structure map NATIH-A.

experience with this process, although it is known that it was tried in the Haft Kel Field in Iran. Therefore, it was extensively tested using numerical simulation, thereby carrying out major sensitivity studies. A typical projection of oil production for Natih under gas injection in the future, including past performance, is shown as Fig. 13. Shown in this graph are the respective contributions of gas/oil gravity drainage and water drive towards total oil production, both in the past and in the future. Note that the relative importance of gas or water displacement is layer specific, but, overall, it can be seen that gas/oil gravity drainage had become the predominant mechanism by the mid 1980s. The predictions shown in Fig. 13 show that, driving the oil rim down into the previously watered-out area, leads to a temporary re-entry of oil into the previously water-invaded matrix (negative contribution by water drive, see dark shaded part below *x*-axis in Fig. 13) which is, however, more than compensated by increased oil production by gas/oil gravity drainage. On the whole, Fig. 13 suggests that gas injection will add significant amounts of reserves in Natih.

Recent field performance in the Natih Field after full gas injection was implemented indicates that the process described above seems to work. In order to illustrate this, the performance of a typical well, N-86, is given in Fig. 14. N-86 produced at both high GOR and high BSW (=water cut) prior to full gas injection, and is seen to be gradually 'oiling in', with the present oil rate being tenfold the previous rate.

The field case presented above invariably begs the question: can matrix wettability, capillary continuity and block-to-block interaction be predicted in advance, thus making it possible to go straight for the optimum process without expensive experimentation in the field, as in the case of Natih? As discussed in the beginning of this paper, it is possible to address the issue of wettability by careful experimentation, thus enabling a proper choice between water and gas injection. In contrast to concepts used in the 1970s, it is now most commonly assumed that capillary continuity between matrix blocks exists. Within Shell, some simple geometrical rules have been developed allowing one to judge the degree of block-to-block interaction, requiring, most importantly, a good geological model of the naturally fractured reservoir. Since it plays an essential role in the gas/oil gravity drainage process, it should, however, always be included in the reservoir modelling.

WATER FLOOD INDUCED FRACTURES

In an effort to understand the mechanism of water flood induced fracturing, a considerable amount of work has been carried out both within Shell and throughout the industry in general (Clifford *et al.* 1991). This effort is driven from two directions, both aiming at cost reduction. First, increasing amounts of produced water have to be handled and disposed off in many of today's water floods. An obvious solution to the water disposal problem is the re injection of the produced water. In addition, one may want to relax some of the very stringent treatment procedures for injection water to save costs. Second, in water floods in tight reservoirs, very small well spacings would be required to allow development at reasonable off takes and injection rates, unless the injectivity and productivity of the wells can be improved. In both cases, limited fracturing of injection wells may have to be allowed.

Various models predicting the development of a water flood

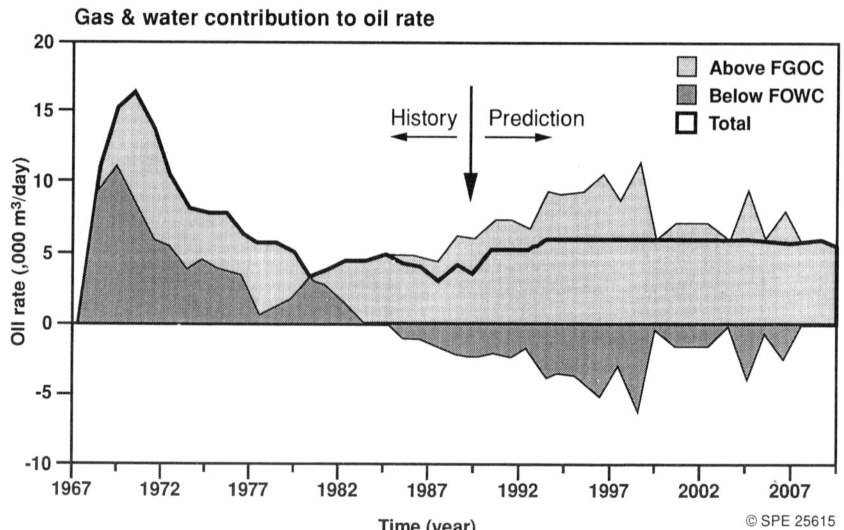

Fig. 13. Relative effectiveness of gas/oil gravity drainage and water/oil displacement during historical period and during oil rim lowering. Natih Field, Oman.

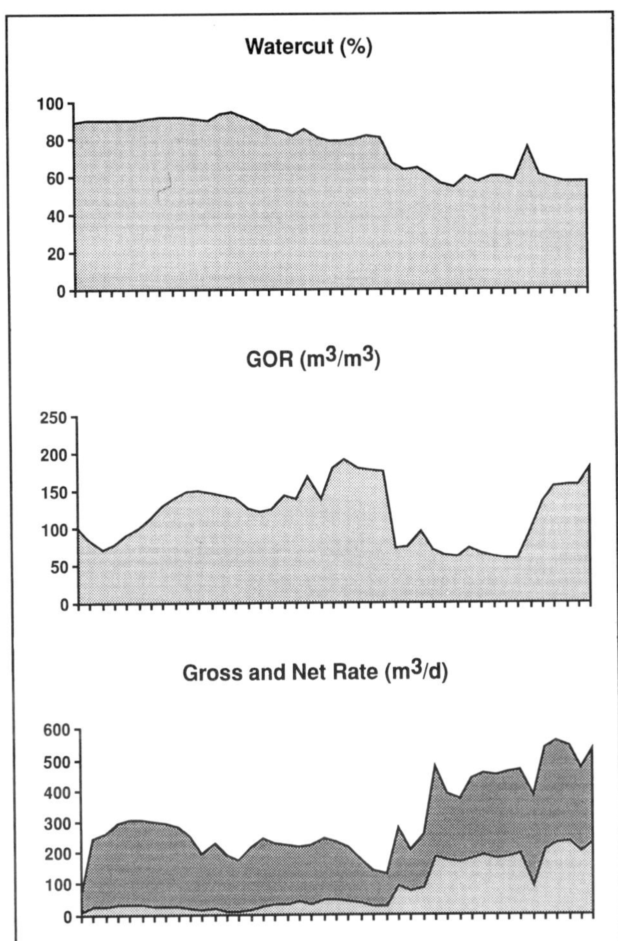

Fig. 14. Typical well response to gas injection: N-86 'oiling in', Natih Field, Oman.

induced fracture were developed that culminated in the development of a tool which uses a thermal reservoir simulator as a host and computes the development of fractures in water injection wells based on sound rock mechanical principles. Complete coupling between fluid/heat flow and the fracture growth was achieved in this simulator, however, the fracture was assumed to cover the full formation height at all times. For an overview on this work, in which also references to more detailed work can be found, see Niko (1992).

An important aspect in the process of fracturing water injection wells are the so called thermo- and poro-elastic back stresses. The thermo-elastic back stresses due to cooling of injection water are particularly important during water injection of highly permeable North Sea sandstone reservoirs, leading to fracturing much sooner than would normally be expected. These thermal fractures are relatively short and, because of a thermal barrier, mostly contained within the formation. Typical fracture half-lengths of 50 feet have been determined in field tests, thus having no appreciable influence on the sweep efficiency of a water injection project.

In a more recent development within Shell, the assumption of constant fracture height has been relaxed, albeit at the cost of losing some degree of coupling between fluid/heat flow and fracture growth. This is illustrated in Fig. 15 which relates to fracture growth in a tight chalk. The idea is to compute the three-dimensional fracture growth first, by using a separate in-house state-of-the-art hydraulic fracture simulator, and second, by feeding the information relating to fracture shape into another simulator such as the one described above. The combination thus allows the assessment of 3D fracture growth on the sweep efficiency of a water flood. Since running two simulators in sequence is rather cumbersome, current in-house efforts are directed towards linking of a simple but consistent 3D fracture growth model into a thermal simulator, thereby also including a fracture wall skin model. The latter is meant to describe the build-up of flow resistance near the fracture wall due to injection of impurities with the injection water.

Although there is no intentional fracturing of water injection wells in Shell's water floods in the North Sea, many cases of thermal fractures have been reported, as in the case of the Fulmar Field (Niko & Ovens 1993). Also within the Shell Group, Petroleum Development Company of Oman operated a pilot test in their Lekhwair Field (Niko & Ovens 1993) with

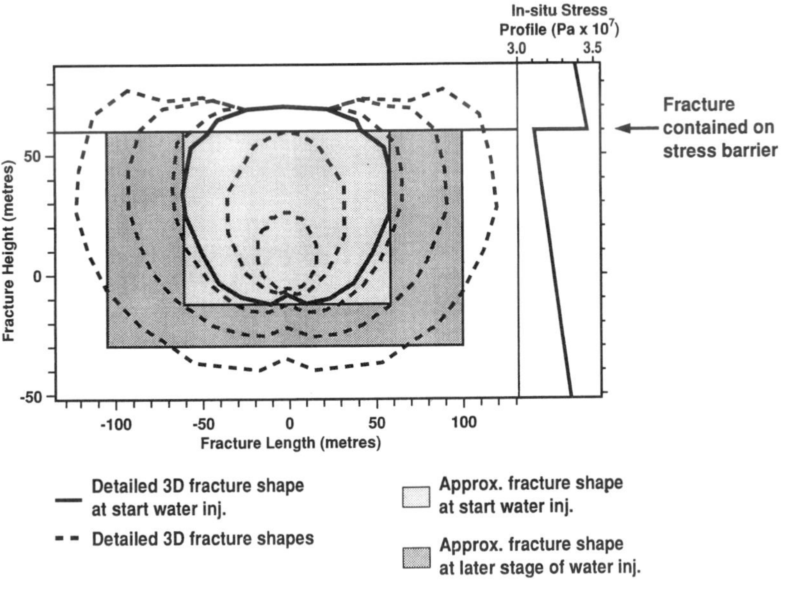

Fig. 15. 3D simulation of fracture growth by state-of-the-art fracture modelling (tight chalk).

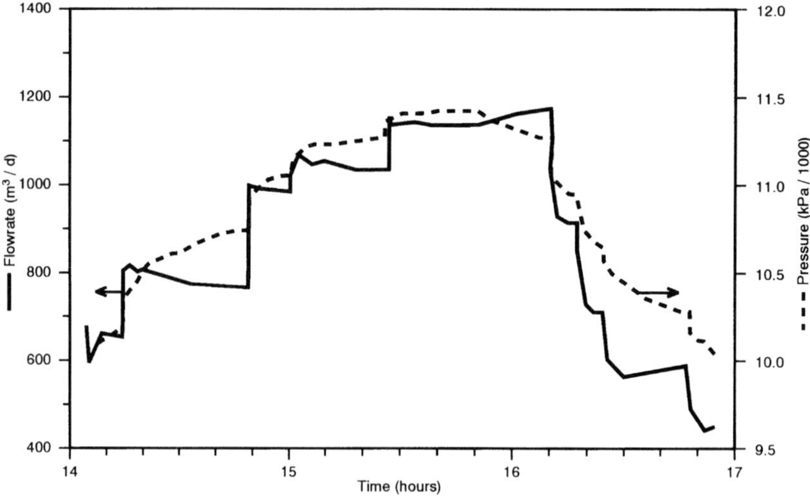

Fig. 16. Lekhwair 54- high rate injection test, initial breakdown. Lekhwair Field, Oman. © Steering Committee European IOR Conference.

the deliberate intention of saving wells by injecting under fracturing conditions into a chalk formation with permeability of 5 mD. Fig. 16 illustrates a formation breakdown test in one of the injectors. Although not all questions, particularly relating to the vertical containment of the fracture, could be resolved satisfactorily, the pilot was basically successful. The Lekhwair water flood presently includes the fracturing of water injection wells in selected areas, in combination with extensive infill drilling, using also horizontal wells, to maximize oil recovery.

RESERVOIR ENGINEERING PRACTICES AND WORKING WITH UNCERTAINTIES

Fast, user-friendly and integrated computer tools are largely responsible for the change in reservoir engineering practices. Our current portfolio of reservoir engineering tools residing on workstations includes pressure transient analysis, phase behaviour, material balance/forecast screening and reservoir simulation. All tools are linked in that output from, for example, the phase behaviour package is directly available as input to compositional simulation. Apart from the links between the different reservoir engineering tools, there are links with data bases and tools of other petroleum engineering disciplines. For example, the output of a detailed geological modelling tool, after appropriate grid block reduction/averaging, can serve as direct input to reservoir simulation.

Working with uncertainty in reservoir engineering can be traced back as far as the early 1970s when the idea of probabilistic reserve estimating was introduced in Shell. Since then, all petroleum engineers, not only reservoir engineers, have to take account of uncertainty throughout their work. The main reason for addressing uncertainty is our incomplete knowledge of the reservoir description and its dynamic behaviour, particularly prior to development when data are sparse. Today, multidisciplinary petroleum engineering teams aim to express reservoir uncertainty in the form of alternative (discrete) reservoir models or realisations, all of which are consistent with the available data. Thus, while alternative development plans/scenarios have historically been evaluated using a single 'base' reservoir model, contemporary development planning exercises address a matrix of reservoir models/realisations and field development concepts. This ensures that upside potential and downside risk are properly addressed and fully evaluated. It also provides a basis for quantifying the value of additional data gathering to reduce uncertainty and risk.

The handling of uncertainty in computerised reservoir engineering tools is still subject to development. One example of such a development is the tracking of uncertainty during reservoir simulation, i.e. showing how the uncertainty in a particular simulator input parameter translates into uncertainty in the production performance. Another example pertains to pressure transient analysis. Rather than reporting a single value for permeability, skin factor and turbulence coefficient, which are obtained by matching predicted pressures (based on a postulated reservoir model) against observed pressures, we have included the associated errors. This adds a degree of 'comfort' or 'discomfort' to the results of a well test analysis, mainly for the benefit of operations reservoir engineers who use the results of well tests as the basis for their workover plans.

CONCLUSIONS

Working in truly integrated teams has given each petroleum engineering discipline the opportunity to increase its impact on the business by contributing its own specific skills. Clear examples where recent advances in the understanding of reservoir processes, have led to successful application in the field and added significant amounts of new reserves, can be found in the fields of assessment of residual oil saturation, horizontal hole development, production from naturally fractured reservoirs and water flooding under fracturing conditions. Apart from these advances, change in working practices (such as working with uncertainty) and the availability of a wide portfolio of computerized tools, has provided the reservoir engineer with opportunities he has never had before.

Apart from the persons listed in the references, the authors of this paper would like to acknowledge all researchers of the recovery processes department in Koninklijke Shell Exploratie en Produktie Laboratorium (KSEPL) and to thank the following of their colleagues in Shell for allowing them to refer to their work and for providing comments: S. Evans, O. van Kessel, S. Coutts, C. van Dijkum, A.M. Schulte, J.W. Roosch, B. Walker and J. Drohm. Moreover, thanks are due to the managements of Shell Internationale Pet. Mij., Shell UK EP Ltd., Shell Gabon, Petroleum Development Company of Oman, and the Ministry of Petroleum and Minerals of Oman, for permission to publish.

LIST OF SYMBOLS

B_o oil formation volume factor
h_o height of oil column
k_v vertical permeability
k_h horizontal permeability
L_w length of a horizontal well
N_p cumulative oil production
q_o oil production rate
t time
Δ_p well drawdown
φ_{eff} effective porosity (=porosity time movable oil saturation)

REFERENCES

BOERRIGTER, P. M., LEEMPUT, L. E. C. VAN DE, PIETERS, J., WIT, K. & YPMA, J. G. J. 1993. Simulation of fractured reservoirs—Case studies. *SPE 25615. Proceedings of the Middle East Oil Technical Conference, Bahrain.*

CLIFFORD, P. J., MELLOR, D. W. & JONES, T. J. 1991. Water quality requirements for fractured injection wells. *SPE 21439.*

DIKKEN, B. J. 1989 Pressure drop in horizontal wells and its effect on their production performance. *SPE 19824. Proceedings of the SPE Annual Technical Conference, San Antonio.*

KOKKEDEE, J. A. & BOUTKAN, V. K. 1993. Towards measurement of capillary pressure and relative permeability at representative conditions. *Proceedings of the 7th European IOR Symposium, Moscow,* 28-34

KONIECZEK, J. 1990. The concept of critical rate in gas coning and its use in production forecasting. *SPE 20722. Proceedings of the SPE Annual Technical Conference, New Orleans.*

NIKO, H. 1992. Reservoir engineering issues of horizontal hole development. *SPE 22387. Proceedings of the SPE International Meeting on Petroleum Engineering, Beijing.*

—— & OVENS, J. 1993. Water flooding under fracturing conditions: from theoretical modelling to field process. *Proceedings of the 7th European IOR Symposium, Moscow,* 109–119.

PERKINS, T. K. & GONZALEZ, J. A. 1982. The effect of thermo-elastic stresses on injection well fracturing. *SPE 11332.*

POPTA, J. V., HEYWOOD, J. M. T., ADAMS, S. J. & BOSTOCK, D. R. 1990. Use of borehole gravimetry for reservoir characterisation and fluid saturation monitoring. *SPE 20896. Proceedings of the European Petroleum Conference, The Hague.*

SEINES, K., LIEN, S. C. & HAUG, B. T. 1992. Two horizontal well tests in Troll demonstrate large production potential from thin oil zones. *SPE 22373. Proceedings of the SPE International Meeting on Petroleum Engineering, Beijing.*

TIEFENTHAL, S. A. 1992. Supercritical production from horizontal wells in oil rim reservoirs. *SPE 25048. Proceedings of the European Petroleum Conference, Cannes.*

WELLINGTON, S. L. & RICHARDSON, E. A. 1994*a*. Redesigned ester single well tracer test that incorporates pH driven hydrolysis changes. *SPE Reservoir Engineering,* 9, 233–239.

—— 1994*b*. Simultaneous use of ester and in situ generated CO_2 as the oil tracer. *SPE Reservoir Engineering,* 9, 240–246.

Challenges in modelling reservoirs in the North Sea and on the Norwegian Shelf

Adolfo Henriquez and Charles Jourdan
Den norske stats oljeselskap a.s. (Statoil), 4035 Stavanger, Norway

ABSTRACT: Reliable predictive models for hydrocarbon production in the high cost Norwegian Offshore have required the development and application to full field simulation of a range of improved 3D geological reservoir modelling techniques. The authors' philosophy for reservoir modelling is outlined and attention focused on the acquisition of the correct type of data, understanding their frequently large inherent uncertainties, and the use of stochastic geological modelling, which, if handling advanced geological concepts and analogue data, enable a proper investigation of the effects of reservoir heterogeneity on fluid flow. Quicker and more trustworthy history matching and 'repairs' of models are made possible. There are challenges to be faced within areas such as geological concepts, estimation of S_w, S_{or}, and permeability, upscaling, ranking of realizations, the use of 3D seismic to determine lithological and fluid compositions and the conditioning of stochastic models using seismic attributes, well test and production data.

KEYWORDS: *Norwegian Shelf, reservoir modelling, stochastic geological methods, uncertainty, data gathering*

INTRODUCTION

During the last 10 years Statoil, and the other Norwegian oil companies, have been actively engaged in detailed 3D geological reservoir modelling. This is because field development in a high cost offshore environment demands generation of the best possible predictions of production profiles and the best possible basis for well/drainage strategies etc. In particular, a lot of attention has been paid to the modelling of reservoir heterogeneity since this has a strong effect on the movements of fluids through the reservoir during production of hydrocarbons. Using different statistical techniques such as truncated Gaussian, Marked Point, semi-Markov and Boolean combined with advanced geological concepts, a whole range of heterogeneities and facies within fluvial and shallow marine depositional environments have been considered as shown in Table 1. Much of the original work is described in a dissertation by Tyler (1994a). Dreyer (1994), in another dissertation, discusses important aspects of the geological input for this type of modelling.

Figure 1 is a visualization of a case where marine background sediments have been stochastically modelled and subsequently truncated by stochastically modelled fluvial channels. The large number of wells used to condition the modelling are visible. This figure gives a good indication of the type of complexities being handled at the facies level. Within the modelled facies, methods such as Gaussian Random techniques can then be used for distributing the petrophysical properties.

The Norwegian oil research community has also put a lot of effort into the consideration of uncertainties within other elements of geological modelling such as in the geophysical control (time picks, depth conversion, etc.), petrophysical input, etc. A stage has now been reached where there is a need to consolidate the modelling and evaluation methods that have been developed. The current software commercialization needs to be completed and more experience gained in its use.

The challenges in modelling reservoirs on the UK–Norwegian Shelf during the next few years will be significantly influenced by several factors. Firstly, the oil price is expected to continue at the present relatively low level for some time yet and the continued, necessary focus on economic returns will keep to a minimum the data gathering which is essential for modelling. Also the giant fields developed in the 1970s and 1980s have reached the end of their plateau production, or are close to this phase. High watercuts and high gas–oil ratios in the producing wells will tax the processing facilities; reduce oil production and influence the timing of tying in existing satellite fields. New fields in the mature production areas will generally be smaller and geologically more complex and unable to sustain an intensive appraisal program, while new fields on frontier areas will tend to be in deeper water.

The challenges will be both of a behavioural and a technical nature. With respect to the former it becomes quite clear that new approaches are necessary—for example, flexible risk-tolerant development plans with phased development for marginal fields where data gathering is related to economic exposure. New technology must be applied quickly, turning problems into advantages wherever possible, for example, in the use of problematical gas production to enhance recovery. In both these cases the reservoir models generated will still have to provide reliable predictions of the right type to help achieve the overall project goals. At the same time the reservoir modelling techniques must be able to continually

First published in *Petroleum Geoscience*, Vol. **1**, 1995, pp. 327–336.

Table 1. *Fields and types of heterogeneities modeled at Statoil in the last few years*

Type of heterogeneity	Field application
Channel facies, channels, channel belts in fluvial settings	Snorre, Gullfaks Sør, Statfjord, Heidrun, Oseberg Gamma
Interfingering deltaic/ shallow marine facies	Statfjord
Geometry/infill of incised valleys	Statfjord, Heidrun
Microscale (laminæ) effects on production from fluvial deltaic and wave-dominated coastlines	Statfjord
Tidal channels/other facies in tide-dominated coastline deposits	Smørbukk, Heidrun
Integration of well test data in characterizing fluvial deposits	Oseberg Gamma

follow the developments within the individual technical disciplines which are integrated within reservoir modelling. For example, new concepts arising within reservoir scale sequence stratigraphy, advances in determining lithology from seismic, etc. Reservoir modelling techniques must keep up with the demands being placed on reservoir management.

AUTHORS' PHILOSOPHY FOR RESERVOIR MODELLING

Before going further we will state our philosophy with regard to reservoir modelling. This can be summarized as follows:

- reservoir modellers should be in a position to influence data collection in advance of the modelling to ensure that the correct data are collected;
- the quality of the data being input to the models should be properly understood;
- relevant information within 3D seismic data sets should be fully utilized;
- stochastic/statistical modelling is frequently necessary to properly evaluate the effects of heterogeneity on fluid production, to help identify the areas of unswept hydrocarbons and to capture uncertainty for effective risk management. These heterogeneities can be surfaces, facies bodies, petrophysical parameters and faults and fractures;
- models derived in this way must be conditioned to well (static), seismic and test/production data and be readily history matched;
- modelling techniques must be applicable to large producing fields with perhaps more than 100 wells, as well as to small fields with limited well and production data;
- upscaling of the models is a critical stage that requires close attention if invalidation of the original geological input is to be avoided;
- modelling will be most successful where well balanced multidisciplinary and cross-functional work groups draw on expert advice and use robust, supporting integrated software.

We are today in a position to model multiple reservoirs in giant fields such as the Statfjord Field where the number of wells being conditioned is considerable (> 100) and the indications are that history matching is being achieved faster than using deterministic methods. This type of modelling will allow us to make rapid repairs to models where we have difficulties in history matching or where new well data becomes available. Despite this progress we still can make advances which, not only have a technical merit, but also a direct economic impact.

THE CHALLENGES

The following sections will deal with several of the technical challenges which are considered to be important within reservoir modelling and which are related to the philosophy already outlined above. These challenges are considered in a series of questions.

Do reservoir modellers influence data gathering?

Many times the fact is bemoaned that operational people never collect the data needed for detailed reservoir modelling. Typically wells never went deep enough to establish a

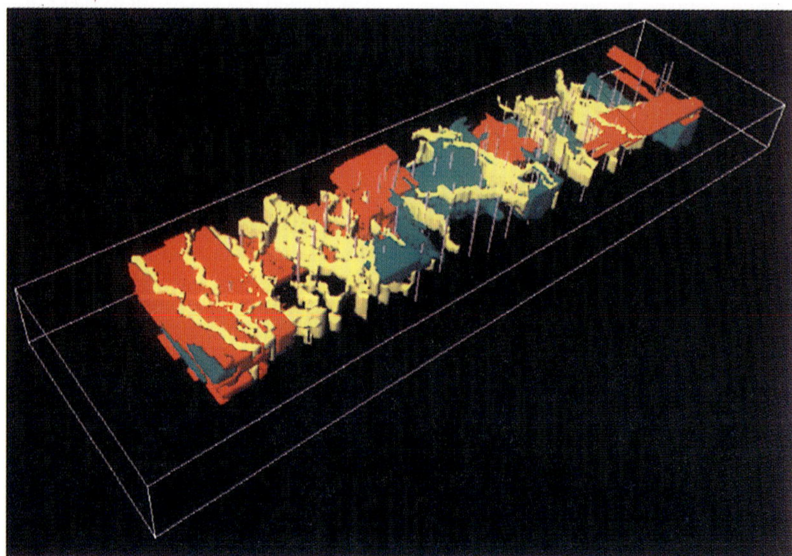

Fig. 1. Example of state-of-the-art reservoir geological model. Notice that the Boolean method used for the channel distribution is conditioned to all the drilled wells.

correlation datum, the cores were cut out because of cost savings, the zones were not individually tested or not tested at all, there should have been a pilot hole in some of the horizontal wells to get zone representative data, etc.

The situation is also being reached where drillers are developing new cost-saving techniques that are warmly welcomed by management but that can be a direct threat to the needs of data gathering. For example, slimhole drilling must not be allowed to exclude the possibility to take logs, cut cores or test drilled sections (– but if they permit extensive coring the advantages will be great!). Branched wells and subsea developments are both areas that provide field development advantages but also threaten the interests of the reservoir modeller if they exclude logging, and especially production logging (PLT). This requires that data gathering must not be unilaterally excluded or remain undeveloped, and that a proper balance is maintained between short-term cost saving and longer-term value creation. A moment which clouds the decision about collecting a given datum is that, given the aim of a flexible and robust plan for development and operations, the early time production should not be affected by small changes in the data. However, later in the mature production phase, evaluation of enhanced oil recovery or unitization exercises may suffer greatly economically by lack of data which should have been taken and evaluated earlier.

There is no other option – reservoir modellers must educate others about the value of data gathering and be willing and able to participate in the decision-making process.

Are the correct data collected and the essential elements modelled?

This can be illustrated by using two simple examples of LWD (logging while drilling) and core description.

(1) A common assumption is that LWD can always replace wireline logging and therefore it is an excellent way to lower operational costs. However, the correct estimation of petrophysical parameters, a key element in reservoir modelling, can be jeopardized if one is not aware of the problems of interpreting LWD data. Figures 2a, b and c (Følo 1994a) are used to illustrate this point. In the case of normal oil accumulations (Fig. 2a), the LWD gives good estimates of porosity, since the measurements of formation density are not significantly affected by invasion. However, where light hydrocarbons are present the amount of invasion at the time of measurement is very uncertain and, consequently, relating the density measurement to porosity is similarly uncertain (Fig. 2b). The problem is further compounded if we then make porosity–permeability extrapolations to uncored sections. In Figs 2b and c the choice of the extent of invasion for a given density can mean the difference between interpreting a low permeability and a moderate permeability reservoir.

Without due care and attention during interpretation, money saved during logging can create negative value later (cost reduction is not always cost effective).

(2) Given an extensively cored section a geologist or petrophysicist could spent a life time describing this material in splendid isolation, and use a lot of money in doing so. This is clearly unsatisfactory given the goals that exist within the oil industry. It is therefore necessary to establish the best practices for what is relevant in particular cases. For example, when should one acquire high-resolution probe permeameter data, which lithological characteristics are critical to our high resolution sequence stratigraphic understanding, which heterogeneities are relevant in water flooding?

BP and Statoil researchers working together have already come some way in this direction and presented their results at several conferences (e.g. Kjønsvik *et al.* 1994; Jones *et al.* 1995). In these studies fluvial reservoir and shallow marine reservoir models generated using stochastic/statistical techniques provided the input to reservoir simulation. It was then possible to investigate the effects of different heterogeneity on reservoir flow performance at several different scale levels and using different recovery methods and well spacings. Figure 3a shows the heterogeneity evaluated in a fluvial model, i.e. cross-bedding types, facies distributions within channels, channel characteristics within channel belts and the relationships between channel belts. Figure 3b shows how for the case being studied (water flood), that it was the small scale and the much larger-scale heterogeneity that has the greatest effect on recovery at breakthrough while the medium scale is not so important. Recognizing the important heterogeneity in each case allows the modelling process to be more effective by focusing on the relevant controlling factors, both in data collection and data handling.

To succeed here requires that the individual technical experts always have a clear view of the quality of the input data, the requirements of the final product and the correct relative value of their own discipline's contribution in this interdisciplinary process of reservoir modelling.

Are well data as certain as one would like to believe?

Given all the complexities that exist when modelling undrilled areas it is not surprising that there is an innate desire to believe that well data have little associated uncertainty. However, when looking at field examples this is not the case. Not only are there problems with navigational data such as depth surveys and deviation surveys, but most of the petrophysical parameters used in reservoir modelling come from indirect wireline measurements that subsequently have to be interpreted as reservoir characteristics such as porosity, saturation, etc. It is essential to be fully aware of the error and uncertainty that come from instrument response, formation property and the interpreter's choice of models and parameters.

The following examples illustrate some of the problems that exist.

Core-to-log correlations

There is often a large uncertainty in determining the best fit to establish the core porosity–log density correlation for a reservoir zone. This best fit is then used to assign porosity values from logs to non-cored intervals. Figure 4 from a North Sea Field shows how the spread of data in an economically important reservoir zone results in substantial uncertainties in HCPV—in this case 5–10% between the different fits shown.

Determination of effective net/gross

The correct determination of net/gross in thin-bedded reservoirs (even with good quality sandstones) and relatively poor quality formations still presents us with a major challenge.

Figure 5 is a data set from reservoirs offshore mid Norway where diagenesis has adversely affected reservoir quality. From this figure it is obvious that the choice of porosity cut-off will be critical in the net pore volumes that form the input to

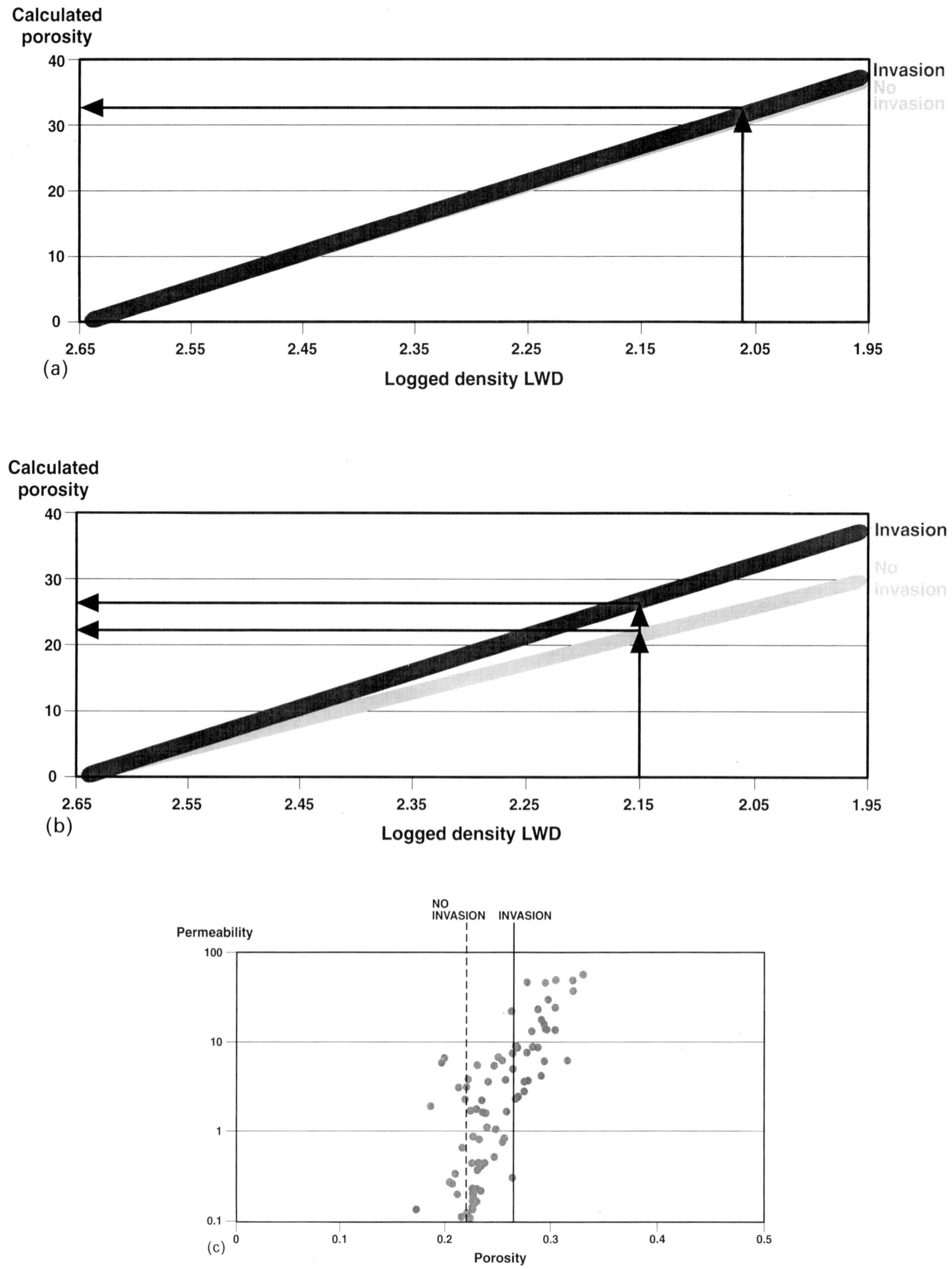

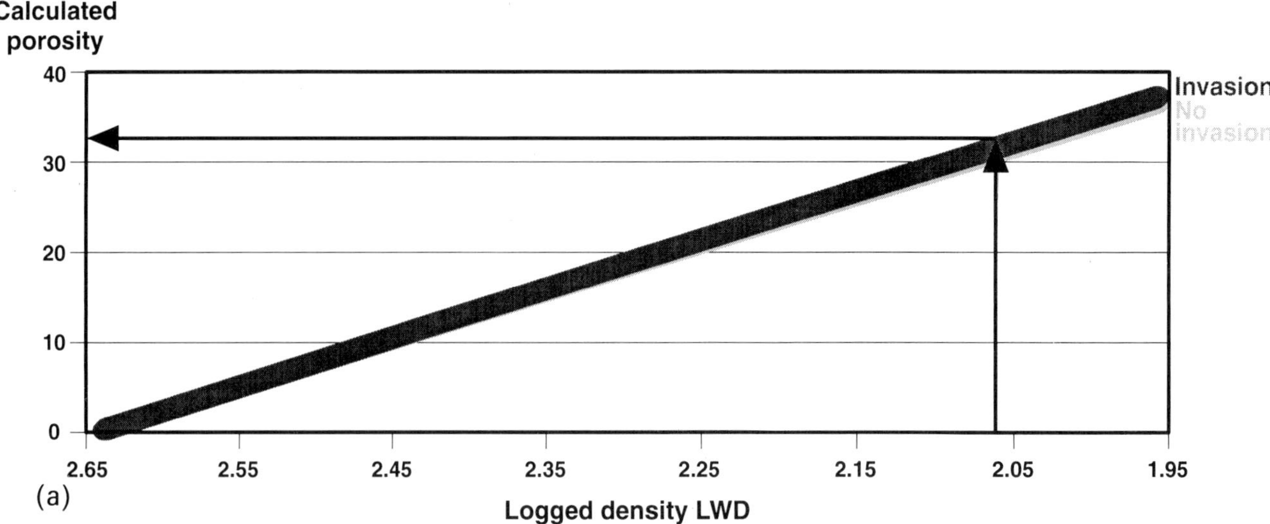

Fig. 2. (**a**) LWD density logging – the effect of invasion in normal oil reservoir on porosity calculations is small (Gullfaks, oil field, oil-based mud); (**b**) LWD density logging – effect of invasion in light hydrocarbon reservoir (Sleipner Øst, gas field, water-based mud); (**c**) the effect of invasion in permeability extrapolations in light hydrocarbon reservoirs may be large. From Flølo (1994a).

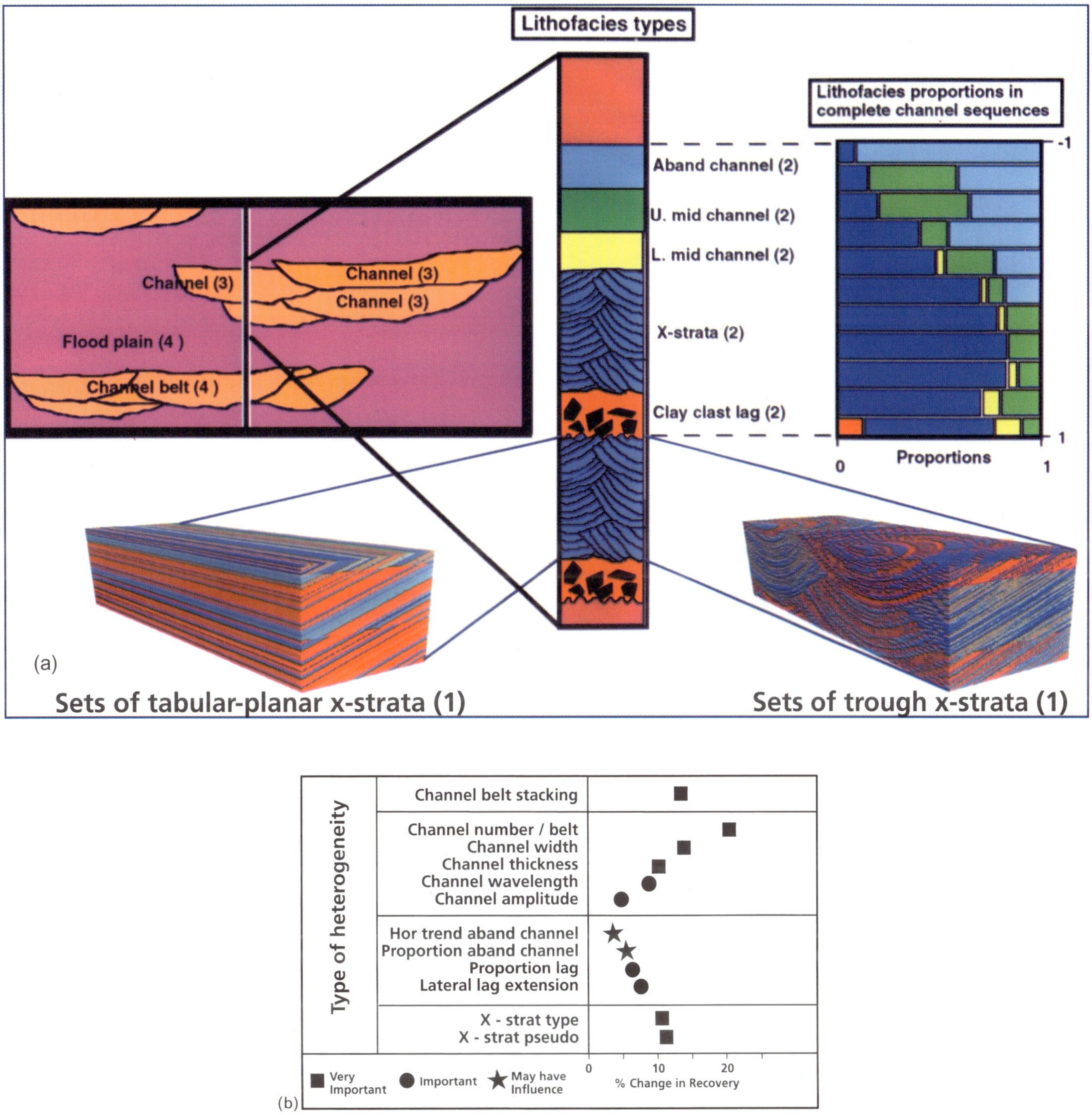

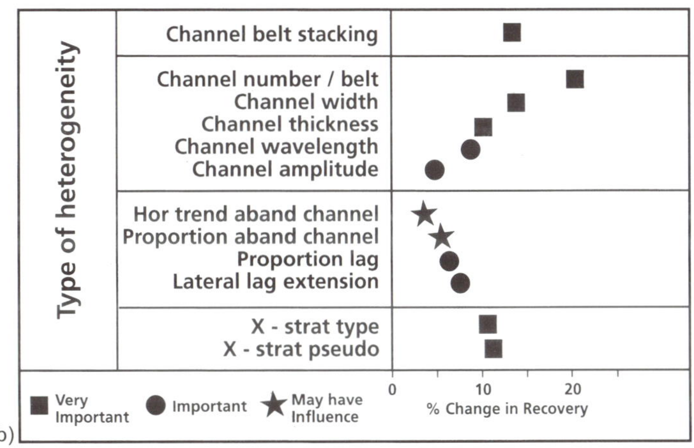

Fig. 3. Heterogeneity in a fluvial reservoir: (**a**) modelling the different levels; (**b**) influence of the different levels on breakthrough recovery (water flood).

the model. In this particular case the average porosity of the formations fall within the steep slope parts of the curves, thus the uncertainty is large. This, of course, affects the predictive ability of the reservoir models and the final results of the modelling should reflect this uncertainty.

What is the correct static net/gross where thin beds are present? Shoulder effects and poor electric log resolution in thin-bedded heterolithic reservoirs may need to be corrected. This is to ensure the correct spatial distribution of hydrocarbons in the model and a better identification of the features that may form important heterogeneity.

Looking back at fields in production it is not uncommon to observe that the original estimates of net/gross from standard petrophysical techniques have not always been very accurate. Figure 6 shows the standard petrophysical interpretation of net/gross in a North Sea well and also the subsequent results from PLT data. In the interval 2087–2094 mMD the original petrophysical interpretation estimated only c. 1 m of net pay. The PLT measurements which are younger towards the right of the figure indicate that hydrocarbons have been displaced over much more of the interval indicating that supposed non-pay has actually contributed to production.

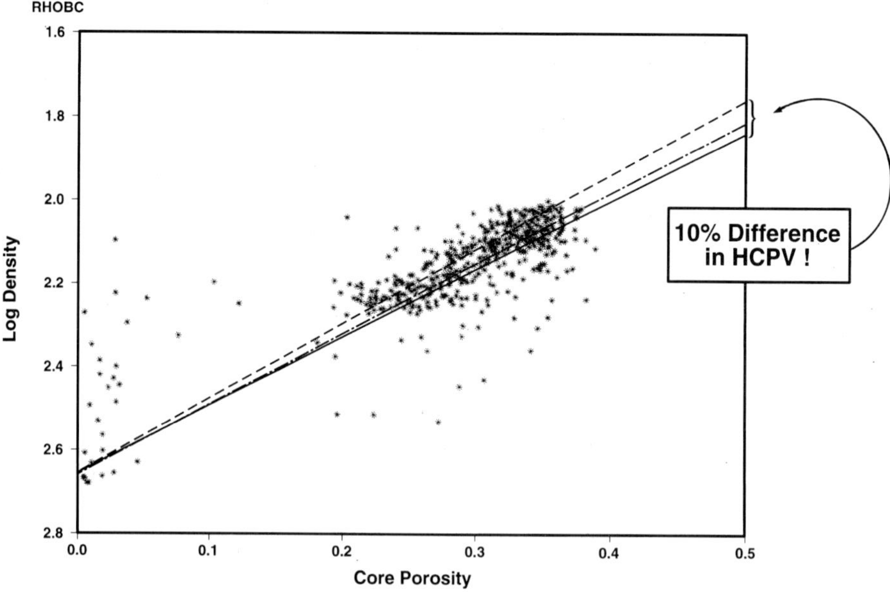

Fig. 4. Volumetric uncertainties in core-to-log-correlations.

Calculation of water saturation

In any reservoir model there will always be a question mark against the values of the orignal water saturations being used since common petrophysical models are based on oversimplified fluid phase geometry and do not take into account bed shoulder effects, etc. Continuing developments in this area show that improvements can still be expected (for example, de Kuijper *et al.* 1995; Stalheim & Eidesmo 1995).

Estimation of permeability from logs

Determination of permeability from logs has always been a hazardous business. Coupled with the need to improve the volumetric parameters in reservoir modelling, as already discussed above, there is no doubt that a better interpretation of permeability from logs will be a great advance. Looyestijn *et al.* (1994) have shown the potential that exists within NMR logging both for improving estimation of permeability and other reservoir parameters. Here, there is a need to focus on new developments that will improve our reservoir description.

From these few examples it should be obvious that for reservoir modelling the large uncertainties that exist in this area need to be rapidly reduced and brought under control. The sum of the effects mentioned here can mean that modellers unwittingly construct models that have fundamentally incorrect initial pore volumes and dynamic parameters. However, even with a reduction in uncertainty, sampling of these petrophysical data should take into account the remaining uncertainty instead of using single values as is often the case today. This means stochastically sampling the data.

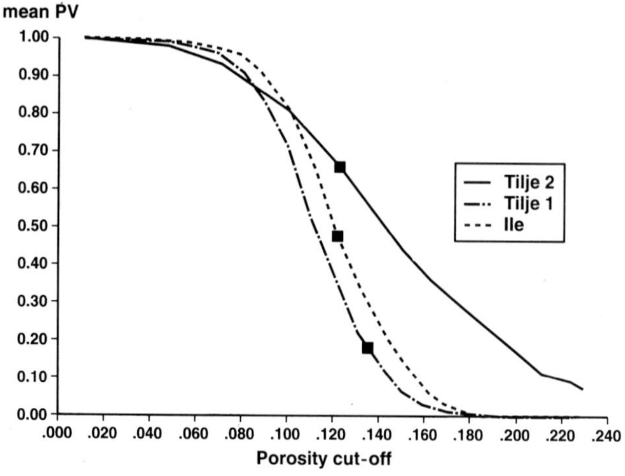

Fig. 5. Variation in mean pore volume with porosity cut-off values (Ile and Tilje formations, Haltenbaken).

What is the basis for estimation S_{or} after flooding?

It is increasingly being recognized that the residual oil, S_{or}, is a complex quantity. It depends on the sedimentary facies, lithology, lamination, mineralogy, recovery process, rates of fluid movement (and therefore position with respect to wells) and on the simulation grid blocks themselves. Furthermore, laboratory measurements are a considerable source of potential error. A lot of sub-optimal investment and potentially unnecessary research has arisen because of the failure to correctly predict this value in reservoir models. It is partly the reason for the large post-production increases in field reserves seen in the North Sea since the sweep has been more efficient than first anticipated, i.e. the S_{or} estimates have been pessimistically high. Here more than other areas, the experiences to date must be carefully reviewed. There are a lot of data from reservoirs that have been flooded by gas and water—what can we learn from these? Figure 7 (Flølo 1994b) illustrates how such data can be collected from fields and used to build up an experience data base which can be compared with the laboratory measurements that have too strongly controlled previous estimates. This figure also shows the need to understand properly the relationship between the history of the flooding and the time at which measurements have been made. In the given example from the Oseberg Field, none of the measurements represent the S_{or}, that is the final result of flooding, which in this case is a gas flood.

During the last few years interesting results have been obtained in petroleum research institutions and oil companies that point to a probable breakthrough in the physical understanding of both residual oil and relative permeability, using network- and pore-modelling. Such a breakthrough will, however, require that future efforts are properly co-ordinated.

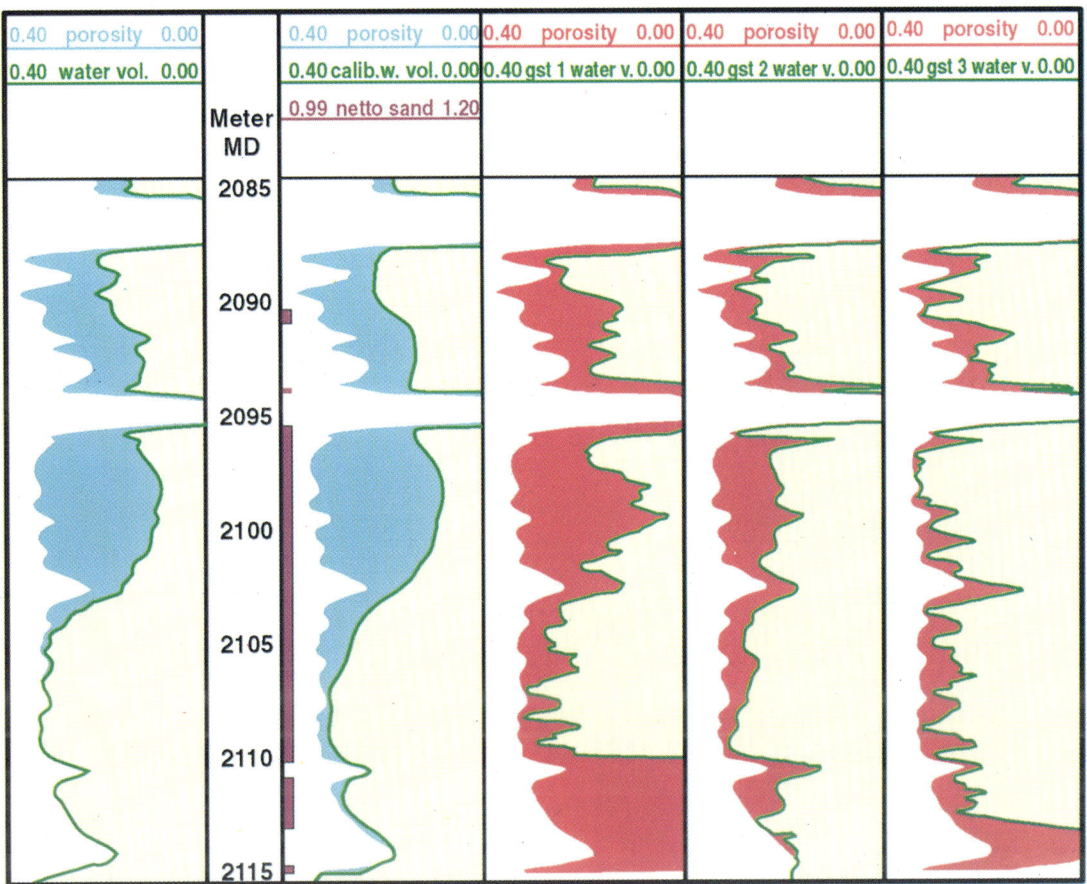

Fig. 6. Production logging highlighting problems in defining effective net/gross.

The aim should be that, given the physical parameters mentioned above; the relative permeability for every grid block should be calculated. It is important to realize that the advances in practical modelling of heterogeneity at all levels in the reservoir make possible the assignation of realistic relative permeability to every sedimentological facies, in contrast to the conventional assignation to whole formations or large regions of the reservoir. Distinguishing between S_{or} as due to rock–fluid interaction or to bypassed oil has also become possible by explicit modelling of the reservoir heterogeneity.

Can better use be made of 3D seismic to extract reservoir data both directly and stochastically?

This is a fundamental area for improvement since seismic data are the best source of reservoir data in the large inter-well

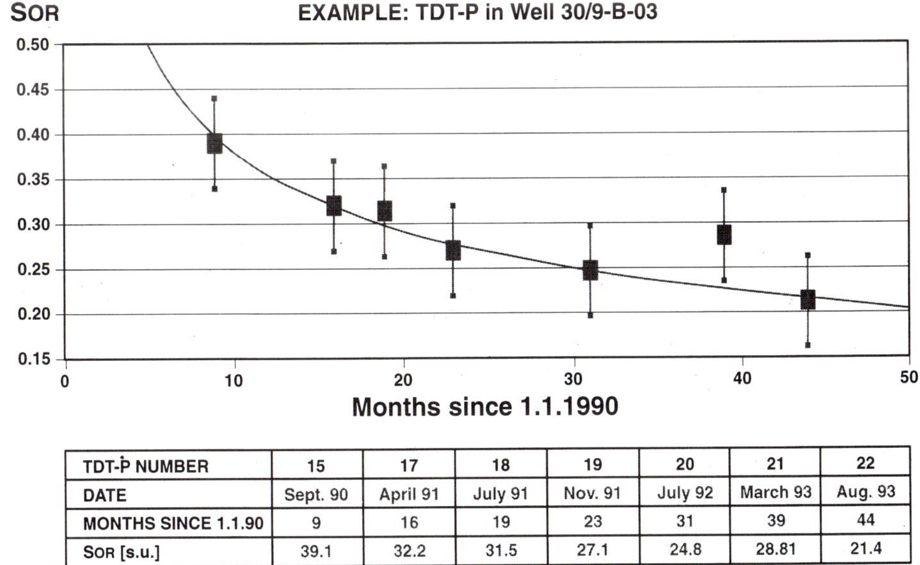

TDT-P NUMBER	15	17	18	19	20	21	22
DATE	Sept. 90	April 91	July 91	Nov. 91	July 92	March 93	Aug. 93
MONTHS SINCE 1.1.90	9	16	19	23	31	39	44
SOR [s.u.]	39.1	32.2	31.5	27.1	24.8	28.81	21.4

Fig. 7. Determination of residual oil saturation (from Flølo 1994b).

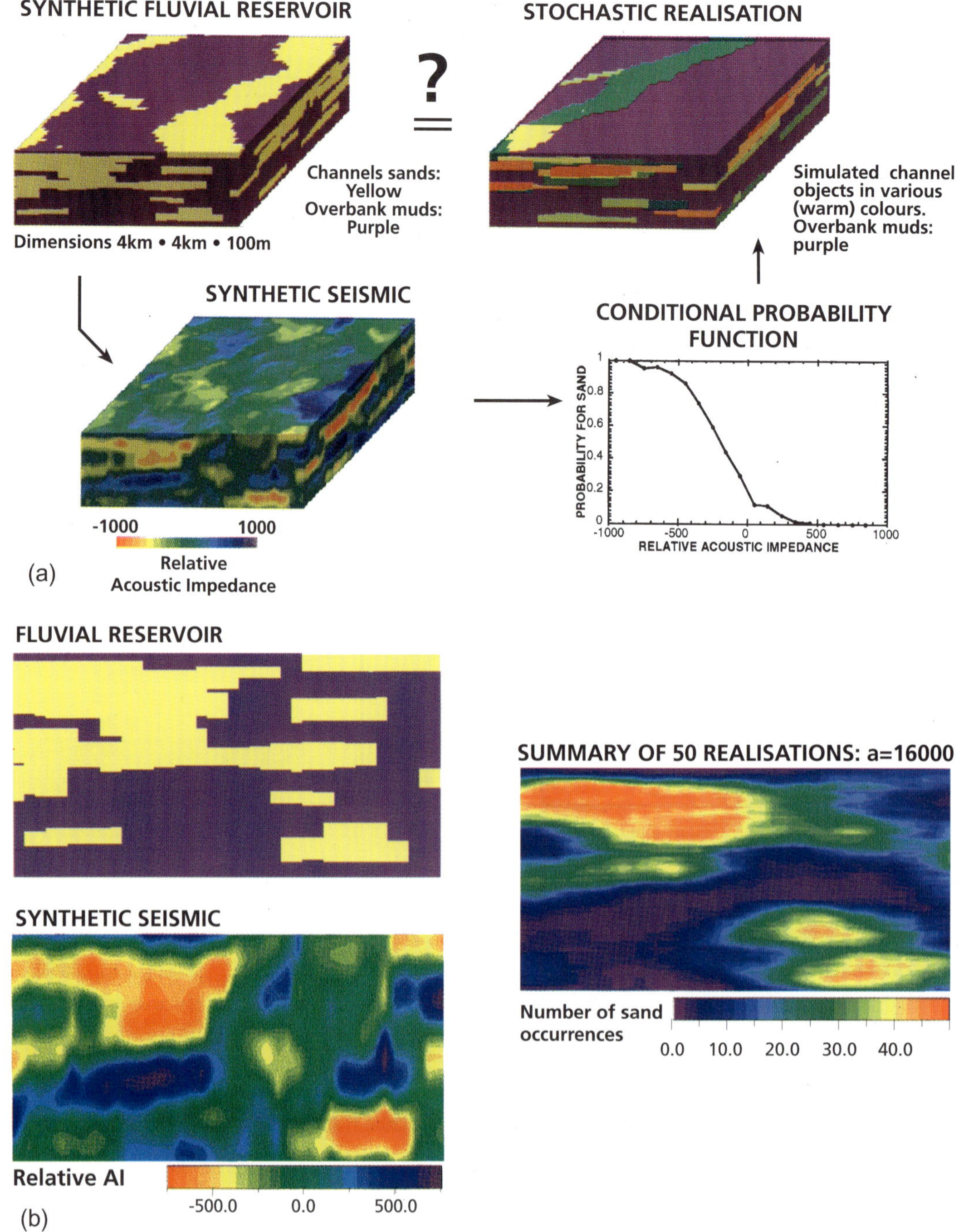

Fig. 8. Conditioning stochastic models with seismic data: (**a**) principles; (**b**) some early results. (From MacDonald *et al.* 1995).

volumes of fields. It is also an area where many companies are focusing their efforts and looking for every small development within seismic processing, data collection and analysis that give a better resolution in both seismic imaging and lithology-fluid determination (Bunch & Dromgoole 1995). Examples are now being published where the ability to determine the distribution of reservoir fluids and the use of time-lapse 3D seismic is giving invaluable information for the mapping of fluid fronts that can calibrate reservoir models (Anderson *et al.* 1995a and b).

Undoubtedly, any advances within the area of lithology and fluid determination from seismic data will be of great interest. Here the role of new offshore shear wave technology may prove to be important. Examples published by Berg *et al.* (1994) show the value of this technology. It is hoped that considerable improvements within the lithology-fluid area will follow which will make significant contributions to reservoir modelling. The potential of shear wave technology within time-lapse 3D seismic based reservoir monitoring is also being considered and will involve close integration with our detailed reservoir models.

Even where it is not possible to establish a 'direct' determination of lithology or fluid from seismic data it must not be forgotten that there is still a lot of useful, although less certain, information that can probably be extracted from modern 3D seismic data cubes. It must be a goal to condition stochastic models using probability functions for lithology based on the seismic attributes that are found to be influenced by lithology; Figure 8 shows details from the work of MacDonald *et al.* (1995). Starting from an initial sedimentary model which provided the reference case (an alluvial system with channel deposits), a seismic impedance cube was generated (Fig. 8a). It was then possible to evaluate whether such impedance data and lithology probability functions could be successfully combined with existing stochastic modelling techniques for channel placement to better constrain the sand distribution in the simulations. Figure 8b shows a summary of 50 simulations and how closely they approximate to the original model. This sort of modelling may become important on the Norwegian Shelf where a significant proportion of undeveloped reserves are reservoired in fluvial systems.

Can geological understanding give a greater predictive accuracy in reservoir models?

One of the reservoir engineer's greatest frustrations is the geologist's ability to always give so many possible answers to each question asked about the geological input to the reservoir model. However, while significant levels of uncertainty will always exist in geological models it is important to still push for reducing (and at least quantifying) the uncertainty level in several areas. This is important with respect to the challenges that small, complex fields and the reduced data gathering will bring. The situation has improved considerably, and can improve further, especially through the advances being made in high resolution sequence stratigraphy, where the stratigraphic concepts are being developed on relevant, surface exposed, reservoir analogues (e.g. Dreyer & Fält 1993). There is now a much better basis for establishing the basic framework of the reservoir and therefore both the lateral and vertical distribution of the flow elements in the model. Software has already been developed which enables modelling of important intra-reservoir bounding surfaces and their uncertainty as well as the relationships of sandbody stacking and heterogeneity distribution to these important sequence stratigraphic surfaces. With improved facies modelling it is also natural that the better physical understanding of relative permeability that exists today will allow these to be combined and therefore again improve reservoir models. Implicit in this assumption is that the problems of upscaling can be overcome (see later).

There is also considerable room for improvement within the sphere of the modelling of structural heterogeneity at a reservoir scale and the area of stress state management. Conditioning to seismic data has already been mentioned but conditioning to well test and production data should also be built into models wherever possible and joint research efforts are currently being directed towards this.

Is it feasible to history match enough stochastically generated models to make better estimates of the uncertainties in production profiles?

Work on Norwegian Fields has shown that it is possible to generate a large number of models, all of them honouring all the known geological data. The question is how to determine those realizations with the most significance and how to streamline the history matching. Tyler (1994b) described a method to rank the realizations by using a fast random walk simulator. Here, a few representative models can be chosen for the most time-consuming finite difference reservoir simulation calculations. Another method (Tyler *et al.* 1995), based on stream-line calculations, can actually reproduce coarsely the production profiles that would be generated by more accurate conventional reservoir simulation. Another interesting method has been developed by Tyler *et al.* (1993) where only the area of interest is studied in detail by refining the simulation grid in this region and taking into consideration the interflows to the rest of the reservoir by using the flux at the boundaries obtained from coarser reservoir simulation models. This allows us to also alleviate the well known problem of upscaling, referred to later in this paper.

Work is currently being carried out to generate automatically a simulation grid with enough refinement around the wells and respecting faults and other geological features. In the near future, it should be possible to develop gridless simulation: it is only pressures, saturations and well rates which are of interest to the reservoir engineer. Voronoi grids (or so called perpendicular-bisection grids), as described first in the petroleum literature by Heinemann & Brand (1989) should be able to solve this problem. It should be possible, by using Voronoi grids, to give as the input: the geological model, the well trajectories, the amount of computing time available (and therefore the approximate number of blocks) and the resolution needed in the neighbourhood of the wells. The output will be the reservoir simulation grid.

Is there a need for many more tools or algorithms for modelling facies and heterogeneities?

Looking back at the progress that has been made during the last five years, a moderate level of maturity has been achieved within 3D stochastic reservoir modelling. Using the tools available it is possible to make acceptable 'hybrid' models with different techniques superimposed upon each other. The existing toolbox gives the necessary flexibility to cope with the fact that each reservoir tends to have its own special features that need individual treatment. There will, however, be some further developments to allow a better modelling of, for example, calcite layers within shallow marine sandstones and, as already noted, there is still a need for developments with respect to using seismic and test and production data to condition the stochastic models.

Apart from these, the greatest need now is to have integrated, well maintained commercial products that are robust and proven within normal operating conditions. A great diversity of commercial products, while attractive to the purist modeller, will not be of help to the companies attempting to purchase stable products. There must be a choice, there must always be room for the small players who challenge those set in their ways, but there must not be chaos in the market place.

Are today's upscaling methods adequate for the geological models being generated?

The methods in use today are too simple for what it is necessary to achieve and in some cases there is a strong suspicion that the errors present in a local upscaling add up substantially as the final coarse grid is approached. More use of adaptive gridding to follow the areas of greatest flux might be a preferred direction of investigation, where a problem to be solved is that these areas change with time and that for different layers the areal grid should be different. How relative permeability, which is related to fine-scale facies, is upscaled remains a big question still to be answered. Jones *et al.* (1995) have shown promising results in this direction.

It is acknowledged that the upscaling problem is essentially insoluble. In general, it is impossible to obtain the same results from a fine reservoir simulation identical to the geological model and a coarser reservoir simulation grid where the input data has been reduced by several orders of magnitude. Pseudo-relative permeability, grids which are conformable to the heterogeneity, introducing more parameters by using tensor permeability are all ways to try to alleviate the upscaling problem. However, only increased computing power with the possibility of simulating the multi-million blocks of the fine geological grid will do away with the upscaling issue. It is therefore also important to remember that reality has only to be achieved to the degree that the results are relevant to the decisions being made.

Is it possible to allow for the geological model crashing?

Experience has shown that while there has been a focus on uncertainty in depth conversion, seismic picks, porosity, etc., not enough consideration has been given to the possibility that the model used is wrong. For example, the field compartmentalization is completely different than first envisaged.

To tackle this problem technical staff must be flexible in their understanding of the possible outcomes and again the actual needs being addressed; much of this comes from experience. At the same time we need to have integrated software systems that allow alternative models to be rapidly generated, verified and subsequently rapidly simulated and history matched. This will be most successful where there is a close linking between the software applications and common databases and where the large processing capacity that is available today is used effectively.

CONCLUDING REMARKS

This paper has hopefully made it clear that 3D reservoir modelling depends on progress in a lot of disciplines and that within these individual disciplines the amount of uncertainty can be large. This requires that efforts are directed to reducing uncertainty and making others aware of this uncertainty. Being an integrated subject, reservoir modelling will to a large degree be as successful as the weakest link in the chain. The chain in this case is the product of reliable software, of well organized databases containing the correct type of data collected at the correct time and of technical experts who understand the particular problems and who can act cross-functionally. This should result in an integrated product which gives a good enough answer for the problem being investigated and at the same time clearly illustrates the total uncertainty in the final predictions. In many cases such a result can only be achieved through an extensive use of stochastic modelling of heterogeneity, whether it be surfaces, facies or petrophysics.

We have come a long way and there is still a long way to go but, importantly, there is still plenty of room for value creation here!

The authors wish to thank Statoil for permission to publish this paper and the organizers of the Symposium 'AD1995 NW Europe's Hydrocarbon Industry' who encouraged us to prepare this overview of the challenges within reservoir modelling. Thanks are also due to the many colleagues who have made important contributions to our thoughts and presentation.

REFERENCES

ANDERSON, R. N., BOULANGER, A., WEI HE, SUN, Y. F., LIQUIN XU *et al.* 1995a. 4D seismic helps track drainage, pressure compartmentalisation. *Oil and Gas Journal*, **93**, 55–58.

——, 1995b. Method described for using 4D seismic to track reservoir fluid movement. *Oil and Gas Journal*, **93**, 70–74.

BERG, E., SVENNING, B. & MARTIN, J. 1994. SUMIC – A new strategic tool for exploration and reservoir mapping. 56th meeting EAEG, Vienna, June 1994, abstract E055.

BUNCH, A. W. H. & DROMGOOLE, P. W. 1995. Lithology and fluid prediction from seismic and well data. *Petroleum Geoscience*, **1**, 49–57.

DREYER, T. 1994. *Applications of quantitative sedimentology in the characterisations of subsurface reservoirs and outcrop analogs*. PhD thesis, University of Bergen.

—— & FÄLT, L-M. 1993. Facies analysis and high-resolution sequence stratigraphy of the Lower Eocene shallow marine Ametlla Formation, Spanish Pyrenees. *Sedimentology*, **40**, 667–697.

FLØLO, L. H. 1994a. LWD or wireline for formation evaluation. NPF 5th Conference on Reservoir Management, Stavanger.

——, 1994b. Residual oil saturation after gas flooding—a TDPT case study. 16th European Formation Evaluation Symposium, Aberdeen.

HEINEMANN, Z. & BRAND, C. 1989. Modelling reservoir geometry with irregular grids. Paper SPE 18412, presented at the SPE Symposium in Houston, February 6–8.

JONES, A., DOYLE, J., JACOBSEN, T. & KJØNSVIK, D. 1995. Which sub-seismic heterogeneities influence waterflood performance? A case study of a low net-to-gross fluvial reservoir. *In*: De Haan, H. J. (ed.) *New Developments in Improved Oil Recovery*. Geological Society, London, Special Publications, **84**, 5–18.

KJØNSVIK, D., DOYLE, J., JACOBSEN, T. & JONES, A. 1994. The effect of sedimentary heterogeneities on production from a shallow marine reservoir—what really matters? SPE 28445, presented at EUROPEC, October 1994.

DE KUIPER, A., SANDOR, R. K. J., HOFMAN, J. P., KOELMAN, J. M. V. A., HOFSTRA, P. & DE WAAL, J. A. 1995. Electrical conductivities in oil-bearing shaly sand accurately described with the SATORI saturation model. SPWLA 36th Annual Logging Symposium, June 1995.

LOOYESTIJN, W., SCHIET, M. & RUNIA, J. 1994. Successful field test of nuclear magnetic resonance (NMR) logging in the Rotterdam Field. 16th European Formation Evaluation Symposium, Aberdeen.

MACDONALD, A., BERG, J. I., SKARE, Ø. & HOLDEN, L. 1995. Constraining a stochastic model of channel models using seismic data. EAPG conference 1995, Glasgow, abstract F052.

STALHEIM, S. O. & EIDESMO, T. Is the saturation exponent 'n' a constant? SPWLA 36th Annual Logging Symposium, June 1995.

TYLER, K. J. 1994a. Heterogeneities: challenges in reservoir management. Skrifter 9/1994, Doktorgradsavhandling, Høgskolen i Stavanger.

——, 1994b. Ranking of production performance from detailed geological models. 4th European Conference on the Mathematics of Oil Recovery, Oslo.

——, Svanes, T. & Omdal, S. 1993. Faster history matching and uncertainty in predicted production profiles with stochastic modelling. Paper 26420, SPE Annual Technical Conference and Exhibition, Houston.

——, Vesterholm, L. & Henriquez, A. 1995. Ranking of stochastic models. In prep.

Reservoir description in the 1990s: a perspective from the flow simulation through layercake parasequence flow units

Patrick W. M. Corbett, Tim Good, Jerry L. Jensen, Jon J. M. Lewis[1], Gillian Pickup, Philip S. Ringrose and Ken S. Sorbie

Department of Petroleum Engineering, Heriot-Watt University, Riccarton, Edinburgh, EH14 4AS, UK
[1] *(Present address: Landmark ????*

ABSTRACT: Reservoir description, in the 1990s, is developing through the increased integration of geology and engineering. This is being accomplished by the integration of a wide range of concepts across the disciplines. Sequence stratigraphy, flow units and engineering architectural considerations are being integrated towards the development of more geologically-realistic simulation models. Sequence stratigraphic concepts are being used to guide upscaling. A new discipline of geoengineering is being forged.

This paper explores this integration in the flow simulation of two case studies. The description and modelling of a relatively heterogeneous shoreface (Rannoch Formation) is contrasted with a more homogeneous shoreface (Lochaline Sandstone). The flow performance of the heterogeneous shoreface (Rannoch) is seen to be sensitive to the geology at various scales. In contrast, the flow performance in the less heterogeneous shoreface (Lochaline) is seen to be relatively insensitive to the geology. The simple models, describing a layercake parasequence flow unit, show how the importance on flow of various aspects of geological heterogeneity can be assessed. The method can be used to identify the most critical aspects of the geology, providing guidance for the description and simulation of reservoirs.

KEYWORDS: *reservoir description, reservoir modelling, geoengineering*

INTRODUCTION

Reservoir description is the description of a (hydrocarbon) reservoir in the subsurface for the purpose of development. More specifically, reservoir characterization is the quantification of that description for the purpose of reservoir simulation (Lake 1989). Reservoir description incorporates the 'hard data' of reservoir characterization and the 'soft data' that is geological experience, concepts and understanding. The purpose of reservoir simulation is to forecast the outcome of various development scenarios for that reservoir, incorporating and estimating the uncertainties that are present in the calculation process.

Finite difference numerical simulators require hard data for porosity, permeability and other saturation dependent functions in a series of grid blocks from which the flow can be simulated in response to pressure changes and flows (injection and production) induced by well operations. Because the size of the grid blocks in a field simulation model are by necessity large, the effective properties are functions of the geological architecture (Kortekaas 1985; Lasseter *et al.* 1986; Ringrose *et al.* 1993) at scales less than the grid block size. The geological architecture is largely 'soft' data, determined by a few geometrical relationships (i.e. those observed in core slabs) and a stratigraphic/sedimentological model. This model can be derived from sequence stratigraphic concepts in shallow marine sediments (van Wagoner *et al.* 1990), from architectural concepts in fluvial systems (Miall 1988; Shanley & McCabe 1994) and from stratigraphic concepts in aeolian systems (Kocurek & Havholm 1993), for example. As it is the effective property that is sought, the integration of hard petrophysical data and soft architectural models is required.

At the larger scale, the architectural arrangement at the interwell scale (Weber & van Geuns 1989) and the flow unit concept (Hearn *et al.* 1984, Ebanks 1987) provide the framework for the development of effective petrophysical properties by the averaging of plug data. Recent usage of the probe permeameter has highlighted some problems with the representivity of traditional plug data within flow units (Hurst 1993).

In a producing field, the tuning of various parameters in the reservoir model until the response of the model matches the production data (water breakthrough, water cut development after breakthrough, field pressures, etc.) is known as history matching. There is no formal procedure for ensuring either the quality or the uniqueness of the history match. The parameter match achieved need not be unique because of the number of poorly defined input data which are available to the simulation engineer as matching parameters.

The method outlined below takes into account the soft geological data (sequence stratigraphic framework) to develop the effective properties of geological building blocks. The approach has the potential to reduce the amount of history matching by focusing adjustments on the appropriate parameters.

From K. Glennie & A. Hurst (eds), 1996, *AD1995: NW Europe's Hydrocarbon Industry,* Geological Society, London, pp. 167–176

A GEOLOGICALLY-DRIVEN METHOD FOR RESERVOIR SIMULATION

The following steps provide a geologically-driven method for the construction of reservoir simulation models.

1. Identify the reservoir flow units in a sequence stratigraphic setting (i.e. incorporating the stratigraphically-significant surfaces and the progradational, aggradational, retrogradational nature of the units). Determine whether these are architecturally layercake, jigsaw or labyrinth.
2. Identify the genetic units: laminae, laminasets, beds, bedsets, channel components, channels, etc., for the respective stratigraphic setting.
3. Map the petrophysics (porosity, permeability and capillary pressure) onto the genetic units. Statistical methods can be used to identify representative element properties and geometries with the help of outcrop analogues.
4. Scale-up the petrophysical data by simulations based on the genetic units and their stacking patterns; we note below that this is by no means a solved problem but, here, we are assuming that scale-up is performed within the limitations of the usual pseudo-ization approaches (Kyte & Berry 1975; Stone 1991 etc.).
5. Generate a flow model, history match with data by returning through 1–4 incorporating the uncertainty in the data to identify critical elements in the model. If the model fails to match, consider other geologically reasonable models which are consistent with the data (e.g. variations on the structural model).

This method follows the concept of 'geoengineering' (Goggin 1988) whereby the geology is 'engineered' for the purpose of reservoir simulation modelling. More formally, we are trying to 'match' the lengthscales of the upscaling process onto the natural geological lengthscales as determined by the depositional process and the fluid mechanics. The 'fluid mechanical' length scales which prevail in a multi-phase fluid displacement process are determined by the capillary/viscous ratio and the viscous/gravity ratio.

As noted above, there are problems with the application of conventional scale-up methods such as those of Kyte & Berry (1975). These relate to limitations concerning the basic assumptions which are made in the calculation of the pseudo-properties, such as pseudo-relative permeability. However, there is also a problem with boundary conditions in all such pseudo-ization or upscaling methods. That is, the effective parameters (pseudo-functions) are not independent of the flows around the local domain which is being upscaled. For example, in the geological context being discussed here, we cannot contend that there is a unique set of pseudo-functions for a particular facies – say, made up of HCS bedforms – since it depends on the flows in adjacent facies (as well as on the balance of forces – viscous/gravity ratio etc.). However, there are two geological lengthscales where something more practical and reasonable can be done in the upscaling as follows:

(i) at the laminaset scale (mm–cm), which might form our smallest representative elements for pseudo-ization, the boundary condition problem can be avoided by assuming capillary equilibrium. Here, the fluid distribution is a unique function of the capillary pressure and can be calculated easily; a steady-state calculation then suffices to determine the pseudo-functions as described elsewhere (Smith 1991; Pickup & Sorbie 1994);

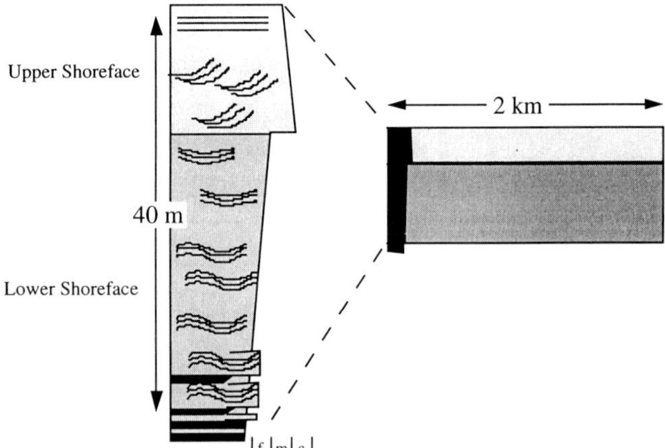

Fig. 1. The geological description of an ideal shoreface reservoir (after Walker 1981).

(ii) at the bedform scale (cm–m), the simplest and most reasonable assumption is that a particular representative bedform is embeddded in similar bedforms (e.g. Corbett *et al.* 1992). Thus, the pseudo-functions are calculated in a representative region which has the bedform geometry surrounded by similar bedforms. This still requires that some decision must be taken about the directions of the external flows. However, in practice, we tend to take 'horizontal' and 'vertical' flows to obtain the limits of the pseudo-functions (noting that the bed itself may not be 'horizontal').

At present, no upscaling method exists which has solved the boundary condition problem for all cases and this is an active area of research. This is a problem of mathematical technique (or algorithm) development and the topic of this paper is more a matter of developing the correct geologically-based approach to upscaling. If the approach is correct, then any one of a wide range of available techniques can be applied by us (e.g. Kyte & Berry 1975; Stone 1991; Pickup & Sorbie 1994).

INTRODUCTION TO CASE STUDIES

Our case studies are taken from shorface reservoirs. In a shoreface reservoir, the flow unit (*sensu* Hearn *et al.* 1984) is at a parasequence scale (*sensu* van Wagoner *et al.* 1990) with layercake characteristics (*sensu* Weber & van Geuns 1989) (Fig. 1). In such a reservoir, simple averages of core plug data have been proposed as the effective reservoir properties (Weber & van Geuns 1989). Plug averages have been shown, however, to be inadequate in reservoirs that are heterogeneous at the small sub-plug and inter-plug scales (Ringrose *et al.* 1993). Recent probe data from shoreface reservoirs show structure at these scales (Corbett & Jensen 1993) and the proposed method has been developed to quantify the effects and, if found to be significant, account for the effects of relatively small-scale geology.

Shoreface reservoirs

Shoreface units are widespread elements in shoreline sediments. A number of sedimentary models for shoreface systems have been developed in recent years (Walker 1981; Brenchley *et al.* 1992; Scott 1992). In these models, several

Fig. 2. A photograph of an analogue for the ideal reservoir in Fig. 1 taken from the Lower Cretaceous Book Cliffs in Utah. The photograph (20 m high) shows one of the parasequences (van Wagoner *et al.* 1990) with lower shoreface at the base of the photograph to upper shoreface at the top.

common features can be identified, and these are summarized as:

- coarsening-up profile;
- low-angle cross-stratification;
- discontinuous shales at the base;
- occasional diagenetic horizons.

These are summarized in Fig. 1, the genetic unit for a shoreface, i.e. the reservoir description at the parasequence scale. Shoreface units similar to the ideal section can be found in the Book Cliffs, Lower Cretaceous of Utah (Fig. 2) where they are known to form repetitive mesoscale stratal elements, deposited in response to cyclical changes in sea-level (van Wagoner *et al.* 1990).

In Fig. 1, the finer-grained lower shoreface, dominated by low angle, hummocky cross-stratification, is overlain by the coarser-grained upper shoreface dominated by trough cross-bedding and parallel lamination. Shoreface systems can be correlated over conventional well distances with aspect ratios in excess of the 50 : 1 illustrated (Bryant & Flint 1993). For the scale of a cross-sectional reservoir model, between typical injector and producer pairs, this description would suggest a Layercake Model (Weber & Van Geuns 1989). Shoreface reservoirs are widespread producing zones around the world with important examples in the North Sea (Rannoch Formation, Beryl Formation, Fangst Formation, Vlieland Formation), West Africa, western US and SE Asia. The Cretaceous Book Cliffs in Utah provide dramatic exposures of shoreface sequences (Fig. 2).

Rannoch Formation

The Rannoch is an important North Sea reservoir and its description and modelling have been addressed in a number of recent papers (Thomas & Bibby 1991; Scott 1992; Corbett *et al.* 1993). The Rannoch shoreface differs from the ideal model in the example given here by having minor shales in the lower shoreface and having a rippled unit between the middle shoreface and the overlying Etive Formation Upper shoreface (Fig. 3). It is recognized that the nature and characteristics of the Rannoch varies thoughout the Brent Province and these changes (e.g. permeability changes due to a diagenetic overprint) will require modification of the Rannoch model shown here (Corbett 1993).

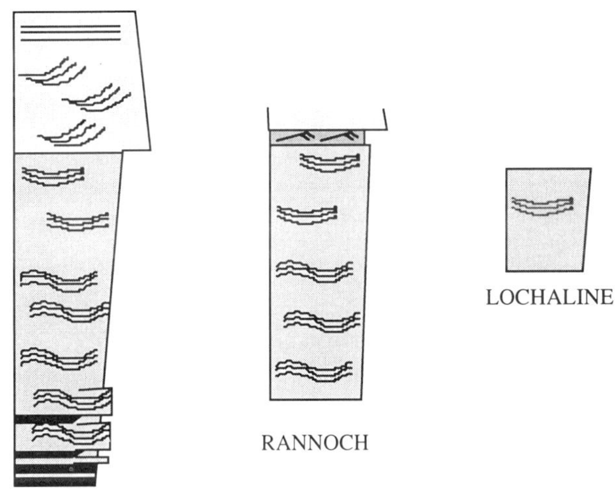

Fig. 3. A comparison of an ideal shoreface with the Rannoch and Lochaline case studies showing the relative thicknesses of the respective units.

Lochaline Sandstone

The Lochaline Sandstone is mined for sand on the west coast of Scotland, providing access to study a sandbody architecture in 3D (Lewis *et al.* 1990). The Lochaline Sandstone also differs from the ideal model in lacking a significant upper shoreface and shales. This unit is relatively thin (Fig. 3). In these respects, neither of our examples is the same as the ideal model, although elements of the ideal model are recognizable and both have aspects that are common to all shoreface sandstones.

APPLICATION OF THE GEOENGINEERING APPROACH

1. Sequence stratigraphic framework

The units identified in the case studies are interpreted as layercake parasequences. The Rannoch Formation is part of a prograding shoreface system in the northern part of the Brent province (Mitchener *et al.* 1992). As such it is regionally and

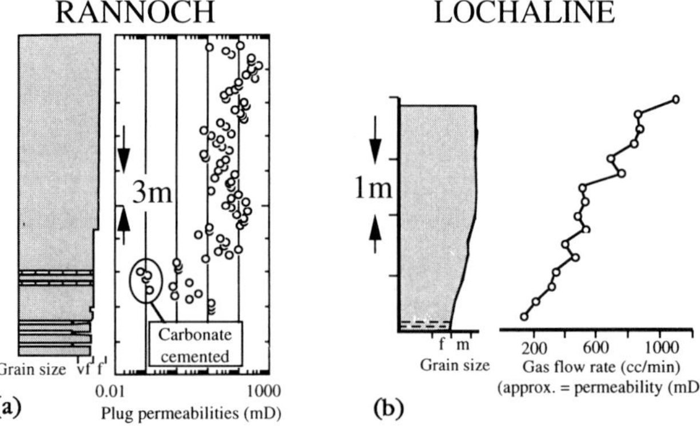

Fig. 4. Permeability profiles in shoreface sandstones: examples from (**a**) the Middle Jurassic Rannoch Formation (from Corbett *et al.* 1993) and (**b**) Cretaceous, Lochaline Sandstone, W. Scotland (from Lewis *et al.* 1990).

locally correlatable and defines a flow unit by being correlatable between wells. The internal structure of the Rannoch is heterogeneous at various scales and these will be discussed further below. The Rannoch/Etive package is bound by time-stratigraphic shales. The Rannoch is considered a separate flow unit from the Etive.

The stratigraphic significance of the Lochaline Sandstone is harder to define regionally, as the Lower Cretaceous sandstone unit is confined to a limited outcrop on the west coast of Scotland. Within the outcrop there is lateral extent over 2 kilometres. Being thin, the unit may not extend over interwell distances in a North Sea context; however, in an onshore context the Lochaline Sandstone would also define a reservoir flow unit of layercake nature at a parasequence scale.

2. Genetic unit identification

Sequence scale

The shoreface parasequence defines the reservoir flow unit and is generally mappable between wells. This is a characteristic of layercake reservoirs. The middle and lower shoreface elements are identifiable within a Rannoch flow unit. At this scale, the gravitational competition between the fluids will interact with the coarsening-up profile. Coarsening-up profiles in well-sorted sandstones are consistent with increasing permeability trends and these are commonly seen in shoreface sands (Fig. 4).

Bed scale

In shoreface sandstones, the lower to middle shoreface are commonly dominated by hummocky and swaley cross-stratified genetic units. The bed scale phenomena, occurring over vertical distances of 1–2 m, are distinctive in outcrop (Fig. 5) and on permeability profiles within wells (Fig. 6). In the Upper Shoreface, trough cross-bedding to parallel lamination are the dominant genetic units.

In the Lochaline Sandstone it is difficult to identify bedding structure (except on weathered outcrop surfaces). Permeability patterns do not reveal any bedding, the sand having been largely homogenized by bioturbation.

In shoreface sequences, discontinuous and continuous shales (shale genetic units) can be present in the lower part of the shoreface unit. The tendency is for the units to become less well preserved and more discontinuous as the shoreface develops (Brenchley *et al.* 1992). The Rannoch and Lochaline sections shown in Fig. 4 are largely devoid of shales. Shales in other shorefaces can be important (e.g. as shown in the Book Cliffs simulations, Ciammetti *et al.* 1995).

In addition, diagenetic units in the form of concretionary carbonate cements are present in shoreface systems (and are noted in the Rannoch section in Fig. 4). In the Rannoch Formation, pressure tests (designed in order to measure the properties of the concretions) in the Cormorant Field showed the cemented zones to be of limited lateral extent in that field (Braithwaite *et al.* 1989). The few diagenetic cemented zones were not considered significant in the Rannoch model. In the Lochaline Sandstone, well cemented (quartz) nodules were also described and, although included in the simulations, were not found to be significant because of their isolated nature. The spatial distribution of these nodules may be random, or alternatively developed along stratigraphic surfaces within the sediment. In the former situation they are not likely to be significant, in the latter case they can form significant flow barriers or baffles.

Lamina scale

Lamina are variably developed in shoreface systems. In the Rannoch Formation, the presence of mica ensures that the laminae are well developed and very visible. In the Lochaline

Fig. 5. Beds and laminaset scale features in a shoreface sandstone, Book Cliffs, Utah.

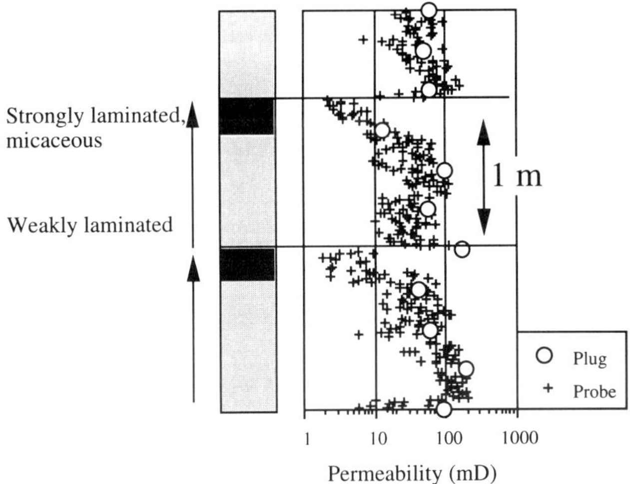

Fig. 6. Beds in shoreface sandstones as detected by probe permeability profiles. Note the poor representivity of the core plug data at this scale.

Sandstone, it is difficult to see any sedimentary structure except on weathered surfaces.

Examples of laminated shoreface elements from the Rannoch Formation are described in Corbett & Jensen (1993). Low contrast, high contrast and rippled laminaset elements are the 'building blocks' of HCS beds.

3. Mapping the petrophysics onto the genetic units

(a) Permeability

Shoreface elements are generally a few metres to tens of metres thick. In thin units, probe data can show trends which describe the sandbody by reflecting the grain size profile (Fig. 6, Fig. 4b). In thicker shoreface intervals, such as the Rannoch Formation, core plugs show the coarsening-up trend (Fig. 4a). The vertical permeability profiles are useful in showing trends at various scales. A statistical measure (the semivariogram) is also useful for identifying the spatial structure (Fig. 7, from Jensen *et al.* 1996*b*). The structure observed in semivariograms of permeability data is an important indication of grain size trends and should be incorporated in flow simulation models as it can impact the flow performance (Jensen *et al.* 1996*a*).

Lateral variograms can be constructed from permeability data measured at outcrop, in a mine (as in the Lochaline Sandstone case) or on cores from a horizontal well. In each case, the lateral permeability profiles of probe data are not necessarily maintained along a stratigraphic horizon and therefore cannot be considered truly horizontal. Probe measurements provide a small support (measuring perhaps the properties of a single lamina) a better measurement would come from plugs or even whole core samples along a transect (showing the variability of bedding at the larger scales). The lateral permeability semivariogram in the Lochaline Sandstone attempted to follow a stratigraphic datum. In the Lochaline lateral semivariogram (Fig. 8), there is evidence of some structure at 60 m lag in an overall trend (increasing variance with increasing lag spacing). The cause of this structure (i.e. hole at 60 m) has not been identified geologically and may be due to uncertainties in the stratigraphic control.

The permeability structure of beds, which are generally 1–2 m thick, are best shown by the probe permeameter data

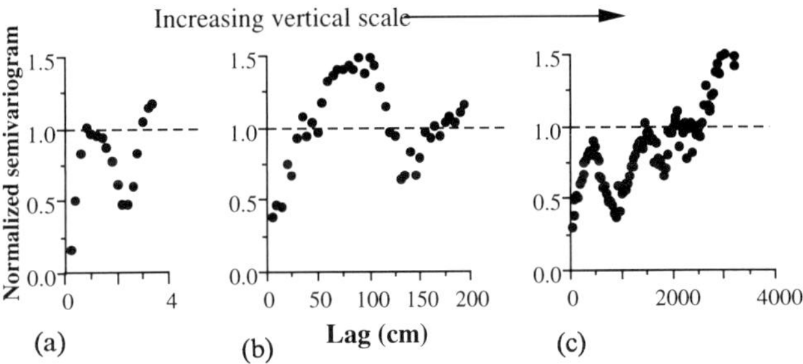

Fig. 7. Semivariograms showing permeability structure in vertical profiles at **(a)** the parasequence (a trend); **(b)** bed (cyclicity); and **(c)** lamina (cyclicity) scales (from Jensen *et al.* 1996*b*).

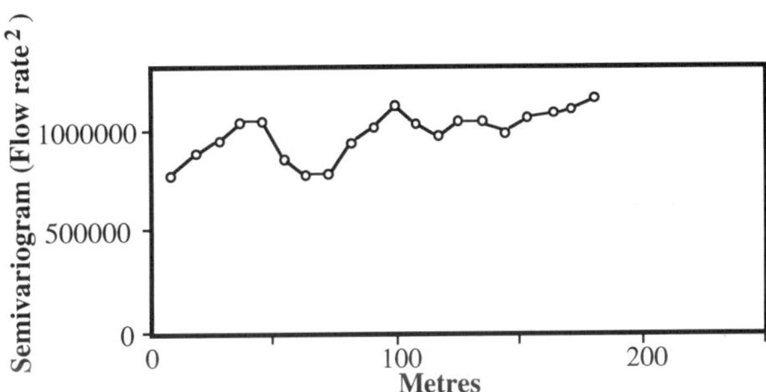

Fig. 8. Lateral semivariogram from the Lochaline Sandstone in the Lochaline mine.

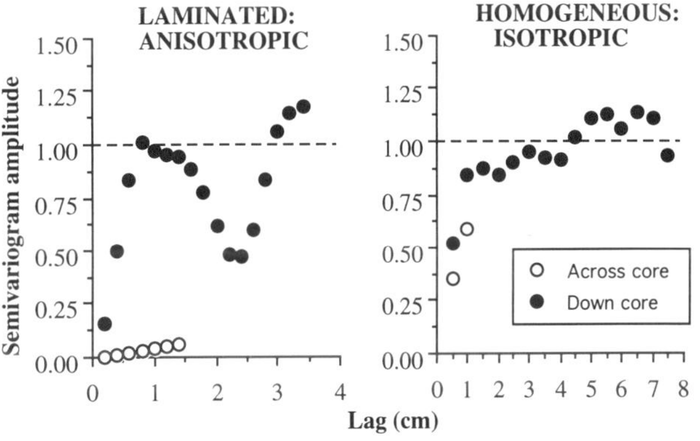

Fig. 9. Vertical and lateral semivariograms for lamina elements in the Rannoch Formation: massive (top), laminated (bottom).

gathered at centimetre spacing (Fig. 4). The variogram picks up the structure in the form of a 'hole' at a lag distance equivalent to the dominant bed thickness (Fig. 7b). The lateral geometry of the shoreface beds cannot be determined from core. However, identifying the shoreface nature of the system, aspect ratios can be measured at analogous outcrops (Corbett *et al.* 1994) and these can be used to model variability at this scale. The applicability of analogue data has to be assessed for each reservoir and this requires more quantification of stratal elements across a wide range of systems so that the variability and representivity of elements can be established.

At the small scale, semivariograms in vertical and horizontal directions from probe data at millimetre spacing (i.e. along and across core) can be used to determine the small-scale structure (Fig. 9).

Table 1. Permeability variation (measured by Cv = standard deviation/mean) for various scales in the Lochaline Sandstone and the Rannoch Formation

	LOCHALINE	RANNOCH
Lamina scale	0.25	0.31–1.52
Bed scale	0.48	0.63–0.93
Parasequence scale	0.59	0.76*

* indicates that the carbonate intervals have been excluded in this analysis.
Note that the heterogeneity increases with scale in the Lochaline Sandstone, only heterogeneous (*sensu* Corbett & Jensen 1992) at the parasequence scale. In contrast the permeability variation in the Rannoch is significantly higher and heterogeneous at all scales. The greatest variability is observed in some of the lamina elements – the ripple lamination has a Cv of 1.52 – however, these do not dominate at the larger scales.

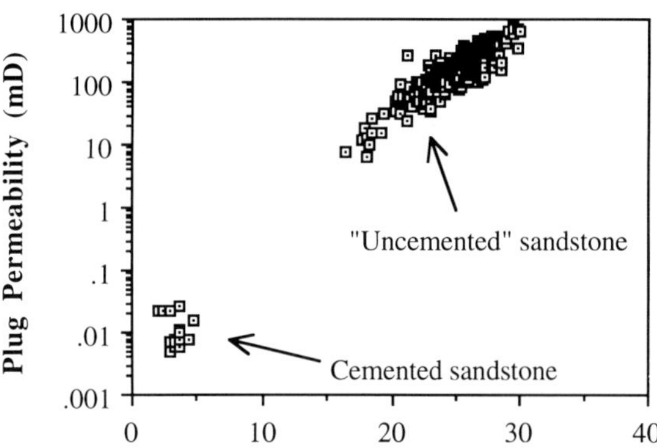

Fig. 10. Porosity versus permeability crossplot for the Rannoch Formation from a North Sea well. In this example, the low permeability cemented material can be easily distinguished from the relatively uncemented sandstone.

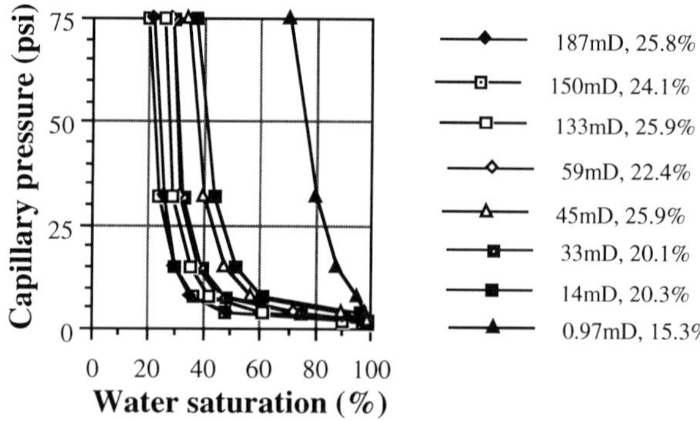

Fig. 11. Family of drainage capillary pressure curves for the Rannoch Formation (from Corbett 1993).

In these examples, the permeability structure can be identified by a series of targetted sampling schemes on the recognized genetic units. The variability (as measured by the coefficient of variation; Cv = S.D./Mean) can be used to quantify the Rannoch and Lochaline shorefaces at the significant scales (Table 1).

The non-reservoir unit properties can sometimes be identified as subsets of the plug data (Fig. 10). In this example, the cemented intervals have very different poroperm characteristics from uncemented Rannoch intervals. The uncemented intervals show significant heterogeneity (Cv 0.76) and cannot be treated as a homogeneous reservoir without further analysis of the heterogeneity.

(b) Other important parameters

1. *Porosity.* Porosity is determined at various scales from wireline logs and core plugs. Porosity is important for the volumetric determination of fluids in place but has little direct impact on flow. In the Rannoch, the porosity variation (15–30%) is much less (factor of 2) than the permeability variation (10–1000 mD; factor of 100) so detailed porosity analysis is not necessary. The pore size distribution will be important in influencing capillary pressure; however, the collection of capillary pressure data in a genetic units framework allows for this. If large porosity variations are present, the porosity data should also be collected in a genetic units framework.

2. *Capillary pressure.* Capillary pressure (Pc) is commonly measured on selected 'homogeneous' core plugs as a measure of the pore size distribution. Often in simulation models, when large grid blocks are employed, Pc is not used other than to initialize the model. This is because the effects of Pc are minimal for large grid blocks (but not necessarily within 'real' reservoir rocks). Figure 11 shows a family of drainage curves (brine displacing air) for the range of permeability classes within the reservoir. These show typical variations for the Rannoch, which are largely a function of permeability. In this case, the appropriate curve can be assigned according to the permeability. For water-flood conditions, imbibition curves (water displacing oil) would be more appropriate, and these can be scaled from the air-brine drainage data with the aid of some careful experimental data. For completeness, water–oil imbibition curves, at reservoir conditions, on carefully selected representative stratal elements could be sought; however, these experiments are not easy and therefore costly. The use of drainage curves is a pragmatic solution to this engineering problem.

3. *Relative permeability.* Absolute permeability is measured by an inert gas (usually in the case of probe and plug permeameters, nitrogen). Gases have to be corrected to an equivalent fluid permeability (i.e. a slippage correction). In the presence of two or more phases (i.e. combinations of oil, gas and water), the relative permeability to oil, water and gas in the presence of another phase depends on the saturation and wettability. Relative permeability is difficult to measure in the laboratory at reservoir conditions, consequently few samples are run.

In the reservoir simulation model, a selected curve or numerical relationship is often used in preference to measured data (Muggeridge, 1991). The relative permeability functions are a critical control on the performance of the flow simulator. The effects of lamination (core-plug) scale heterogeneity upon relative permeability curves have been recently reviewed by Ringrose *et al.* (1994). Rather than averaging out the relative permeability performance for the different samples measured in the laboratory, the differences should be understood in terms of sub-sample size, geological structure and wettability differences. This may require taking larger samples (e.g. whole core) to obtain relative permeability curves above the lamination scale, to take smaller samples within the laminae, or use numerical models based on pore geometry and wettability to simulate the lamination scale wettability behaviour.

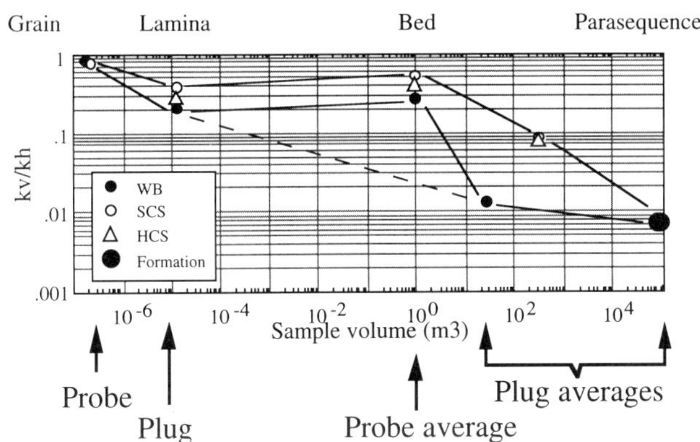

Fig. 12. Permeability anisotropy with increasing scale in the Rannoch Formation. (Subdivision of facies in the Rannoch shown: WB wavy bedded; HCS hummocky cross-stratification; SCS swaley cross-stratification, refer to Corbett 1993.)

4. ***Permeability anisotropy ratio.*** There are few methods for systematically measuring the anisotropy ratio in reservoirs. Vertical (k_v) and horizontal (k_h) plug permeabilities can be used to determine the ratio at the plug scale within a facies. Plug data must be carefully averaged for larger-scale features. Inter-bed flow can be severely affected by minor layers (e.g. bounding surfaces) not detected by probe or plug. For this reason the horizontal permeability is often accurately estimated by the arithmetic average; however, the vertical permeability tends to be overestimated by the harmonic average (Corbett 1993). The degree of anisotropy in layered sedimentary rocks tends to increase with scale as shown in Fig. 12 for the Rannoch Formation. Measurements on cubic plugs in Rannoch material generally support the probe and plug estimates presented here (Halvorsen pers. comm. and Braithwaite *et al.* 1989) at the appropriate scale. Pressure tests and specially-designed well tests can be used to give measures of vertical permeability at larger scales; however, these dynamic data are not available in this case.

The anisotropy ratio is an important parameter in reservoir simulators and, because conventional plug measurements were so poor at describing the 'effective' parameter, the anisotropy ratio typically became the parameter to adjust until a history match was achieved. Careful measurement of the permeability at the lamina and bed scale can be exploited to give improved estimates of anisotropy at the parasequence scale. These will further constrain the simulation model.

4. Simulation and scale-up of genetic units

Effective permeability is a function of geometry (refer to a recent review in Pickup *et al.* 1994). For simple geometries, statistical averages will suffice: arithmetic average for flow parallel to layer flow, harmonic average for flow across layers. The geometric average is the correct flow for a random permeability field in 2D. For tilted layers or beds, a tensor permeability may be required, allowing for flow normal to the pressure gradient.

Capillary pressure and relative permeability functions (in the simulator and the laboratory) are sensitive to the recovery process (i.e. water-flood, gas-flood), the flow rate, the reservoir geometry and are non-linear functions of saturation. Scale-up of these saturation-dependent properties for reservoir simulation grid blocks is not as 'simple' as averaging permeability. An accepted scale-up technique for these parameters is 'pseudo-ization' leading to the use of pseudo-functions (Kyte & Berry 1975; Lake *et al.* 1989). Pseudo-properties of grid blocks are numerical functions that reproduce the correct flows and pressure drops determined from a high resolution simulation for the same volume. A coarse grid model with pseudo-properties will perform the same way as a very fine grid simulation, reducing the number of blocks needed for a given reservoir volume, allowing large volumes to be accurately simulated. If the simulations are based on the geologically-representative elements, the

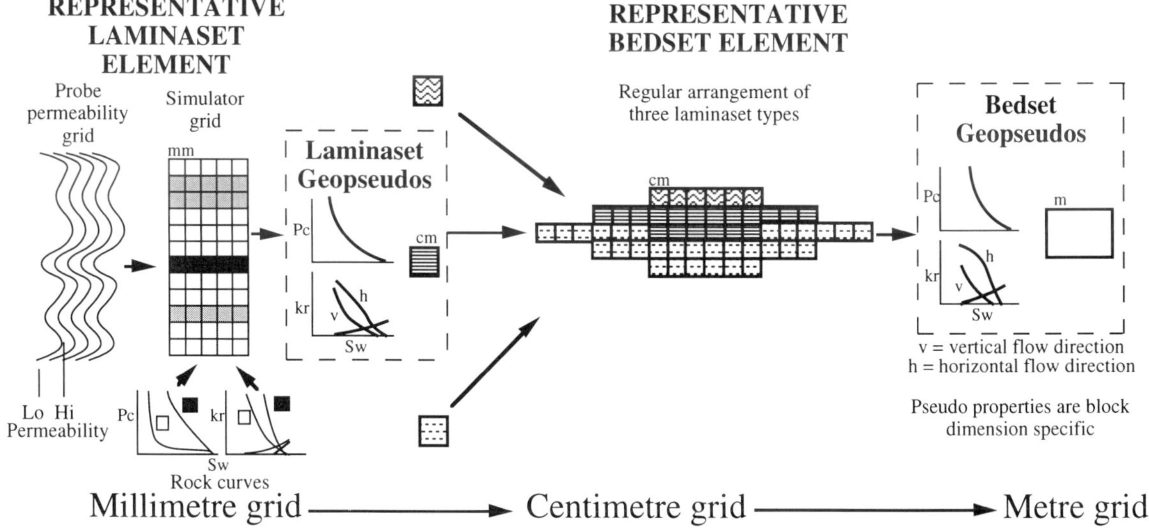

Fig. 13. Summary of the geopseudo method for deriving effective flow properties at the laminaset and bedset scales. On the left are the absolute permeability variations measured by the probe permeameter. To the right of the permeability field, the simulator grid is depicted with the appropriate relative permeability and capillary pressure curves for each grid block. Following simulation at the fine scale the pseudo-properties of the laminasets are generated by the simulator. The centre shows the laminaset pseudo-properties being used in a bedset model at the next scale. Finally on the right, the pseudo properties for a metre-scale block, which include capillary effects and sedimentary structural effects are generated. These genetic unit pseudos (geopseudos) can be used for the appropriate elements in the larger scale models (from Corbett *et al.* 1992).

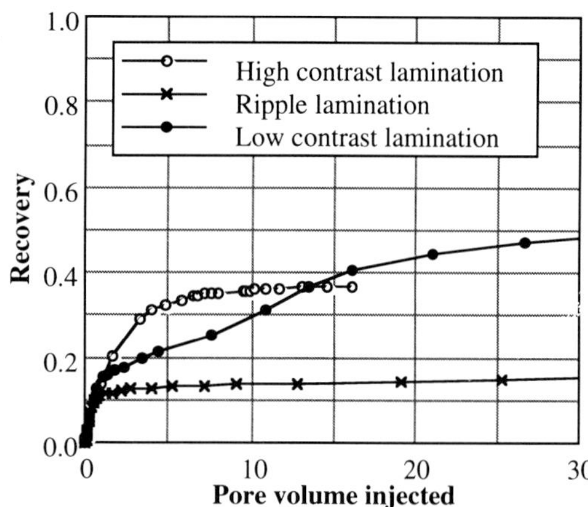

Fig. 14. Horizontal flow performance of three Rannoch Formation laminasets. Low contrast lamination is the most uniform and ripple lamination the most variable. From Corbett & Jensen (1993).

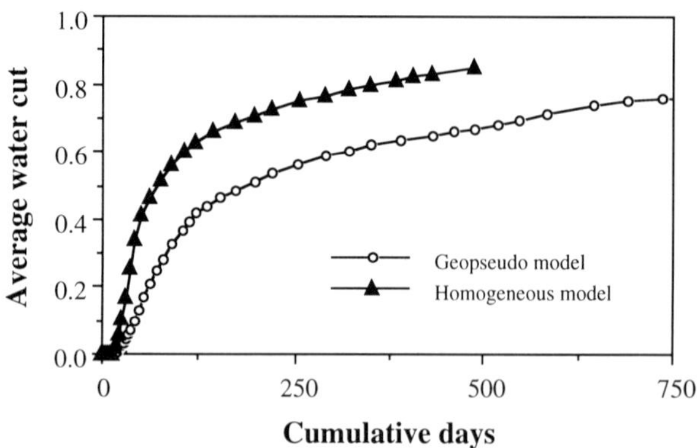

Fig. 15. Comparison of recovery for a scaled-up geopseudo model for the Rannoch compared with a homogeneous case (from Corbett 1993). The effect of the beds and lamina is to reduce the overall recovery.

pseudos are known as 'geopseudos' (Corbett *et al.* 1992). The geopseudo procedure is illustrated in Fig. 13.

As an example, the flow performance (shown as water cut development through time in a horizontal flood) of representative lamina elements from the Rannoch are shown in Fig. 14). The differences are due to permeability contrasts in layered systems (low contrast; $150 < k < 400$ mD and high contrast; $4 < k < 220$ mD) and the isolated geometry of ripple laminated systems ($1 < k < 60$ mD). The low contrast laminated interval has a Cv = 0.31 and effectively performs like a uniform rock. These elements are incorporated in scaled models and applied at the parasequence scale (Fig. 15), where the performance can be compared with the rock curves (i.e. relative permeability curves scaled for the appropriate grid block size but not accounting for the sub-grid block heterogeneity) showing the effects of the heterogeneity. The k_v/k_h ratio has been appropriately scaled by the use of horizontal and vertical pseudo-functions allowing for this property to change with saturation (if appropriate). At the lamina, bed and formation scales the Rannoch showed some sensitivity to the geological structure (Figs 14 & 15).

In the Lochaline Sandstone, flow at lamina and bed scales was insensitive to the low variability (Cv < 0.5). Only at the unit scale in the Lochaline Sandstone were any effects of the heterogeneity seen (Fig. 16). The effects were very minor

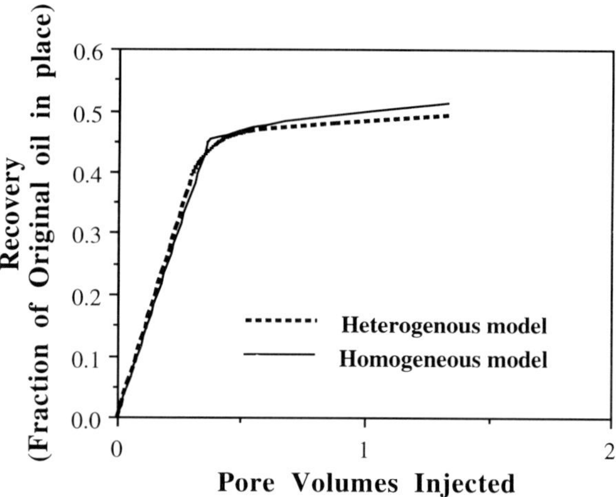

Fig. 16. Comparison of recovery for the Lochaline Sandstone comparing a model with fine-scale geology (heterogeneous case) with a model of uniform permeability (homogeneous case).

because of the relatively thin (4 m) sandstone and the moderate heterogeneity (Cv = 0.59).

5. Performance prediction and history matching

The geopseudo scale-up method outlined above has been carried out on a North Sea Rannoch Formation reservoir with encouraging results (refer to Corbett *et al.* 1992 and Corbett 1993, for further details). The model is based on a 'digital' petrophysical model of the Rannoch Formation in which the genetic units and stratal elements are accounted for in the scale-up. While the approach followed is very deterministic, uncertainty and variability could be introduced at each step of the scale-up using a mixture of object and continuous geostatistical modelling techniques. This would allow the performance to be expressed in a range of outcomes, quantifying the uncertainty and/or variability.

In the Rannoch example considered, production data can be reasonably matched without any tuning of parameters (Fig. 17). Additional fine tuning of the model could be carried out by adjusting the parameters at each scale within the bounds of the data (using geostatistical techniques) and our understanding of the shoreface geological model. If the numerical model cannot be sufficiently matched at this stage, the effect of other possible influences (e.g. faulting) can be further investigated.

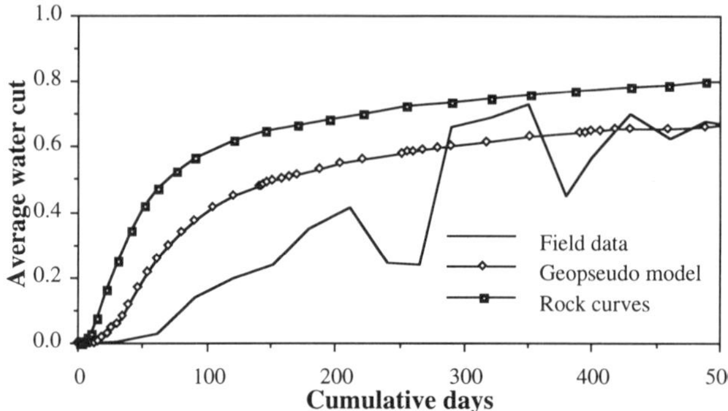

Fig. 17. The geopseudo model simulation presented in Fig. 15 compared with production data from the a Rannoch Field on which the model was based (from Corbett *et al.* 1992). The match with the geology built in to the relative permeability curves is an improvement over the rock curve model. The effect of the small-scale geology can thus be assessed by comparison with 'homogeneous' models and real data.

DISCUSSION

The importance of small-scale structure

Recent studies have shown that the small-scale heterogeneities can be important in reservoirs (Kjonsvik *et al.* 1994) and in laboratory floods (Honarpour *et al.* 1995; Huang *et al.* 1995). In this work, the small-scale heterogeneity was shown to be significant, only after reservoir connectivity has been established. In layercake shoreface reservoirs, connectivity of the flow units (in the absence of faulting) is not a major problem. In these reservoirs, the small-scale structure (at the lamina and bed scale) is an important control on flow performance. These small-scale effects need to be scaled if they are to be represented in larger-scale simulations. The approach presented here is designed to allow the incorporation of a geological model in the scale-up. An important point about the effect of lamination (where it is important; Ringrose *et al.* 1993), is that it affects **two** important features of the pseudo-function, namely (i) the two-phase anisotropy which is **much** higher than the single-phase anisotropy and cannot be accurately represented as a simple k_v/k_h multiplier; and (ii) the phase trapping or levels of residual oil can be very different in laminated systems which may radically alter the end point saturations in the pseudo-relative permeabilities (Corbett *et al.* 1992; Ringrose *et al.* 1993; Huang *et al.* 1995).

The use of geological analogues

The two shoreface sandstones shown differed in a number of respects from each other. Both had a common element, the coarsening-up, and differences – the Rannoch has more variability at the small scales. High resolution probe data allowed these differences to be quantified, with respect to permeability. The quantification of further shoreface outcrops will enable representivity of shoreface systems to be better determined in future.

In addition, detailed simulations, incorporating the geology as best as current technology allows, of what is directly observed in outcrops is important to the understanding the effects of the various geological heterogeneities in shoreface systems (Ciammetti *et al.* 1995). Other workers have published similar uses of outcrop data (Høimyr *et al.* 1993), and many more of these studies are needed in order to determine what level of geological description is needed for flow simulation.

Petrophysical sampling

Petrophysical sampling needs to be carefully planned to address the description of the various stratal elements. Measuring the flow properties of the basic elements (laminaset and bedset scales) and calibrating the scaled effective properties against dynamic data (e.g. well tests) provides an alternative to 'statistically valid' 1 ft spaced samples). Simulation of geological elements, as presented in this paper, forces the collection of petrophysical data fit for the purpose of simulation.

Ease of application

The scale-up as outlined here is currently rather labour intensive as the models need to be hand-crafted. The development of appropriate software will encourage the correct scale-up of relative permeability curves (Pickup 1994) which are fundamental to the performance of a numerical model. Presently, the demands of the fine scale modelling (largely time-related) are holding back the widespread application of the presented approach, even though the significance of the small-scale effects are now well established (Alvestad, pers. comm.).

For a range of scales, geostatistical methods allow rapid handling of large data sets and handling of uncertainties in properties, geometries and stacking patterns (e.g. Rossini *et al.* 1994). These tools are becoming increasingly available to address various modelling issues and can be incorporated into the above method.

Finally, having established the above geologically well-founded approach to multiphase fluid flow upscaling, we recognize that there are other mathematical problems which are still outstanding. As discussed above, problems relating to boundary conditions and other numerical and mathematical issues must also be solved in order to provide us with a fully reliable 'toolbox' for upscaling.

CONCLUSIONS

The modelling of shoreface reservoirs with the benefit of a geological model, that has been used to drive both the sampling and scale-up, is an appropriate way to combine geology, petrophysics and reservoir engineering.

Geoscientists find a geologically-based approach to reservoir simulation more consistent with the developments in stratigraphic concepts. This is particularly so in shoreface systems where these concepts are best understood and the effects of the stratal elements can be shown to effect performance.

The development of 'geoengineering' as a discipline with the role of building geologically-reasonable flow simulation models is a foreseen development during the 1990s although there are still some outstanding mathematical problems which must be solved for the most general application of such methods to be reliable.

The use of geologically-realistic flow modelling has wider applications for the modelling of many geofluid phenomena (e.g. oil migration, diagenetic studies, contaminant transport, etc.) especially where the small-scale geology impacts the macroscopic properties. The methodology can be extended to the scale-up of other petrophysical data (e.g. acoustic impedance) for modelling different (seismic) properties.

The work presented here draws from a number of research projects carried out in recent years by the Reservoir Description Group at Heriot-Watt University. These projects have been generously funded by BP, Chevron, Deminex, Mobil, Shell, Esso, Elf, Department of Trade and Industry, Total, Statoil, Norsk Hydro, Enterprise, British Gas, Amerada Hess. We also acknowledge Statoil's support of the Lochaline study whilst JJML was at Imperial College. We would like to acknowledge the help from, and discussions with, various colleagues that have helped during the development of this work: Ben Lowden, Richard Pelling, Malcolm Arnot, John Underhill and Sandy Tudhope. Geoquest RT have generously provided the ECLIPSE simulator, providing an opportunity for geoscientists to do flow simulation and the insights gained thereby have been critical to the development of methods presented in this paper.

The authors would finally like to thank Andrew Hurst and Dan Schwartz for their painstaking review of the first draft which greatly helped distinguish the supportable arguments from pure speculation – the final manuscript is an appraisal of where we are, and perhaps where we'd like to go.

REFERENCES

BRAITHWAITE, C. I. M., MARSHALL, J. D. & HOLLAND, T. C. 1989. Improving recovery from the Dunlin Field, U.K. Northern North Sea. Paper SPE 19878, presented at the 64th Annual Technical Conference and Exhibition, San Antonio, Texas, October 8–11.

BRENCHLEY, P. J., FLINT, S. S. & STROMBERG, S. G. 1992. Quantitative facies discrimination and the application of sequence stratigraphy to bed length modelling of shallow marine heterolithic facies, *In:* Montadert, M. & Eschard, R. (eds) *Subsurface reservoir characterization from outcrop observation.* Editions Technip, Paris.

BRYANT, I. D. & FLINT, S. S. 1993. Quantitative clastic reservoir geological modelling, problems and perspectives. *In:* Flint, S. S. & Bryant, I. D. (eds), *The Geological modelling of hydrocarbon reservoirs and outcrop analogues.* International Association of Sedimentologists Special Publication, **15**, Blackwell Scientific, Oxford, 3–20.

CIAMMETTI, G., RINGROSE, P. S., GOOD, T. R., LEWIS, J. J. M. L. & SORBIE, K. S. 1995. Waterflood recovery and fluid flow upscaling in a shallow marine and fluvial sandstone sequence. Paper SPE 30783, Presented at the 70th SPE Annual Technical Conference & Exhibition, Dallas, October 22–25, 845–858.

CORBETT, P. W. M. 1993. *Reservoir characterisation of a laminated sediment, Rannoch Formation, Middle Jurassic, North Sea.* PhD thesis, Heriot-Watt University.

——— & JENSEN, J. L. 1992. Estimating the mean permeability: How many measurements do you need? *First Break*, **10**, 89–94.

——— & ——— 1993. An application of probe permeametry to the prediction of two-phase flow performance in laminated sandstones (lower Brent Group, North Sea). *Marine and Petroleum Geology*, **10**(4), 335–346.

———, RINGROSE, P. S., JENSEN, J. L. & SORBIE, K. S., 1992, Laminated clastic reservoirs–The interplay of capillary pressure and sedimentary architecture. Paper SPE 24699, presented at the 67th SPE Annual Technical Conference and Exhibition, October, Washington, 365–376.

———, STROMBERG, S. G., BRENCHLEY, P. J. & GEEHAN, G. 1994. Laminaset geometrics in fine grained shallow marine sequences: core data from the Rannoch Formation (North Sea) and outcrop data from the Kennilworth Member (Utah, USA) and Bencliff Grit (Dorset, UK). *Sedimentology*, **41**, 729–745.

EBANKS, W. J., Jr. 1987. Flow unit concept–integrated approach to reservoir description for engineering projects. *American Association of Petroleum Geologists Bulletin*, **71**, 551–552.

GOGGIN, D. J. 1988. *Geologically-sensible modelling of the spatial distribution of permeability in eolian deposits: Page Sandstone (Jurassic) Northern Arizona.* PhD thesis, University of Austin, Texas.

HEARN, C. J., EBANKS, W. J. Jr., TYE, R. S. & RANGANATHAN, V. 1984. Geological factors influencing reservoir performance of the Hartzog Draw Field, Wyoming. *Journal of Petroleum Technology*, **36**, 1335–1344.

HØIMYR, O., KLEPPE, A., & NYSTUEN, J. P. 1993. Effects of heterogeneities in a braided stream channel sandbody on the simulation of oil recovery: a case study from the Lower Jurassic Statfjord Formation, Snorre Field, North Sea. *In:* Ashton, M. (ed.) *Advances in Reservoir Geology*, Geological Society, London, Special Publication, **69**, 105–134.

HORNAPOUR, M. M., CULLICK, A. S., SAAD, N. & HUMPHREYS, N. V. 1995. Effect of rock heterogeneity on relative permeability. Implictions for scaleup *Journal of Petroleum Technology*, **47**, 980–986.

HUANG, Y., RINGROSE, P. S. & SORBIE, K. S. 1995. 'Capillary Trapping Mechanisms in Water-Wet Laminated Rocks' Paper SPE 28942, presented at the SPE Reservoir Engineering Conference, November, 1995.

HURST, A. 1993. Sedimentary flow units in hydrocarbon reservoirs: some shortcomings and a case for high resolution permeability data. *In:* Flint, S. S. & Bryant, I. D. (eds), *The Geological modelling of hydrocarbon reservoirs and outcrop analogues.* International Association of Sedimentologists Special Publication, **15**, Blackwell Scientific, Oxford, 191–204.

JENSEN, J. L., CORBETT, P. W. M. & RINGROSE, P. S. 1995. Permeability semivariograms, geological structure and flow performance. *Mathematical Geology*, **328**, 419–435.

———, LAKE, L. W., CORBETT, P. W. M. & GOGGIN, D. J. 1996. *Statistical Analysis for Geoscientists and Engineers.* Prentice Hall.

KOCUREK, G. & HAVHOLM, K. G. 1993. Eolian sequence stratigraphy–a conceptual framework. *In:* Weimer, P. & Posamentier, H. W. (eds) *Siliclastic sequence stratigraphy, Recent developments and applications.* American Association of Petroleum Geologists, Memoir, **58**, 393–409.

KJONSVIK, D., DOYLE, J., JACOBSEN, T. & JONES, A. 1994. The effects of sedimentary heterogeneities on production from a shallow marine reservoir–What really matters? Paper SPE 28445, presented at Europec, 25–27 October.

KORTEKAAS, T. F. M. 1985. Water-oil displacement characteristics in cross-bedded reservoir zones. *Society of Petroleum Engineers Journal*, **25**, 917–926.

KYTE, J. R. & BERRY, D. W. 1975. New pseudofunctions to control numerical dispersion, *Society of Petroleum Engineers Journal*, **15**, 269–276.

LAKE, L. W. 1989. *Preface to Reservoir Characterisation – 1.* SPE Reprint Series No. **27**, Society of Petroleum Engineers, Richardson, Texas.

———, KASAP, E. & SHOOK, M. 1989. Pseudofunctions–The key to practical use of reservoir description, *In:* Buller A. T. *et al.* (eds.) *North Sea Oil and Gas Reservoirs II.* Norwegian Institute of Technology, Graham & Trotman, London, 297–308.

LASSETER, T. J., WAGGONER, J. R. & LAKE, L. W. 1986. Reservoir heterogeneities and their influence on ultimate recovery. *In:* Lake, L. W. & Carroll, H. B. (eds), *Reservoir Characterisation.* Academic Press, Orlando, Florida, 545–560.

LEWIS, J. J. M., LOWDEN, B. D. & HURST, A. 1990. Permeability distribution and heterogeneities in sub-surface exposures of a shallow marine sandbody. *Field Guide for 13th International Sedimentological Congress, Nottingham, UK, 1–3 September.*

MIALL, A. D. 1988. Reservoir heterogeneities in fluvial sandstones: lessons from outcrop studies. *American Association of Petroleum Geologists Bulletin*, **72**, 682–697.

MITCHENER, B. C., LAWRENCE, D. A., PARTINGTON, M. A., BOWEN, M. B. J. & GLUYAS, J. 1992. Brent Group: sequence stratigraphy and regional implications. *In:* Morton, A. C., Haszeldine, R. S., Giles, M. R. & Brown, S. (eds) *Geology of the Brent Group.* Geological Society, London, Special Publication, **61**, 45–80.

MUGGERIDGE, A. H. 1991. Generation of effective relative permeabilities from detailed simulation of flow in heterogeneous porous media, *In:* Lake, L. W., Carroll, H. B. J. & Wesson, T. C. (eds) *Reservoir Characterisation II.* Academic Press, Orlando, Florida.

PICKUP, G. E. 1994. The Geopseudo Atlas: geologically based upscaling of multiphase flow. *Proceedings of the Society of Petroleum Engineers European Computer Conference, Aberdeen, 15–17 March*, 277–289.

———, RINGROSE, P. S. & CORBETT, P. W. M. 1994. Geology, geometry and effective flow, *Petroleum Geoscience*, **1**, 37–42.

——— & SORBIE, K. S. 1994. Development and application of a new two phase scaleup method based on tensor permeabilities. Paper SPE 28586, presented at the 69th SPE Annual Technical Conference & Exhibition, New Orleans, 25–28 September.

RINGROSE, P. S., JENSEN, J. L. & SORBIE, K. S., 1994, 'The use of geology in the interpretation of core-scale relative permeability data'. Paper SPE 28448 presented at the 69th Annual Technical Conference & Exhibition, New Orleans, 25–28 September, 881–890.

———, SORBIE, K. S., CORBETT, P. W. M. & JENSEN, J. L. 1993. Immiscible flow behaviour in laminated and cross-bedded sandstones. *Journal of Petroleum Science and Engineering*, **9**, 103–124.

ROSSINI, C., BREGA, F., PIRO, L., ROVELLINI, M. & SPOTTI, G. 1994. Combined geostatistical and dynamic simulations for developing a reservoir management strategy: A case history. *Journal of Petroleum Technology*, **46**, 979–985.

SCOTT, E. 1992. The palaoenvironments and dynamics of the Rannoch-Etive nearshore and coastal succession, Brent Group, Northern North Sea. *In:* Morton, A. C., Haszeldine, R. S., Giles, M. R. & Brown, S. (eds) *Geology of the Brent Group.* Geological Society, London, Special Publication, **61**, 1–17.

SHANLEY, K. W. & McCABE, P. J. 1994. Perspectives on the sequence of continental strata. *American Association of Petroleum Geologists Bulletin*, **78**, 544–568.

SMITH, E. H. 1991. The influence of small-scale heterogeneity on average relative permeability. In: Lake, L. W., Carrol, H. B. Jr. & Wesson, T. C. (eds) *Reservoir Characterisation II.* Academic Press, San Diego, 52–76.

STONE, H. L. 1991. Rigorous Black Oil Pseudo Functions. Paper SPE21207, presented at the SPE Symposium on Reservoir Simulation, Anaheim, CA, 17–20 February.

THOMAS, J. M. D. & BIBBY, R. 1991. The depletion of the Rannoch-Etive sand unit in Brent sand reservoirs in the North Sea. *Proceedings of the 3rd International Reservoir Technology Conference, US Deptartment of Energy and National Institute for Petroleum and Energy Resources*, paper 3RC-28.

VAN WAGONER, J. C., MITCHUM, R. M., CAMPION, K. M. & REHMANIAN, V. D. 1990. *Siliclastic sequence stratigraphy in well logs, cores and outcrops*, AAPG Methods in Exploration Series, No. 7.

WALKER, R. G. 1981, *Facies Models. Geoscience Canada*, Reprint Series 1, Geological Society of Canada, Hamilton, Ontario.

WEBER, K. J. & VAN GEUNS, L. C. 1989. Framework for constructing clastic reservoir simulation models. *Journal of Petroleum Technology*, **42**, 1248–1297.

Petrophysical innovations and their influence on exploration and production in NW Europe

R. A. Skopec,[1] J. Foot[2] and R. Marion[2]

[1] *University of Aberdeen, Department of Geology and Petroleum Geology, Meston Building, King's College, Aberdeen AB24 3UE, UK. (Present address: Texaco E & P Technology Department, Houston, Texas, USA.)*
[2] *BP Exploration Operating Company Limited, Farburn Industrial Estate, Dyce, Aberdeen AB2 0PB, UK.*

ABSTRACT: The current economics of the petroleum industry necessitates more cost-effective petrophysical practices and technology. The need to evaluate complex lithologies and unconventional reservoirs with reliable petrophysical tools is fundamental to successful reservoir characterization and field development. Up to recently, wireline logging developments have consisted primarily of evolutionary improvements in current technology, e.g., nuclear, acoustic, and electromagnetic methods.

Three tools responsible for major technological advancements in petrophysics include: a new generation of nuclear magnetic resonance (NMR) logging tools, sonic logging tools capable of making direct shear-wave measurements, and low-invasion gel coring for preservation of *in situ* rock properties. NMR is making a dramatic impact on the estimation of reserves by differentiating mobile and immobile reservoir fluids, better defining fluid contacts, determining porosity independent of lithology, and improving estimates of permeability. Acoustic-array wireline logging tools using dipole sonic sources have improved the quality of borehole evaluations and have contributed to advancements in surface- and borehole-seismic interpretation and data integration. The goal of obtaining core which is representative of the formation while minimizing physical and chemical alteration of the rock during coring and handling is now possible using low-invasion, gel encapsulation. This technology has added a new dimension to analysis of rock subject to wettability alteration and mechanical disaggregation.

KEYWORDS: *petrophysics, wireline logging, nuclear magnetic resonance, acoustics, coring, N. Sea*

INTRODUCTION

Exploration and development in the North Sea and continental Europe has entered a mature stage, where most discoveries are marginal to moderate in size and quality. Petrophysical protocol can no longer include cost-ineffective and sometimes redundant evaluations. The need to reduce operating costs has led to improvements in the way current logging tools are operated. Significant time can be lost tripping in and out of the borehole and in running wireline tools at reduced logging speeds. Because of the high cost of rig time and the increase in the drilling of high-angle and horizontal wells, which may require pipe-conveyed logging systems, versatility in the data acquisition process has become critical. In many high-angle wells, coiled tubing conveyed logging is being used to reduce trip time and reduce operating costs. Stacking, or combining, several wireline tools to acquire data in a single trip is made possible through advanced multiplexing methods and more powerful well-site computers. The use of digital signal processing and adaptive digital filters has increased data capacity and telemetry capabilities from 80 to 660 Kbytes/sec in a standard seven-conductor wireline, enabling the development of array and down hole imaging tool designs (Adoumieh *et al.* 1990). Without sacrificing data quality, logging speeds can now be increased to minimize wireline data acquisition time. A new generation of slimhole logging tools is emerging to accommodate smaller diameter, higher temperature and higher pressure completions. New logging-while-drilling methods are providing the petrophysicist with options for difficult logging conditions and high-angle wellbores.

The philosophy of when to run particular open-hole wireline logs or cut full-diameter core is best described in terms of oil field development. The extent of the petrophysical programme changes during the exploration, appraisal and development stages to meet engineering and geoscience objectives (Fig. 1). During the exploration phase, when little or no data are available and risk is high, a diverse logging programme is executed to maximize the reservoir description process. To reduce uncertainty in the reservoir evaluation, continuous full-diameter core is cut for stratigraphic definition and determination of physical reservoir parameters. Core can be used to establish a petrophysical basis for calibration of indirect reservoir evaluation tools, for example, wireline logs and seismic data.

During appraisal, the issue is not of whether hydrocarbons are present, but of how much recoverable oil is present, how it will be economically recovered, and how it is spatially

From K. Glennie & A. Hurst (eds), 1996, *AD1995: NW Europe's Hydrocarbon Industry*, Geological Society, London, pp. 177–184

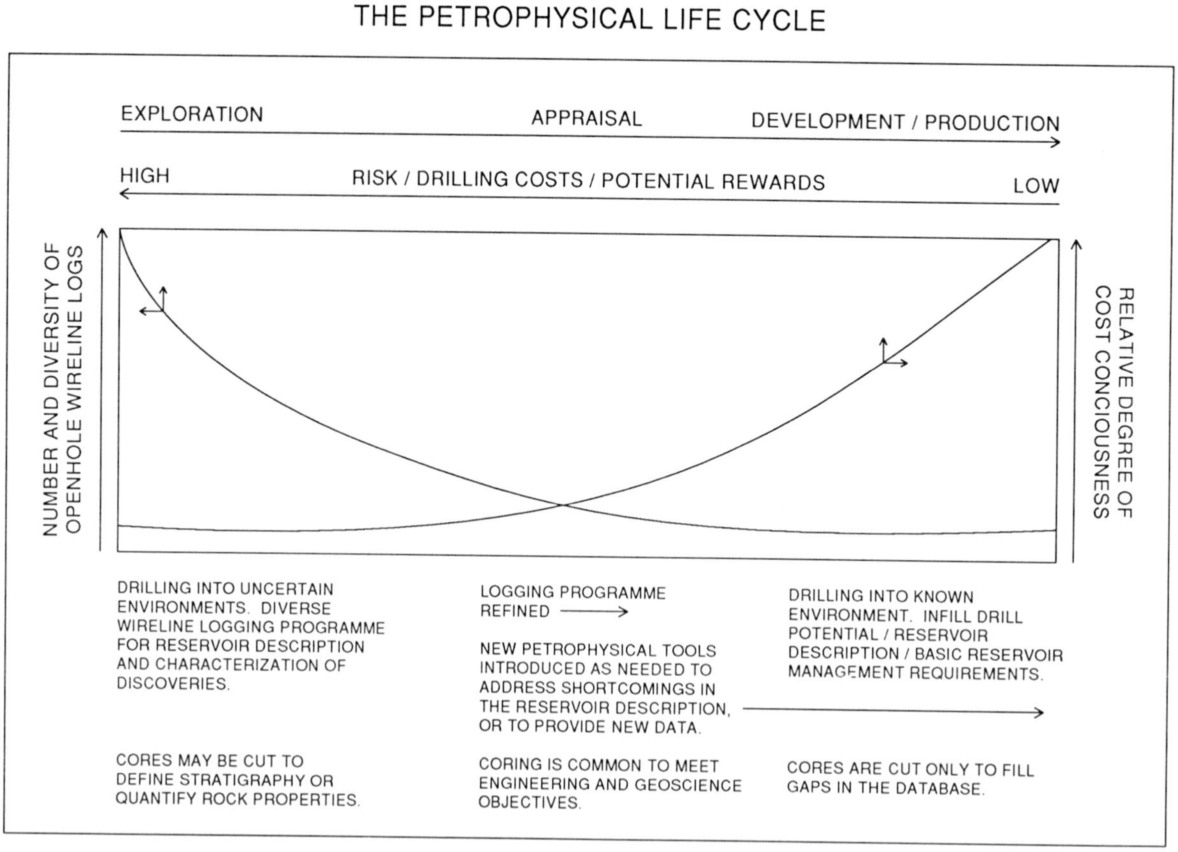

THE PETROPHYSICAL LIFE CYCLE

EXPLORATION APPRAISAL DEVELOPMENT / PRODUCTION

HIGH RISK / DRILLING COSTS / POTENTIAL REWARDS LOW

NUMBER AND DIVERSITY OF OPENHOLE WIRELINE LOGS

RELATIVE DEGREE OF COST CONCIOUSNESS

DRILLING INTO UNCERTAIN ENVIRONMENTS. DIVERSE WIRELINE LOGGING PROGRAMME FOR RESERVOIR DESCRIPTION AND CHARACTERIZATION OF DISCOVERIES.

LOGGING PROGRAMME REFINED →

NEW PETROPHYSICAL TOOLS INTRODUCED AS NEEDED TO ADDRESS SHORTCOMINGS IN THE RESERVOIR DESCRIPTION, OR TO PROVIDE NEW DATA. →

DRILLING INTO KNOWN ENVIRONMENT. INFILL DRILL POTENTIAL / RESERVOIR DESCRIPTION / BASIC RESERVOIR MANAGEMENT REQUIREMENTS.

CORES MAY BE CUT TO DEFINE STRATIGRAPHY OR QUANTIFY ROCK PROPERTIES.

CORING IS COMMON TO MEET ENGINEERING AND GEOSCIENCE OBJECTIVES.

CORES ARE CUT ONLY TO FILL GAPS IN THE DATABASE.

Fig. 1. The petrophysical life cycle during the exploration, appraisal and development/stages for hydrocarbon reservoirs.

distributed. When a project enters the appraisal stage, the logging programme is refined and the number of logs run usually decreases as the reservoir is delineated and rock and fluid properties are quantified. If the reservoir is petrophysically complex and heterogeneous, additional core may be cut to perform advanced laboratory testing. This may include recovery-orientated data related to multiphase flow and fluid distribution. As appraisal proceeds, new logging tools may be introduced to address shortcomings in the reservoir description, or to provide new information that cannot be obtained with existing petrophysical technology.

When a project is determined to be economically viable, risk is a minor factor, and geoscientists and engineers deal with the development and production phases of the field. The theme of reducing operating costs is now at a maximum and the minimum number of logs and cores will be acquired during this stage. The petrophysical programme may be enhanced for infill reservoir description, but only after careful review of the existing database. Alliances with service companies are common at this stage to attempt further reductions in operating expense.

The various stages of field development and the life cycle of the petrophysical programme evolve in a field-specific manner and no two programmes are identical. In every project, the value of petrophysical data is maximized when acquired early and by its integration with other data sets. Comprehensive petrophysical evaluation is dependent on the integration of data acquired from wireline logs and continuous full-diameter core.

INNOVATIONS

Since the mid-1980s, wireline logging technology has focused on development of the following:

- improvements in thin bed resolution;
- borehole imaging devices;
- direct shear wave sonics;
- nuclear magnetic resonance tools;
- sampling and testing tools;
- advanced production (through casing) methods;
- logging while drilling;
- nuclear tool development;
- enhanced data acquisition and presentation formats.

Laboratory petrophysics and core acquisition advances include:

- low-invasion non-damaging coring methods;
- non-destructive imaging techniques;
- outcrop and laboratory data acquisition hardware e.g. probe permeametry;
- scale-up procedures for fluid flow simulation.

For the sake of brevity, this paper reviews three of the most innovative petrophysical technologies, NMR logging, multipole sonic logging and down-hole core preservation. For a comprehensive review of recent innovations in petrophysical technology, the reader is referred to Prensky (1994a,b).

NUCLEAR MAGNETIC LOGGING

Perhaps the most significant logging tool to be developed in the last decade is the nuclear magnetic resonance Log (Magnetic Resonance Imaging Log, MRIL® and Combinable Magnetic Resonance Log, CMR®). Nuclear magnetic resonance (NMR), documented for the first time about fifty years ago, was first applied to chemical analysis for the investigation of molecular structures (Bloch *et al.* 1946). NMR logging has been one of the great hopes for significant improvement in petrophysical evaluation since its first application in the 1960s. Magnetic resonance imaging (MRI) opened new frontiers in diagnostic medicine and is used routinely to examine soft body tissues. MRI can be used to non-destructively image several physical and chemical properties of porous rocks and multiple fluid phases contained in their pores with spatial resolution at the submillimetre level (Edelstein *et al.* 1988).

Early versions of NMR logging tools, known as the nuclear magnetism log (NML) using free induction decay (FID) technology were bulky, logging speeds were slow, and doping of the drilling mud with magnetite was required to nullify the borehole signal (Hilchie 1988). Recently, a second generation of wireline logging tools using pulsed, spin-echo technology have been introduced (Miller *et al.* 1990). Modern versions of NMR logging devices have been designed to overcome many of the shortcomings of the first generation tools. Nuclear magnetic resonance provides a way to manipulate certain atomic nuclei for statistical analysis. A measurement is made of the energy released by protons (hydrogen nuclei), expressed as a function of time, as they are released from a temporary magnetic field. The nuclei of many atoms spin, are charged, have magnetic moments, and act as small magnets which can be controlled by an induced magnetic field. The NMR wireline tool in its simplest form has a large coil which produces an electrical field when electrical current is passed through it (Fig. 2). The magnet creates a large static magnetic field (B_0) which aligns the nuclear spins. A radio frequency (RF) coil supplies an oscillating field (B_1) which is orthogonal to B_0. In MRIL logging, the tool is tuned to evaluate hydrogen nuclei (e.g. protons) in the fluid of the rock's pore space. The total signal is proportional to the total number of protons present in both the free and bound fluids.

The MRIL system (Fig. 3) consists of the following:
(a) surface equipment
- power supplies, data acquisition/processing computer, and data display computer;

(b) down-hole equipment
- gamma ray tool for depth control;
- MRIL electronics package;
- magnet and antenna probe.

NMR theory

Hydrogen nuclei can be viewed as behaving like spinning tops that bear charges and produce a magnetic field. The strength and direction of the magnetic field are expressed in terms of a vectorial quantity known as the nuclear magnetic moment. In the absence of any external field, the nuclear moment is randomly orientated. In the presence of a static, uniform magnetic field B_0, the spin system becomes polarized and the magnetic moments align such that their net vectorial sum, the bulk magnetization (M), is orientated with the magnetic field. If the magnetization is tipped away from the direction of the magnetic field, it will start to precess about this direction, like a spinning top that is non-vertical (Fig. 4). The precession frequency, known as the Larmor frequency, is the product of magnetic field strength and the gyromagnetic ratio, γ, a physical constant with a unique value for each nucleus.

® MRIL, mark of Numar Corporation.
® CMR, mark of Schlumberger Wireline Services.

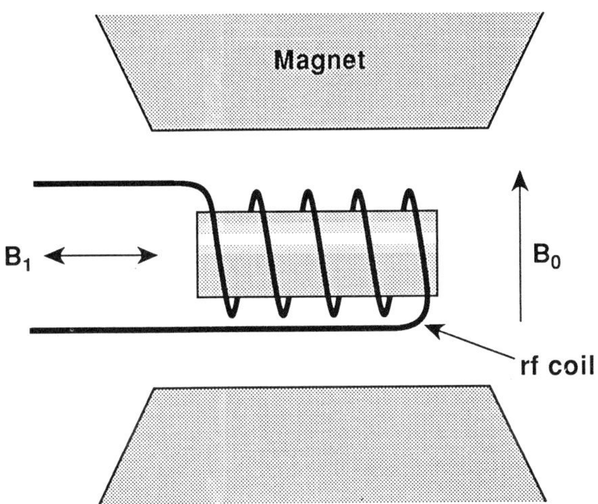

Fig. 2. The nuclear magnetic resonance (NMR) magnet and RF coil. The magnet creates a large static magnetic field, B_0, and the RF coil supplies an oscillating field, B_1, which is orthogonal to B_0 (after Coates *et al.* 1993).

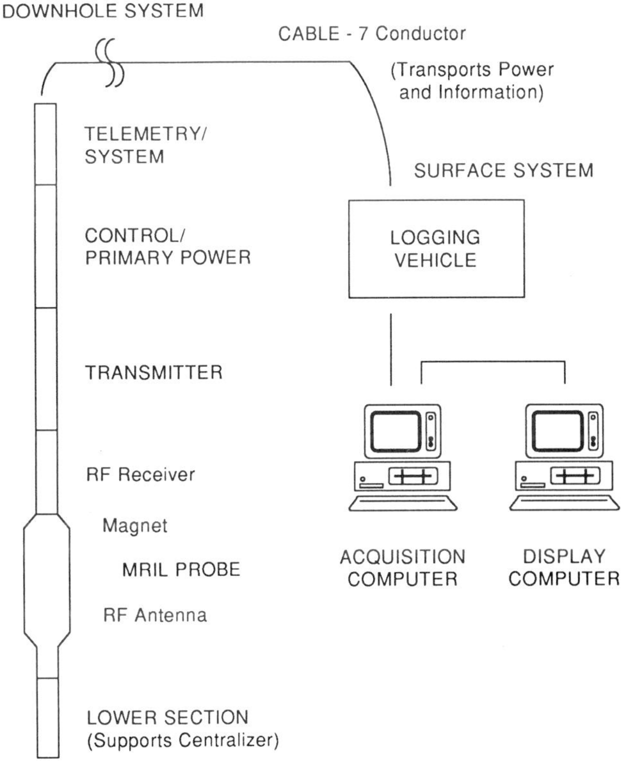

Fig. 3. The Magnetic Resonance Imaging Log (MRIL) down-hole and surface operating system (after Miller *et al.* 1990).

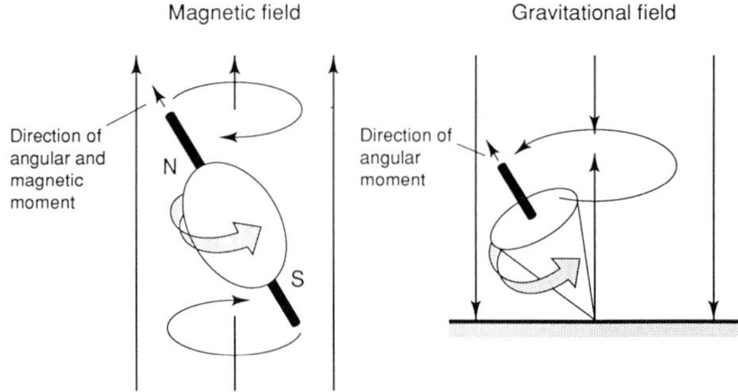

Fig. 4. Comparison between a spinning top precessing in a gravitational field and a spinning nucleus precessing in a magnetic field.

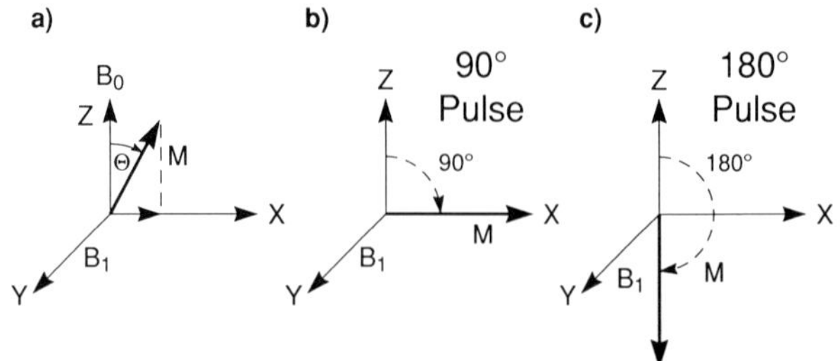

Fig. 5. The principles of pulsed nuclear magnetic resonance measurements with respect to spin tipping. (**a**) shows the thermal equilibrium condition where the spin magnetization is aligned along B_o, (**b**) a 90° pulse will tip the spins into an equatorial plane, and (**c**) a 180° pulse will invert the spins. The 180° pulses are repeated periodically leading to a series of spin echoes.

Tipping of the magnetization is the principle behind pulsed (spin echo) NMR where short RF pulses at the Larmor frequency are used to orientate nuclei in a precisely controlled manner. The amount by which the magnetization is tipped is proportional to the amount of energy delivered to the spins and is dependent on the strength of the RF field and its duration. Figure 5a shows the thermal equilibrium condition where the spin magnetization is aligned along B_o. The RF pulse along the B_1 axis exerts a torque on the nuclear spins which causes rotation from the B_o direction. Thus, a 90° pulse will tip the spins into the equatorial plane, and a 180° pulse will invert the spins (Figs 5b,c). The 180° pulses are repeated periodically leading to a series of spin echoes, known as the Carr-Purcell (CP) echo sequence (Carr & Purcell 1954).

The most significant parameter estimated by the NMR log is the free-fluid index (FFI). The initial magnetization of the formation is a function of the number of aligned and orientated protons, which is directly linked to FFI. This means that the Irreducible Water Saturation (S_{wir}) in water-wet systems or immobile fluid saturation can be estimated directly from NMR logs and estimates of **net pay** can be improved. Conventional water saturation tools and models estimate only that portion of the reservoir pore volume containing water; free and bound water are not differentiated. NMR logging data can help determine whether hydrocarbons will be produced free of water and when the free water level (FWL) and fluid contacts are unknown or difficult to define using conventional logging tools. The NMR measurement of porosity is independent of lithology and other intrinsic rock properties. NMR measurements are classified according to their relaxation times, spin-lattice (longitudinal, T1) and spin-spin (transverse, T2). Both T1 and T2 are closely related to the dimensions of rock pores (surface-area-to-volume) ratio; hence, permeability can be derived from these measurements (Howard *et al.* 1990). Productive versus non-productive intervals can be delineated and a general estimation of reservoir quality is possible using NMR logging.

NMR rock physics

The time behaviour of the NMR relaxation parameters, T1 and T2, characterizes the pore-size distribution of the formation as well as the total porosity. The smallest pores with bulk volume irreducible fluid (BVI), are distinguished from larger pores containing movable fluids by the free fluid index (FFI). The effective pore space is designated MPHI or Φ_e. Because of short relaxation times, clay-bound water contained in micropores cannot be identified–this means that NMR measures the effective porosity, Φ_e, and not the total porosity, Φ_t.

The following equations apply to NMR petrophysical interpretation:

$$\mathrm{BVI} = (S_{ir} - S_{cbw}) \times \Phi_t \leq \mathrm{BVW}$$
$$\mathrm{FFI} = (1 - S_{ir}) \times \Phi_t$$
$$\mathrm{MPHI} = (1 - S_{cbw}) \times \Phi_t$$
$$\Phi_e = \mathrm{BVI} + \mathrm{FFI}$$
$$\Phi_{cbw} = \Phi_t - \Phi_e$$

Where S_{ir} is irreducible fluid saturation–generally water in water-wet rocks and oil in oil-wet rocks; BVW is bulk volume water, and S_{cbw} is clay (atomically-bound) water saturation.

Figure 6 illustrates the MRIL rock model (Coates *et al.* 1991). The bulk irreducible fluid volume (BVI) does not include clay-bound water, although this fraction of fluid is generally at irreducible conditions. Clay-bound water is not considered to be part of the effective porosity. In many instances, capillary-bound fluids are representative of ineffective pore space. Since the NMR method cannot identify the clay-bound fluid fraction, this is estimated by material balance. The producible water fraction is equal to the total volume of water minus the capillary-bound and clay-bound water.

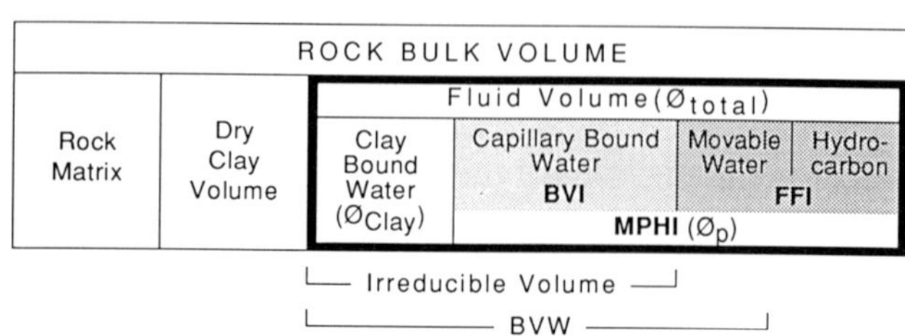

ROCK BULK VOLUME					
		Fluid Volume(Øtotal)			
Rock Matrix	Dry Clay Volume	Clay Bound Water (ØClay)	Capillary Bound Water BVI	Movable Water	Hydro-carbon
			MPHI (Øp)		FFI
		Irreducible Volume			
		BVW			

Fig. 6. The MRIL rock model (adapted from Coates *et al.* 1991).

Rock permeability is derived from established relationships and NMR-derived surface area, pore size distribution, and porosity values. Rocks with equal values of porosity can have vastly different permeability because of pore size, tortuosity and connectivity variations. NMR provides more realistic estimates of permeability because of its unique ability to measure surface area and pore size characteristics.

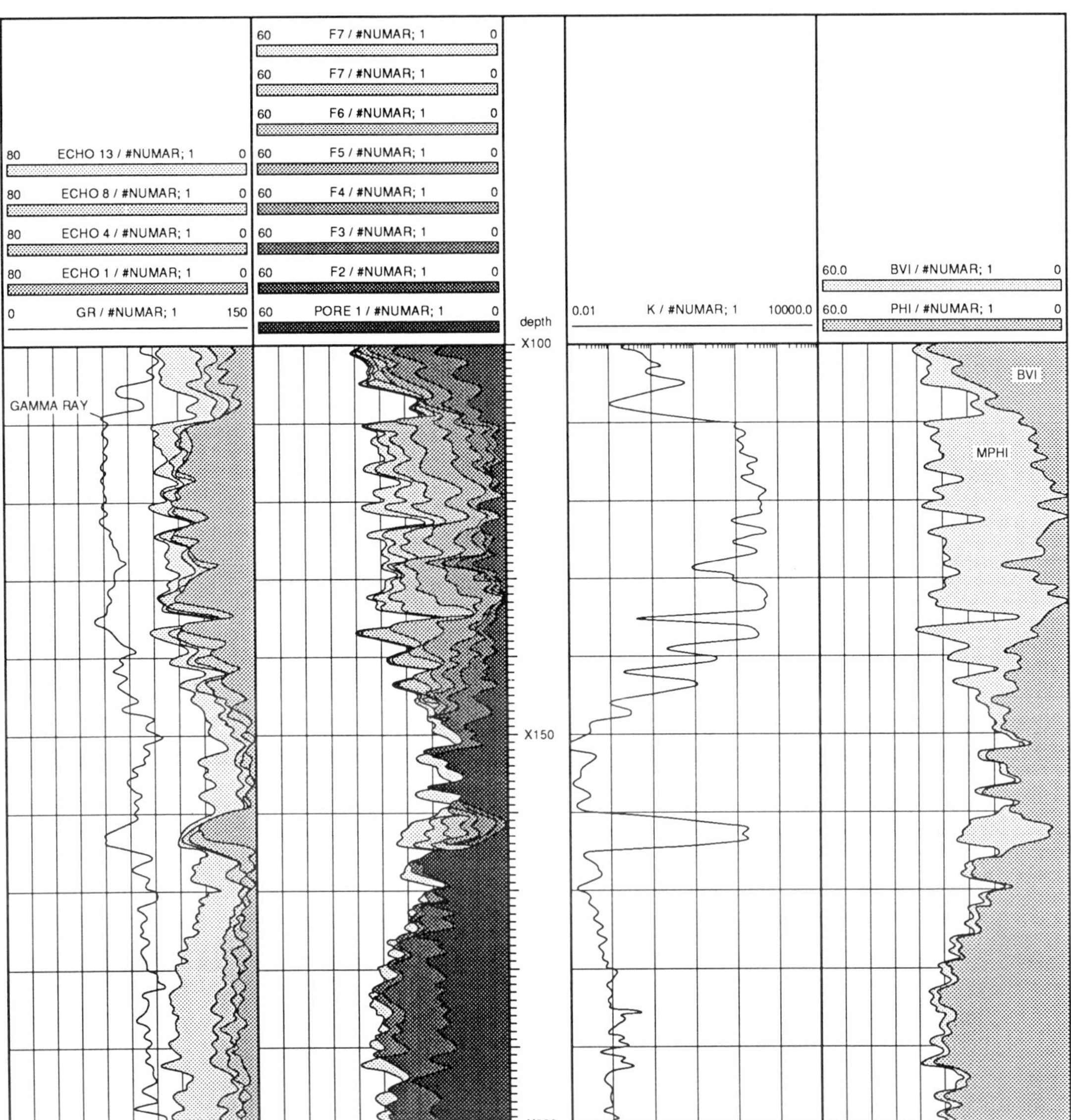

Fig. 7. An MRIL log example. Track 1 includes the gamma ray for depth correlation and pre-selected echoes of 1, 4, 8, and 13 msec which are related to pore size distribution. Track 2 contains the T2 relaxation time bins used to calculate various petrophysical parameters. Track 3 contains an empirically-derived permeability value. The bulk volume irreducible fluid (BVI) and the total effective porosity (MPHI) are shown in Track 4.

MRIL interpretation

The Numar MRIL log (Fig. 7) will be used to explain some of the petrophysical parameters derived from NMR logs. Track 1 includes pre-selected echoes (raw NMR pulses equivalent to relaxation measurements) of 1, 4, 8, and 13 ms which are related to pore size distribution (Coope 1994). The separation between echo numbers qualitatively represents the fractional proportion of pore sizes for the pore class defined by the echo number. As indicated by darker shading, the effective pore size increases as the echo number increases. The echo number gives a direct indication of reservoir quality based on the free fluid capacity of the rock. The section from X100 to X150 contains large pores with a fairly narrow range of pore sizes. The standard gamma ray trace is also included in Track 1. Track 2 contains the T2 time bins (relaxation values) used to calculate various petrophysical parameters.

Track 3 contains an empirically-derived permeability value based on a bound water model (Coates *et al.* 1991). Higher permeability values are calculated in the upper section where large pores are abundant. The Bulk Volume Irreducible fluid (BVI) represents the small pores filled with non-movable fluids, water bound by capillary forces, minus the clay-bound water saturation, multiplied by the total porosity. The NMR signal cannot discriminate the clay-bound water, thus, the effective porosity is measured rather than the total porosity. An excellent estimator of productivity is the BVI/FFI ratio, which is a measure of the surface-to-volume ratio. Track 4 shows both the BVI and the total effective porosity (MPHI). The area between the two traces is equal to the FFI. The FFI is greatest in the upper section where irreducible fluid is at a minimum.

MULTIPOLE ACOUSTIC LOGGING

The need to obtain better shear wave data is becoming increasingly more important in petroleum exploration and production. Recent advances in seismic processing by

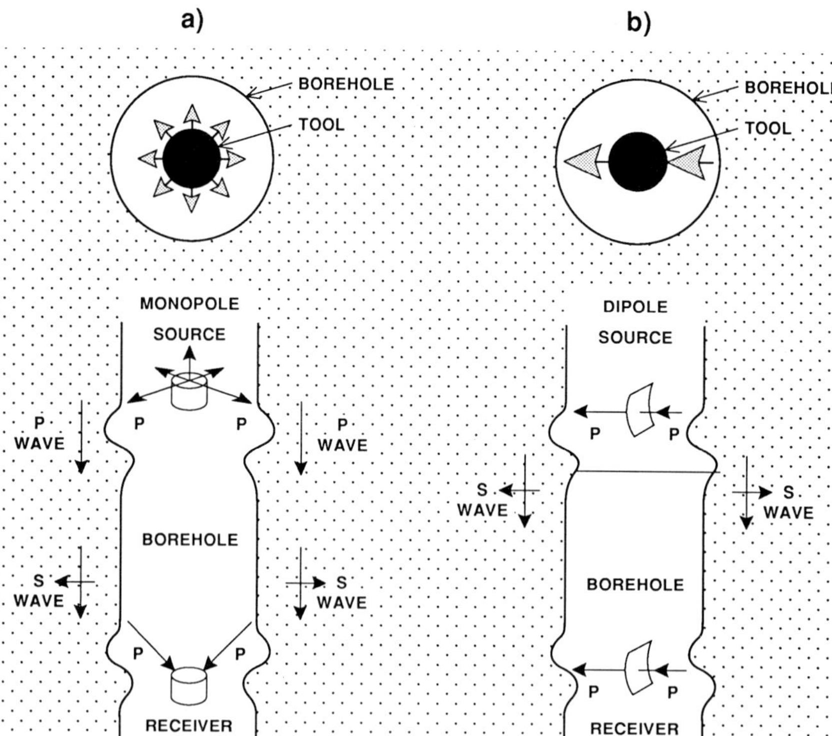

Fig. 8. P- and S-wave generation with the (**a**) conventional monopole logging tool, and (**b**) an asymmetric (dipole) source and receiver using a directed flexural wave.

amplitude-versus-offset (AVO) analysis and borehole seismic evaluation have been the driving forces behind wireline multipole sonic log development (Skopec & Ross 1994). Shear waves are also used in the calculation of rock mechanical properties, for example, formation strength and elastic moduli, detection of fractures using tube (Stoneley) waves, lithology determinations, gas detection and permeability estimates. The dipole energy source and receiver array provide shear measurements for both soft and hard formations, regardless of the shear velocity. Monopole sonic tools can only detect shear velocities that are faster than the borehole fluid velocity.

The Dipole Shear Sonic Imager (DSI®) and Multipole Array Acoustilog (MAC®) tools incorporate dipole-based technology as well as monopole sources for acquisition of compressional- and shear-wave data. In monopole sonic tools, an omni-directional pressure source creates a compressional-wave pulse in the borehole fluid which propagates out into the formation. As this pulse enters the formation, it creates a slight uniform bulge around the borehole wall (Fig. 8a). This, in turn, excites both compressional- and shear-waves in the formation. Head waves form as they travel up the wellbore, rather than the more desirable direct compressional- and shear-waves.

A dipole tool uses a directional source that behaves much like a piston, creating a pressure increase on one side of the borehole and a decrease on the other. The directed wave causes a small flexing of the borehole wall and this excites compressional- and shear-waves in the formation (Fig. 8b). Unlike monopole tools, the dipole will always record shear (flexural) waves regardless of shear-wave speed (Harrison *et al.* 1990). Recent versions of the dipole tool contain upper- and lower-dipoles for improved borehole compensation and evaluation of acoustic anisotropy.

® DSI, mark of Schlumberger Wireline Services.
® MAC, mark of Western Atlas Wireline Services.

DOWN-HOLE CORE PRESERVATION

A major issue confronting the coring and core analysis industry has been damage to the core during acquisition and handling. The goal of coring and core preservation should be to obtain rock which is representative of the formation while minimizing physical and chemical alteration of the rock during coring and handling. Standard down-hole coring assemblies do not preserve *in situ* reservoir properties because no provision is made for core preservation prior to surface handling. Low-invasion coring systems help minimize drilling fluid invasion, but rock wettability and saturation can still be altered by drilling fluid imbibition and/or diffusion before core analysis begins. The quality of many laboratory petrophysical measurements is dependent on the saturation state of the core.

New technology using high-viscosity gel for down-hole core encapsulation and preservation is an alternative to operator-intensive well-site methods (Skopec *et al.* 1995). The viscous, zero water loss core preservation gel is a high molecular weight polypropylene glycol which is insoluble in water and environmentally safe. Because the gel comes in direct contact with the core during and immediately after it is cut, further exposure to drilling fluid is minimized. Once at the surface, exposure of the core to air is significantly reduced. For poorly consolidated rocks with sufficiently high compressive strengths, the high viscosity gel stabilizes and enhances the mechanical integrity of the core. Surface handling is improved and core damage is reduced during core transportation to the laboratory. Figure 9a shows the down-hole core preservation assembly before coring begins. The gel is contained in the inner-barrel prior to coring and is distributed by a core-activated floating piston valve after core begins to enter the core head (Fig. 9b).

Excess gel is displaced from the inner-barrel by the core through the extended pilot catcher shoe, out the throat of the bit, and past the cutters where it mixes with, and is

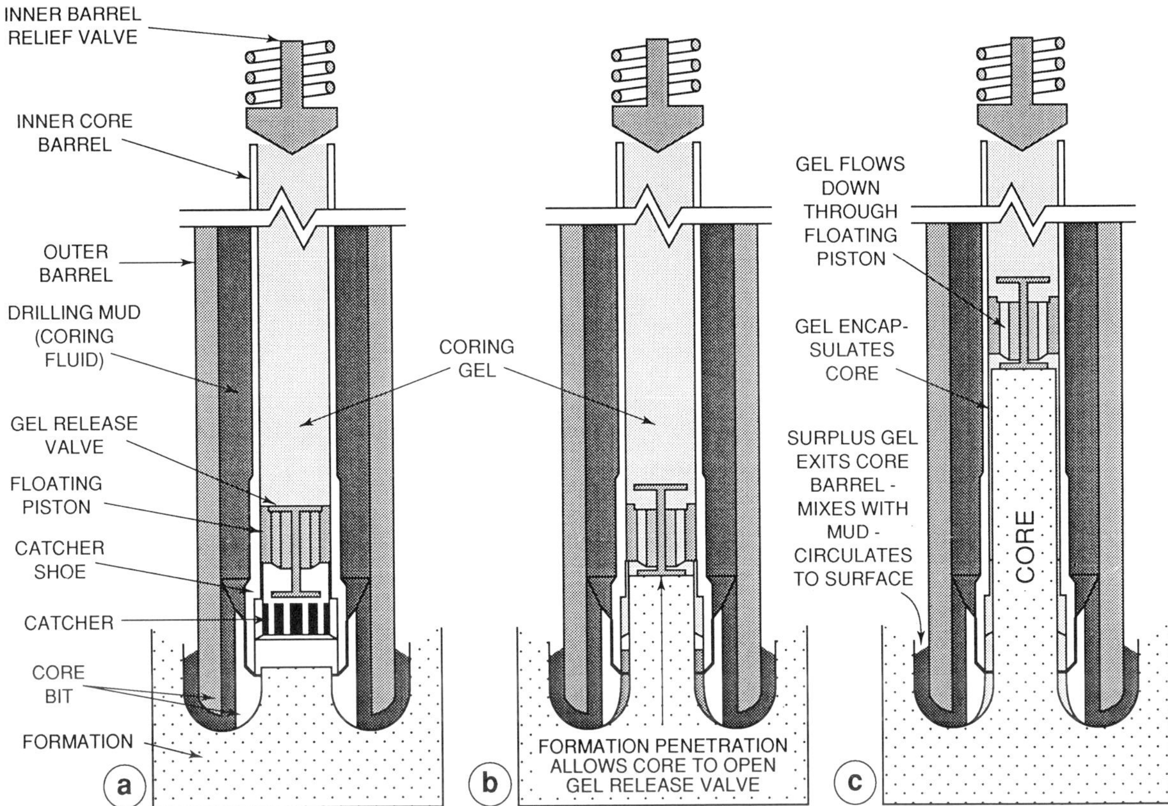

Fig. 9. The down-hole core preservation assembly (**a**) inner barrel piston closed prior to gel encapsulation of core, (**b**) gel release valve opens, and (**c**) gel encapsulation and core preservation (from Skopec *et al.* 1995).

dispersed by, the drilling fluid. At the cutter-to-rock contact, the gel displaces drilling fluid and protects the core from flushing and drilling fluid filtrate invasion. Static filtrate invasion of the core by over-balanced drilling fluids in the inner-barrel, an intrusive phenomenon common to conventional coring, is eliminated. Approximately two gallons of gel remains in the inner-barrel annulus and forms a thin protective layer over each 30 ft section of core. Figure 9c shows a gel-encapsulated core prior to surfacing.

Gel encapsulation is not known to affect surface gamma logging or non-invasive core imaging methods such as X-ray and MRI scanning. The preservation gel can be peeled, scraped, or wiped from the core surface. Gel on the outer surface and in the near-surface pore layers of the core generally presents no more problem to the core analyst than mud cake in conventional analysis. Down-hole core preservation can be used in combination with pressure-retained core barrels if residual oil saturation determinations or special core studies are to be conducted.

CONCLUSIONS

Petrophysical data acquisition in NW Europe has been affected by dramatic economic and technical developments. All petrophysical tools to be run in a borehole must be justified on the basis of cost effectiveness and technical gain. The petrophysical life cycle affects the types and number of logs to be acquired and influences whether conventional full-diameter core will be taken. Several new wireline logging tools, in particular NMR and multipole acoustic logging, are expanding the potential for measurement of reliable *in situ* petrophysical properties. With the advent of low-invasion down-hole preservation methods, cores with preserved rock properties are improving reservoir evaluation.

The authors wish to acknowledge valuable discussions with D.F. Coope of Numar UK Limited. We thank the management of BP Exploration Operating Company for permission to publish this paper. All figures were prepared by B. Fulton, University of Aberdeen.

REFERENCES

ADOUMIEH, R., BERNEKING, D., OLSEN, B., KUMAR, R., & PHILLIPS, J. 1990. The Maxis system-imaging for reservoir characterisation. *Schlumberger Oilfield Review*, **2**(2), 31–42.

BLOCH, F., HANSEN, W. & PACKARD, M. 1946. The nuclear induction experiment. *Physical Review*, **70**, 474-485.

CARR, H. Y. & PURCELL, E. M. 1954. Effects of diffusion on free precession in NMR experiments. *Physical Review*, **94**, 630.

COATES, G. R., MILLER, M., GILLEN, M. & HENDERSON, G. 1991. The MRIL in Conoco 33-1: an investigation of a new magnetic resonance imaging log. *Transactions of the SPWLA 32nd Annual Logging Symposium*, paper DD.

COOPE, D. F. 1994. Petrophysical applications of NMR in the North Sea. *Transactions of the Sixteenth European Formation Evaluation Symposium, SPWLA*, paper Z.

EDELSTEIN, W. A., VINEGAR, H. J., TUTUNJIAN, P. N., ROEMER, P.B. & MUELLER, D. M. 1988. NMR imaging for core analysis. Paper SPE 18272 presented at the SPE 63rd Annual Technical Conference and Exhibition.

HARRISON, A.R., RANDALL, C.J., ARON, J. B. *et al.* 1990. Acquisition and analysis of sonic waveforms from a borehole monopole and dipole source for the determination of compressional and shear speeds and their relation to rock mechanical properties and surface seismic data. Paper SPE 20557 presented at the SPE 65th Annual Technical Conference and Exhibition.

HILCHIE, D. W. 1988. NML-The hope for production prediction. *Journal of Petroleum Technology*, March, 273–275.

HOWARD, J. J., KENYON, W. E. & STRALEY, C. 1990. Proton magnetic resonance and pore size variations in reservoir sandstones. Paper SPE 20600 presetned at the SPE 65th Annual Technical Conference and Exhibition.

MILLER, M. N., PALTIEL, Z., GILLEN, M. E., GRANOT, J. & BOUTON, J. C. 1990. Spin-echo magnetic-resonance logging–porosity and free-fluid index. Paper SPE 20561 presented at the SPE 65th Annual Technical Conference and Exhibition.

PRENSKY, S. E. 1994*a*. A survey of recent developments and emerging technology in well logging and rock characterisation. *The Log Analyst*, 35(2) 15–45.

——— 1994*b*. Supplement to a survey of recent developments and emerging technology in well logging and rock characterisation. *The Log Analyst*, 35(5) 78-84.

SKOPEC, R.A., COLLEE, P. & SHALLENBERGER, L. 1995. High viscosity gel encapsulation for core preservation and improved reservoir evaluation. *Journal of Petroleum Technology*, 47(5)

——— & ROSS, C. P. 1994. Amplitude-Versus-Offset interpretation, scaling factors, and other challenges associated with acoustic data integration. Paper SCa 9417 presetned at the Society of Core Analysts Annual Meeting.

Part Four: Future Challenges

The remaining resource of the UK North Sea and its future development

M. C. Daly, M. S. Bell & P. J. Smith

BP Exploration Operating Co. Ltd, Farburn Industrial Estate, Dyce, Aberdeen AB2 0PB, UK

ABSTRACT: Prediction of the remaining North Sea oil and gas resource is crucial to the UK offshore industry. Based on these resource estimates, government must ensure the optimum exploitation of the nation's hydrocarbons. Oil and gas companies, through strategic positioning based on these estimates, wish to pursue profitable exploration and production programmes.

A conventional view of the total resource of the UK North Sea yields an estimate of 48×10^9 bbl oil equivalent, of which 17.2×10^9 bbl oil equivalent has already been produced. Of the remainder, 13.2×10^9 bbl lies in existing producing fields and discoveries with proven reserves. The remaining 17.6×10^9 bbl is roughly equally split between possible reserves associated with, as yet, unappraised discoveries (9.3×10^9 bbl) and an estimate of 'yet-to-be-found' oil and gas (8.3×10^9 bbl). Clearly, the greatest room for error in these figures, and therefore the greatest opportunity for resource base growth, is in the volumes 'yet to be found'. A significant change in our perception of the 'yet-to-find' (YTF) resource will influence both government policy and oil company strategy.

In this paper, we discuss our most likely estimate of the resource, and present an alternative that compares this view with the experience of a similar estimation in the Gulf of Mexico, a basin with a much longer exploration and production history than the North Sea. The implications of this unconventional future are then discussed in terms of remaining resource, and the technology and economic environment the North Sea required to realize a long-term sustainable future.

KEYWORDS: *North Sea, yet-to-find, resource prediction, infrastructure*

INTRODUCTION

In 1995 there exists a widely held perception that the North Sea has reached a stage of maturity from which the way forward will be an inexorable and fundamental exploration decline, that the giant fields have all been found, and that the resource base is decreasing in attractiveness, proving increasingly difficult to find and increasingly unprofitable to develop and to produce.

To date, the UK North Sea, defined as the northern, central and southern basins (Fig. 1) has produced some 17.2×10^9 barrels of oil equivalent (boe). A further 13.2×10^9 boe lies in proven reserves in producing fields and appraised or mature discoveries. A remaining 9.3×10^9 boe of possible reserves exists in unappraised or immature discoveries. Over and above these figures is a prediction of the 'yet-to-find' volumes.

The most uncertain of these figures is the 'yet-to-find volume', yet it is on the perception of this volume and its characteristics (expected field sizes, phase and API gravity, and geographic distribution) that the UK government must plan its petroleum policies and oil and gas companies must develop their investment strategies.

The necessity to plan for the future has, over the years, led to 'guestimates' of our unappraised and undiscovered reserves on both local and global scales. Among the many, reference should be made to Campbell & Ormaasen (1987) who suggest that a high proportion of Europe's undiscovered reserves lie in the Norwegian offshore, while Leckie & Chew (1991) consider the reserves of NW Europe in a global setting. In this volume, Kassler examines the distant future with a discussion of both pessimistic and optimistic scenarios. Against this background, the outlook presented is one of an optimistic future.

PLAY FAIRWAY ANALYSIS AND RESOURCE PREDICTION

At BP we base our resource assessment on play fairway analysis. This analysis underpins our view of prospectivity and, therefore, the volume, phase and distribution of predicted resource. A play fairway is defined as an area of potential reservoir facies with the possibility of a cap rock and a petroleum charge. Reservoir, known or postulated, is generally seen as the prime criterion, as it represents the drill target.

The nature of play fairway analysis varies throughout the life of a basin. Once a play has been recognized, which may take the drilling of many wells, an aggressive search begins for the largest fields in the play. This typically lasts up to five years, establishes the basin's resource base and characteristically discovers the largest fields. This phase represents the high-value growth phase of a basin and is here referred to as Sector 1 (Fig. 2). This search for the early giant pools dominates all other considerations as it is in this phase where the big prize is won, irrespective of oil price and fiscal regime.

Once the largest fields have been discovered, value can still be created but at a progressively lower rate as field sizes decrease and more and more difficult reservoirs are exploited. This consolidation phase is labelled Sector 2.

From K. Glennie & A. Hurst (eds), 1996, *AD1995: NW Europe's Hydrocarbon Industry*, Geological Society, London, pp. 187–193

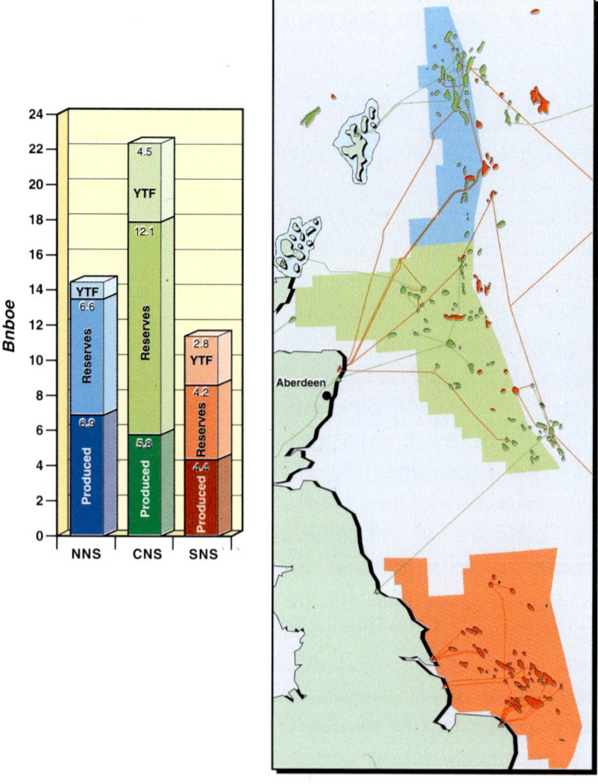

Fig. 1. Map of the North Sea showing the locations of the three major producing regions in the UK North Sea, the northern North Sea (NNS), the central North Sea (CNS) and the southern North Sea (SNS) and the resource attributed to them. The green and red shapes are the major oil and gas fields respectively, the lines are the pipelines.

During this period the growth of infrastructure encourages the exploration for smaller pools. Eventually, it is possible to over-invest in a fairway by pursuing smaller and less cost-effective accumulations, ultimately destroying the value previously created. This part of basin life we call Sector 3. It is during the Sector 2 phase in a basin's life that an accurate appreciation of the remaining resource is critical to avoiding value destruction.

The boundaries between the three sectors are clearly definable only with the benefit of hindsight. They are also not fixed through time. The boundary between Sector 1 and 2 is a function of field size, involving the discovery of a finite number of giant fields ($>250\times10^6$ boe). In contrast, the boundary between Sector 2 and Sector 3 is an economic cut off with respect to what is profitable and value adding. The change from Sector 2 to 3 is characteristically marked by increasingly higher finding and development costs and decreasing success. Clearly, the onset of value destruction will vary as cost structures decrease, either through the use of new technology or through more productive ways of doing business.

In 1995 in the UK Continental Shelf (UKCS), the West of Shetlands Tertiary play fairway is clearly in Sector 1, where giant fields are being discovered and the expectation is that there are more to come. In contrast, in the North Sea, most play fairways are firmly in Sector 2. For some companies in certain fairways, the line between Sector 2 and Sector 3 may already have been crossed.

Predicting the future potential of play fairways is a notoriously contentious issue. The method(s) chosen are governed by the degree of historical discovery data available to give a perspective on the maturity of the play, and the level of geological knowledge focused on what there might be 'yet to find'. In the North Sea today, there is clearly a high degree of both.

NORTH SEA RESOURCE PREDICTION

BP's assessment of the North Sea resource base has been developed around some ten play fairways that are analysed independently on the criteria of historical success and geological knowledge. Figure 3 is a cartoon chronostratigraphic diagram showing the play fairway characterization used. Figure 4 shows the cumulative discovery curves for the Brent and Upper Jurassic shallow marine play fairways. Each of these have fundamentally different characteristics.

The Brent profile (Fig. 4a) may be interpreted as one of classical maturity, where, in the last 15 years, only 10% of the

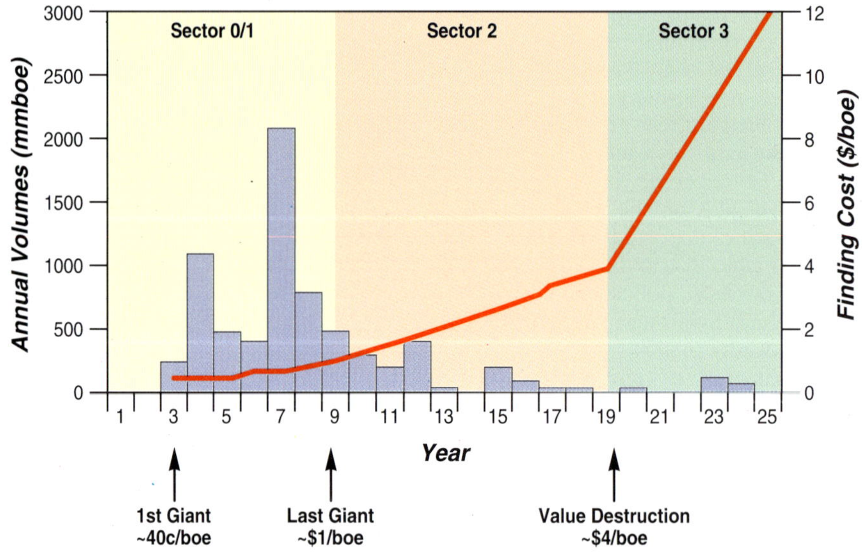

Fig. 2. Diagram showing the sector nomenclature and the three-fold life of a successful petroleum play or basin with associated trends in finding cost and annual discovered volumes.

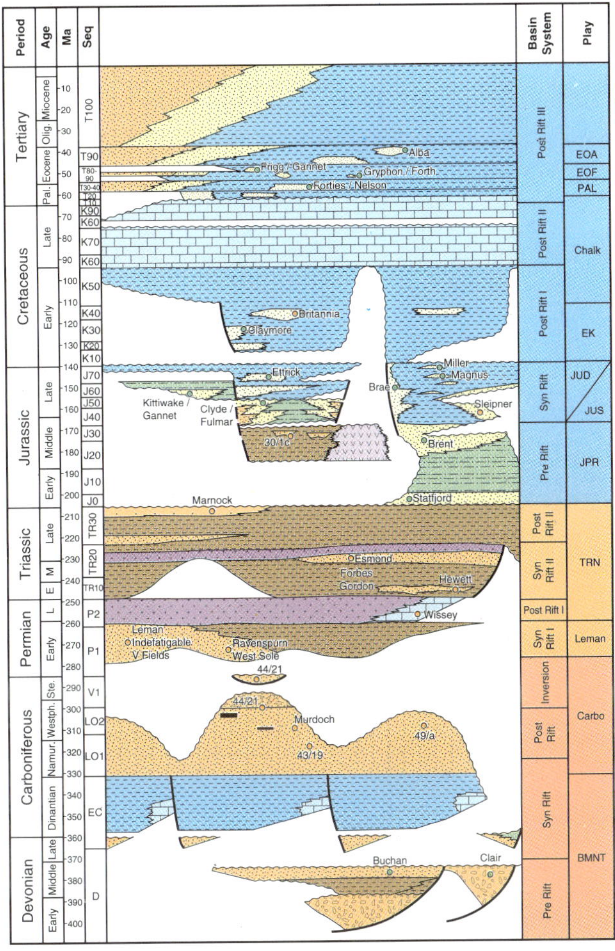

Fig. 3. Chronostratigraphic diagram of the North Sea outlining the various plays and their relationship to the stratigraphy (after Rattey & Hayward 1993).

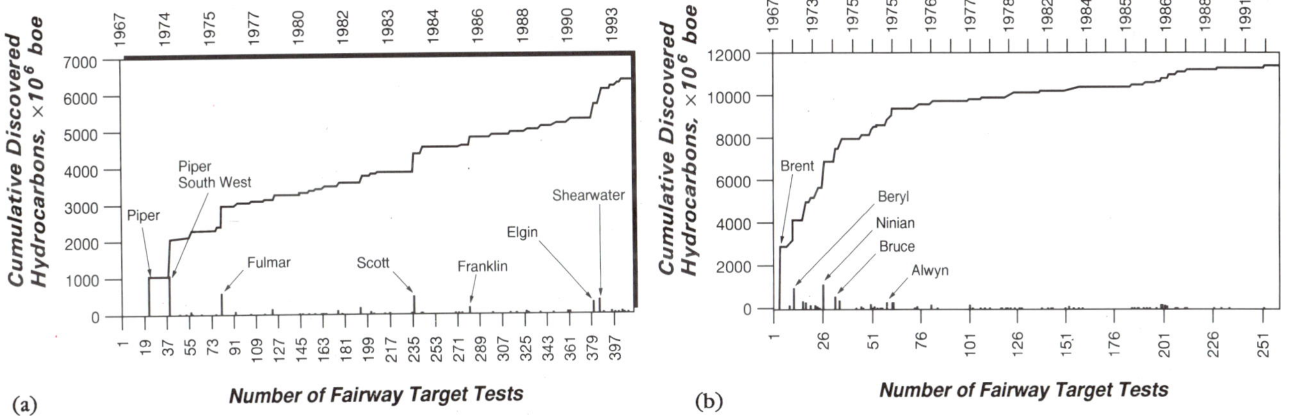

Fig. 4. Discovery versus target tests and time for the: **(a)** Middle Jurassic Brent and **(b)** Upper Jurassic shallow marine fairways.

resource (1×10^9 boe) has been added by 80% of the wells (200) targeted in the fairway, a finding rate of approximately 5×10^6 bbl per well. In contrast, the first 50 Brent wells experienced finding rates of over 150×10^6 boe per well. Clearly, the time to be exploring the Brent play fairway was in the mid 1970s.

The play fairway of the Upper Jurassic shallow marine facies (Fig. 4b) shows a grossly similar distribution but with major breaks as giant fields have been found late in the life of the fairway. Most recently the Elgin and Shearwater discoveries have rejuvenated this play fairway. Both these discoveries are part of the central North Sea high pressure/high temperature province and represent a further characteristic of maturity – that of addressing the difficult plays late in the life of a basin. In this case, due to technology developments that allowed the safe testing of these difficult prospects and a dearth of alternative targets. Such discoveries, after more than 350 wells in the fairway, leave the door open to further large finds and blur a sharp change from Sector 1 to 2.

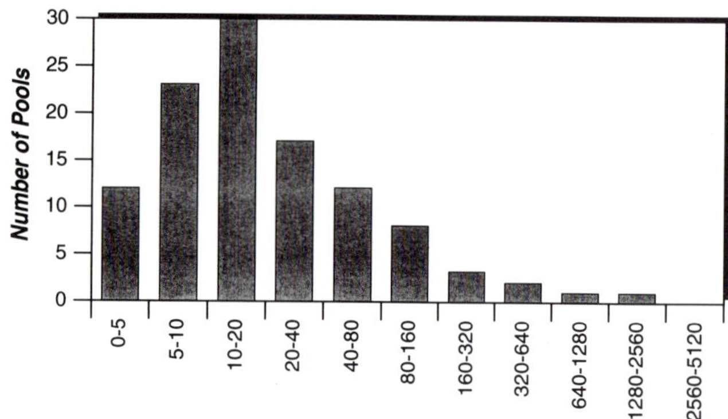

Fig. 5. Field size distribution of the Leman gas play in the southern North Sea. Note that the gap in the small pool part of the profile reflects an economic cut-off rather than a true lack of pools of this size range.

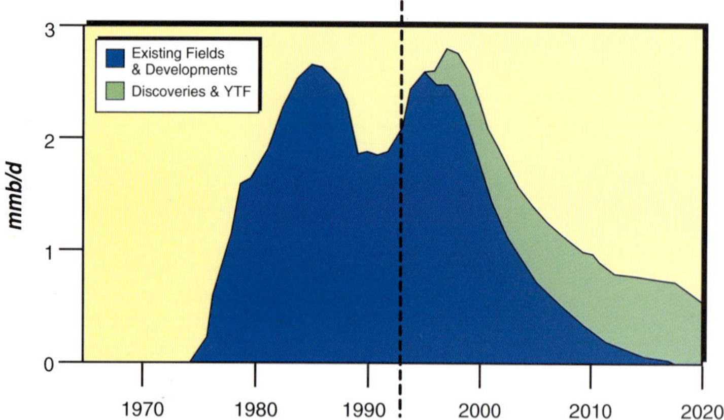

Fig. 6. Production profile for the UKCS with a future profile prediction based on our analysis of the past performance of the basin and out estimation of 'yet-to-find'.

Having examined the maturity of a play, the next step is to understand the distribution of field size in the play. Figure 5 shows the field size distribution for the Leman gas play of the southern North Sea. This approximately log normal shape is argued to be the fundamental shape of reserves distributions. If so, it can immediately be used to define missing field sizes and therefore a 'yet-to-find' for the fairway. However, the assumption of log normal distribution is not proven and, at best, is just another guide.

An important feature of this type of analysis is that the small pool sizes are under represented, because, in the past, we have not specifically looked for $10–20 \times 10^6$ bbl pools. This 'small pool' area of such plots defines one of the major opportunities of the future in the maturing North Sea basin, provided they can be found at high rates of success and developed at low unit cost.

The two approaches described above, discovery curves and field-size distributions, often based on detailed statistical analysis, form the basis of the historical look at the yet-to-find resource. Coupled with a geologically based understanding of the remaining prospect density, play and prospect risk, and a notional volume for prospects we are currently unable to image seismically, these data allow us to make predictions on 'yet-to-find' resource in the North Sea play fairways.

Our 1993 median prediction for the 'yet-to-find' of the UK North Sea play fairways is 8.3×10^9 boe. This figure has been built up from small geographical areas of common geological risk and are here simply aggregated to give a total UK North Sea 'yet-to-find'. This figure, however, represents a conventional view, arrived at by conventional analysis.

An examination of North Sea reservoirs shows marked differences in the apparent exploration maturity of the different reservoir systems. Analysis of the discovery profiles of the various reservoir units leads to the conclusion that, with one or two notable exceptions, the future discoveries in the North Sea will be of less than 50×10^6 bbl, will be in the high ($>40°$) or low ($<25°$) API range, and will be concentrated in and around the already discovered accumulations.

Figure 6 shows our current conventional, view of the resource base in terms of past and future production. Current fields and developments, undeveloped discoveries and 'yet-to-find' potential, are all played out as a future production profile. This future is only one model of how the resource might be produced over time.

Two features of the chart are striking. Firstly, we are less than half way through the production profile of the North Sea, under conventional wisdom. Secondly, when the 'yet-to-find' resource is added to the proven and probable resource,

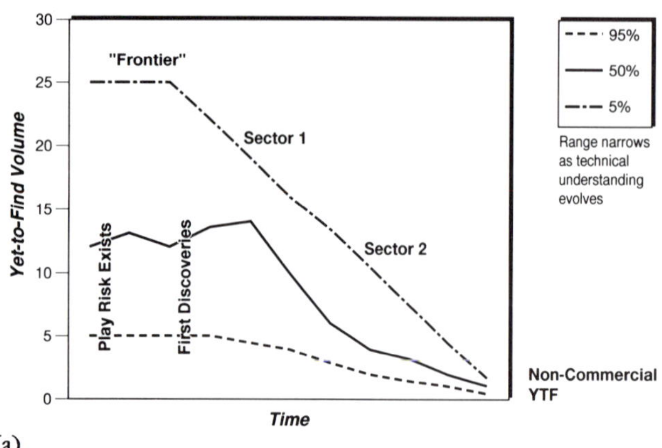

(a)

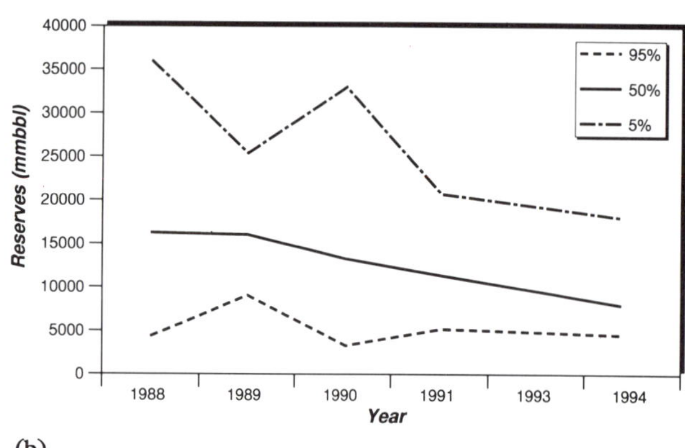

(b)

Fig. 7. Yet-to-find ranges through time: (**a**) theoretical curve showing the narrowing of the range as a play matures; (**b**) 'yet-to-find' ranges for the UK North Sea during the past five years.

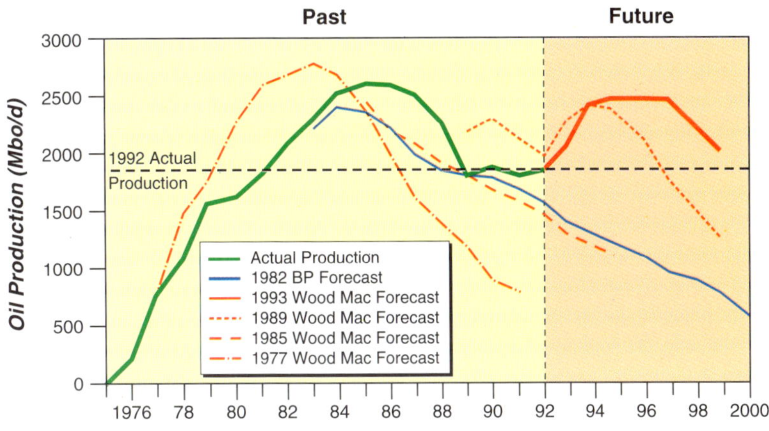

Fig. 8. Past predictions of oil production profiles for the UK North Sea.

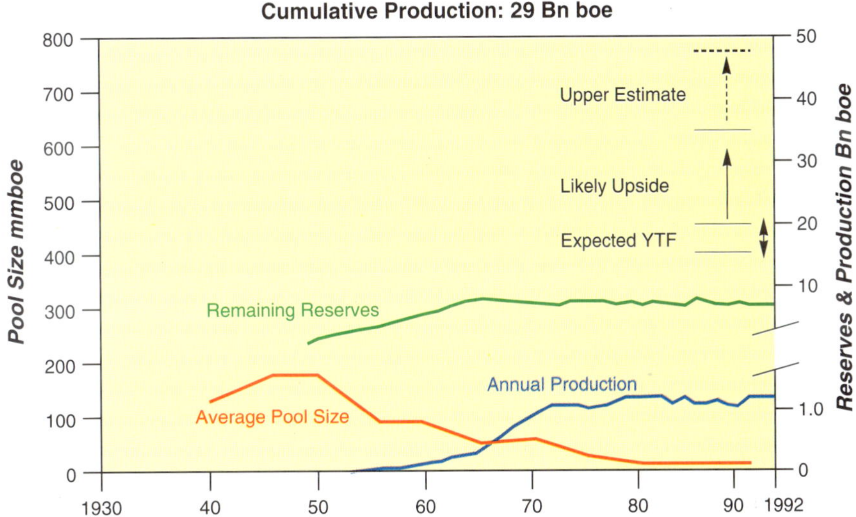

Fig. 9. Production profile of the Gulf of Mexico up to 1992. Note the sustained production rate and remaining reserves against a background of reducing pool size.

the basin life is extended to about 2020, but at production levels of about 25% of present production. This observation tends to confirm the rapid decline and implies a future life span of little more than 25 years in the traditional North Sea basins.

PAST PREDICTIONS

Predictions of 'yet-to-find' reserves may be expected to improve with increasing basin maturity and knowledge. Figure 7a shows this idea graphically by a theoretical narrowing of the range placed on our 'yet-to-find' resource with time. Figure 7b shows little evidence of this happening in our estimates based on a five-year time span over which we have directly comparable data. Clearly, our ability to predict resource is suspect.

A different way to view this expected future is to compare predictions about the North Sea production profile. Figure 8 shows predictions of UK oil production made during the last fifteen years. The notable feature of all the past predictions is that they have consistently underestimated the actual production out into the future. Clearly this is a form of 'planners droop' and leads to the obvious question: 'How reliable is our current estimate of the future?'

THE GULF OF MEXICO EXPERIENCE

The North Sea is not the only basin for which predictions have proved inaccurate. An examination of the Gulf of Mexico, a mature, offshore, oil and gas basin, demonstrates that a mature basin can continue to expand over time and continually exceed expectations.

Figure 9 shows how, as the basin has been exploited, annual production has risen and then stayed constant at about 1×10^9 boe per year, while the average pool size within the basin has continually decreased. Furthermore, the remaining reserves have stayed constant at nearly 10×10^9 boe every year for the past 30 years.

The performance of the Gulf of Mexico is impressive, and demonstrates that, with the right dynamics, a basin can continue to expand and exceed expectations for many years. Possibly there is a lesson here for the soothsayers of the North Sea, to be wary about underestimationg the possible future.

This leads us to the question 'what are the right dynamics in the UK for a long-term sustainable future, and what are the possibilities for the North sea resource base?' The question perhaps leads us away from our current preoccupation with 'when does the game end?'

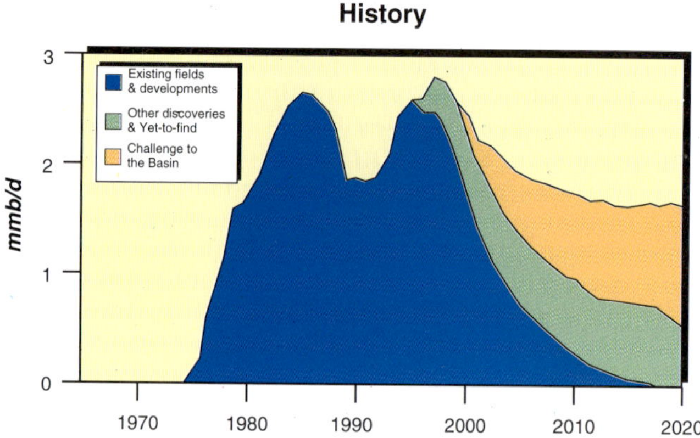

Fig. 10. A new prediction of the UK North Sea's future based on a sustainable resource base, continued technological improvements, a viable economic climate and a focus on mutually beneficial relationships within the industry.

DISCUSSION

By analogy with the Gulf of Mexico, we propose an alternative outcome to the North Sea resource base outlined above. Figure 10 shows this outcome as a production profile containing significant additional resource volumes to those conventionally identifiable today. The thesis is that additional resource will continually become available to the industry through time, given the right technological developments and business dynamics. This vision is not substantiated by anything we are able to identify today. It is more an assertion that we don't know what we don't know, and our track record is telling us there is always more resource accessible than we perceive at any one time. We will term this additional resource the 'North Sea Challenge'.

To allow this resource growth, there are four broad elements that must be in place for a mature basin to sustain itself: resource, technology, infrastructure and the appropriate relationships between operators, contractors, government, and society.

Regarding the resource base, the North Sea Challenge volumes will continue to increase as technology improves and operating decrease. The industry standard recovery factor is now at 45%, with many fields approaching 60% recovery (Forties Field is currently 59%). Moreover, small fields ($5–30 \times 10^6$ boe) are becoming increasingly commercial. These volumes sit below previous economic cut-offs. However, our increasing ability in subsea completion technology and potential tie-back distances will allow their connection to existing infrastructure. Increased 3D seismic coverage will also dramatically increase their number as imaging of the subsurface improves.

The second element that must be in place is that cost-effective technology must continue to evolve and become available throughout the industry. In the North Sea this will focus on enabling the development of the deeper, smaller and presently non-commercial pools. Industry standardization as outlined in the 'Cost Reduction In the New Era' (CRINE) initiative will go a long way toward achieving these developments at cost-effective rates. The number of subsea trees will increase by an order of magnitude in the next 6 years. Improved subsea technology will see the almost total phasing out of new manned platforms.

The third element that must be in place is that core infrastructure must be maintained to allow for the exploitation of incremental opportunities that would otherwise be uneconomic. This applies to both infrastructure required for subsea tie-backs and also the pipelines and terminals required for export and processing.

Finally, the fourth element critical to sustained basin life is the right business relationships, These must be based on an alignment of interests between contractors and operators, government and society. Only with this alignment will the industry meet the challenges ahead, will the government continue to enjoy a flourishing industry and will society allow us to operate.

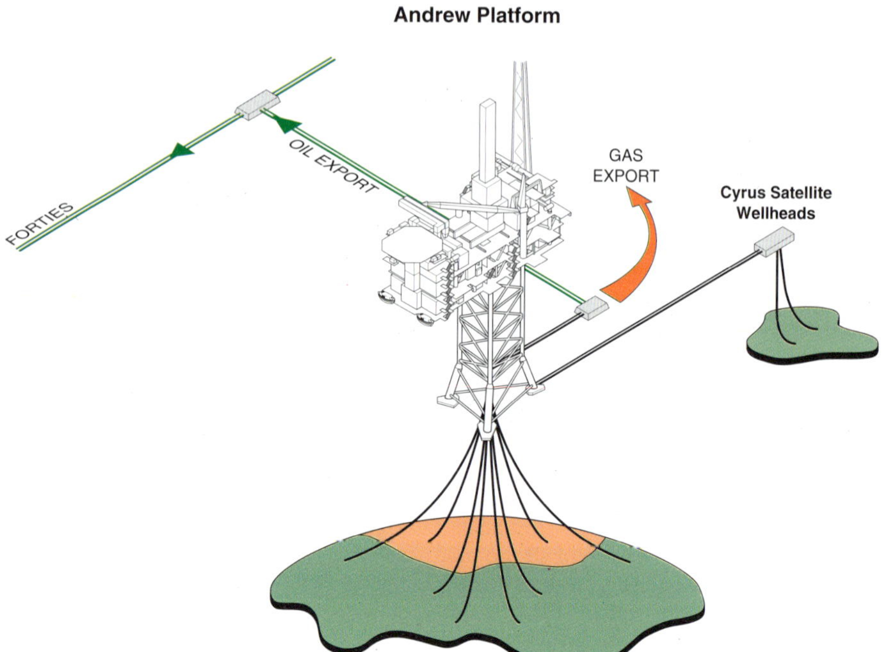

Fig. 11. A cartoon of the Andrew/Cyrus development showing the duality of the development and the crucial proximity to existing Forties infrastructure.

THE ANDREW/CYRUS DEVELOPMENT – AN EXAMPLE

The four elements can be illustrated in BP's Andrew/Cyrus development (Fig. 11) in the Central North Sea, which is due to produce first oil late in 1996. Andrew was discovered in 1974, but languished as 'uneconomic' until 1990, when the first field developments utilizing horizontal well technology led to suggestions that the reservoir could be developed profitably. The 10% additional recovery from a thin oil column sandwiched between a gas cap and a large aquifer was sufficient to make Andrew an attractive investment. Imaging technology, based on 3D seismic data, was crucial in developing the depletion philosophy for Andrew, utilizing a few horizontal wells that produce at high rates.

Proximity to the Forties pipeline allows for optimal use of existing infrastructure. The high off-take rates, coupled with a managed and saleable gas profile, makes the most of the limited resource.

Finally, the new relationships available with the contracting industry through alliancing has allowed the cost base to be driven down to the mutual benefit of all stakeholders in the project.

These four elements – resources, technology, infrastructure, and relationships – reinforce each other to ensure a steady-state production profile supported by reserves replacement at commercially attractive cost structures. In this scenario, the UKCS will be producing large volumes of hydrocarbons for several decades to come.

The authors would like to thank the University of Aberdeen for inviting this contribution to the Quincentennial celebrations. They would also like to thank BP for permission to publish the review and wish to acknowledge the many BP colleagues who have contributed to this work. However, the views expressed are entirely the responsibility of the authors. The lack of specific references reflects the wide ranging source material, much of which is contained in BP internal documents.

REFERENCES

CAMPBELL, C. J. & ORMAASEN, E. 1987. The discovery of oil and gas in Norway: an historical synopsis. *In:* Spencer, A. M. *et al.* (eds) *Geology of the Norwegian Oil and Gas Fields*. Graham & Trotman, London, 1–37.

KASSLER, P. 1996. Global energy challenges over the next 50 years. *This volume.*

LECKIE, G. G. & CHEW, K. J. 1991. The discovered hydrocarbon reserves of western Europe. In: Spencer, A. M. (ed.) *Generation, Accumulation, and Production of Europe's Hydrocarbons*. Oxford University Press, 1–23.

RATTEY, R. P. & HAYWARD, A. B. 1993. Sequence stratigraphy of a failed rift system: the Middle Jurassic to Early Cretaceous basin evolution of the Central and Northern North Sea. *In:* Parker, J. R. (ed.) *Petroleum Geology of NW Europe: Proceedings of the 4th Conference*. Geological Society, London, 215–249.

Global energy challenges over the next fifty years

Peter Kassler

Group Planning, Shell International Petroleum Company Ltd, Shell Centre, London SE1 7NA

ABSTRACT: Two very different energy scenarios have been developed by Shell planners as an aid to strategic thinking about the long-term future. They both arise from the fundamental political and economic changes of the 1980s and early 1990s. 'New Frontiers' is a world in which global liberalization leads to high growth in energy demand, particularly in developing countries. In 'Barricades', countries and institutions reject globalization, and economic growth and energy use are much lower. The environment is a strong theme in both scenarios, but 'New Frontiers', because of its higher energy prices, is ultimately more favourable to the development of renewable energy sources.

KEYWORDS: *energy futures, energy investment, renewables, environment*

INTRODUCTION

We have come to celebrate 500 years of the University of Aberdeen. It was 1495 when Bishop Elphinstone persuaded the Pope and King James IV of Scotland that something had to be done for the benefit of the North-East Scots, who were 'uncultured, ignorant of letters, and almost barbarous'. The solution was the foundation of the Roman Catholic King's College in Old Aberdeen. Then in 1593, George Keith founded the Protestant Marischal College in New Aberdeen. Aberdeen is clearly fertile ground, since it sustained two universities for the next 267 years – for much of that time equal to the total for England and Wales – until they came together in 1860 to form the University of Aberdeen. And once again, Aberdeen has two universities, following the elevation of the Robert Gordon Institute to university status.

I can testify to the advantages of having two parents. It was 1907 when Royal Dutch and 'Shell' Transport and Trading came together to form the Royal Dutch/Shell Group of companies, which is today one of the largest enterprises in the world. Shell companies are major players in the North Sea oil and gas industries, and our strong presence here bears witness to Aberdeen's status as the operational centre of Britain's offshore oil industry.

Shell companies also operate in over 100 countries around the world. We try to anticipate what the future might hold for us using scenarios, and I shall be describing to you our current global scenarios. These are not forecasts. Instead they tell what I hope will be two interesting, challenging and very different stories, mainly about the next thirty years, but with some thoughts extending even further into the future.

The purpose of the scenarios is to draw managers away from thinking about the future in terms of extrapolation of present trends. The energy history of the past century has been punctuated by surprises, discontinuities and sudden changes – the antithesis of extrapolation. Scenarios help us to understand these sea-changes better, by providing deeper interpretations of today's events, and of the underlying forces that will shape the future. They are intended to challenge and change managers' perceptions of the world ('mental maps'), deeply rooted as these often are. Figure 1 is one of a long series of maps which show that the early explorers were convinced that California was an island. We now know that they were wrong, but their belief was a persistent one which no doubt led some of them to make travel plans that were inappropriate. Similarly, in the mid 1980s very few oil industry people were prepared for the 1986 oil price collapse and the radical structural shifts in the energy industry to which it led. This is a good example of mental maps which were not useful aids to business navigation. Those whose mid 1980s mental maps included even vague indications of the rocky shoals ahead of them were in a position to respond better to the situation. Some others were holed below the waterline and sank without trace, or were boarded and taken into port by friendly salvagers.

'NEW FRONTIERS' AND 'BARRICADES'

Our current scenarios start by describing a series of almost revolutionary political and economic reforms which were introduced during the 1980s. 'New Frontiers' is a world in which these reforms are effectively implemented, leading to a transformation of today's poorer countries. In 'Barricades', however, those with entrenched interests in the status quo mount a campaign against change, bringing about a world of divisions and barriers.

The 1980s was a period of profound, even revolutionary change in many parts of the world (Fig. 2). The obvious changes, visible on a map of the world, have been connected with the fall of the Soviet Empire and the disintegration of the Soviet Union. These ended the Cold War and created entirely new assumptions about international relationships, all over the world. The political and military results of these changes are still being worked through, sometimes violently, in ex-Yugoslavia, the Caucasus and elsewhere. At the same time, political liberalization has resulted in free elections and civilian governments in, for example, much of Latin America

From K. Glennie & A. Hurst (eds), 1996, *AD1995: NW Europe's Hydrocarbon Industry*, Geological Society, London, pp. 195–201

Fig. 1. Mental maps.

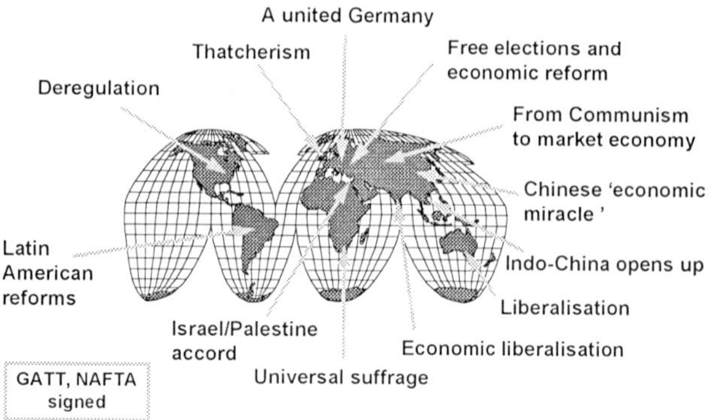

Fig. 2. An era of revolutions.

and parts of East Asia. Even in Africa, there has been progress towards pluralism, in South Africa and elsewhere.

Equally profound changes have occurred in the world of economies and markets. The Soviet collapse has been widely interpreted as a failure not only of authoritarian political regimes but also of centrally planned economies. Liberalism has now been adopted as the received economic wisdom almost everywhere, and has become the basis of economic policies. Among the larger economies, privatization and deregulation have been introduced, with differing degrees of commitment, in the UK, the USA, Japan, Australia and France. Even more far-reaching changes have occurred in some developing countries. In Latin America, for example, where most countries have introduced democratic reforms, together with privatization of state companies and deregulation of markets. Economic reforms have been initiated in the world's most populous countries, China and India.

A further theme of change is that of globalization. It has become trite to speak of the global village, implying that communications and the media now have a world-wide reach. The same, however, is also true of financial transactions and of the market operations of many companies. The path towards globalization will be smoothed by the working of international trade agreements like the General Agreement on Trade and Tariffs (GATT), and the North American Free Trade Association (NAFTA) and even the European Single Market, provided always that these are seen as steps towards an open global trade regime and not as exclusive citadels to be defended by their garrisons.

Responses to the liberalization and other reforms described above define the two scenarios (Fig. 3). 'New Frontiers' is a story of growth, turbulence and change. Economic and political liberalization are expected to work, in the sense of improving societies' ability to create wealth for their members. These processes, however, create enormous upheavals as long-standing barriers are dismantled and poor countries begin to assert themselves, claiming a larger role on the world's economic and political stage.

In 'Barricades', liberalization is held back, creating a world of divisions and barriers with limited room for manoeuvre. People resist liberalization because they fear they might lose the things they value most – their jobs, power, autonomy, religious traditions and cultural identity. This resistance leads to an increasingly divided world of rich against poor, regions against regions, insiders against outsiders and ethnic and religious groups against their neighbours.

ENERGY IN 'NEW FRONTIERS'

The challenge

The reforms of 'New Frontiers' lead to rapid economic growth in the countries which adopt them wholeheartedly, and to high and increasing levels of demand for natural resources, including energy. These changes are not particularly apparent in the immediate future, because of the long

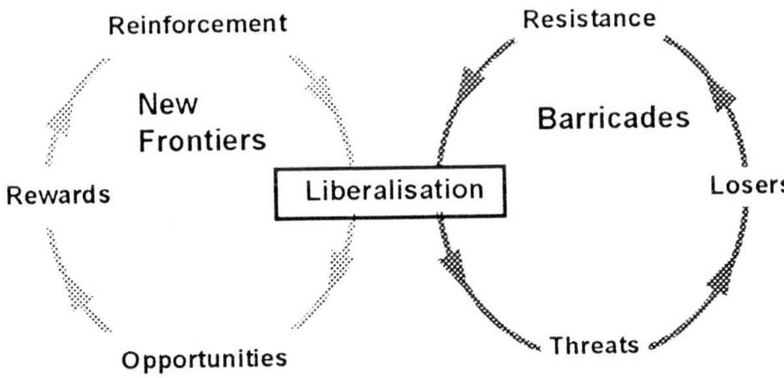

Fig. 3. Two responses.

time and patience required to turn reforming ideas into real wealth creation. From about 2000 onwards, however, the existing economic divergence between low growth rates in the industrialized countries (symbolized by the Organisation for Economic Co-operation and Development (OECD)), and high ones in much of the rest of the world, starts to increase. Liberalization leads to growth rates of 5–6% in non-OECD countries, similar to those of the 1960s and 1970s. If this is sustained, non-OECD countries could account for 70% of world output by 2020, compared with less than half in 1992 (Fig. 4). New opportunities emerge for trading and investment. Companies from developing countries enter global markets, and many traditional industries shrink in rich countries as they grow in poor ones. These changes have powerful implications for energy demand.

By 2020 the number of vehicles in the world will have doubled, with almost all the growth occurring in developing countries. Although the growth of the global economy increases the demand for energy, energy growth is slower than economic growth, because of efficiency improvements. Increased competition, faster technology transfer and reduction in energy subsidies result in more efficient cars and electricity generation. Liberalization and increased consumption lead to a shake-out of heavy industry in such countries as Russia, China and India. The price mechanism, rather than regulation, is used to regulate environmental impact, with energy taxes set to internalize the environmental costs of its use. In rich countries, services and information-based industries replace more energy-intensive manufacturing.

Notwithstanding the developments described above, which mostly favour conservation and decreases in energy intensity,

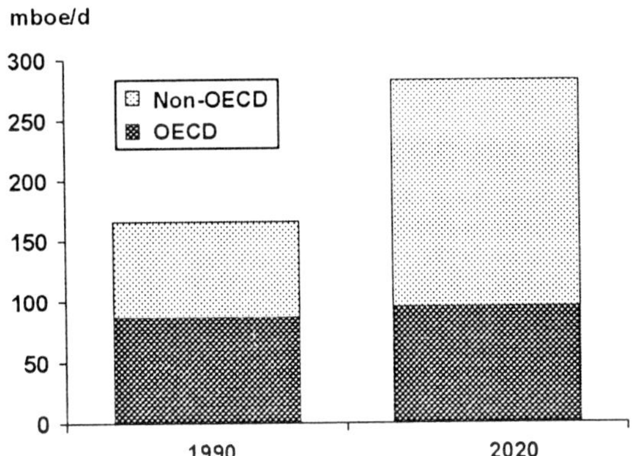

Fig. 5. Demand by primary energy type.

'New Frontiers' is a scenario of very high growth in energy demand. As poor people become more affluent, they use more energy for personal transport, consumer durables like air conditioners and refrigerators, and agricultural improvements such as irrigation pumps and tractors. The very poorest people switch from wood to kerosene, LPG and electricity. As poor countries industrialize, they produce a growing share of the world's energy-intensive products – metals, paper and cement. This implies a major call on the world's energy resources (Fig. 5).

The response

In 'New Frontiers' the demand for 'oil' (liquid petroleum fuels which can include liquids from gas) is estimated to exceed 110 x 10^{-6} bbl/d by 2020. More than half of this would come from the OPEC nations in which production costs are generally low compared to revenues, so there is considerable profit encouraging investment in new production capacity. Demand growth is mainly in developing countries, creating a more diversified crude oil market. A central assumption in 'New Frontiers' is that the state of international affairs in the Middle East would become more relaxed, in particular following the settlement of the Arab–Israeli conflict and increased inter-regional trade and investment, greatly decreasing the risks of market disruption.

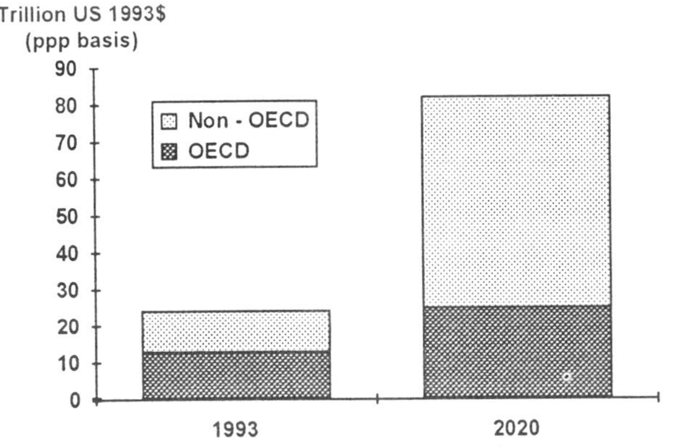

Fig. 4. World GNP (New Frontiers).

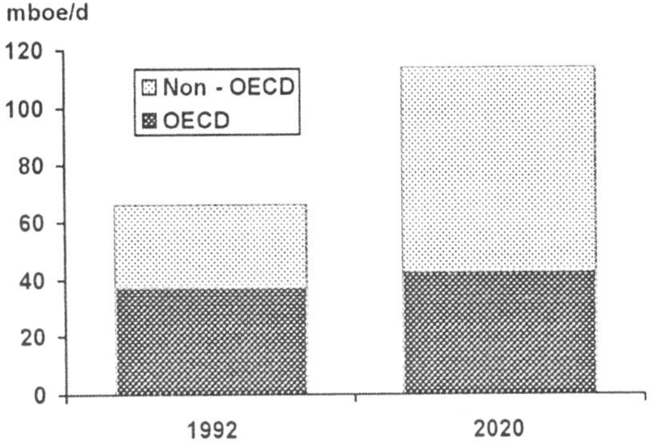

Fig. 6. Oil supply (New Frontiers).

Oil production, therefore, continues to grow (Fig. 6). By 2020 this production will have made serious inroads into reserves, even of some major producing countries. Sustaining acceptable reserve-to-production ratios will require the development of more costly resources such as very heavy oil and gas-to-liquids. This cannot be achieved without some real growth in oil prices, but not necessarily above the $25–30/bbl or so at which unconventional liquid petroleum fuels will be able to be profitably produced. Given such prices, the world's liquid fuel resources are considered sufficient to fuel 'New Frontiers' demand levels up to and beyond 2020.

Demand for gas rises faster than that of any other fuel during the 1990s. Its competitive position relative to coal and oil is strongly enhanced by the efficiency and cleanliness of its use in combined-cycle power generation, supported by an extensive resource base.

By the turn of the century, various new supply developments aimed at the OECD countries will cause these markets to be well-supplied at prices around today's levels. Low economic growth means low total energy demand. Russian industry becomes restructured, leaving gas production capacity in excess of demand. A more interlinked and open pipeline system throughout Europe also increases deliverability of gas and a spot market develops.

Towards 2010 European gas prices are expected to move up in real terms, due mainly to the cost of new, long-distance projects.

Meanwhile, in the developing world, gas demand grows strongly, particularly where there are local resources. In 'New Frontiers', international trade grows significantly, partly by way of new international pipelines moving Middle East and Latin American gas to foreign markets. New technologies such as 'mini-LNG', methanol and gas-to-liquids have potential for commercial development of remote gas resources at reasonable prices.

Coal still has a future in 'New Frontiers', although European and North American demand eventually declines due to removal of subsidies in Europe, and environmental pressures everywhere. In many coal-rich parts of the developing world, however, coal remains an important cheap and indigenous resource. Environmental pressure results in a move towards cleaner coal technologies such as gasification, encouraged by the World Bank and other international agencies.

In many countries, public sentiment and economic realities continue to be against nuclear, and existing plants are retired more rapidly than new ones are constructed. If concerns about global warming are sustained, there could in later years be a modest revival of acceptance of nuclear energy, permitting the industry, for example, to build new plants on existing sites. There may also be patchy nuclear construction in some developing countries which are short of alternatives. Putting all this together, however, nuclear electricity production in 2020 is unlikely to exceed that in 1990.

'New Frontiers' is an environment of technological innovation (as well as of energy demand growth) which is quite favourable to the development of renewable energy. Significant cost reductions should be achieved in biomass gasification, wind power and, especially, photovoltaics which rapidly spread into niche markets, including local use in developing countries. Many buildings will have photovoltaic arrays on their roofs. Finally, a word about who will be doing all this work. 'New Frontiers' contains many elements of change, but one of them will certainly be a high degree of restructuring of traditional energy industries under conditions of intense competition, deregulation and technical change.

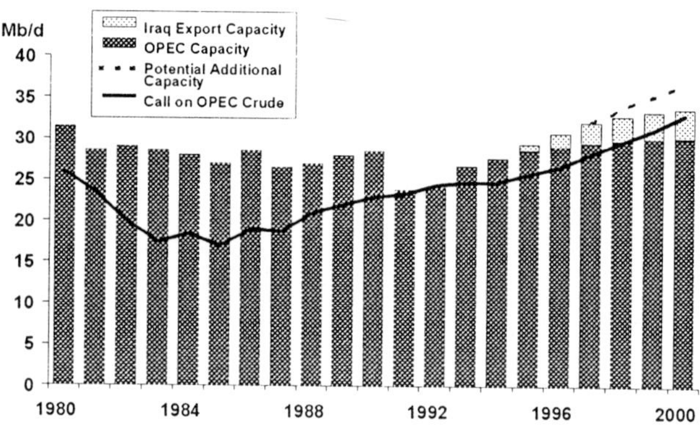

Fig. 7. Crude production capacity vs. call on OPEC.

Monopoly positions and ownership structures will be broken up, both among State oil companies and old-fashioned monolithic utilities. New competitors will arrive on the scene, often aggressive enterprises breaking out of their (now liberalized) home markets. The boundaries between the oil, gas and electricity industries will become blurred and new alliances of companies will operate across them. Different companies will play by different rules, and as new market positions and alliances are built up, pressures on margins will increase. The prizes will go to those who can meet ever tougher requirements and deliver the best value at lowest cost. The energy consumer should flourish in 'New Frontiers', as long as governments can restrain the urge to seize the benefits of competition in the form of energy taxation.

ENERGY IN 'BARRICADES'

The challenge

In the 1990s economic output and energy demand in much of the 'Barricades' world will grow at about the same rates as in the 1980s. One exception is the former Soviet empire where the downturn resulting from the initial economic reforms takes a number of years to reverse. Existing heavy industry can no longer be kept afloat and the international oil industry's privately funded energy projects are slow to show results.

The net effect is that world energy demand grows quite slowly through the 1990s, although for a few years major producers continue to add productive capacity to secure their negotiating positions in the OPEC share-out. With the eventual return of Iraq to the market, there is ample capacity for the rest of the decade (Fig. 7). Oil prices remain weak and acrimonious discussions continue among the OPEC countries. These particular political tensions are made worse by the lack of resolution of more general Middle East problems, such as the Palestinian issue, the sharing of water supplies and other ethnic and religious disagreements. Further instability is added by conflicts spilling over into the Middle East from the patchwork of new republics of ex-Soviet Central Asia. Military expenditure takes precedence over oil investment in the increasingly cash-starved and insecure Gulf producing states. Towards the end of the decade, further investment in 'useless' capacity stops.

In the early years of the twentyfirst century, the world passes rapidly from a situation of excessive complacency about energy supplies to one of potential danger. Even in the

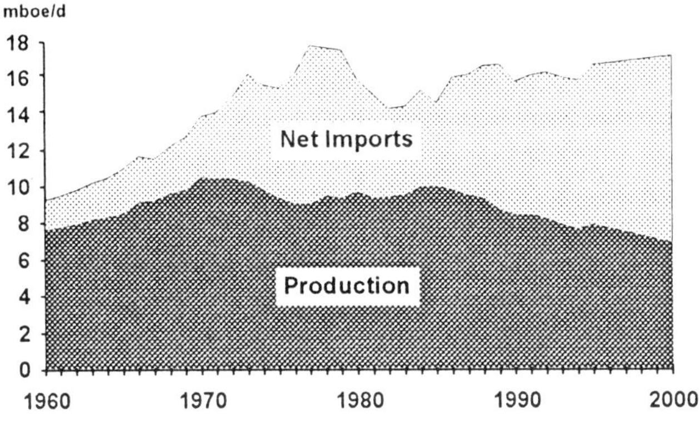

Fig. 8. US oil dependency (Barricades).

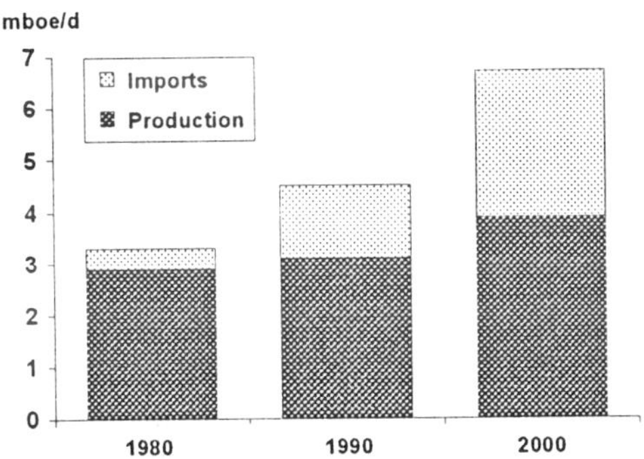

Fig. 9. Natural gas: W. Europe (Barricades).

relatively low-growth world described above, the demand for oil has grown to levels which consume the former OPEC surplus production capacity. Constrained by weak prices and the resulting revenue shortages, the producers have neglected to invest in their oil fields since the middle 1990s. The Middle East has again become a political and military tinder-box. This is a major concern in the USA which has become increasingly reliant on oil imports (Fig. 8).

In the 'Barricades' scenario, we have imagined that at some time after 2000 one or another of the many Middle East disputes ignites the tinder, and leads to an oil crisis. This could be an event in the Gulf itself, in the Straits of Hormuz or Central Asia, and could be a rekindling of one of the age-old ethnic disputes of the region. The crisis itself is probably short-lived. The oil price shoots up to over $40/bbl briefly, but the supply disruption can be dealt with in weeks or months, given the high state of flexibility of the world's oil industry after many years of political uncertainty. It is, nevertheless, another nasty shock for the oil-consuming world, and one which has profound psychological consequences. The crisis sparks off a set of attitudes among governments and consumers which we have nicknamed 'Energy is Bad', causing a major discontinuity in the energy industries.

There are three elements in the story about 'Energy is Bad':

- concerns about supply security in large energy-importing industrial economies: USA, Japan, Europe. This applies to imported gas as well as to oil (Fig. 9);
- concerns about the environment, particularly vehicle-related pollution and global warming;
- and the shock itself with its attendant disruption of everyday life as visible evidence of insecurity of supplies.

The reaction is swift and dramatic. Energy-importing countries scramble to free themselves from dependence on imports, egged on by their own 'green' constituencies. The response is largely by way of regulation, such as:

- laws mandating strictly regulated energy conservation;
- regulations encouraging further development of alternatively-fuelled vehicles (such as electric vehicles) and promoting their use (Fig. 10);
- support for nuclear plant construction;
- encouragement of biofuel production;
- even higher taxation of oil and gas fuels, especially if imported.

The result of these policies is that the growth of oil and gas demand in OECD countries is severely limited, and oil

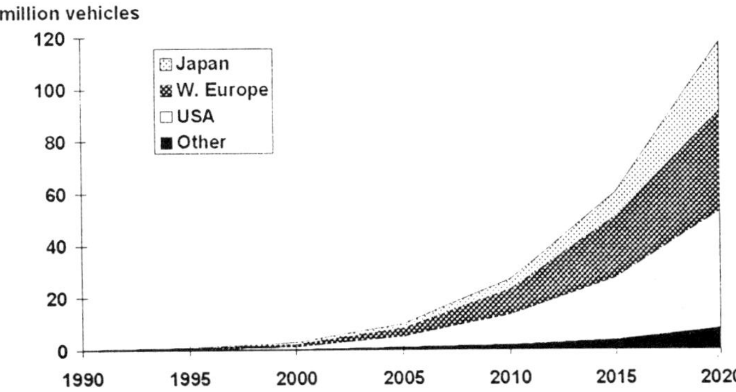

Fig. 10. OECD electric vehicles (Barricades).

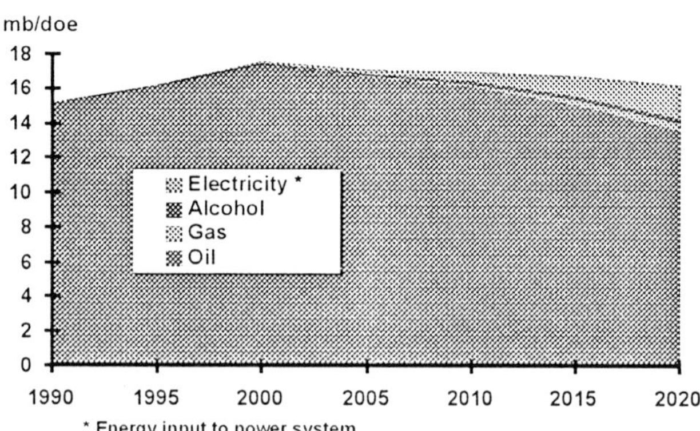

Fig. 11. OECD road transport fuels (Barricades).

consumption may even decrease in some important markets such as transportation (Fig. 11). Overall oil demand does not recover, despite the price falling back to its pre-shock level. The new tilt of the energy playing fields will give a powerful impetus to competing fuels such as nuclear and renewables, whose most important characteristic is localness – energy can be produced within the barricades, more or less independently of the outside world. The 'Barricades' scenario is not good news for international oil and gas companies.

The response

Low oil prices in the 1990s create a difficult financial environment and investment climate for oil-producing companies, cash-starved in an environment of increasing technical costs of new production and political risks. After the crisis there is a strong drive towards oil self-sufficiency, activated by subsidies for domestic exploration and production, preferably by domestic oil companies. The USA is a key participant in the oil market and will move mountains to protect whatever new marginal production can still be developed within its borders. Economic nonsenses proliferate in this environment, but because of low demand growth, the world manages to muddle along.

'Barricades' will be more benign for gas than for oil because of its environmental advantages. The USA will use subsidies and incentives to bring on new domestic supplies, some of which will replace oil for transportation (in the form of compressed natural gas or methanol). As gas exploration is much less mature than that for oil, many countries including LDCs have potential to develop their own new supplies, even if these are modest on the global scale. In Europe, major gas storage projects are considered a useful insurance against supply disruptions caused by political upsets in Algeria or Russia. New international gas pipelines, on the other hand, are unlikely to be considered good investment prospects in 'Barricades'.

The future of coal in this scenario varies from place to place. Green lobbies will contest its use in many OECD countries, unless accompanied by very large investments in pollution-eliminating equipment. For countries having domestic coal supplies, on the other hand, its use will be a natural response to the call for self-sufficiency. China and India, for example, will certainly continue to use their coal supplies on grounds of cheapness and availability. Overall, these factors lead to a modest growth of world coal demand over the scenario period.

Other energy sources include nuclear and renewables. Reluctant government support for nuclear will lead to some new investment on the grounds that it is the least undesirable of several unattractive alternatives. This will depend partly on the state of opinions in the global warming debate at that time. New markets for renewables such as solar and biomass are largely in poor countries which do not offer the kind of market opportunities required to drive expensive R&D programmes. If governments provide the very high subsidies required, agricultural land, especially in Europe and the USA, could be used to grow crops for liquid fuels substitution, especially for diesel vehicles. Otherwise renewables will be developed locally, cheaply and on an *ad hoc* basis. Their time will come later, as I shall suggest in a few minutes.

THE ENVIRONMENT

In discussing both scenarios, there has been an underlying theme that concern about the environment is common to both – it has now been a major world issue for more than thirty years and will and must continue to be so, under any conceivable scenario. What will vary are the relative priorities to be assigned to various environmental questions and the methods proposed for their solution. In general, 'New Frontiers' will favour market solutions. In 'Barricades', the same concerns are addressed by regulations and laws.

One major dimension of environmental concern is that of problems related to poverty. The very poor have to worry about basic questions of water purity, sanitation, overcrowded cities, smoke from wood and coal fires and consumption of tree resources for firewood in homes. These are basically local problems which can be alleviated by way of the processes of wealth creation, leading to investment in housing and other infrastructure, and ability to pay for less polluting fuels such as LPG. This process will be favoured in 'New Frontiers' in parallel with the overall economic growth in that scenario, resulting from the liberalization reforms. By implication, the responses in 'Barricades' will be slower.

As levels of wealth increase, new areas of concern emerge. Rich countries today worry about emissions and local air quality, recycling standards, biodiversity and the ozone hole. These or other matters may turn out to be long-term worries, but in any case the rich countries seem likely to have different problems from the poor ones. Those affecting poor people are immediate and physical, while rich people's concerns are of a less tangible and longer-term nature.

The moral of this story is that, while poor countries may echo polite worries about the problems of the rich, it is unfair and unrealistic to expect them to be as concerned about these as about the unpleasantness of their own immediate environments. It also implies that as extreme poverty is gradually conquered as a consequence of 'New Frontiers' policies, and progress is made on many local environmental problems including air and water pollution, developing countries' attention will tend to shift towards today's rich country problems, such as CO_2 emissions.

It is hard to resist a value judgement between the market-based environmental management of 'New Frontiers' and the draconian regulation of 'Barricades' under the 'Energy is bad' regime. One feels instinctively, however, that the 'New Frontiers' approach is likely to lead to better quality energy management than the other, largely because the market mechanism is more cost-effective than a command structure, which has been discredited in so many countries over the past few years.

ENERGY AFTER THE SCENARIO PERIOD (POST 2020)

All the evidence on evolution and replacement of existing energy systems suggests that their infrastructure tends to have a lifetime measured in decades, and that for this reason technological transformation is an extremely slow process. We are therefore easily reassured that industries like ours will not be overwhelmed by technological surprises, because the time required to replace the hardware will give us time to adjust to the new technology, whatever it may be. We are also well aware, on the other hand, that past attempts to anticipate long-term technology developments have been pathetically unsuccessful, which may inhibit us from trying to do so in the future.

This brief postscript to my description of the energy scenarios is mainly to argue that, nevertheless, it is possible to make some general observations which are relevant to thinking about the long-term future. There have been major transitions in the energy world. Since the Industrial Revolution, wood has been replaced by coal as market leader, then by oil. Natural gas is now increasing its share of primary energy supplies and electricity has become the principal energy carrier. Such transitions have occurred at least every seventy years in the past, and each has added to the complexity of the energy system. We know of no reason why this change process should now slow down, stop, or go into reverse.

A number of non-fossil-fuel technologies are available for the future. These include wind, biomass, geothermal, solar thermal, photovoltaic, tidal and wave energies. The cost of these energies, although still much too high, has been falling steadily following a learning curve broadly in line with cumulative production. In these early stages, they find a place through the exploitation of market niches (such as photovoltaic electricity production for specific applications far from the grid in sunny countries). A mature, established energy like oil has reached a much flatter part of its learning curve and costs will fall more slowly.

On the basis of these learning curves, we would expect new renewable forms of energy to start to be competitive with fossil fuels around 2020 or 2030. After this, use of the new energies would grow. It is not necessary, for this argument, to identify which renewable energy has the best prospects. Technologies will compete, but the market will decide.

On the demand side, the need for energy will depend on wealth and lifestyles at that time. This subject stretches the imagination as much as anything in energy planning. One major question is whether the very long-standing geometric growth of personal mobility (measured in average km travelled per person per year) will continue. This trend appears to have been persistent since the early nineteenth century, and correlates closely with the world's consumption of transportation fuels. If one considers what might disrupt it, one candidate might be virtual reality. Virtual music and virtual entertainment are now common-place, and it is not implausible to argue that the powerful and vivid images which can be conveyed may replace part of the desire or need to travel, on holidays (replaced by virtual holidays) or business (replaced by high-tech tele-meetings) for example.

Of course, we cannot predict the future. We certainly cannot forecast what will happen over one hundred years. It is our view, however, that the two developments outlined above – the progressive reductions in the cost of new renewable energies through normal market mechanisms, and the impact on energy demand of new lifestyles and new technologies – suggest a likely decline in fossil fuel consumption around the middle of the next century. Since this means that the present CO_2-emitting energy forms would be replaced by non-CO_2-emitting ones, it has powerful implications for the global climate change debate, which need to be explored carefully and in more detail. One thing we can be pretty sure of is that the pattern of energy supply will continue to increase in complexity.

CONCLUSION

My observations have looked to the next 50 years. The long history of the University of Aberdeen shows how a sound institution can adapt and survive over periods which make even these time horizons look short. But it is important not to take the future for granted. Who hears today of the University of Fraserburgh? It too was founded in Aberdeen (in 1597 – just four years after Marischal College), but it did not survive for long. Successful adaption is not easy, nor guaranteed, and our scenarios are one of the ways we try to prepare ourselves for an uncertain future. The University of Aberdeen has clearly succeeded in weathering many fundamental changes in its environment, and I wish it continued success in the years, even centuries, to come.

Based on work carried out by members of Group Planning, Shell International Petroleum Company.

Designing a portfolio management programme to optimize cash-flow

D. Fassom

Fina Exploration Limited, Fina House, Ashley Avenue, Epsom, Surrey KT18 5AD, UK

ABSTRACT: The design and implementation of any portfolio management programme must, by definition, be tailored to the drivers and particular objectives of the company owning the assets. This paper will concentrate on one of the most important driving forces, namely managing cash-flow. Five key steps are required to achieve an effective portfolio management programme:

1. establish targets/goals;
2. describe and value the assets in your company's portfolio;
3. identify and catalogue potential 'customers';
4. construct appropriate deal structures and other strategies to achieve your targets;
5. work hard and do deals.

Portfolio management must be a team effort. Without clear direction from management, the targets cannot be properly set. Without the expertise of geoscientists, engineers, economists and tax experts the assets cannot be accurately valued, and without the lawyers the deals can never be consummated in law. Experience indicates that it is the technical understanding and valuation of the company's assets that is most important, and asset valuation will always be the most accurate where high-calibre technical staff are allowed to work together, and where they are given time and resources to complete the task. Without a good knowledge of asset valuation, portfolio management is at best a liability and at worst a potential company wrecker. Portfolio management armed with reliable valuations can, on the other hand, achieve its objectives and significantly contribute to a company's health and success.

KEYWORDS: *portfolio management, cash-flow, profit, asset valuation*

INTRODUCTION

Portfolio management is a process which aims to reconfigure a company's asset base to achieve a wide range of possible corporate objectives. Portfolio management is not confined to the exploration theatre but extends to the whole portfolio including appraisal, producing and mature fields. As such, portfolio management, relates to the re-organization of projects and assets rather than simply prospects. Some objectives that portfolio management can achieve are as follows:

- To optimize investment by:
  acquiring attractive assets in areas of company-owned infrastructure or a newly discovered 'play fairway' (by swaps, purchases or farm-ins);
  ensuring there are sufficient projects in the portfolio for a healthy investment and reward profile.

- Rebalancing or 'hedging' the risk/reward profile of a company's portfolio by:
  increasing or decreasing the ratio of high and low risk opportunities to suit the company's risk profile (via swaps, farm-outs, farm-ins etc.).

- Accelerating investment activity on the portfolio by:
  - reducing the number of partners and hence improving partner dynamics in an asset;
  - forming joint ventures with cash-rich companies;
  - farming out projects which would otherwise be delayed.

- Optimizing cash-flow by:
  - selling fields which require significant investment;
  - farming out selected projects;
  - withdrawing from costly or poor-quality blocks and projects.

The design and implementation of any portfolio management Programme must, by definition, be tailored to the drivers and particular strategic objectives of the company. This paper will concentrate on one of the most important drivers, namely managing cash-flow. In this paper the author will define cash-flow, examine its importance and look at how pressures on cash-flow arise in the oil sector. The measures designed to optimize cash-flow will be described although such methodology could equally be applied to achieving asset growth, focusing on core areas, or indeed dealing with any other

From K. Glennie & A. Hurst (eds), 1996, *AD1995: NW Europe's Hydrocarbon Industry*, Geological Society, London, pp. 203–207

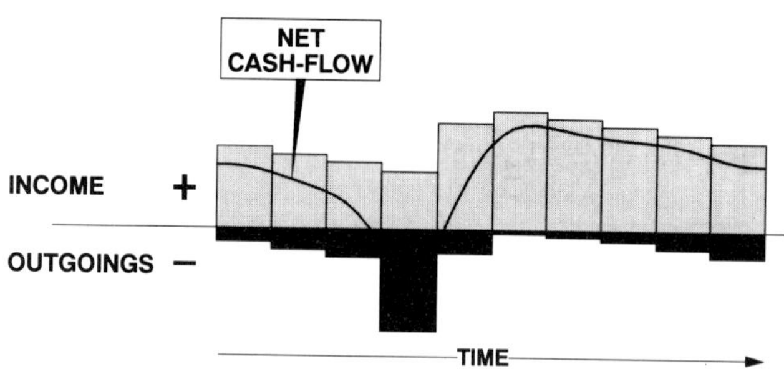

Fig. 1. Net cash-flow is the difference between income and outgoings.

aspect of your portfolio your company wants to change and improve.

WHAT IS CASH-FLOW?

Cash-flow and profit can be confused. The profit (or loss) of an oil and gas company is calculated by using a set of approved accounting procedures which allow the accountants to spread huge expenditures over project life rather than in the year it is spent or borrowed. As such, the profit and loss figures can be quite misleading if one year is taken out of context. So let us be quite clear, profit is not the sum of money your company has in its pocket at the end of the year.

On the other hand cash-flow is a very tangible, and instantaneous representation of your company's position; it is, quite simply, the difference between cash physically coming in and cash going out. Cash coming in will be from your company's producing fields and any asset sales; cash going out would be the cost of asset acquisitions, tax, interest payments, salaries, exploration costs, overheads etc. The example in Fig. 1 shows a hypothetical company's cash-flow profile. In the early years cash-in exceeds cash-out and a positive cash-flow prevails. In the middle years however, the spectre of negative cash-flow is looming. Why is this happening? Perhaps the field or fields production are all declining at the same time; the price of oil may have fallen; exploration costs may be high. At the same time funds for a new field development will be needed. Unfortunately the incoming cash-flow cannot support the expenditure so money must be borrowed. The later years show a return to a positive cash-flow.

WHY IS CASH-FLOW SO IMPORTANT AND HENCE A DRIVER?

Whilst a company with poor profitability, a weak asset base or a string of dry holes may wither and eventually cease to trade, a company with negative cash-flow and unsympathetic banks can be shut down instantly. So cash-flow is very important not only for the company but also for its employees. A steady and regular cash-flow is also important from the investors' points of view. Remember they want to receive a steady dividend every year if possible. For this reason the reader can understand the impatience of the 'Euro-tunnel' investors – their cash flow problems will mean no dividends until the next century!

In reality, no oil and gas company wants to approach the fringes of a nightmare scenario where the banks could foreclose. Hence companies try and keep a tight grip on their cash-flow. At certain times, however, even the most far-sighted company can find themselves in a low cash-flow position. In these situations sometimes the only way out is the re-organization of the company's assets. Portfolio management does just that.

WHAT ELSE CAN CAUSE PRESSURES ON CASH-FLOW TO ARISE IN THE OIL SECTOR?

The previous example was of a company that had either not planned its cash-flow properly or had experienced an unexpected loss of revenue. What else can cause cash-flow pressures on our upstream business? To examine this, the author cites an example from the perspective of his own company, the PetroFina Group. The PetroFina Group is one of Europe's leading multi-nationals with a market capitalization of almost $8 billion with extensive interests in both upstream and downstream activities throughout the world, but particularly in Europe and the United States. Hence Fina's UK upstream business is often competing with other upstream affiliates or with the downstream, for capital from the parent company. Because the upstream business is so capital-intensive the demands for cash can be quite large and, at certain 'crucial' times (for instance, when there is not enough cash to fund projects), one has to make a decision to

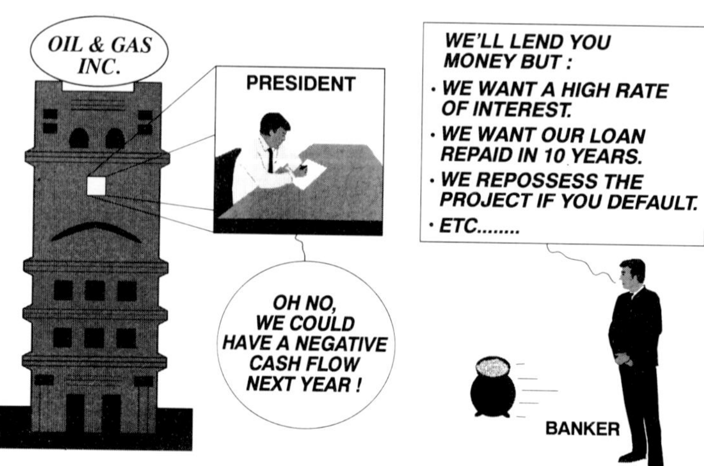

Fig. 2. Overborrowing.

Fig. 3. Successful portfolio management.

either delay a project, borrow more money or cut back elsewhere. Sensible companies certainly do not want to delay good-quality developments (because these are the life-blood of a successful organization) and they do not want to over-borrow for obvious reasons (Fig. 2). Because of these considerations, financial pressures can often be placed upon our 'controllable expenditure' such as exploration etc. This is not always popular – but understandable.

When a company is faced with such a problem it can either defer or cut back its exploration expenditure or it can find a solution which maintains activity and keep its assets working. Keeping assets working with minimal expenditure is one aspect of successful portfolio management (Fig. 3).

PORTFOLIO MANAGEMENT – 5 KEY STEPS

The design and implementation of a successful portfolio management programme requires 5 key steps to be followed.

1. Establish strategic direction(s), goals and any corporate constraints.
2. Describe and value your portfolio of assets.
3. Identify and catalogue your potential 'customers'.
4. Construct appropriate deal-structures.
5. Work hard and do deals.

Establish targets/goals

Before embarking on the time-consuming exercise of portfolio management, the first thing that is needed is for the portfolio manager and the rest of the staff to understand the prime objectives of the programme. If the main driver is to optimize cash-flow, he or she must be clear on how far this drives the programme. Are there other secondary drivers or limiting factors such as conserving certain assets, core areas that must be protected, maintaining a good balance of gas and oil projects, etc? All of these issues must be clarified otherwise the portfolio could be decimated by an over-enthusiastic or misinformed portfolio manager.

Describe and value your portfolio

This sounds quite obvious but it is imperative to look at the risks, the range of possible values and other attributes of each project before the programme starts.

Fina utilizes a number of technical and economic indicators which have been developed in-house and which provide us with a description of each project. We start with a thorough in-house review of the project interpretation, assessing risks and the range of possible reserves. This is followed by interactive economic analysis to determine the probability of making a commercial discovery and, in the event of a commercial discovery, what size we can expect. We then balance this reward and risk against the cost of the exploratory work to see if we have a drillable project or if we need to await higher product prices, new infrastructure or new technology, or as I will mention later, get someone else to take all the risk via a farm-out. The principal method of expressing risk and reward and then ranking projects in Fina utilizes the principal of expected monetary value or EMV.

The EMV is simply a mathematical representation of risk,project value and the amount of risk-capital required. There are two sides to the calculation.

(a) The positive (or reward) element is calculated as follows:

Probability of commercial success (Ps) × net present value (NPV) of the expected discovery size in the event of a commercial success (in Fina we call this the mean success value or MSV).

i.e. $Ps \times NPV_{MSV}$

(b) The negative (or capital exposure) element is calculated as follows :

Cost of the exploration or appraisal well (the abortive costs) × probability of not making a commercial success (1–Ps).

i.e. Abortive costs × (1–Ps).

Therefore the EMV can be expressed as $(Ps \times NPV_{MSV}) - (\text{Abort. cost} \times [1-Ps])$.

If the EMV is negative then the project is either too risky for the potential reward or, the cost of the exploration well is too high for the risked reward the project offers. Conversely, if the EMV is positive then it may be an attractive drilling opportunity.

It must also be noted that the EMV is just one of numerous indicators used in project ranking etc. The EMV number is only a guide and the author recommends that the user should never lose sight of the basic technical and commercial data which are used to calculate this and other 'numbers'.

Identify and catalogue our potential customers (Fig. 4)

It may seem obvious but even if you have established your objectives and what your assets are worth, you still would get nowhere by simply sitting waiting for the phone to ring. Many companies may not even be aware that they should be doing business with you. Look at brokers' reports, network with the industry and establish a list of likely companies and build a convincing case why a party should do business with you. Time spent doing this will yield a much higher 'success rate' on your deals.

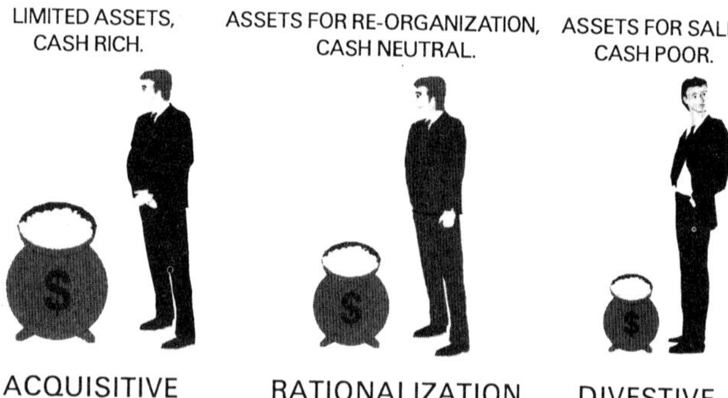

Fig. 4. Identify your 'customers'.

Construct appropriate deal structures and other strategies to achieve your targets

The number of deal-structures to optimize cash-flow is limited only by your imagination and legislation. The four main methods are as follows :

Selling assets/buying assets

The selling of assets is often controversial but it does provide instantaneous cash-flow for the seller, and if it is a field under appraisal there is an added benefit of avoiding costly development expenditure. Obviously if you do not get a good price you haven't done a very good job.

From the buyer's perspective it may be that they have excess cash, and a purchase will add future cash-flow at a higher rate of return than they can get on a deposit account in a bank.

Relinquishing acreage

This is always a difficult decision to make, but remember, some annual licence costs can run into hundreds of thousands of pounds. There is a point where tough decisions have to be made to avoid these blocks bleeding your company dry.

Swaps

In a swap, unless there is an impending well or other high-cost expenditure, there is normally no immediate cash-flow effect. Indeed, because the deal involves no E&A work, the benefit of a swap is often purely strategic. The author is not denigrating swaps but simply saying make sure its the best action for your company bearing in mind your 'driver'. Over-focusing on geographical 'core areas' for instance can reduce your spread of risk, and reward in the basin.

Farm-outs/farm-ins

By comparison with swaps, farm-outs have a much more immediate effect.

1. By avoiding the drilling expenditure there is an immediate cash-flow saving equivalent to the well cost. This can run into millions of pounds.
2. The project is actually drilled (rather than staying on the shelf) and if it is a discovery the farm-out company keeps some of the interest in the project and has added these reserves at a finding cost of zero. This is obviously an excellent method of reserve replacement for companies with a poor cash-flow.
3. If the project had not been farmed out and was dry, this expenditure counts as an expense in accounting terms and must be deducted from the company's profits (not very welcome to shareholders). Furthermore the abortive costs will impact directly on the company's cash-flow.

Obviously it is down to your own imagination and ability as a negotiator to get the best farm-out terms for your company, but one thing to remember is that farming-out a prospect which results in a discovery can be very good business, particularly if the prospect may not have otherwise been drilled. If it was dry then your company has just saved a substantial sum of money. Also remember that one has got to do business in the future with the farminees so try and negotiate a deal in which both sides feel they have benefited. Above all, in the author's opinion, be clear and honest about your company's motives – it pays off in the long run.

Work hard and do deals

Since the inception of Fina UK's portfolio management programme in 1991, the company has continued to fund a large number of wells. In addition, it has also farmed-out a substantial amount of wells. The farm-out programme has yielded one discovery to date at a predictable finding cost of zero. Had Fina funded these wells its finding costs would have soared to over $6 a barrel.

At the end of 1995 Fina's portfolio of UK blocks had been reduced from 125 to 72. In the process, however, Fina has increased its average equity in what it considers high quality blocks and has increased its operatorship to 19 blocks, several of which contain promising discoveries.

Perhaps the most painful aspect of portfolio management are disposals. Whether it be by relinquishment or sale, the loss of a block from a portfolio can sometimes be a difficult occasion for a company. In Fina UK's case, substantial receipts have been made from such sales and this has gone someway to ease the sense of loss. Such disposals have not only transformed Fina UK's cash-flow position but they have also reduced its debt repayments to the extent that Fina is now in a much better position to pursue its exploitation efforts in the North Sea and other areas.

CONCLUSION

Portfolio management must be a team effort (Fig. 5). Without clear direction from management, the targets cannot be

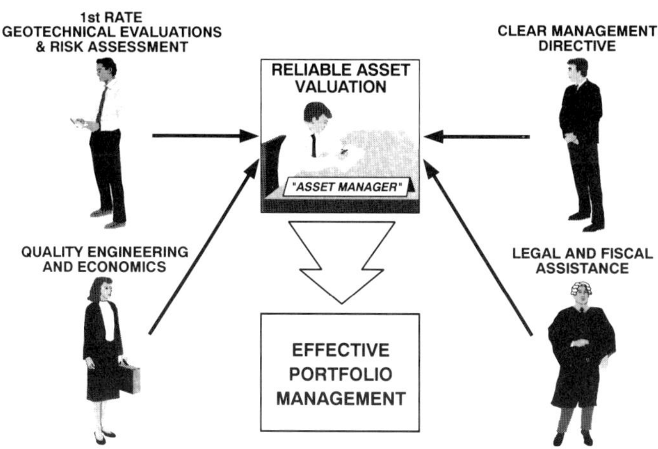

Fig. 5. Effective portfolio management is a team effort.

properly set. Without the expertise of geoscientists, engineers and economists the assets cannot be accurately valued, and without the lawyers the deals can never be consummated in law. In the experience of the author it is the valuation of the assets that is most important, and asset valuation will always be the most accurate where high-calibre technical staff are allowed to work together, and where they are given time and resources to complete the task.

Without a good knowledge of asset valuation, portfolio management is, at best, a liability and, at worst, a potential company wrecker. Portfolio management armed with reliable valuations can, on the other hand, achieve its objectives and significantly contribute to the company's health and success.

As a final comment the role of the Department of Trade and Industry and, in particular, the Licensing Group must be recognized. Without their advice, efficiency and hard work many of these important transactions could have suffered delay or may not have been possible to complete.

Overall management of risk in oil field development

R. L. Bruce[1], A. M. Minty[2] and C. A. J. Gregory[3]

[1] *Vectra Technology Ltd, 310 Europa Boulevard, Westbrook, Warrington WA5 5YQ, Cheshire, UK (Present address: Risk Management Research Institute 48, Princess St., Manchester M1 6HR, UK)*

[2] *Risk Management Research Institute, 48, Princess St., Manchester M1 6HR, UK*

[3] *Mobil North Sea Ltd, Grampian House, 65 Union Row, Aberdeen AB1 1SA, UK*

ABSTRACT: In the last decade, and particularly since 1988, significant effort and expenditure has been applied in the offshore sector to the reduction of risk. This risk has been measured in terms of fatilities and injuries to the workforce and the effectiveness of the investment to reduce risk has been assessed on the basis of potential lives saved or fatalities averted. Regulations, notably SI2015, have demanded that risks to personnel be reduced to levels that are 'as low as is reasonably practicable' or ALARP. The regulations state that operators must demonstrate that the risks have been reduced to ALARP levels and advise that this can be done using cost-benefit analysis. ALARP levels are achieved, they state, when the cost of further risk reduction is 'grossly disproportionate' to the benefits achieved. Although the ALARP criterion has caused confusion it has provided a sound philosophical basis for changing the nature of regulations from prescriptive to goal-setting. It could be inferred from this criterion, also, that an underlying principle for regulation could be that the goals/objectives of the regulators, operators and asset owners are entirely consistent. The authors of the paper believe that this concept can be applied to the overall management of risk of an organization. Oil field development and insurance purchase strategies can be cited as examples of how the techniques can be applied. The paper will seek to show how risk, decision-making and asset management can be integrated under an 'ALARP' style concept.

KEYWORDS: *risk, management, safety, decision-making*

INTRODUCTION

With ever-increasing pressures on all industries, not least the offshore sector, to reduce costs and improve profitability, it is more important than ever to adopt the best methods available for achieving this while remaining within constraints of current safety legislation. This is all set against a background of conflicting pressures. Safety requirements are becoming more stringent, exploration is being forced into deeper, more inhospitable waters, oil prices remain low compared to the levels of the early 1980s and insurance premiums, a major component of the opex, are increasing out of all proportion to other costs.

There are many initiatives to reduce costs, both within organizations and across the industry. One of the latest is the CRINE (cost reduction in a new era) initiative of the UK Offshore Operators' Association. Laudable though this is, it does not get to the root of the problem. It is not cost **reduction** *per se* which should be sought, but cost **effectiveness**. Many things can be done more cheaply, but if at the same time safety, availability or reliability is compromised then the apparent cost saving could back-fire in the form of unacceptable downtime, accident clean-up and asset damage costs and business interruption costs.

This paper will describe a decision framework which can be utilized to assist operators in allocating their financial resources in the most optimal fashion between risk prevention measures and risk financing measures. Risk, in this context, constituting the whole spectrum of business risk affecting the venture, safety and personnel risk simply being a subset of those broader risks.

In addition to the more obvious risks affecting offshore installations such as the major hazards of fire and blast scenarios, there are the normal business risks such as whether or not to develop a new field, what type of production system to use, how much capacity to build into the system to meet unknown future recovery rates, whether to upgrade a field with, say, an injection platform. In addition the financial arm of the corporation have to make decisions in the face of uncertainty in the money markets. How should excess capital be invested in the light of risk in the stock market *vis-a-vis* the risks associated with a specific project.

All of these situations have a common thread. The organization is faced with making asset management decisions (assets being both capital and hardware) while facing uncertainty as to future outcomes (the price of oil, the recoverable reserves, the price of Yen against Sterling). Risk, decision-making and asset management is then a theme which underpins the whole of business life, and any decision support tools that may help optimize these crucial decisions must be worthy of investigation.

From K. Glennie & A. Hurst (eds), 1996, *AD1995: NW Europe's Hydrocarbon Industry*, Geological Society, London, pp. 209–213

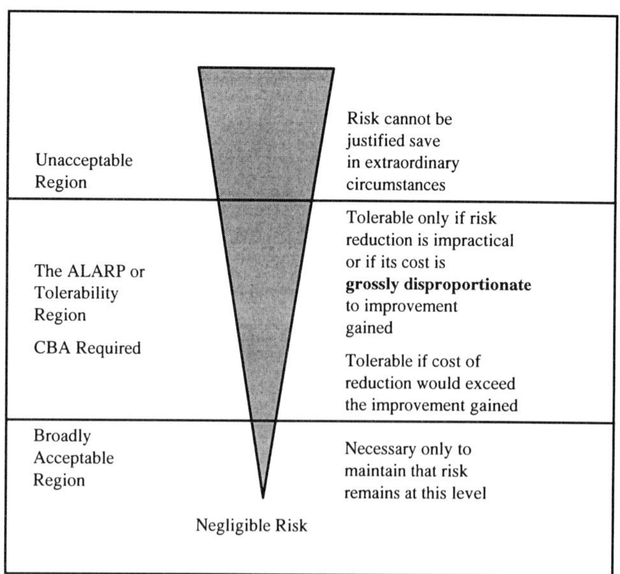

Fig. 1. Schematic representation of the ALARP principle.

THE TREATMENT OF RISK

Whatever the form of risks being faced, there are many ways in which operators can manage those risks. They can protect themselves from the potential consequences or they can make provision against those consequences occurring. Means of protection can be further subdivided into prevention, control, mitigation and avoidance (avoidance could include opting to use a Not Normally Manned Installation). Making provision (usually financial) against the consequences usually means insuring against the risk. They could also, of course, choose not to do any of these and simply carry the risk themselves. The various means of managing risk can be categorized as follows:

- Avoidance
- Prevention
- Control
- Mitigation
- Transfer
- Retention

If financial constraints were our only concern we could begin now to discuss the optimization problem, namely how can we best allocate finance between the above six mechanisms so as to optimize the long-term profitability for the operator. Before we look at that however, we must remember the safety requirements which must be the overriding design constraints.

In the UK, and increasingly in many other parts of the world, personnel safety is being stipulated within the As Low As Reasonably Practicable (ALARP) principle. A maximum level of individual risk is defined (typically 1 in 1000 chance of death per year per individual). This is the **intolerable** level and any design which, following an appropriate risk assessment, exceeded this level would be unacceptable. In addition, however, even when risks are below the intolerable level they must be further reduced to levels which are ALARP. This is achieved through cost benefit analysis, where modifications are considered for reducing levels of risk and must be implemented unless it can be shown that the cost would be **grossly disproportionate** to the levels of benefit produced. Figure 1 shows how different levels of risk must be treated under the ALARP principle.

Of the risk management mechanisms discussed earlier, the first four – prevention, control, mitigation and avoidance – are all effective in reducing personnel risk (in addition, of course to reducing the risk of asset damage and lost production). The latter two, however – transfer (insurance) and retention – are merely financing mechanisms (retention usually takes the form of using a captive insurer or occasionally a contingency fund). Figure 2 shows schematically how the six mechanisms fit into the ALARP safety requirements. The first four mechanisms must be used to 'design out' risk to achieve an 'ALARP' design. The 'residual risks' must be 'ALARP'. The problem which must be addressed is how can we use all six mechanisms in an optimum manner to maximize long-term profitability, within the constraints of the residual risks meeting the ALARP criteria.

The answer lies in the use of mutli-attribute decision theory or MADT which is a form of life cycle cost (LCC) minimization approach only more sophisticated. In fact, LCC can almost be thought of as a subset of MADT. The following section explains how cost benefit analysis is currently applied, in increasing levels of refinement, MADT being a sophisticated approach to cost benefit analysis (CBA).

COST BENEFIT ANALYSIS

At its simplest level cost benefit analysis (CBA) is based on a simple comparison between the 'cost' of a design upgrade option and 'benefit' accrued from the upgrade. The cost is usually expressed as the discounted equivalent annual cost and the benefit is usually the reduction in probable loss of life. The ratio of the two is sometimes called the implied cost of an averted fatality or ICAF.

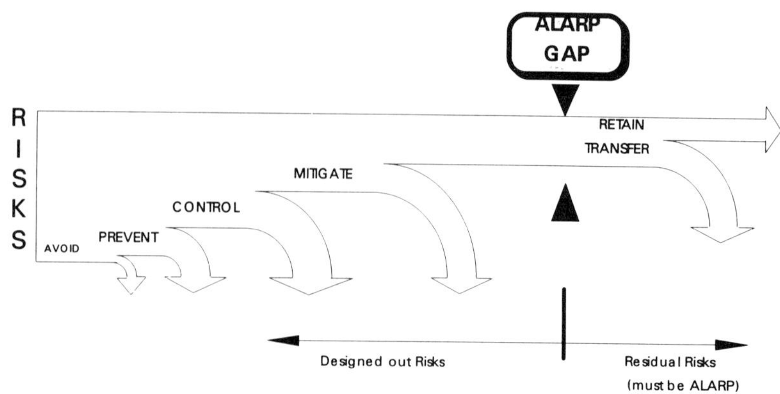

Fig. 2. Schematic representation of risk control measures and ALARP.

$$ICAF_C = \frac{C - \Delta O}{\Delta PLL.K}, \quad (1)$$

where C is discounted annual cost of upgrade; ΔO is reduction in operating costs (e.g. probable asset damage costs, insurance premiums etc.); PLL is probable loss of life (per annum);

PLL is $\sum_{i=1}^{N} f_i \, n_n$ = sum over all *N* scenarios of frequency times number of fatalities.

ΔPLL is PLL $_{\text{before modification}}$ − PLL $_{\text{after modification}}$. The subscript 'c' on $ICAF_C$ devotes the **calculated** ICAF as opposed to some stipulated **allowable** ICAF denoted $ICAF_A$.

K is known as the gross disproportionality factor referred to earlier. It is used to artificially deflate the $ICAF_C$ in recognition of a number of points, although there appears to be little consensus as to the true rational for including K. The reasons given include:

- to allow for omission of other benefits over and above the reduction in PLL, namely reduced business interruption and asset damage (ΔO is not always included in the calculation);
- to allow for risk aversion – PLL as defined above makes no allowance for the aversion generally felt for multi-fatality accidents – ΔPLL is therefore a lower bound on the benefits;
- to allow for uncertainty in the risk assessment calculation.

Contemporary interpretation of UK safety legislation holds that the gross disproportionality factor should be greater in cases when risk levels are closer to the intolerable level.

MULTI-ATTRIBUTE DECISION THEORY

As mentioned above, multi-attribute decision theory (MADT) is a form of life cycle costing. In LCC, average or 'expected' life cycle cost is calculated. This can include known fixed costs but more importantly it is made up of the cost of each credible scenario multiplied by the probability of that scenario occurring. Two important modifications are made to the LCC approach to arrive at MADT. Firstly, instead of calculating the expected value of **cost** we calculate the expected value of the **utility** of that cost. The utility is a measure of the effect of a particular cost on an organization. This is not necessarily a linear function of actual cost. The second modification is that all of the consequences of each scenario are combined into a single **multi-attribute** utility function. $U(c_i, f_i)$. $U(c_i, f_i)$ represents a function of both cost 'c_i' and fatalities 'f_i' incurred in hazard scenario '*i*'. The concept can be further extended to include other consequences such as environmental damage (French 1988; Keeney & Raiffa 1993).

Figure 3 shows a typical multi-attribute utility function for costs and fatalities. As can be seen the utility is at a maximum for zero cost and zero fatalities. This represents trouble-free operation of the installation throughout the year. Each of the hazard scenarios identified in the quantified risk analysis (QRA) would plot on this surface at some value of cost and fatalities. Taken together with its probability and summed over all scenarios gives the **expected utility**. The expected utility function is calculated both with and without a proposed modification in place and the decision – whether or not to implement the modification – is based on the highest expected utility.

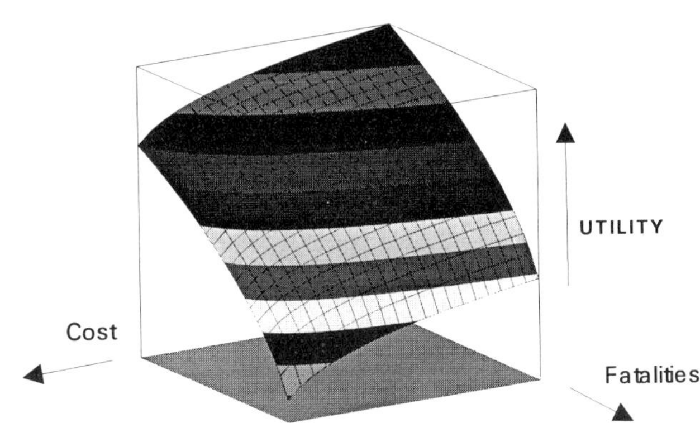

Fig. 3. Utility curve as a function of both cost and fatalities.

The decision framework described here avoids most of the shortcomings which effect the simple CBA approach. All the attributes (measures of consequences) of each scenario are included in the utility function (although of course they could, if so desired, have been included in the CBA). Risk aversion to multiple fatalities is implicitly incorporated into the utility function through its concave shape (Fig. 3). The surface becomes increasingly steeper for higher number of fatalities representing a greater aversion to multiple deaths. Risk aversion to large costs is modelled in a similar fashion.

The methods described in this section are under further development and being applied to several areas of risk and asset management in industries affected by major hazards. One of the most fruitful areas currently being developed is that of insurance strategy optimization. The optimum level of insurance to purchase and the equitable level of risk premium to pay are being determined through MADT techniques.

APPLICATION TO EXPLORATION RISKS

Decision trees, as described above, can be used to make crucial exploration decisions. The decision nodes would be – do we drill at all?; if so, how many exploration wells and, what pattern of drilling is the optimum strategy? (balancing chance of a successful find against cost of operation if wells are dry). The probabilities needed for a decision of this type cannot be estimated so easily as in the case where quantified risk assessment (QRA) data is available. They can be estimated, however, through a combination of expert judgement and historical data (e.g. hit rates for different types of geological prospects). Uncertainties which need to be modelled clearly include the size of the reserves which may be found, if any, but also include the future price of oil, tax legislation, operating costs, safety legislation, environmental constraints etc.

Much of the foregoing has been applied by exploration companies for many years, there is nothing new in estimating expected returns on risky ventures. A key element not commonly applied, however, is the use of utility and value functions as opposed to raw costs at the end of the decision tree branches. Rational treatment of uncertainty surrounding the estimated costs and probabilities is also often inadequately addressed. It is important using a decision support tool not just to look at the first answer which drops out and take that as the recommended decision. Sensitivity and uncertainty analyses should be performed to determine whether the optimum decision is particularly sensitive to one or more key input variables.

In addition the value of acquiring additional information should be examined. For example, the exploration decision is not limited to 'drill' or 'not drill' but also includes 'obtain more data first', e.g. more seismic information (French 1988). Such 'value of information' analyses can be performed within the framework of the decision analyses methodology proposed herein.

APPLICATION TO PURCHASE OF MAJOR HAZARDS INSURANCE

The management of risk in those industries exposed to major hazards involves complex issues. These span many disciplines from engineering through to economics. It is rare however to find an integrated approach to risk management in these industries. Typically the safety management, engineering design and operational personnel will co-operate, (either willingly or through the necessity to produce regulatory safety documentation) in order to (attempt to) optimize the design of the facility from a cost-benefit point of view. The costs are usually immediate costs and, where operating costs are included they rarely consider the insurance implications of differing design options. The risk manager on the other hand often takes over where the design team finishes and attempts to optimize the financing of the residual risks which could not be (easily) designed out. The primary options open to the risk manager are whether to self-insure or purchase market insurance and whether to opt for some form of partial cover using deductibles, coinsurance or first loss policies. The option of establishing and using a captive insurer is also a favoured mechanism in many industries, not least offshore.

In many situations the design/safety team and the risk manager operate quite independently. Occasionally, negotiations with brokers and underwriters will centre around the need to install additional safety measures into the facility. In these situations the two departments will collaborate. Such collaboration however is unlikely to achieve true optimization of corporate objectives, since a rigorous overall decision model has not been developed.

It is through the application of QRA data, in the form of probabilities and consequences of all potential scenarios, that the 'missing link' can be established between cost-effective, safe design and cost-effective risk financing. An overall life cycle cost benefit model can be established, termed an holistic model by the authors, utilizing techniques borrowed from the field of operational research and decision theory. This report develops the concepts needed to apply these techniques.

A principal reason for extending risk quantification studies to include CBA is to provide direct inputs to aid consistency in resource allocation decisions. In addition to risk parameters concerning both 'critical' individuals and population groups, such analyses should address the likelihood of occurrence of financial costs arising from accidents involving material damage both on- and off-site and production losses during shutdown periods for repairs. This may then be employed in evaluating the worth of further safety measures, where statistically expected reduction in such costs can reflect an 'insurance' value.

Although it might be posited that QRA initially, followed by CBA, can be applied to determine an optimum level of expenditure on insurance premiums, this, in reality is too simplistic a claim. CBA used to evaluate the worth of safety expenditures, may exclude other factors upon which the utility of the expenditure to the firm is assessed; the effect of whether quantitative risk calculations are over-conservative would also have a bearing upon the utility of a decision. These, coupled with the requirement that safety expenditure must be implemented on an 'ALARP' basis, add further difficulty to understanding the application of QRA to insurance premium application. These and similar issues are currently being investigated under a multi-client research study by the authors.

The prime objective of the current study is to develop techniques by which QRA-derived data (probabilities and consequences for all perceivable hazards) can be utilized to allow rational decisions to be made regarding the management of the relevant risks. In particular whether each risk should be avoided, reduced, transferred or retained, and if transferred whether that should be in full or in part (i.e. what deductible?). Optimization of this decision must also be subject to the requirement that personnel risks have been reduced firstly to a **tolerable** level and thereafter to an **ALARP** level.

One criticism often levelled at these risk-based techniques is how do we know what the risks are – what are the confidence bounds on our estimates of both probabilities and consequences? Many people would argue that these uncertainties will swamp the analysis and the only safe option left open to us is to cover ourselves against the worst credible scenario. This means designing against the worst event and insuring against the maximum credible loss.

Two points should be made here which counter such criticism. Firstly it is not the purpose of risk-based techniques to remove conservatism. That is left to the specialists in each technology. The purpose of risk-based techniques is to understand the uncertainties affecting a decision and treat them rationally. Secondly a key element in these techniques is the derivation of a decision-maker's attitude to risk (risk averse or risk prone) and the 'value or utility function' (Fig. 3), which expresses severity of different consequences to an organization. In fact, it is only through concepts such as these that a truly rational approach to risk and asset management can be achieved. Adopting a policy of full insurance cover or any level of cover, without regard to the likelihood of different accident scenarios is highly irrational.

Figure 4 shows how a decision tree approach can be utilized to determine the optimum decision regarding insurance purchase. A safety upgrade system is being considered. This may or may not have a bearing on insurance premiums, it will, however, reduce consequences and/or probabilities associated with accident events and thereby reduce the expected cost of operation.

The decision tree shows a square box denoting a decision node, i.e. whether or not to install the new system. This is followed by a second decision node on what level of insurance to purchase. Finally, there is a chance node which represents the probabilities of all credible outcomes. The end point of each branch represents the value of that sequence of decisions and outcomes. More correctly, the utility of that outcome should be entered here from Fig. 3. The decision tree is evaluated from right to left. At chance nodes the expected utility of all the chance outcomes is evaluated (i.e. probability × utility of outcome). At decision nodes the decision leading to the highest expected utility is chosen. The optimum decision in Fig. 4 is shown to be 'install the system and do not purchase insurance'.

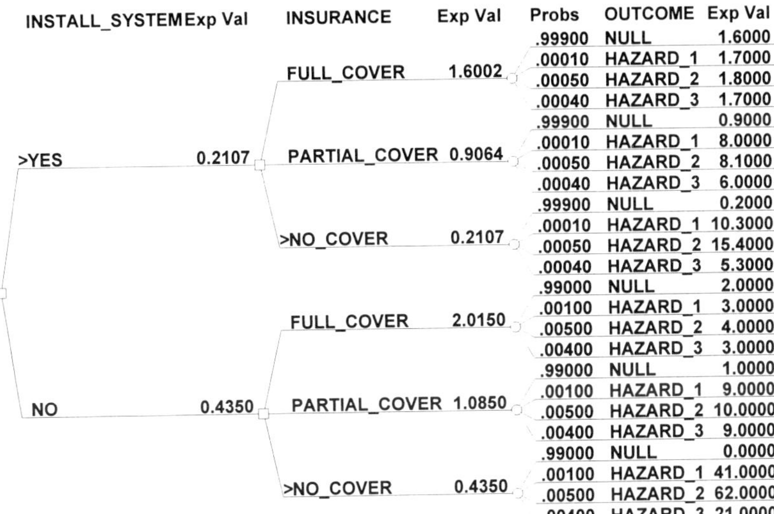

Fig.4. Decision tree analysis of insurance optimization.

CONCLUSIONS

This paper has discussed the various sources of risk, both technical and commercial, that affect organizations. Although written from the point of view of the offshore industry the concepts described are applicable to all branches of industry and business sectors. The overriding tenet of the paper is that all decisions which are made in the face of uncertainty are most rationally dealt with using a framework which takes cognisance of that uncertainty and the effect of potential outcomes on the organization. The non-linear relationship between actual costs (and revenues) and the 'effect on' or 'value to' the organization should be a key part of the decision model. Probabilities associated with potential outcomes must be estimated as best as possible. Where QRA or actuarial data is available, so much the better. Where this is not the case, subjective probabilities should be used. Where there is a perceived need to be prudent with a particular decision (i.e. err on the safe side) this is best dealt with through the value function rather than by arbitrarily inflating estimates of probabilities or assuming worst case outcomes whenever a chance outcome is faced.

Good decisions don't guarantee good outcomes. They do, however, guarantee that in the long run the profitability of an organization is as optimized as possible within the constraints of the information available.

In areas of decision-making, such as major hazards insurance where prudence has always been a hallmark of the risk manager and full cover is the normal decision, it may transpire that such decisions are not necessarily optimal. In such cases part of the money previously earmarked for insurance purchase could be reallocated to extra safety hardware leading to multi-million pound savings, with an added bonus of increased personnel safety into the bargain.

REFERENCES

KEENEY, R. L. & RAIFFA, H. 1993. *Decision with Multiple Objectives–Preferences and Value Trade-Offs*, Cambridge University Press, Cambridge.

FRENCH, S. 1988. *Decision Theory–An Introduction to the Mathematics of Rationality.* Ellis Horwood, Chichester.

Prospective developments, production and revenues from the UKCS 1995–2000: a financial and regional simulation

A. G. Kemp and L. Stephen

University of Aberdeen, Department of Economics, Edward Wright Building, Dunbar Street, Aberdeen AB9 2TY, UK

ABSTRACT: This paper examines the development and production prospects for oil and gas from the UKCS in the period 1995–2000 using a financial simulation model applied to a large database. Emphasis is given to the prospects in the different regions of the UKCS. Oil production is likely to peak in the period and decline slowly thereafter. Gas production will increase substantially throughout the period to 2000. The relative importance of the central North Sea as a producing region will increase substantially in the period, while the northern North Sea will show a decrease. Investment patterns will reflect these trends. Revenues from the UKCS will continue to make a substantial contribution to the economy. The UK should be more than self sufficient in oil to beyond 2000. Potential gas production should also exceed UK gas demand well beyond 2000.

KEYWORDS: *financial simulation, production, investment, revenues, cost savings*

INTRODUCTION

Activity levels in the UKCS have been subject to several changing influences in recent years. The oil price has fluctuated considerably with the average value remaining far below the real levels experienced in the period 1974–1985. There are major uncertainties facing both supply and demand, with the former generally being the more difficult to predict. OPEC strategy and cohesion are uncertain. The prospects for Russian production and Iraq being allowed to export freely on to the world market are also highly uncertain.

On the demand side it is not clear to what extent consumption is responding to the relatively low crude oil prices. There is some evidence, especially in advanced countries, that the price elasticity of demand for petroleum products is not symmetrical. The negative responsiveness of consumption to 'high' oil prices is not producing a corresponding positive response in consumption when oil prices are 'low'. Improvements in energy efficiency and conservation which were developed when oil prices were 'high' are unlikely to be reversed when 'low' prices prevail. A further complication is the enhanced use of excise duties on petroleum products by governments – especially in Western Europe. This is producing a larger wedge between crude oil prices and product prices. Product prices may not appear to be especially 'low' and in the UK, for example, prices of key products are currently increasing in real terms.

The uncertainties in the oil market are even more uncertain on the supply side. Non-OPEC production as a whole continues to be more buoyant than was thought likely a few years ago. This reflects a combination of the fruits of past geological successes in some countries and the technological progress in the industry which has resulted in new projects becoming more attractive to investors. The consequence has been that instead of falling in recent years as was widely predicted, aggregate non-OPEC production has grown. This is in spite of substantial falls in some countries, especially the FSU and USA. Non-OPEC production may continue to increase in the near future. In the medium-term much depends on whether the large output falls in Russia can be reversed. This is clearly possible on the basis of proven reserves, but the continued political uncertainty and the associated inadequate investment leave the production prospects unclear.

Within OPEC the continued uncertainty about supplies from Iraq has been a dominant feature for some time. This has been a negative influence on the world price. To some extent, therefore, the effects of the removal of the UN sanctions are incorporated in present oil prices. The actual volumes which Iraq can put on to the world market are not so clear. This depends on both production and export capacities. The reaction of OPEC in terms of strategy to the re-emergence of Iraq as a major exporter is also unclear. The effect on the oil price of full Iraqi exports will depend critically upon the extent to which OPEC is able to redetermine quotas in a manner which is perceived as credible by the oil traders. The removal of one of the market uncertainties has a positive effect as well.

In recent years the effective surplus of OPEC capacity has been quite low compared to the position in the 1980s. This by itself has a positive effect on price. The phenomenon has also encouraged the view that in the medium term there is likely to be a medium-term shortage of productive capacity with the consequence that oil prices will rise very substantially. Globally the industry is still investing substantially, and the willingness to invest is clearly still obvious, given the appropriate encouragement. On this basis a hike in prices from the operation of normal market forces is not very likely. The intervention of political factors can overwhelm

From K. Glennie & A. Hurst (eds), 1996, *AD1995: NW Europe's Hydrocarbon Industry*, Geological Society, London, pp. 215–225

these normal forces and can inhibit long-term investments. This could lead to unpredictable price hikes as have occurred in the past.

The net effect of all the factors discussed above is for an oil price outlook subject to considerable volatility depending in large part on the influence of events relating to OPEC behaviour. The fundamentals of supply and demand point to no obvious upward trend in real oil prices for some years.

A recent feature of activity in the UKCS has been dramatic improvements in the application of new technology. Major examples are 3D seismic and horizontal drilling. Wider application of these has enabled projects to proceed at considerably lower cost. Another notable recent feature has been the CRINE initiative (cost reduction initiative for the new era). This, especially in conjunction with the application of new technology, has already produced significant cost savings, particularly in the development costs of new fields. In turn this has influenced the volume of new activity. The industry is hoping that CRINE will continue for the foreseeable future and produce further benefits. Currently there is much emphasis on reducing operating costs with many platforms undergoing substantial demanning.

Activity levels have been affected by the enhanced employment of the existing infrastructure of pipelines and platforms to facilitate the development of new fields. This has made more projects viable and, in some cases, accelerated the cycle time between discovery and development. In the 1980s, cycle times on average increased, but currently there are notable examples of some remarkably fast lead times.

Discovery rates have continued to be very satisfactory by international standards. The overall success rate has held up remarkably well given the maturity of much of the acreage. The reserves discovered per well drilled have, of course, fallen very substantially in the 1980s compared to the 1970s. Interestingly, however, this fall has been arrested in recent years with the reserves discovered per exploration and appraisal well being relatively constant at around 10×10^6 bbl oil equivalent. Recent successes West of Shetland are likely to cause this to increase in the near future.

This paper examines the possible levels of development and production activity over the period 1995–2000. Recent years have seen a change in the geographic emphasis of exploration activity in the UKCS. There has been an increased interest in the Irish Sea and West of Shetland. This paper puts these new developments in perspective by providing a regional analysis of prospective activity in 5 distinct areas, namely the Southern Gas Basin, central North Sea, northern North Sea, West of Shetland and the Irish Sea.

METHODOLOGY AND ASSUMPTIONS

The analysis was undertaken using a large computerized financial model developed at the University of Aberdeen to simulate future activity levels in the UKCS. Primary inputs into the model include all the publicly available information on currently producing fields relating to their historic and expected production rates, investment costs, operating costs and abandonment costs. From a variety of sources, information has been gathered on all new discoveries which have not yet been developed, or even fully appraised in some cases. Estimates were made of the time periods at which such fields would be ready for development work and production to commence. For these future fields, estimates were made of the likely investment costs, operating costs and abandonment costs. In making such calculations some judgements were made regarding the technological progress that will be required, and the cost assumptions have, to an extent, assumed that such progress will, in fact, be made. However, in other respects the bank of future fields can be described as conservative as it does not include any new discoveries from further exploration successes.

With regard to field developments the distinction is made between probable and possible developments. Probable developments are defined as those which are currently being appraised, or for which reasonably accurate estimates of reserves are available, and development seems likely in the next few years. Possible developments refer to fields where the reserves estimates are not well defined at present and where some uncertainty exists regarding possible development concept, timing, and production profiles. In these cases a best estimate of recoverable reserves and development expenditure has been employed.

Key inputs into the financial model are assumptions regarding future oil and gas prices. Three price scenarios have been chosen. The low one has an oil price of $14 per barrel in constant real terms and a gas price of 15 pence per therm in constant real terms. The medium oil price is $18 per barrel in constant real terms and the gas price 19 pence per therm, also constant in real terms. The high oil price scenario starts at $20 per barrel in 1995 and rises linearly to $25 in real terms in 2000 and $28 in 2005. The corresponding gas price is 20 pence in 1995 and rises to 25 pence in real terms by 2000 and 28 pence, again in real terms, in 2005. Exceptions to these gas prices relate to contracts signed in the 1960s and 1970s. These generally contain prices well below those applicable to new contracts. For these 'old' fields, contract prices which reflect the (estimated) values have been chosen as the basis of projections.

The general inflation rate assumed averages 3.5% per year. There is thus a considerable increase in prices in money-of-the-day terms. The price scenarios adopted are not forecasts. They have been chosen to examine the effects of a range of realistic price prospects over the next 5 years.

The financial simulation model operates by calculating the pre-tax and post-tax returns to the projects under the different oil and gas price scenarios. When a post-tax real rate of return of at least 10% (13.5% in money-of-the-day terms) is achievable, the project is deemed to be acceptable. If this minimum return is not attainable the field's development is postponed until such time as this return is achievable. Another scenario when the threshold rate of return was 20% in real terms (*c.* 23.5% in MOD terms) was also examined.

It was assumed in the modelling that fields for which Annex B approval had already been obtained would be developed irrespective of the effects of lower oil prices. For all other fields it was assumed that their development would depend upon the prospective attainment of the specified minimum rate of return. This statement applies to fields where development is currently being contemplated.

It should be emphasized that, with respect to gas, the development of a new field is presumed to be triggered when a post-tax return at least equal to the threshold rate of return is in prospect. It is presumed that contracts for the sale of the gas can be concluded. In the present market situation in the UK, the time is soon approaching when the production potential from the UK sector of the UKCS will exceed the demand for gas in the UK. The timing of the development of new gas fields will depend upon the availability of new export markets as well as contracts in the UK. Difficulties and possible delays with respect to marketing have not been examined further here.

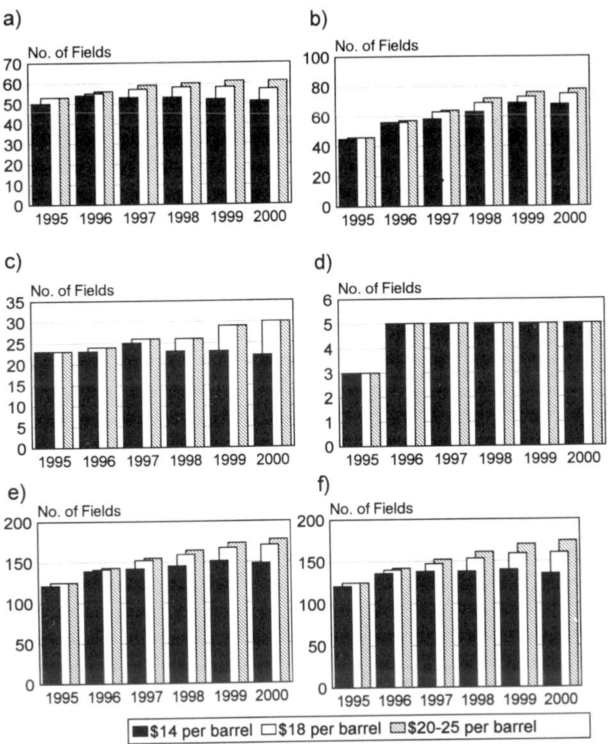

Fig. 1. Total number of fields: (**a**) southern North Sea, Threshold IRR 10% (real); (**b**) central North Sea, Threshold IRR 10% (real); (**c**) northern North Sea, Threshold IRR 10% (real); (**d**) Irish Sea, Threshold IRR 10% (real); (**e**) UKCS, Threshold IRR 10% (real); (**f**) UKCS, Threshold IRR 20% (real).

RESULTS

Numbers of producing fields

The financial model calculates the prospective post-tax returns to new fields which become available for development over the period. It also calculates the period at which continued production from existing fields is no longer economically viable. When these calculations are combined the total numbers of fields in production are obtained. This exercise was conducted for the 5 regions distinguished in the study. Each is discussed in turn.

(1) Southern North Sea

In Fig. 1a the prospective position in the southern North Sea is shown. Under the 19 pence gas price case the total number of fields in production increases from 53 in 1995 to 58 in 1998 and then falls to 57 in 2000. Under the high gas price scenario, the number increases throughout the decade from 53 in 1995 to 61 in 2000. The higher price triggers off the development of more new fields and postpones the abandonment date of existing ones. Under the 15 pence price scenario the number increases from 50 in 1995 to 54 in 1996 but thereafter falls back to 51 in 2000.

The results indicate the considerable price sensitivity of activity even over a relatively short period. By 2000 the variation in the number of producing fields is as much as ten. The bank of discovered fields in the southern basin is substantial, but under the price scenarios examined prospective returns are generally moderate with a noticeable number of marginal prospects.

(2) Central North Sea

In Fig. 1b the possible numbers of new fields in the central North Sea are shown. It is seen that under all price scenarios the numbers increase substantially over the period. Under the $18 price case the numbers rise from 46 in 1995 to 75 in 2000. This reflects a very brisk pace of development over the next few years. Under the high price case the number reaches 78 in 2000, while under the $14 case the corresponding number is 68. The sensitivity of the pace of new developments to oil price behaviour is thus very substantial. Under the $14 price there is actually a reduction in the number of producing fields between 1999 and 2000 reflecting the abandonment of fields as well as a lower number of new developments.

(3) Northern North Sea

The results for the northern North Sea are shown in Fig. 1c. Under the $18 price case the number of fields in production increases from 23 in 1995 to 30 in 2000. With the high price case the number reaches 30, but under the $14 case there are only 22 producing fields in that year. This is less than the number in 1995. The results indicate not only the sensitivity of activity in the East Shetland basins to oil price behaviour but also the much lower expected number of new projects, even under high oil prices

(4) West of Shetland

In the West of Shetland region under all price scenarios, two fields are confidently expected to be in production before 2000, with the first starting in 1996. Under the $18 and high price cases a third will be developed and should come on stream by 2000. It is possible that other developments will be under way but production will not commence until after 2000.

(5) Irish Sea

In Fig. 1d it is seen that there will be 5 fields on stream in the Irish Sea by 1996 under all the price scenarios shown. These fields will all still be producing in 2000. Other prospects exist but production may not take place until after 2000.

(6) Aggregate numbers of fields in production

The resulting aggregate numbers of fields in all sectors are shown in Fig. 1e. The total number increases substantially from 1995 levels to 177 under the high price, 170 under the $18 case, and 148 under the $14 scenario. Under the last scenario, the number in 2000 falls from its 1999 level. The case discussed so far is one where the threshold return is 10% in real terms (*c.* 13.5% in MOD terms). With relatively low oil prices capital rationing becomes a more pressing problem. Investors may react to this by increasing their threshold rate of return. The results of employing a rate of 20% are shown in (Fig. 1f). The differences compared to the base case become significant by 2000. In that year there are 174, 160, and 135 fields in production under the three price scenarios. The difference is obviously greater under the low price case, mainly because the lower level of profitability results in more new fields becoming uneconomic at the higher discount rate.

Production trends

Prospective production trends under the three price scenarios are now considered. These trends reflect the combined effects of new developments and the continued production from existing ones. The latter calculation incorporates the abandonment of some fields where they pass the economic cut-off point. Each region is considered in turn.

(1) Southern North Sea

In the southern North Sea gas production is expected to reach a peak in 1995 and gradually decline thereafter (Fig. 2). Production could exceed 1.6×10^{12} standard cubic feet (scf) in 1995 and under the 19 pence price case fall to 1.37×10^{12} scf by 2000. Under the high price case, output could be 1.4×10^{12} scf in 2000 and 1.27×10^{12} scf under the 15 pence case. From Fig. 2 it is seen that the fall only becomes significant from 1998 onwards.

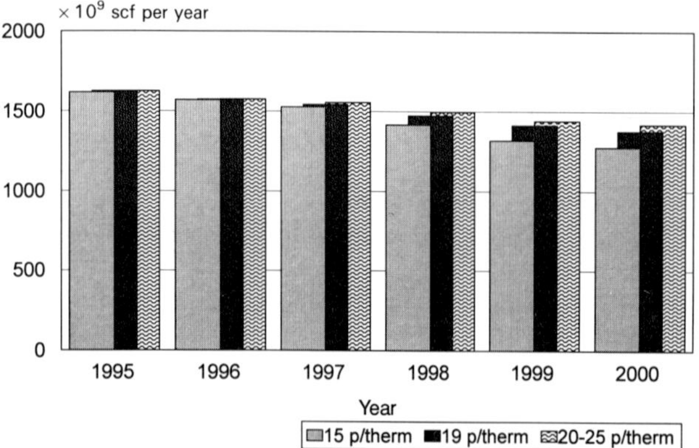

Fig. 2. Total gas production, southern North Sea, Threshold IRR 10% (real).

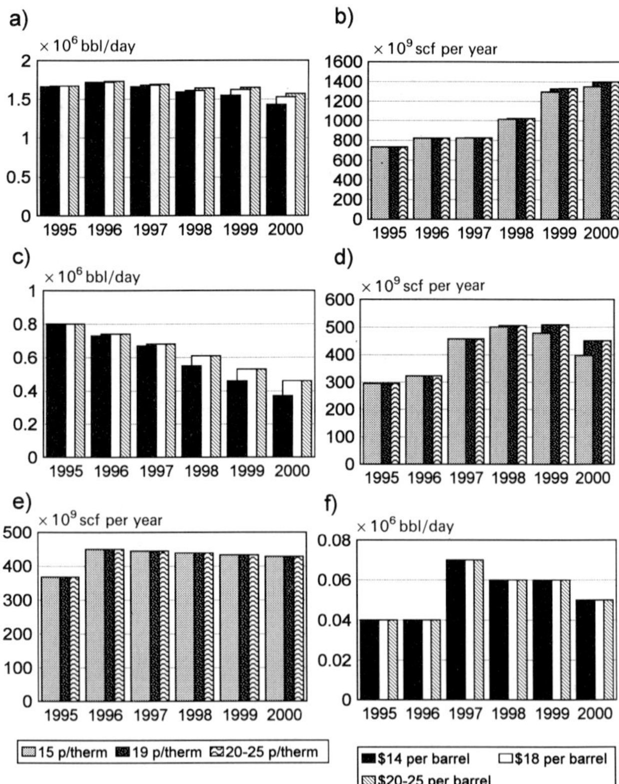

Fig. 3. Total production: **(a)** oil production, central North Sea, Threshold IRR 10% (real); **(b)** gas production, central North Sea, Threshold IRR 10% (real); **(c)** oil production, northern North Sea, Threshold IRR 10% (real); **(d)** gas production, northern North Sea, Threshold IRR 10% (real); **(e)** gas production, Irish Sea, Threshold IRR 10% (real); **(f)** oil production, Irish Sea, Threshold IRR 10% (real).

(2) Central North Sea

In the central North sea oil and gas are now both important and it is useful to distinguish between them. In Fig. 3a oil production prospects are shown. It is seen that output may increase to reach a peak of around 1.73×10^6 bbl/day in 1996 and fall at a gentle pace thereafter. Under the high price case it may be around 1.57×10^6 bbl/day in 2000 and 1.43×10^6 bbl/day under the $14 case. The price sensitivity is perhaps not dramatic. This reflects the dominance of output from fields which are already in production in the total picture in 2000.

In Fig. 3b the expected trends in gas production from the central North Sea are shown. There is a strong contrast with the oil prospects. It is seen that gas production from this sector may nearly double over the period 1995–2000. The estimate for 1995 is 0.73×10^{12} scf and under the 19 pence price it could reach 1.4×10^{12} scf in 2000. This means that by 2000 output from the central sector could be as large as in the Southern Basin (Fig. 2). In subsequent years central sector gas production will exceed that from the Southern Basin. The growth in output is seen to accelerate from 1998 onward with the coming on stream of large new fields. It is seen that production rises sharply under all price scenarios and the difference in 2000 between the high and low price cases is not large. This reflects the dominance in the overall total of production from existing fields plus those whose development is more or less guaranteed.

(3) Northern North Sea

In Fig. 3c oil production prospects are shown for the northern North Sea. It is seen that output falls continuously from a 1995 figure of 0.8×10^6 bbl/day. Though some new developments do occur under the two higher prices they cannot compensate for the fall in output from the existing producing fields. In 2000 output could be around 0.46×10^6 bbl/day under the 2 higher price scenarios and 0.37×10^6 bbl/day under the low price. This represents a fast pace of decline and indicates why concern has been expressed about the future of the Sullom Voe terminal.

The prospects for gas production in the northern North Sea are somewhat better (Fig. 3d). This shows that gas output will increase from nearly 0.3×10^{12} scf in 1995 to 0.5×10^{12} scf in 1998. By 2000 it will have fallen, however, to reach 0.45×10^{12} scf under the 2 higher price cases and just under 0.4×10^{12} scf under the low case.

(4) West of Shetland

The oil production prospects West of Shetland to 2000 are shown in Fig. 4. It is seen that output commences in 1996, and under the two higher price scenarios, rises to around 0.18×10^6 bbl/day in 2000. Further increases are expected in succeeding years. In addition to the three oil fields whose development is likely by 2000, several other discoveries have been made, and further oil developments are probable in the

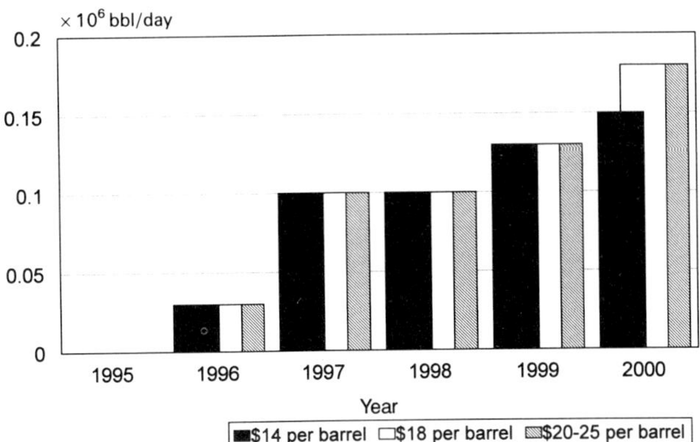

Fig. 4. Total oil production, West of Shetland, Threshold IRR 10% (real).

period after 2000. Some gas fields have been discovered, but their commercial viability requires considerably higher gas prices than presently prevail. Gas developments are thus unlikely until well beyond 2000.

(5) Irish Sea

The Irish Sea is currently receiving considerable new development activity. Gas and oil production are both set to increase. In Fig. 3e prospective gas production could rise from 0.37×10^{12} scf in 1995 to 0.45×10^{12} scf in 1996. Output at this level is nearly maintained to the year 2000. Prospective oil production is still comparatively small, but in Fig. 3f it is seen that it may reach 70 000 bbl/day in 1997, after which it falls somewhat.

(6) Total UKCS

The aggregate picture for all regions in the UKCS is now considered. In Fig. 5a, it is seen that output will reach 2.55×10^6 bbl/day in 1996 and 1997 under the high price case after which it falls gently to a level of 2.25×10^6 bbl/day in 2000. The trend under the \$18 price is not very different, but under the \$14 scenario output falls off more sharply reaching 2.01×10^6 bbl/day in 2000.

The production prospects in the scenario where the threshold rate is 20% in real terms are shown in Fig. 5b. In the near future there is no difference compared to the case with 10% threshold rate. By 2000 the difference becomes more noticeable especially under the low price scenario. Production is 1.9×10^6 bbl/day compared to 2.01×10^6 bbl/day under the 10% threshold case. This is still not a large difference. It reflects the fact that total production in 2000 is overwhelmingly dominated by output from existing fields and new ones where a development commitment has been made. For years beyond 2000 the difference in results becomes progressively greater.

Under all price scenarios there is a steady increase in gas production from a 1995 figure of around 3×10^{12} scf (Fig. 6a). By 2000 output under the 19 pence scenario is 3.7×10^{12} scf. Under the high price case production is slightly higher while under the low price it is 3.45×10^{12} scf. There is not so much price variation as with oil production.

Under 20% real threshold rate total gas production by 2000 shows only a small difference compared to the 10% threshold case (Fig. 6b), but with the low price production becomes 3.16×10^{12} scf rather than 3.45×10^{12} scf. For years beyond 2000 the difference becomes progressively greater.

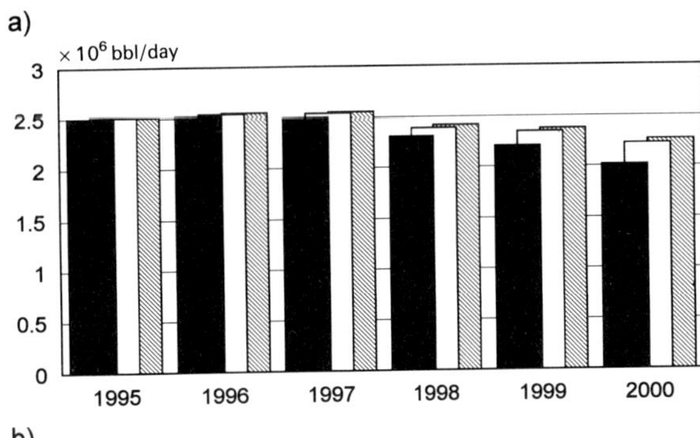

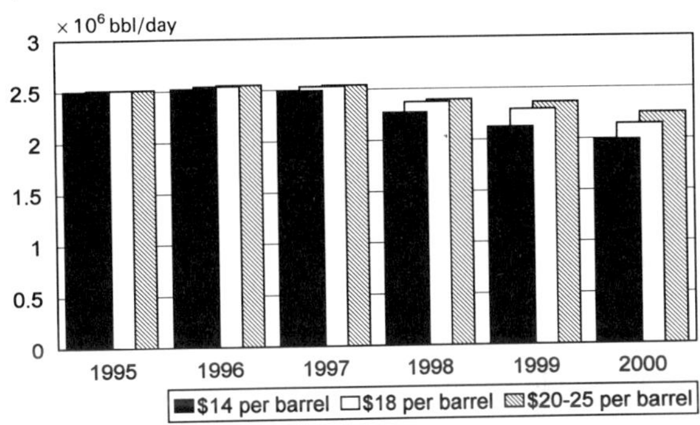

Fig. 5. Total oil production UKCS: **(a)** Threshold IRR 10% (real); **(b)** Threshold IRR 20% (real).

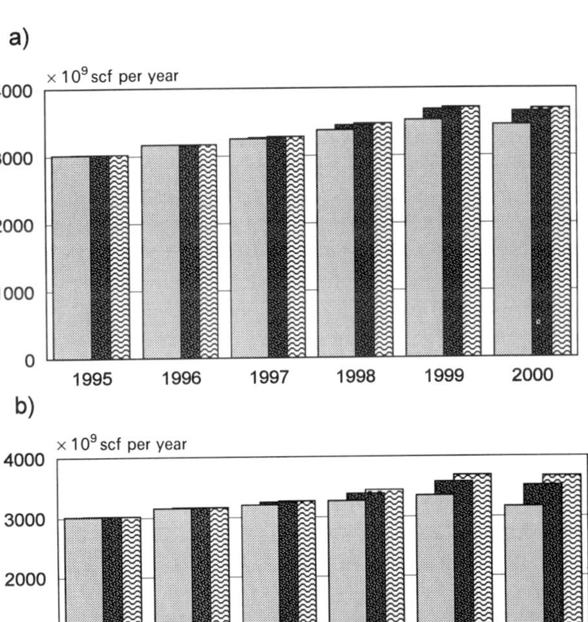

Fig. 6. Total gas production UKCS: **(a)** Threshold IRR 10% (real); **(b)** Threshold IRR 20% (real).

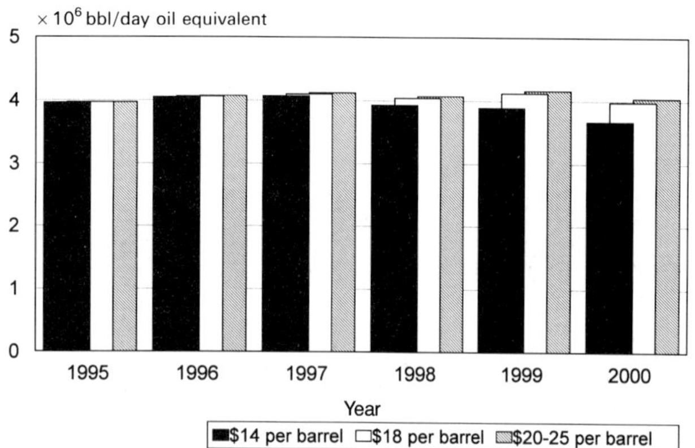

Fig. 7. Total oil and gas production, UKCS, Threshold IRR 10% (real).

In Fig. 7 total prospective production of hydrocarbons is shown under the 10% real threshold case. It is seen that under the high price case output exceeds 4×10^6 bbl/day oil equivalent through to 2000. Under the low price it falls from 1997 onwards to 3.6×10^6 in 2000. Under the 20% threshold case under the low price scenario the production is 3.4×10^6 bbl/day oil equivalent in 2000.

Development expenditure

Prospective trends in development expenditures are now considered. These are overwhelmingly on new fields but some relate to refurbishment of existing fields. It should be stressed that in this study they exclude expenditure on terminals. In Fig. 8a total development expenditure prospects for the whole of the UKCS are shown with the 10% net threshold. It is seen that, at 1995 prices, the levels exceed £4 billion in 1995 and 1996 under the medium and high price scenarios and fall to around £3 billion in 2000. Under the low price case the level falls to £2.35 billion in 2000.

Trends in development activity are quite sensitive to the choice of threshold rate (Fig. 8b). The figures are noticeably lower under the $18 and $14 price scenarios with the difference becoming apparent as early as 1996. In 2000 with the $18 case the level falls to £2.4 billion rather than £3 billion. Under the $14 scenario expenditure is only £1.7 billion compared to £2.35 billion with the 10% threshold.

Development expenditure has capital and drilling components. The expected trends in these do not follow the same pattern (Figs 8c and d). It is seen that capital expenditure falls from 1996 onwards on all price scenarios. Under the high price case it falls from over £3 billion to £2 billion in 2000 and with the low price the decrease is from £2.8 billion in 1995 to £1.4 billion in 2000. Development drilling expenditure keeps up much better over the period. Under the medium and high prices expenditure is as large in 1999 as in 1995.

The different trends reflect two factors. The first is the success of the industry in reducing the costs of producing facilities for new fields, including further use of the existing infrastructure. The second is the continued need for production drilling on all new projects, including on those where the capital costs have been reduced.

The distribution of total development expenditure among the 5 regions is shown in Fig. 9a for the $18 price. The striking feature is the continued dominance of the central North Sea. Expenditure in this sector is currently increasing in absolute terms but, even when it falls after 1998, its relative importance is still striking. The other noteworthy feature is the decrease in importance of the northern North Sea both in absolute and relative terms. By the end of the decade investment in the Southern Basin may exceed that in the northern North Sea, though it is currently very much higher in the latter sector.

Operating expenditures

Prospective aggregate operating expenditures are shown (Fig. 9b) for all sectors of the UKCS. Under the 2 higher oil prices they are seen to increase from £4.3 billion in 1995 to £4.7 billion and £4.6 billion in 2000. Under the low price they fall to £3.9 billion in 2000. The increase is due to the growth in the number of fields over the period. This has more than offset the effects of cost reductions achieved to date, and confidently expected. With the low price scenario, the lower number of fields in production results in a decrease in aggregate operating costs after 1997. The behaviour of operating costs under the 20% threshold is not very different from the 10% case. This is because the overwhelming bulk of these costs emanates from fields which are already in production or under development.

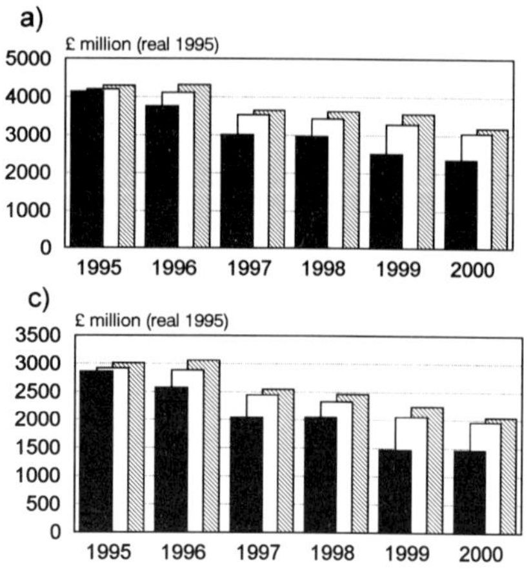

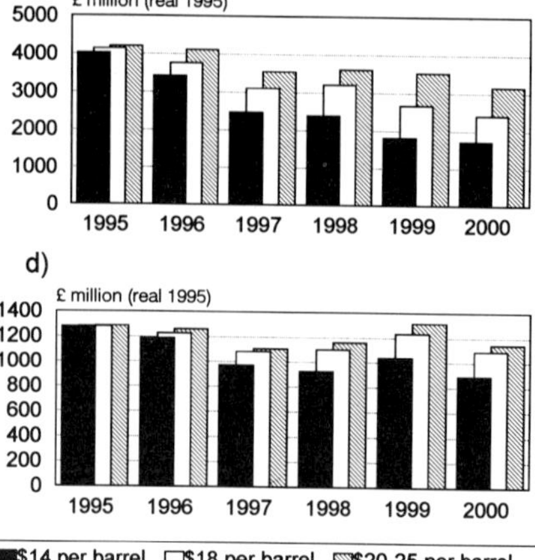

Fig. 8. Total expenditure UKCS: (**a**) total development expenditure, Threshold IRR 10% (real); (**b**) total development expenditure, Threshold IRR 20% (real); (**c**) total capital expenditure, Threshold IRR 10% (real); (**d**) total drilling expenditure, Threshold IRR 10% (real).

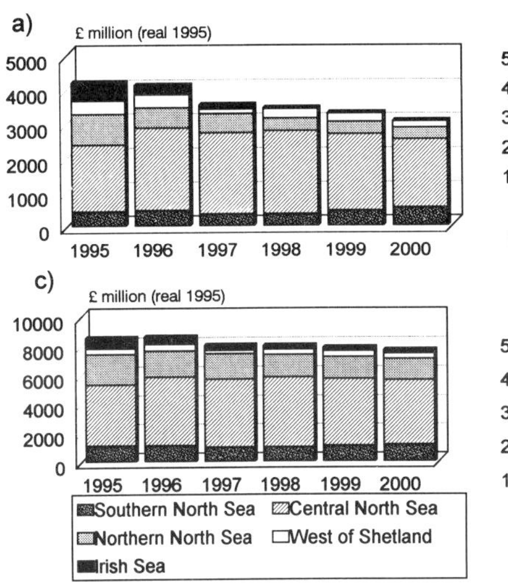

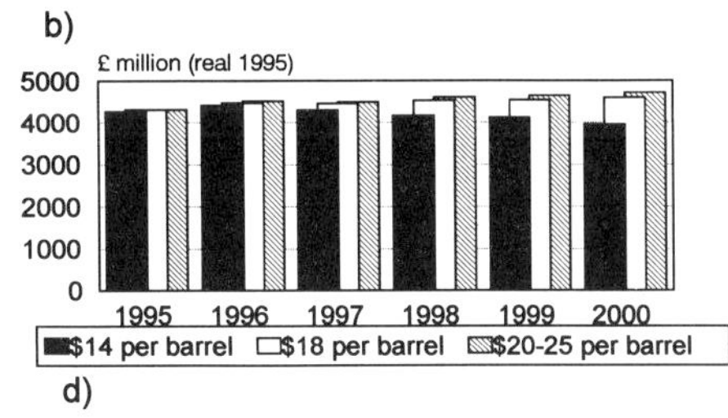

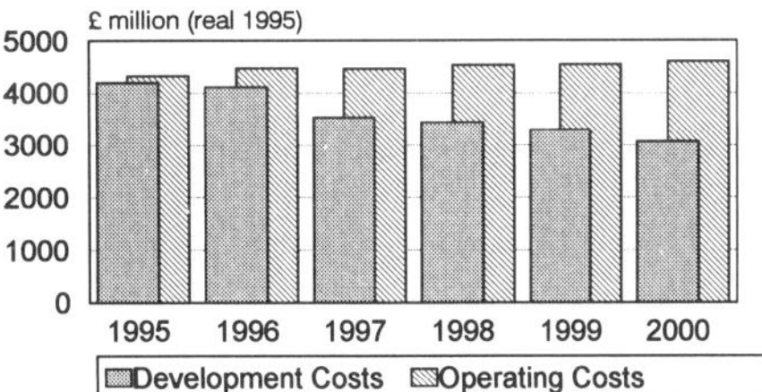

Fig. 9. Expenditure UKCS, Threshold IRR 10% (real): (**a**) development expenditure, $18 real oil price; (**b**) total operating expenditure; (**c**) total expenditure, $18 real oil price; (**d**) development and operating expenditure, $18 real oil price.

Total expenditure

The aggregate expenditures across all sectors are shown in Fig. 9c under the $18 price. The total increases in 1996 to reach almost £8.6 billion, but falls at a gentle pace thereafter. The importance of the central North Sea is reinforced. The share of total expenditure going to this sector will actually increase in the period to 2000. The decrease in the importance of the northern North Sea over the period is also noteworthy. In this case, however, it is seen that even in 2000 expenditure in that sector still exceeds the level in the Southern Basin. This is because of the much higher operating costs in the northern North Sea.

The relative importance of development and operating costs is shown for the $18 price (Fig. 9d). Historically, investment costs have exceeded operating costs. This has now changed and over the next few years operating costs will exceed investment costs by an increasing margin. In 2000, development costs may be £3 billion while operating costs are £4.6 billion.

Gross revenues

The potential gross revenues from oil and gas production are now examined. In Fig. 10a they are shown for oil and gas across all sectors under the 3 price scenarios and the 10% real threshold rate. They indicate the gross contribution of oil and gas to the balance of payments and to the national economy generally. This contribution is obviously very substantial under all the price scenarios considered. Under the $18 price case gross revenues are around £16.5 billion in 1995. In real terms they will stay at broadly this level in the period through to 2000.

The price sensitivity of the revenues is very marked. This reflects, not only the direct impact, but also the indirect effects through the influence on the pace of new field developments and the timing of field abandonment. In 1995 the difference of £5 billion between revenues under the low and high prices is solely due to the direct price effect, but in 2000 the difference of over £10 billion reflects these indirect effects on production as well. The price effect remains the stronger one.

The relative importance of oil and gas revenues is shown in

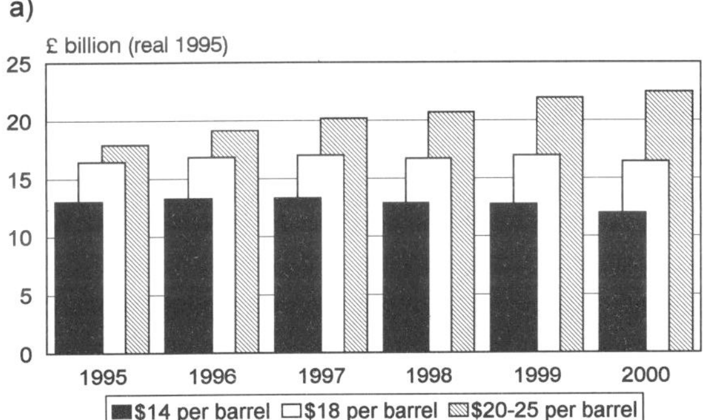

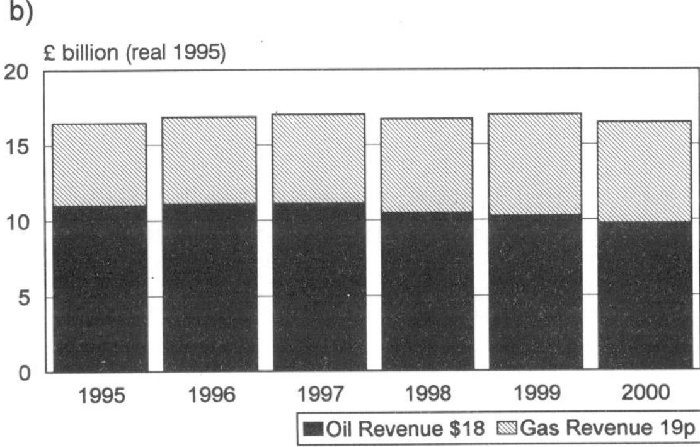

Fig. 10. Production revenues UKCS, Threshold IRR 10% (real): (**a**) gross production revenues; (**b**) oil and gas production revenues, $18 oil price.

Fig. 10b for the $18 and 19 pence price cases. Over the period to 2000 the relative contribution of gas is seen to increase. In 1995 it accounts for 33% of total revenues and in 2000 for 41%. The contribution of gas will become even more important next century.

Effects of cost savings

In recent years much attention has been given by the industry to the reduction of costs through the CRINE

a)

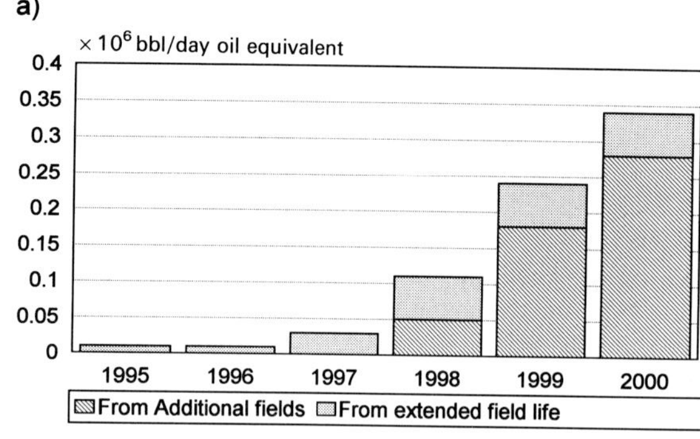

b)

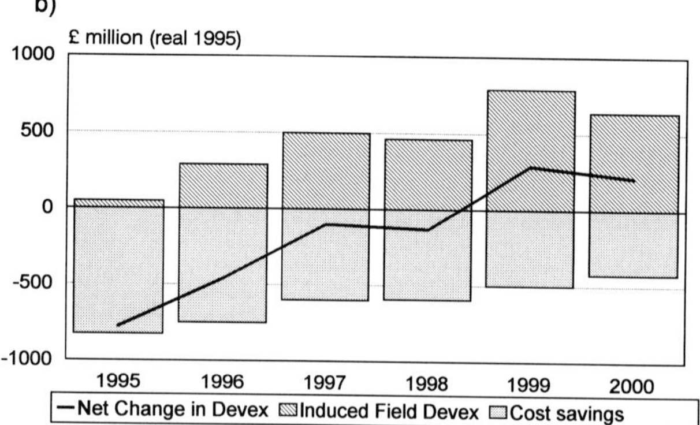

Fig. 11. Change with 20% reduction in development and operating costs, $14 real oil price: (**a**) additional oil and gas production; (**b**) additional total development expenditure.

a)

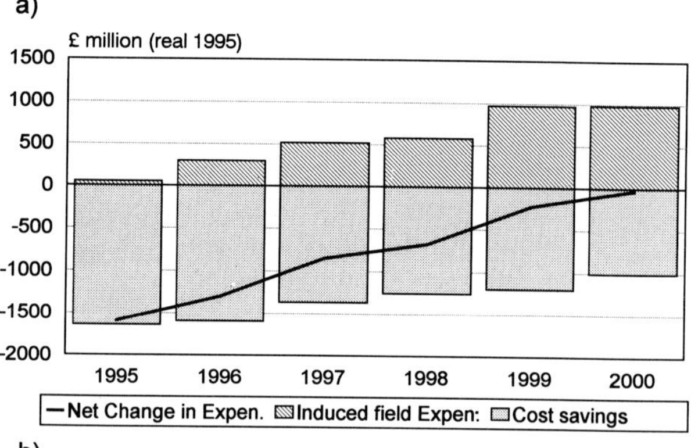

b)

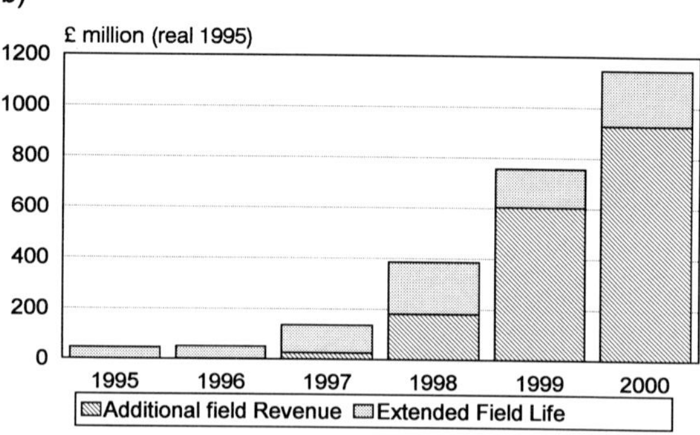

Fig. 12. Change with 20% reduction in development and operating costs, $14 real oil price: (**a**) additional total expenditure; (**b**) additional revenue.

initiative (cost reduction initiative for the new era). The consequences of further success in reducing costs are now examined. These effect (influence) activity levels in various ways. They can facilitate the development of fields which would otherwise be uneconomic. The lives of existing fields can be extended. There will thus be positive as well as negative effects on expenditure levels.

The effects of cost savings on overall activity levels will be greater under relatively low oil price conditions. This is because at higher oil prices more of the new projects would proceed without cost savings. In the present study, emphasis is given to the effects under the $14 price scenario. A case is examined where cost savings of 20% in investment and operating costs are made.

(1) Extra new field developments and field life extensions

Under the 10% real threshold return criterion it was found that in the period 1995–2000 no less that 22 extra new field developments were triggered. These would not have been economic without the cost savings. In addition the lives of all existing producing fields are extended. The extent of this effect depends upon the decline rates of the production profiles in individual fields.

(2) Extra production

The extra production of oil and gas which may be obtained in the period to 2000 is seen to climb to a figure of 400 000 bbl/day oil equivalent in 2000 (Fig. 11a). This is a very substantial effect for a comparatively short time period. The contributions in the early years are mostly from extensions to the lives of existing fields. The contribution from new fields obviously takes some time to become apparent. By 2000 this is the larger source of the extra production. This becomes even more noticeable in subsequent years. The great bulk of the increased output is in the form of oil. In 2000, 300 000 bbl/day extra oil production is obtained.

(3) Change in investment expenditure

Cost savings obviously entail a reduction in expenditure but, because they may lead to more new fields being developed, there is also increased, induced expenditure. It is illuminating to consider the **net** effect. In Fig. 11b the effects are shown of a reduction in development costs by 20% under the $14 price scenario. The negative figures show the size of the savings in development costs on all projects which are proceeding or are expected to be undertaken in the future. The figures relate to projects which in any case are going ahead. The positive figures show the investment expenditure on fields whose development is induced by the cost savings. Without these the projects would not go ahead.

The dramatic finding is that within a comparatively short time the net effect of the cost savings is either zero or positive. For the very near future the investment is already committed, and so the net effect of cost savings will be negative, but the finding that from 1997 onwards the induced investment matches or exceeds the cost savings is remarkable. The net positive effect continues to be positive after 2000. This finding is not present under the higher oil price scenarios because more of the future projects would, in any case, be viable without additional cost savings.

Investors are endeavouring to reduce operating costs as well as investment costs. In Fig. 12a the effect of 20% cost savings in all costs from 1995 onwards are shown. In this case

a)

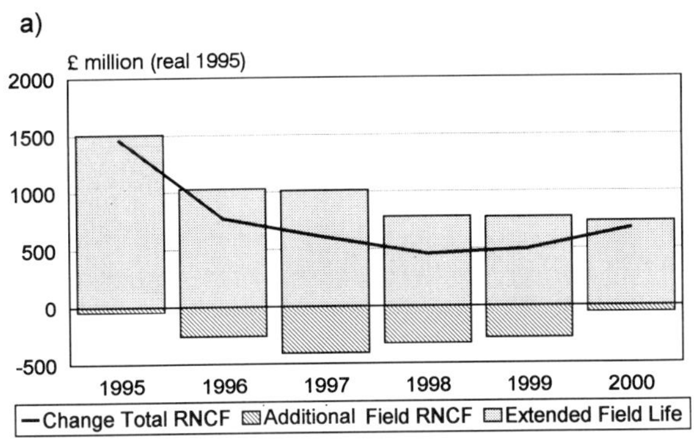

b)

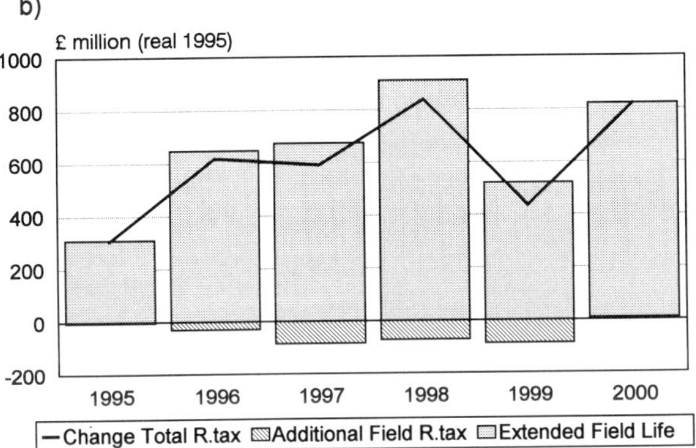

Fig. 13. Changes with 20% reduction in development and operating costs, $14 real oil price: (**a**) additional real net cash flow; (**b**) additional real taxes.

the value of the cost savings greatly exceeds any induced expenditure in the near term. This is because the cost savings apply to all the existing producing fields. It is seen, however, that by 2000 the value of the total induced expenditure exceeds the value of the cost savings. This phenomenon becomes more noticeable after 2000. It is on the basis of the above results that it may be claimed that cost savings in the UKCS can be a positive sum game for all participants.

(4) Extra gross revenues

The gross benefit to the national economy from cost savings is conveniently measured by the extra gross revenues procured. These are shown in Fig. 12b for the $14 price and cost savings of 20% across the board. The results indicate that these grow substantially over the period to reach nearly £1.5 billion in 2000. In the early years, the bulk of the extra revenues comes from extensions to the lives of existing fields, but by 2000 the revenues from induced new field developments become considerably more important. Beyond 2000 this phenomenon is even more pronounced. The value of the extra revenues is, of course, a function of the oil price as well as the output, and higher values could thus be obtained from this source.

(5) Improved industry net cash flow

Successful cost reduction initiatives will improve the petroleum industry net cash flows. This has been examined for the $14 price scenario (Fig. 13a). The results show the changes to the industry cash flows. The positive part of the chart shows the post-tax effect of the cost savings made on existing fields and those future ones which in any case are going ahead. In the negative part of the chart the change to industry cash flow of the new developments is shown which are induced by the cost savings. This net result includes the combined effect of the induced investment and operating costs plus any revenues and taxes. Over the period to 2000 the effect is (understandably) negative given that several new developments are triggered. The negative, real net cash flows are relatively small, however, and are always more than compensated by the value of the cost savings on projects already going ahead. The savings on such projects are more than adequate to finance the induced new field developments. The continuous line shows the overall net effect.

(6) Increased tax revenues

The cost savings increase the taxable capacity of the activities. The government thus obtains a direct benefit in the form of enhanced revenues. On 'old' fields (those receiving development consent before April, 1982) the marginal rate of tax is 70.7% as it includes royalty, Petroleum Revenue Tax and corporation tax. For fields developed between April, 1982 and March, 1993 the marginal rate is 66.5%. For fields developed after that date the rate is 33%.

In Fig. 13b the changes to tax revenues are shown. The lightly shaded part of the chart indicates the extra tax revenues obtained as a result of the cost savings on existing producing fields and those future ones that would in any case be developed. The dark shaded part of the chart shows the net effect on tax revenues of the development of fields whose developments are induced by the cost saving. This is mostly negative because the dominant effect is the tax relief for the investment expenditure. It is noticeable that even by 2000 there are positive tax revenues from the induced developments. These become bigger in subsequent years.

The net impact of the cost savings is seen to be substantially positive throughout the period. Government is clearly a very substantial beneficiary from achieved cost savings in the UKCS. The tax relief on new investments is generally at the rate of 33%. The extra take over the period to 2000 is at a much higher rate reflecting the much higher marginal rates of tax applied to existing fields.

Changing nature of producing systems in UKCS

The petroleum industry has had to adapt continuously to the ever-changing environment since gas and oil exploitation commenced in the late 1960s. This adaptation manifests itself in many ways. One very visible change has been in the type of producing systems employed. In the early years fixed platforms were normally employed. In more recent years there has been much greater emphasis on sub-sea completion systems utilizing existing infrastructure. Currently there is increased interest in floating systems.

The historic trend and prospects in choice of producing systems are shown in Fig. 14. The data refer to fields, not individual producing systems. The type of producing system chosen to represent the field is the main one employed. The results indicate that sub-sea systems became more common from the mid-1980s onwards. This trend will continue strongly in the period to 2000. In that year they could account for 36% of the producing systems employed in the UKCS.

The employment of floating systems will also increase. In absolute terms, fixed platforms will still dominate the total stock in 2000. There could be around 98 fields employing

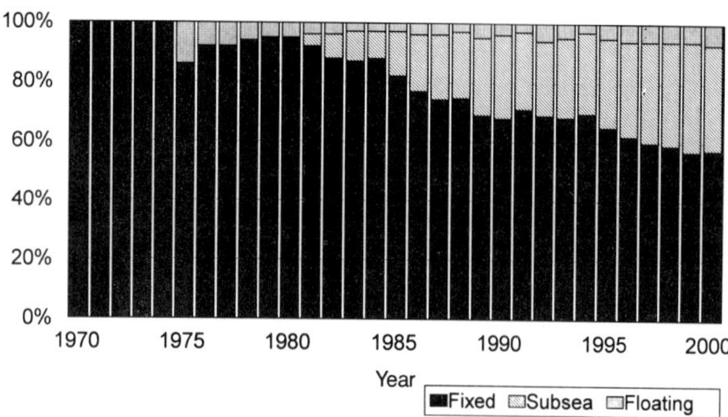

Fig. 14. Development types as % of total, UKCS $18 oil price, Threshold IRR 10% (real).

fixed platforms, and 61 employing sub-sea schemes as the main producing system. The number of fields where floating platforms constitute the main system will be around 10, though that number should grow considerably after 2000.

UK oil and gas production and demand

In Fig. 15a, recent official projections for oil demand are shown alongside the production possibilities from the UKCS emanating from this study. They indicate that there should be a substantial surplus of production over UK demand for oil-based energy. It should be noted that the demand figures shown show the highest and lowest estimates. They exclude non-energy uses of oil. The projections indicate that even with the lowest production possibility and the highest demand, a significant production surplus is still in prospect in 2000.

The corresponding projections of UK gas demand and potential production from the UKCS indicate that over the period there is generally a surplus of potential production over UK demand (Fig. 15b). In 2000 itself, the position is more complex. If gas prices are low and UK demand grows at the highest possible rate envisaged in the official projections, there would be no production surplus, but a deficit. In this context it should be noted that there are good prospects for further substantial new field developments after 2000 which would create a significant surplus. Further, it should be remembered that the study does not take into account further exploration successes.

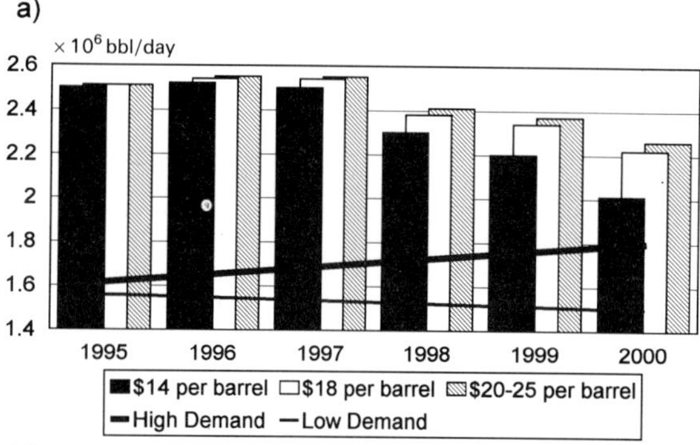

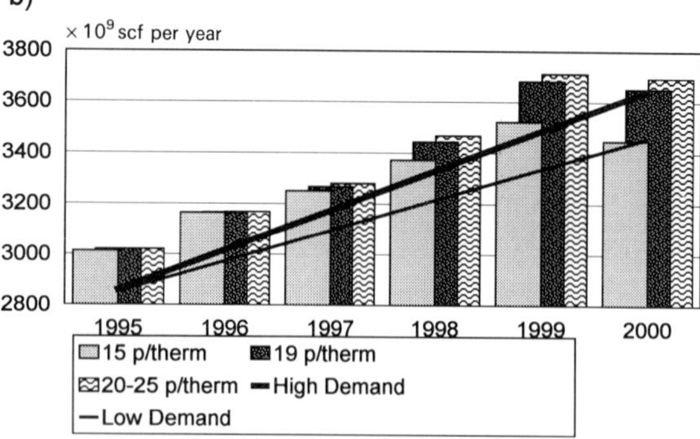

Fig. 15. Total production and consumption UKCS, Threshold IRR 10% (real): **(a)** oil; **(b)** gas.

CONCLUSIONS

The findings in this paper have widespread consequences for the UK oil and gas sectors, especially for the gas market. The prospect of potential UK gas production exceeding consumption became a reality in 1995 (Fig. 15b) with the result that spot gas prices fell dramatically to 8–9 pence per therm. The volumes traded at such prices have been very small and reflect the absence of other outlets for the gas, but the consequences are far-reaching.

Gas buyers, particularly British Gas, had entered into long-term contracts with producers at prices greatly in excess of 1995 spot prices. The average gas acquisition cost of British Gas was around 20 pence per therm in 1995. The whole viability of these contracts has been brought into question as a result of the dramatic developments in 1995.

The expectation that the 'bubble' may continue for some years has been responsible for the decisions by investors to delay some new field development projects. The overall market situation highlights the importance of the gas interconnector project from Bacton to Zeebrugge. The promoters of the project hope that it will commence operations in 1998. With a capacity of 2×10^9 cubic feet per day, exports to the continent via the interconnector could have a major effect on the overall UK supply/demand balance. Much uncertainty still prevails over the extent to which exporters from the UK can break into continental gas markets which are to a considerable extent still dominated by monopsony buyers and monopoly transmitters.

If the very low spot prices continue consumption, particularly in power generations, could be given a further stimulus. Much depends on the possibilities of substitution of gas for coal in existing power stations when current coal supply contracts come to an end.

A further implication of the findings in this study relates to Norwegian imports. From a purely physical viewpoint there is obviously no requirement for further imports for some years. Whether such imports do take place in the near term depends on factors such as revisions to the Frigg treaty and the price at which the gas can be supplied.

So far as general activity levels in the UKCS are concerned the prospects continue to be attractive to 2000. Production of hydrocarbons will continue at very high levels. The success of the cost saving initiatives and the development of new technologies should ensure that even with low hydrocarbon prices, new field developments and exploration continue at healthy levels.

FURTHER READING

BARKER, T., EKINS, P. & JOHNSTONE, N. (eds) 1995. Routledge, London, particularly the following papers:

ATKINSON, J. & MANNING, N. 1995. A survey of international energy elasticities, 47–105,

INGHAM, A. 1995. Responses of energy demand in UK manufacturing to the

energy price increases of 1973 and 1979–80, 121–158,
HODGSON, D. & MILLER, K. 1995. Modelling UK energy demand, 172–184,
BARKER, T., UK energy price elasticities and their implications for long-term CO_2 abatement, 227–253,
GRUBB, M. 1995. Asymmetrical price elasticities of energy demand, 305–310.
DTI. 1995. *Energy Paper No. 65, Energy Projections for the UK: energy use and energy-related emissions of carbon dioxide in the UK, 1995–2020,* HMSO, London.
DTI. 1995. *The Energy Report Vol 2 1995 oil and gas resources of the United Kingdom* , HMSO, London.
ETEBAR, S. 1995. 'Captain Innovative Development Approach', Paper SPE 30369, *presented at the Offshore Europe 1995 Proceedings,* pp 115–123.
KEMP, A. G., ROSE, D. & DANDIE, R. 1992. Development and production prospects for UK oil and gas post-Gulf crisis: a financial simulation. *Energy Policy,* 20(1), 20–29.
——— & STEPHEN, L. 1995. 'Sustaining the Viability of the UKCS in the New World Environment', Paper SPE 30359 *presented at Offshore Europe 1995.*
SHIVERS, III R. M. & BRUBAKER, J. P. 1995. 'Development Planning for the HPHT Erskine Field', Paper SPE 30370, *presented at the Offshore Europe 1995 Proceedings,* pp. 125–134.
UNITED KINGDOM OFFSHORE OPERATORS ASSOCIATION, 1994. *Crine Report, Cost reduction initiative for the new era.* The Institute of Petroleum.
WOOD MACKENZIE, Stockbrokers, Edinburgh. *North Sea Report,* various issues.

Environmental aspects of the offshore industry in the North Sea

D. Tromp

Rijkswaterstaat, North Sea Directorate, PO Box 5807, 2280 HV Rijswijk, The Netherlands

ABSTRACT: In the North Sea area a common policy has been developed with regard to the prevention of marine pollution. This policy is based on a number of international conventions, global as well as regional, regulating various pollution sources, such as dumping of wastes, shipping, 'land-based sources' and discharges from 'man-made' structures such as platforms. These conventions were mainly brought into force in the early 1970s. Their implementation in the North Sea area received political impetus from regularly held ministerial conferences on 'the protection of the North Sea'; the first of these conferences was held in 1984 in Bremen (Germany). As a result, the input of polluting substances in the North Sea has decreased very considerably and some activities, like the dumping of industrial waste into the sea, the dumping of sewage sludge (at the latest by the end of 1998) and the incineration of waste at sea, have been phased out completely.

In the same period that this environmental policy took shape, the development of the offshore oil and gas industry took place. Discharges from these activities are also regulated by one of the conventions (the Paris Convention) and by national legislation based on these conventions. As a result, the input of polluting substances, especially of oil, has also decreased considerably since about 1985, mainly as a consequence of the phasing out in some countries of the use of oil-based muds.

Due to the ageing of production wells, special attention is needed in the future with regard to increasing oil pollution. It is clear that progress has been made with regard to the reduction of discharge of oil and chemicals, but some developments give cause for concern. Measures that need to be taken by the oil and gas industry to minimize or prevent pollution must be in balance with those measures taken by other 'users' of the sea.

KEYWORDS: *offshore industry, environment, North Sea, pollution, Paris Convention*

INTRODUCTION

Parallel with the development of the North Sea oil and gas industry grew an interest in the quality of the environment in this marine area. The Conference on the Human Environment which took place in Stockholm, June 1972, gave a stimulus to this growing awareness. Since that year, a number of international conventions have been concluded with the aim of preventing marine pollution. The work carried out in the framework of these conventions received a political impulse in 1984 when the first Conference on the Protection of the North Sea took place in Bremen at ministerial level. Since then, these conferences have been held regularly. From these activities an environmental policy with regard to the North Sea has emerged. In this paper, the development of this policy is described on the basis of the activities around the various international conventions and ministerial conferences. Following this, those activities of the oil and gas exploration and exploitation which have an environmental impact are compared with this policy.

THE OLDEST SOURCE OF (OIL) POLLUTION: SHIPPING

In 1954 the 'International Convention for the Prevention of Pollution of the Sea by Oil' was concluded. It came into force in 1958. The responsibility for the administration of this convention was, from the beginning of its existence in 1958, carried out by the UN body regulating shipping activities: the Intergovernmental Maritime Consultative Organisation, IMCO (in 1982 the name changed to the International Maritime Organization, IMO). The most important provision with regard to discharge of oily mixtures, contained in this convention, was that in certain zones it prohibited the discharge of '... any oily mixture, the oil of which fouls the surface of the sea'. This meant that in these areas only discharges with an oil concentration lower than 100 ppm were allowed. The North Sea was such a 'prohibited zone'.

This convention was amended in 1962 and 1969. These amendments moved in the direction of more stringent discharge provisions. Presumably because of incidents where very large amounts of oil were spilled into the sea (Torry Canyon, 1967), the convention was again amended in 1971 to include provisions for (oil) tankers relating to tank arrangements and the limitation of tank sizes. The aim of these provisions was to influence the construction of tankers in such a way that, when an accident occurred, the amount of oil spilled as a result of that accident would be limited.

At that time it was realized that oil was not the only pollutant originating from shipping, and in 1973 the 'Inter-

From K. Glennie & A. Hurst (eds), 1996, *AD1995: NW Europe's Hydrocarbon Industry*, Geological Society, London, pp. 227–235

national Convention for the Prevention of Pollution from Ships' (MARPOL) was concluded. This convention, whose aim was: '... to achieve the complete elimination of intentional pollution of the marine environment by oil and other harmful substances and the minimization of accidental discharge of such substances., has a number of Annexes dealing with the discharge or disposal of respectively:

- oil;
- noxious liquid substances (transported in bulk);
- harmful substances carried by sea in packaged forms;
- sewage;
- garbage.

MARPOL 73 was an ambitious convention, replacing the 1954 Convention with regard to oil. It built on the experience gained with the 1954 convention but went considerably further to include a large number of provisions with regard to the construction of ships.

A State could become a Contracting Party to MARPOL 73 only if it adopted at least the two 'mandatory annexes' dealing with oil (Annex I) and noxious liquid substances (Annex II). It turned out that the implementation of Annex II, in particular, was extremely difficult. As a consequence, ratification of the MARPOL 73 Convention went very slowly.

In 1978, IMCO adopted a Protocol to the MARPOL Convention which made it possible for States to become a party to MARPOL while adopting, for the time being, only Annex I. This approach had success; in October 1983 Annex I of MARPOL 1973/1978 came into force. After amendments had been adopted in 1985, Annex II came into force in 1987.

Two of the other (optional) annexes have also been ratified by enough states to form the quorum necessary to make it binding; (harmful substances in packaged form and garbage). At this moment, negotiations are taking place within IMO on a new annex to MARPOL 73/78 dealing with air pollution from ships. Marpol 73/78 is a global convention. This means that its provisions also apply to the North Sea area.

So far as this paper is concerned, the most relevant discharge provisions for oil or oily mixtures are:

(a) for an oil tanker
  - discharge is allowed only if the tanker is more than 50 nautical miles from the nearest land;
  - the instantaneous rate of discharge of oil content does not exceed 30 litres per nautical mile;

(b) for other ships
  - the oil content of the effluent without dilution does not exceed 15 parts per million.

POLLUTION BY DUMPING OF WASTE AND OTHER MATTER

During preparations for the 1972 World Conference on the Human Environment, negotiations on both a regional and global level took place on conventions to regulate the dumping of wastes and other matter into the marine environment. The first results were achieved on a regional level: in February 1972, in Oslo, the 'Convention for the prevention of marine pollution by dumping from ships and aircraft' (the 'Oslo Convention') was signed; it came into force in 1974. All Western European coastal states, from Finland to Portugal, became parties to this convention (Fig. 1), which introduced a totally new concept in environmental policy, namely the principle that some substances are so hazardous that they should not be introduced into the environment.

Fig. 1. The Western European coastal states that are party to the 'Oslo Convention'.

For the elaboration of this principle, the convention contained a so-called 'black list' and a 'grey list'. On the black list were mentioned the substances and materials whose dumping is prohibited (Table 1); the grey list contained substances and materials that required 'special care' and for which a 'specific permit' was required. (The convention included the provision that for all dumping operations a permit was required from the appropriate national authorities.)

The Oslo Convention has served as a model: since 1972 a great number of conventions dealing with the aquatic environment has been concluded, for marine as well as for fresh waters, and nearly all these conventions have adopted the approach of the 'black' and 'grey' list.

Negotiations were also concluded in 1972 on a global convention concerning the dumping of wastes into the seas,

Table 1. *The 'black list' of substances*

1. Organohalogen compounds and compounds which may form such substances in the marine environment, excluding those which are non-toxic, or which are rapidly converted in the sea into substances which are biologically harmless;
2. Organosilicon compounds and compounds which may form such substances in the marine environment, excluding those which are non-toxic, or which are rapidly converted in the sea into substances which are biologically harmless;
3. Substances which have been agreed between the Contracting Parties as likely to be carcinogenic under the conditions of disposal;
4. Mercury and mercury compounds;
5. Cadmium and cadmium compounds;
6. Persistent plastics and other persistent synthetic materials which may float or remain in suspension in the sea, and which may seriously interfere with fishing or navigation, reduce amenities, or interfere with other legitimate uses of the sea.

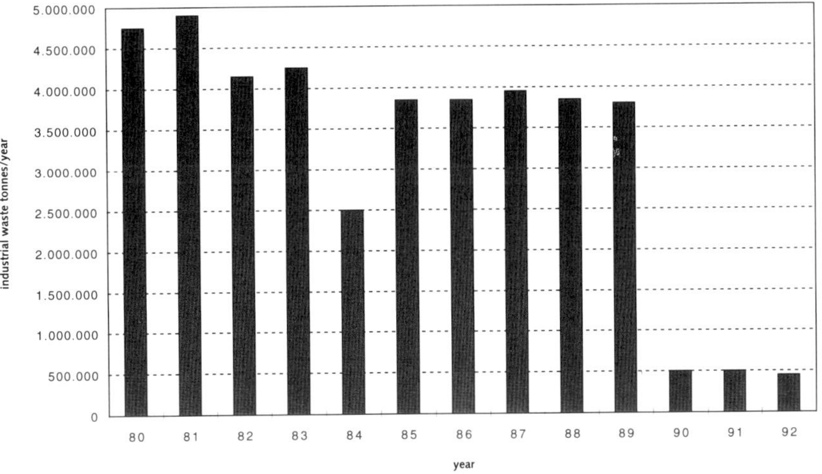

Fig. 2. Amount of waste incinerated on ships in the North Sea (tonnes annually) from Germany, The Netherlands, Belgium, France, United Kingdom, Spain, Sweden, Norway, Ireland and others in the period 1980 until 1992.

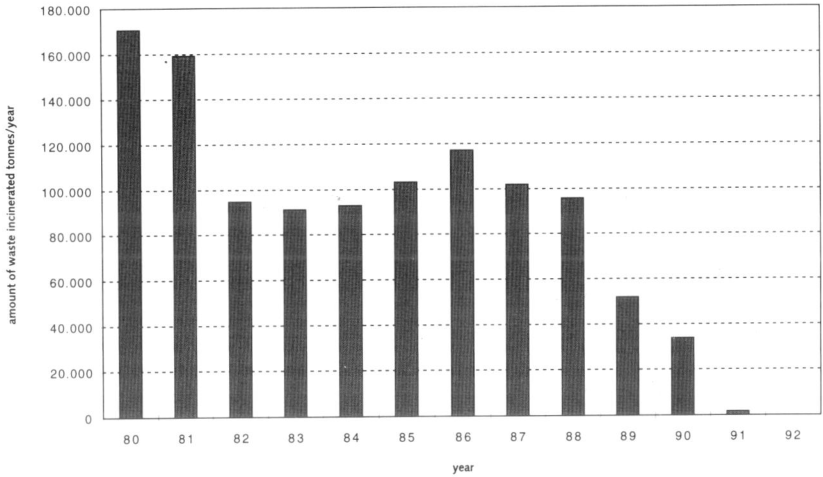

Fig. 3. Dumping of industrial waste into the North Sea (in 1990 fly ashes and mine waste are excluded) by Germany, The Netherlands, Belgium and the United Kingdom from 1980–1992.

and in December of that year, the 'Convention on the prevention of marine pollution by dumping of wastes and other matter' (the London Dumping Convention, or, since 1972, called the 'London Convention 1972') was signed. It came into force in 1975.

The activities carried out in the framework of these two conventions can be described as: harmonizing the policies of Contracting Parties in the field of the dumping of wastes and other matter into the sea with the aim to reduce and to prevent marine pollution from this source.

Since the second half of the 1960s, substances (mainly organic chlorine substances) were incinerated in special ships operating most of the time in the North Sea. Both the global and the regional convention were amended to take account of this development, and provisions with regard to the operation of these incineration-ships were included. Incineration operations in the North Sea came to an end in 1992. Figure 2 shows the amounts of waste burnt within the incineration area in the North Sea.

The material dumped into the North Sea can be divided roughly into:

- industrial waste (see Fig. 3);
- sewage sludge (about 5×10^6 tonnes per year);
- dredged spoil (about 40×10^6 tonnes per year).

POLLUTION FROM LAND-BASED SOURCES

At the time when negotiations on the Oslo Convention took place, it was already recognized that the pollution caused by the dumping of wastes was not the most important source of pollution of the marine environment. It was tackled first presumably because, as a well-defined activity, it could be regulated relatively easily.

The input of pollutants via rivers was known to be of far greater importance, but it was also realized that regulation of that input would be a difficult task because the pollution load of rivers originates from numerous sources of great variety. In 1973, however, the French government took the initiative to start negotiations with the aim of reaching agreement on a convention regulating activities causing pollution of the marine environment, apart from dumping. In 1974, in Paris, the 'Convention for the prevention of marine pollution from land-based sources' (the Paris Convention) was concluded.

Pollution from land-based sources is defined as (art. 3): '... pollution of the maritime area

I. through watercourses,
II. from the coast, including introduction through underwater or other pipelines,
III. from man-made structures placed under the jurisdiction of a Contracting Party within the limits of the area to which the present Convention applies'.

The maritime area mentioned is the same area as that covered by the Oslo Convention (Fig. 1) and includes the North Sea. Also, this convention follows the approach of the 'black' and the 'grey' list, which in character is the same as that of the Oslo Convention.

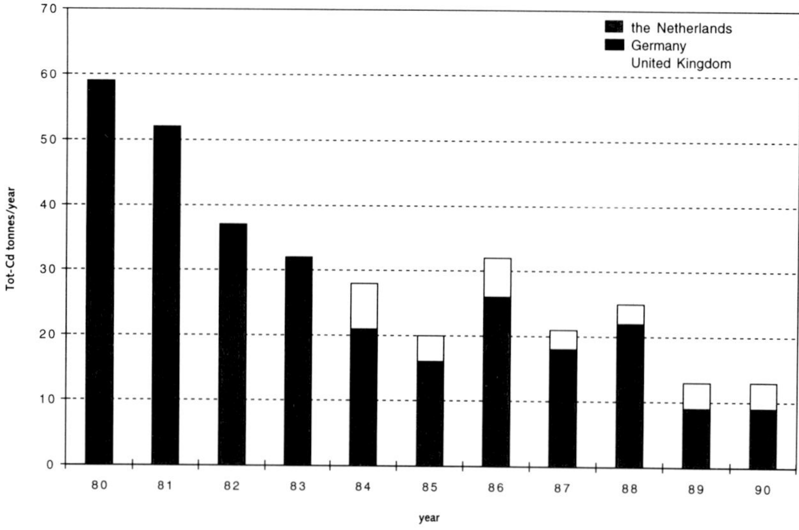

Fig. 4. Mean net cadmium load (tonnes per year) to the North Sea via Dutch, German and English river basins.

Contracting Parties undertake to 'eliminate' pollution of the maritime area by black-list substances and 'to limit strictly' pollution by the grey-list substances. In order to carry out this undertaking the Contracting Parties shall implement programmes and measures 'which shall include, as appropriate, specific regulations or standards governing the quality of the environment, discharges into the maritime area, such discharges into watercourses as affect the maritime area, and the composition and use of substances and products' (art 4). From the definition of 'pollution', it is clear that the discharges from oil and gas exploration and exploitation platforms are covered by this convention.

The Convention came into force in 1978. Apart from Finland, all Contracting Parties to the Oslo Convention are also Parties to the Paris Convention. In addition, the Commission of the European Economic Community, now European Union, became a Contracting Party to the Convention. There is close co-operation between the Commissions of the Oslo and Paris Conventions. They have, for example, a common secretariat based in London.

The wide scope of the convention is reflected in the recommendations and decisions which the Paris Commission has taken. They deal not only with direct discharges of substances in watercourses and the sea (for example mercury in discharges from the chlor-alkali industry) but also with the use of specific toxic substances. As examples, the use of mercury and cadmium in batteries, the phasing out of the use of PCBs and the use of certain pesticides can be mentioned.

In general, the work of the Commission has focused on:

- the reduction of the input of specific hazardous substances via direct discharges;
- the reduction of the input of hazardous substances via discharges of waste streams from certain industrial sectors (for example the pulp and paper industry which discharges chlorinated organic substances);
- the phasing out of the use of certain (black list) substances (PCBs for example);
- atmospheric emissions of black and grey-list substances. In 1989, a protocol to amend the convention came into force which included within its scope pollution of the maritime area via the atmosphere;
- reduction of the input of nutrients (phosphate and nitrogen) in certain 'problem areas' of the maritime area.

The Oslo and Paris Conventions were negotiated in the early 1970s, a period of growing environmental awareness in Western Europe. Activities related to the prevention or reduction or marine pollution represent only a part of the picture in which this growing awareness expressed itself. But the two conventions with the black and grey list approach, offered a direction that could be followed by other organizations dealing with the reduction or prevention of water pollution. The EEC, for example, also started its activities on the environment in the early 1970s, and in 1976 (after lengthy negotiations) a 'framework directive' was adopted: 'Directive 76/464/EEC on pollution caused by certain dangerous substances discharged into the aquatic environment of the Community'. The lists annexed to this Directive are very similar to those of the Paris Convention, and they serve the same purpose; since 1976 the EEC has adopted a number of directives with emission limits and quality objectives for 'black-list substances'.

The same approach was adopted by some 'river commissions': international commissions established in a river basin to tackle the problem of water pollution. The best known of such commissions is probably the International Commission for the Protection of the Rhine.

That the efforts of the national authorities, Paris Commission, the EEC (EU) and the river commissions have had effect is illustrated in Fig. 4, which shows the input into the North Sea of cadmium via rivers and other watercourses for the years 1980–1990 (NICMM 1993).

By monitoring, the Oslo and Paris Commissions jointly assess the state of the marine environment and the effects on this environment of the measures taken. In this task the Commissions work closely with the International Council for the Exploration of the Seas (ICES).

MINISTERIAL CONFERENCES ON THE PROTECTION OF THE NORTH SEA

In 1980, in the Federal Republic of Germany, a report was published on the 'Environmental Problems of the North Sea' (RSUS 1980). One of the (many) conclusions of this report, drafted by a high-level expert group, was that the implementation of the Paris Convention proceeded too slowly. This was one of the reasons why the Government of the Federal Republic of Germany organized a conference at ministerial level with the aim of giving political impetus to the work in the North Sea states to improve the environmental situation of the North Sea area.

The first ministerial-level International Conference on the

Protection of the North Sea took place in Bremen in 1984. Since then, such conferences have been held in London (1987) and The Hague (1990). An intermediate ministerial meeting was held in 1993 in Copenhagen (Ministerial Declaration 1984, 1987, 1990; Intermediate Ministerial Meeting 1993); the last ministerial conference was held in Esbjerg in June 1995.

These conferences certainly fulfilled their purpose: work to improve the quality of the North Sea environment was speeded up, significant results were achieved.

Examples include:

- between 1985 and 1995 to reduce 'substantially' (in the order of 50%) the input to the North Sea from rivers and estuaries of substances that are persistent, toxic and liable to bio-accumulate (Ministerial Declaration 1987, 1990);
- for substances that cause a major threat to the marine environment, and at least for dioxines, mercury, cadmium and lead, to achieve reductions between 1985 and 1995 of total inputs (via all pathways) of the order of 70% or more (Ministerial Declaration 1990);
- between 1985 and 1995 to reduce substantially (in the order of 50%) the input of nutrients (phosphate and nitrogen) into areas of the North Sea where these inputs are likely, directly or indirectly, to cause pollution (Ministerial Declaration 1987);
- to phase out the dumping of industrial waste into the North Sea by 31 December 1989;
- to phase out incineration of wastes at sea by 31 December 1991;
- to phase out the dumping of sewage sludge, at the latest by the end of 1998;
- to request the Oslo and Paris Commissions and ICES to prepare and publish Quality Status Reports of the North Sea at regular intervals. (In 1993 the *North Sea Quality Status Report* was published, prepared by the North Sea Task Force, jointly established by the Commissions and ICES (OPC 1993).
- the continuous reduction of discharges, emissions and losses of hazardous substances moving towards the target of their cessation within one generation (Ministerial Declaration 1995).

The conferences directed many requests for further actions to the EEC and the Oslo and Paris Commissions; with regard to shipping, many requests were made to the IMO.

The decisions taken by the conferences had effect also outside the North Sea area. The decision to phase out the dumping of industrial waste, for example, was taken over by the Oslo Commission and, later, also by the Consultative Meeting of the 'London Convention 1972'.

Another example is the 'Precautionary principle' (action should be taken when there is reason to assume that certain damage or harmful effects on the living resources of the sea are likely to be caused by such substances, even where there is no scientific evidence to prove a causal link between emission and effects), first described in some detail in an international document in the Ministerial Declaration of the Conference in London. This principle has since been adopted by other international organizations (for example the London Convention) and been incorporated in the texts of (new) conventions (for example the OSPAR Convention; see par. 6).

Some of the decisions in the Ministerial Declarations are, from a legal point of view, written in rather imprecise language. For example, with regard to the input of certain substances into the North Sea, it is mentioned that the participants of the Conference agree to take measures, '... with the aim of achieving a substantial reduction (in the order of 50%) in total inputs from these sources between 1985 and 1995'. Such provision would not be acceptable in a legally binding document such as a convention. But the Ministerial Declarations do not contain legally binding obligations; they contain provisions reflecting the political will of the participants. The participants expect from each other that they will give effect to this political will and implement the agreed provisions. Practice has shown that this method can be an effective one.

THE OSPAR CONVENTION: CONVENTION FOR THE PROTECTION OF THE MARINE ENVIRONMENT OF THE NORTHEAST ATLANTIC (PARIS 1992)

As already mentioned, the Oslo and Paris Commissions are closely linked together. They have a joint Secretariat, for example, serving both Commissions and their working groups. There were also some joint working groups. In 1990 when, at the Ministerial Conference, it was decided to phase out the dumping of industrial waste, it became clear that the work with regard to dumping into the sea would decrease, and the work linked to the reduction and prevention of pollution from land-based sources, would increase. Taking into account that the conventions had been drafted twenty years previously, it was decided to replace the two Conventions by a single agreement.

Negotiations started, and in September 1992 in Paris, ministers signed a new convention: the Convention for the Protection of the Marine Environment of the Northeast Atlantic, called the 'Ospar' Convention (Ministerial Meeting 1992). This Convention and its Annexes deals with the prevention and elimination of pollution from:

- land-based sources (Annex I);
- dumping or incineration (Annex II);
- offshore sources (Annex III).

A separate Annex deals with the 'Assessment of the Quality of the Marine Environment'. The 'Polluters Pay Principle' and the 'Precautionary Principle' are both incorporated into the text of the Conventions.

With regard to dumping, the convention has adopted the approach of 'reversed listing': all dumping is prohibited except for those specific (listed) categories of waste.

The offshore annex deals with emissions as well as with the abandonment of platforms and pipelines.

The scope of the Ospar Convention is somewhat wider than that of the two 'parent' conventions, in that it mentions not only pollution but also 'the adverse effects of human activities' on the marine environment. A new element is that Switzerland and Luxembourg are also expected to become Contracting Parties to this convention. The Ospar Convention will come into force thirty days after all Contracting Parties to the Oslo and Paris Conventions have ratified.

The 'black and grey lists', introduced by the Oslo and Paris Conventions, do not appear as such in the new convention. The essence of this concept is maintained, however, by the provision that Contracting Parties, when adopting programmes and measures to prevent and eliminate pollution from land-based sources, shall use a number of the criteria listed in the convention. These criteria include persistency, toxicity and tendency towards bio-accumulation, etc.

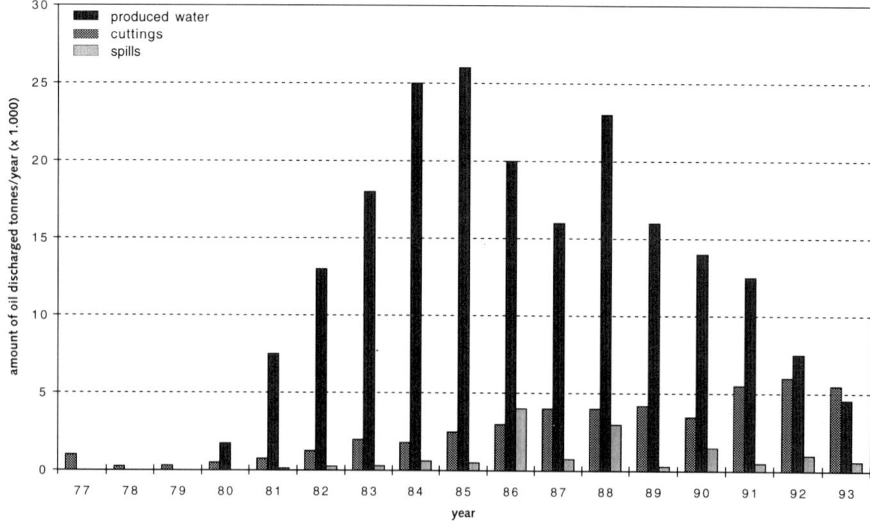

Fig. 5. Amount of oil discharged via produced water, cuttings and spills.

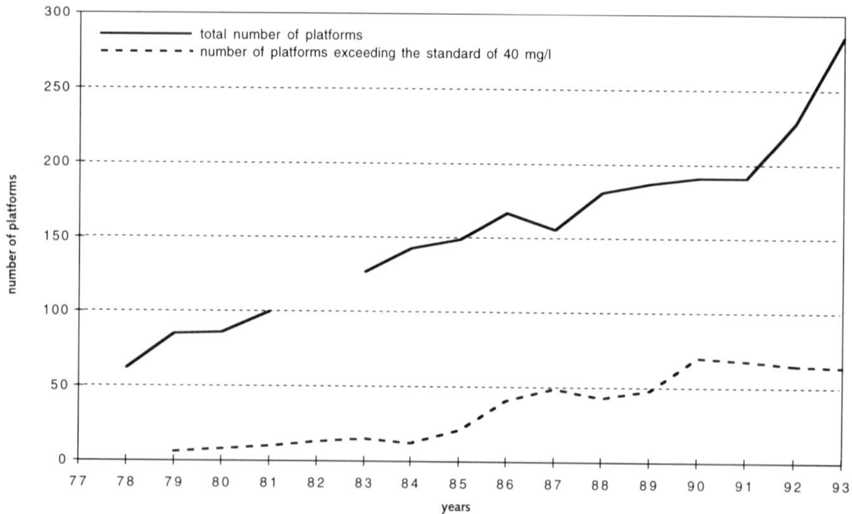

Fig. 6. Number of platforms in the North Sea, and number of platforms exceeding the target standard of 40 ppm.

DISCHARGES FROM THE OFFSHORE OIL AND GAS INDUSTRY

The environmental policy with regard to the North Sea is laid down in a number of conventions and international agreements. The North Sea is an important asset for the surrounding countries, but the exploitation of this resource has to take place within the constraints of that environmental policy. The aims of this policy are to prevent and eliminate 'pollution' and to protect the marine environment against the adverse effects of human activities so as to safeguard human health and to conserve marine ecosystems and, when practicable, restore marine areas which have been adversely affected (art. 2 Ospar Convention).

'Pollution' is defined in the 'Ospar' Convention as follows:

> Pollution means the introduction by man, directly or indirectly, of substances or energy into the maritime area which results, or is likely to result, in hazards to human health, harm to living resources and marine ecosystems, damage to amenities or interference with other legitimate uses of the sea.

Some important elements of this policy are:

- a strict limitation on the input of substances which are toxic, persistent and liable to accumulate, if necessary by the phasing out of its use;
- a limitation of the input of other substances using the best environmental option;
- the reduction of the input, via various sources, of oil and oily mixtures.

The discharges and emissions of the offshore oil and gas industry in the North Sea have to be looked at within the context of this agreed environmental policy.

Both the Paris Commission and the North Sea Conferences have discussed the environmental aspects of the offshore oil and gas industry. These discussions can roughly be divided into three phases. In the first phase attention was focused on the discharge of oil via production water (plus displacement and drainage water). Later it was recognized that more oil was discharged via drilling mud and cuttings. Presently, in the third phase, much more attention is given to the discharge of chemicals than in the past.

DISCHARGE OF OIL VIA PRODUCTION, DISPLACEMENT AND DRAINAGE WATER

The first data gathered in the framework of the Paris Commission with regard to the discharge of oil refer to the year 1977. The total number of installations in the North Sea

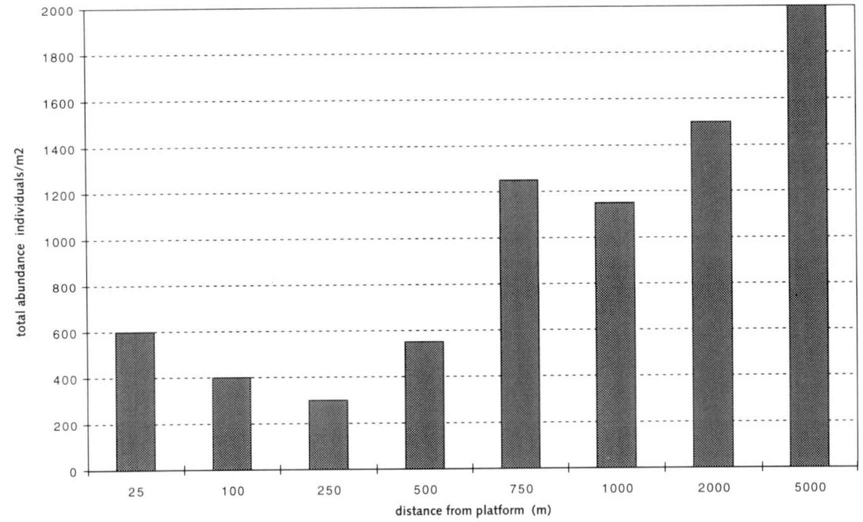

Fig. 7. Total macrofauna abundance at P6b (residual current transect) six years after discharge of cuttings contaminated with oil-based mud.

area was then about 60. Figure 5 shows the amount of oil discharged via produced water for the years 1977–1993; this amount increased by about a factor 10 between 1980 and 1992 (OPC 1992; E & P Forum 1994).

This increased amount is the result of three factors:

(a) the number of platforms had increased during that time span (Fig. 6);
(b) in general, when a production well becomes older it produces more water. When the concentration of oil in the water remains the same, the total amount of oil discharged will increase;
(c) abatement measures on platforms are taken to reduce the oil concentration.

The abatement measures were stimulated with the adoption by the Paris Commission, already at its First Meeting in 1978, of a '(provisional) target standard' for discharges from offshore installations of 40 mg oil per litre (aliphatic hydrocarbons).

In the beginning, the treatment methods applied were mainly gravity separation and dispersed air flotation. More recently, hydrocyclones and centrifuges have also been installed. Although the matter has been discussed a number of times (also at the request of the North Sea Conferences) Contracting Parties to the Paris Commission decided not to change the target standard to a firm standard. Uncertainty about the efficiency of the treatment methods with increasing amounts of produced water is the main reason for this.

Figure 6 shows the number of platforms which did not meet the target standard of 40 mg/l.

A possible solution for the disposal of produced water is injection in the sub-seabed. However, in addition to technical and financial aspects, the question whether it is acceptable to bring, for example, (synthetic) persistent compounds into the seabed has to be considered.

OIL DISCHARGED VIA CUTTINGS

That oil reached the marine environment via the discharge of drilling mud and cuttings was reported to the Paris Commission by the Norwegian authorities for the first time in 1980. With the data for that year, Norway reported that presumably 700 tonnes of oil was discharged via cuttings. This focused the attention of the authorities of the other countries concerned, and in subsequent years it turned out that the oil input via cuttings was far greater than the input via produced water. The matter was discussed several times by the Paris Commission which adopted a number of decisions, the last one, in 1992, suspending the earlier ones (PARCOM Decision 92/2). The essential aspects of these decisions are:

- the use of diesel-based mud was prohibited from 1 January 1987. Only in well-defined, exceptional cases may diesel oil be added to drilling mud;
- whole oil-based muds shall not be dumped or discharged (1988);
- from 1 January 1997 the average oil content of cuttings discharged shall not exceed 10 g oil per kg dry cuttings; meanwhile the target standard of 100 g kg^{-1} shall not be exceeded;
- national authorities shall require prior authorization wherever oil-based mud is used offshore.

In Denmark a voluntary agreement exists between the authorities and the industry that no oil-based mud and cuttings will be discharged. In Norway and The Netherlands, the discharge of oil-based mud cuttings has been prohibited since 1993. If such mud has to be used for drilling operations, the cuttings must be brought ashore for treatment and disposal. (For the amount of oil discharged in the North Sea area via cuttings see Fig. 5.)

From research carried out to assess the environmental impact of oil discharges via cuttings, it has become clear that the biological effect of these discharges can be detected years after the discharge has stopped (Fig. 7; Dean & Mulder 1993, 1994).

DISCHARGE OF CHEMICALS

A large number of chemicals is discharged from platforms into the marine environment via drilling mud or via produced water. These chemicals are either added during the production process (for example corrosion inhibitors, bactericides etc.) or are part of the oil or gas brought to the surface. At present there is no clear overall picture of the amounts and characteristics of the chemicals added and discharged. This is one of the items currently being discussed by the Paris Commission. The aim of the discussion is to ensure that only

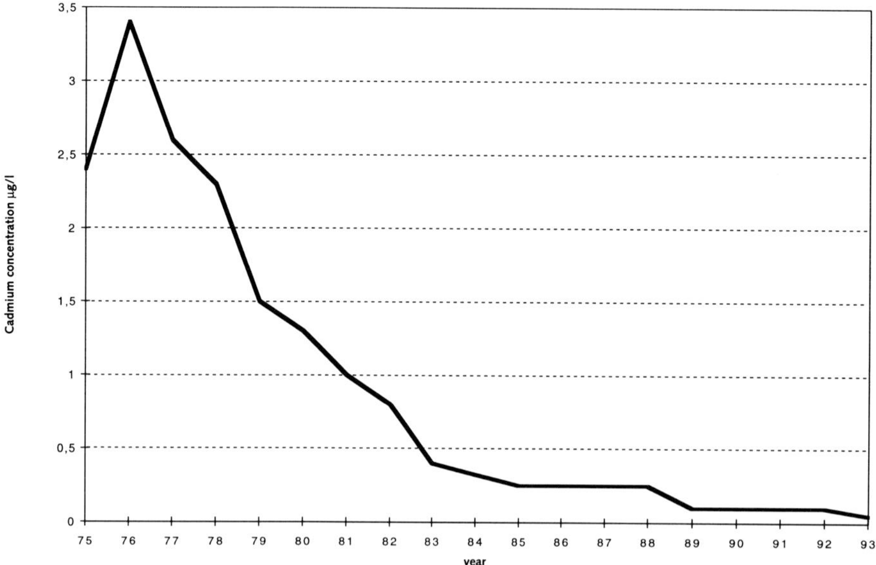

Fig. 8. Total cadmium concentration in Rhine water at the German–Dutch border.

harmless substances will be discharged into the marine environment. It is expected to achieve this aim by operating a notification scheme and an evaluation scheme for chemicals, the CHARM model: Chemical Hazard Assessment and Risk Management model (Schobben *et al.* 1994). There is close co-operation with the E & P industry in the development of this model, which is expected to be operational in 1996; it can then be incorporated into the national regulations of the various countries bordering the North Sea.

Table 2. *Amounts of chemicals discharged in 1989 on the Dutch Continental Shelf via water-based muds and water-based cuttings (Nogepa report 1992)*

Function	Tonnes
Weighting agents	12407
pH-controllers and salinity chemicals	4676
Viscosifers	220
Shale control inhibitors	1800
Fluid-loss control	1022
Gelling products	550
Others (cementing)	131
Thinner, dispersant	101
Lost circulation materials	82
Corrosion inhibitors	38
Deformers	7
Lubricants	5
Emulsifiers	3
Biocides	1
Total	23030

Until that time, Parcom operates, on the basis of a decision it took in 1994, using lists of chemicals: a list of 'adequately tested substances/preparations used and discharged offshore', and a list of substances/preparations the discharge of which is prohibited. Criteria for the placement of a substance on such a list are being discussed with the industry.

One of the main reasons why progress in the control of chemicals is relatively slow is that many of the preparations used are known only by their trade names. The exact composition on the basis of chemical nomenclature is kept confidential because of commercial interests.

Many organic and inorganic substances are present in produced water depending on the geological formation the gas or oil is extracted from. Heavy metals and toxic and/or persistent organic compounds are the most relevant from the environmental point of view.

It is expected that in the coming years, both nationally in the countries around the North Sea and in the Paris Commission, discussion will take place about the necessity or desirability, and also the feasibility, of treating produced water (apart from the treatment it undergoes to reduce oil discharges). This process is stimulated by a request from the Intermediate Ministerial Meeting (1993) to the Oslo and Paris Commission 'to undertake an appraisal of the significance and of the possibilities to reduce emissions of PAHS (Polycyclic Aromatic Hydrocarbons) from offshore installations'. That the amounts of chemicals discharged are not insignificant is illustrated by Table 2.

DISCUSSION

Some 40 years ago, measures were agreed to reduce the input of oil (originating from the operational discharges of ships) into the sea. Around 20 years ago these measures were intensified. In addition, measures to reduce the input of other polluting substances were agreed and implemented. Activities to reduce the pollution of rivers and other inland water courses also helped to decrease the amount of polluting substances reaching the marine environment. Attention was especially focused on black-list substances: mercury, cadmium, persistent and toxic organohalogens, and oil.

That all these efforts have been successful is illustrated by Fig. 8, which shows the cadmium content in the river Rhine near the German–Dutch border; a remarkable decrease from the beginning of the 1970s onwards. One could say that this figure shows the general picture with regard to polluting substances in the countries around the North Sea: a decrease of the input into the marine environment, starting somewhere in the 1970s.

The offshore industry shows a somewhat different pattern.

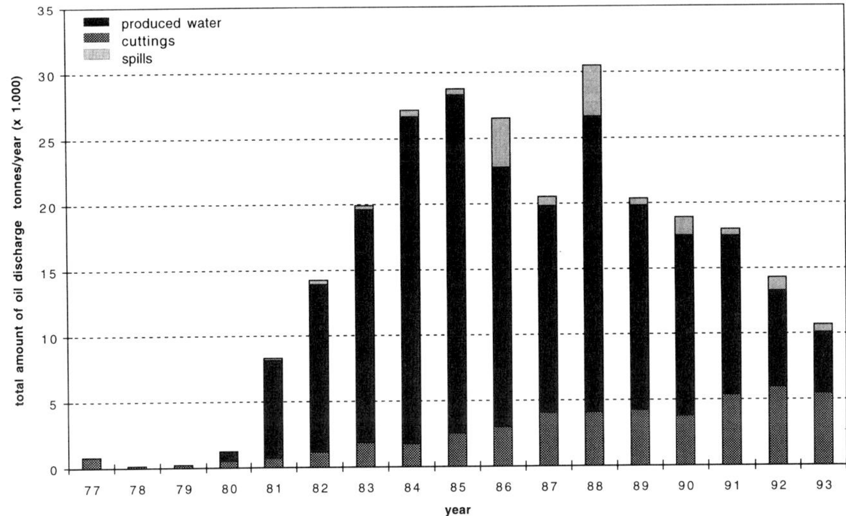

Fig. 9. Total amount of oil discharged via produced water, cuttings and spills.

In 1978 the Paris Commission agreed a (provisional) target standard of 40 ppm oil in produced water. However, the number of platforms which discharged produced water with a concentration of oil higher than the target standard increased steadily. The input of oil via cuttings, according to the information submitted to the Paris Commission, increased between 1981 and 1985 by a factor of 3 to 4; only from 1989 onwards did the amount of oil discharges via this route show a considerable decrease (Fig. 9).

Twenty years after the coming into force of the Paris Convention, the industry is still not able to provide reliable data about the discharge of chemicals or the emissions of polluting substances into the air. However, one could say that from about 1989 onwards, the discharges from the offshore industry followed the downwards trend which on land had started some years earlier: the total input of oil decreases from that year onward. It can be expected that this decrease will continue in the coming years because of the implementation of PARCOM-Decision 92/2 (from 1 January 1997 the oil content on cuttings should be lower than 10 g kg^{-1}).

The situation with regard to chemicals will improve with the implementation of the CHARM model. It is expected that in the near future, attention will be focused on chemicals and on the amounts of oil discharged via produced water. With the production wells becoming older it is likely that the total input of oil via this route will continue to increase, forming a source of chronic oil pollution for a considerable area of the North Sea. At the same time, much money and effort is spent on attempts to reduce the other source of chronic oil pollution: shipping.

It is clear that the activities of the oil and gas industry, which have an impact on the quality of the marine environment, must be in balance with the measures taken by other 'users of the sea' to minimize or prevent marine pollution.

REFERENCES

Annual Reports of the Paris Commission. Secretariat of the Paris Commission, London

DAAN, R. & MULDER, M. 1993. *Long-Term Effects of OBM Cutting Discharges at a Drilling Site on the Dutch Continental Shelf.* NIOZ-Rapport 1993-15, Netherlands Institute for Sea Research/NIOZ, Texel, The Netherlands.

——— & ——— 1994. *Long-Term Effects of OBM Cutting Discharges in the Sandy Erosion Area of the Dutch Continental Shelf.* NIOZ-Rapport 1994-10, Netherlands Institute for Sea Research/NIOZ, Texel, The Netherlands.

E. & P. FORUM, 1994. *North Sea Produced Water: Fate and Effects in the Marine Environment.* Report nr 2.62/204, London.

Intermediate Ministerial Meeting. 1993. *Statement of Conclusion from the Intermediate Ministerial Meeting 7–8 December 1993 in Copenhagen.*

Ministerial Declaration. 1984. *Ministerial Declaration of the First International Conference on the Protection of the North Sea.* Bremen, 31 October and 1 November 1984.

Ministerial Declaration. 1987. *Ministerial Declaration of the Second International Conference on the Protection of the North Sea.* London, 24–25, November 1987.

Ministerial Declaration. 1990. *Ministerial Declaration of the Third International Conference on the Protection of the North Sea.* The Hague, 7–8, March 1990.

Ministerial Meeting. 1992. *Ministerial Meeting of the Oslo and Paris Commissions.* Paris 21–22 September, 1992. The Oslo and Paris Commissions, London.

NICMM. 1993. *De belasting van de Noordzee met verontreinigingde stoffen 1980–1990.* (The input of polluting substances into the North Sea 1980–1990). Rapport DGW- 93.037 National Institute for Coastal and Marine Management/RIZA Institute for Inland Water Management and Waste Water Treatment/RIZA. North Sea Directorate.

Nogepa Report. 1992. *Het gebruik van Generieke boorspoelingen op het Nederlands Continentaal Plat* (The use of generic muds on the Dutch Continental Shelf). Nogepa Report 01-92, The Hague.

OPC 1992. *The discharge of oil in the convention area.* Oslo and Paris Commissions, London.

——— 1993. *North Sea Quality Status Report 1993.* Oslo and Paris Commission, London.

RSUS. 1980. *Umweltprobleme der Nordsee.* Der Rat von Sachverständigen für Umweltfragen. Sondergutachten Juni 1980. Verlag W. Kohlhammer GMBH, Stuttgart und Mainz.

SCHOBBEN, H. P. M., KARMAN, C. & SCHOLTEN, M. C. Th. 1994. *Chemical Hazard Assessment and Risk Management of Offshore E and P Chemicals: second phase 1994.* TNO Rapport R94/315. Netherlands Organization for Applied and Scientific Research/TNO.

Index

Page references in *italics* are for Tables and Figures